COUNTEREXAMPLES IN MEASURE AND INTEGRATION

Often it is more instructive to know 'what can go wrong' and to understand 'why a result fails' than to plod through yet another piece of theory. In this text, the authors gather more than 300 counterexamples – some of them both surprising and amusing – showing the limitations, hidden traps and pitfalls of measure and integration. Many examples are put into context, explaining the relevant parts of the theory, and pointing out further reading.

The text starts with a self-contained, non-technical overview on the fundamentals of measure and integration. A companion to the successful undergraduate textbook *Measures, Integrals and Martingales*, it is accessible to advanced undergraduate students, requiring only modest prerequisites. More specialized concepts are briefly summarized at the beginning of each chapter, allowing for self-study as well as supplementary reading for any course covering measures and integrals. For researchers, the text provides ample examples and warnings as to the limitations of general measure theory.

RENÉ L. SCHILLING is Professor of Probability Theory at Technische Universität Dresden. His research focuses on stochastic analysis and the theory of stochastic processes.

FRANZISKA KÜHN is Research Assistant at Technische Universität Dresden, where she finished her Ph.D. in 2016. She is interested in the interplay of probability theory and analysis, with a focus on jump processes and non-local operators.

COUNTEREXAMPLES IN MEASURE AND INTEGRATION

RENÉ L. SCHILLING
Technische Universität Dresden

FRANZISKA KÜHN
Technische Universität Dresden

CAMBRIDGE
UNIVERSITY PRESS

University Printing House, Cambridge CB2 8BS, United Kingdom

One Liberty Plaza, 20th Floor, New York, NY 10006, USA

477 Williamstown Road, Port Melbourne, VIC 3207, Australia

314-321, 3rd Floor, Plot 3, Splendor Forum, Jasola District Centre, New Delhi - 110025, India

103 Penang Road, #05-06/07, Visioncrest Commercial, Singapore 238467

Cambridge University Press is part of the University of Cambridge.

It furthers the University's mission by disseminating knowledge in the pursuit of education, learning and research at the highest international levels of excellence.

www.cambridge.org
Information on this title: www.cambridge.org/9781316519134
DOI: 10.1017/9781009003797

First published 2021

A catalogue record for this publication is available from the British Library

ISBN 978-1-316-51913-4 Hardback
ISBN 978-1-009-00162-5 Paperback

Contents

Preface

A **counterexample** /ˈkaʊntərɪɡˌzɑːmpl/ is *an example that opposes or contradicts an idea or theory.*[1] It is fair to say that the word 'counterexample' is not too common in everyday language, but rather a concept from philosophy and, of course, mathematics. In mathematics, there are proofs and examples, and while an example, say, of some $x \in A$ satisfying $x \in B$ does not prove $A \subseteq B$, the *counter*example of some $x_0 \in B$ such that $x_0 \notin A$ disproves $A \subseteq B$; in other words, it proves that $A \subseteq B$ does not hold. This observation shows that there is no sharp distinction between example and counterexample, and we do not give a definition of what a counterexample should or could be (you may want to consult Lakatos [94] instead), but assume the more pragmatic point of view of a working mathematician. If we want to solve a problem, we look at the same time for a proof and for counterexamples which help us to capture and delineate the subject matter.

The same is also true for the student of mathematics, who will gain a better understanding of a theorem or theory if he knows its limitations – which may be expressed in the form of counterexamples. The present collection of (counter-)examples grew out of our own experience, in the classroom and on `stackexchange.com`, where we are often asked after the 'how' and 'why' of many a result. This explains the wide range of examples, from the fairly obvious to rather intricate constructions. The choice of the examples reflects, naturally, our own taste. We decided to include only those counterexamples which could be dealt with in a couple of pages (or less) and which are not too pathological – one can, indeed, destroy almost anything by the choice of the underlying topology. We intend the present volume as a companion to our textbook *Measures, Integrals and Martingales* [MIMS], which means that most examples are from elementary measure and integration, not touching on integration on

[1] Oxford dictionaries `https://en.oxforddictionaries.com/definition/counterexample`, accessed 11-May-2019.

groups (Haar measure) or on really deep axiomatic issues (e.g. as in descriptive set theory, see Kechris [89], and the advanced constructive theory of functions, see Kharazishvili [91, 92]).

This book is intended as supplementary reading for a course in measure and integration theory, or for seminars and reading courses where students can explore certain aspects of the theory by themselves. Where appropriate, we have added comments putting the example into context and pointing the reader to further literature. We think that this book will also be useful for lecturers and tutors in teaching measure and integration, and for researchers who may discover new and sometimes unexpected phenomena. Readers are assumed to have basic knowledge of functional analysis, point-set topology and, of course, measure and integration theory. For novices, there is a panorama of measure and integration which gives a non-technical overview on the subject and can serve, to some extent, as a first introduction. The overall presentation is as self-contained as possible; in order to make the text easy to access, we use only a few standard references – Schilling [MIMS] and Bogachev [19] for measure and integration, Rudin [151] and Yosida [202] for functional analysis, and Willard [200] and Engelking [53] for topology.

Some of the counterexamples are famous, many are more or less well known, and a few are of our own making. When we could trace the origin of an example, we have given references and attached names, but most entries are 'standard' examples which seem to have been in the public domain for ages; having said this, we acknowledge a huge debt to many anonymous authors and we do apologize if we have failed to give proper credit. The three classic counterexample books by Gelbaum & Olmsted [65], Steen & Seebach [172], and Stoyanov [180] were both inspiration and encouragement. We hope that this book lives up to their high standards.

It is a pleasure to acknowledge the interest and skill of our editor, Roger Astley, in the preparation of this book and Cambridge University Press for the excellent book design. Many colleagues have contributed to this text with comments and suggestions, in particular M. Auer, R. Baumgarth, G. Berschneider, N.H. Bingham – for the famous full red-ink treatment, C.-S. Deng, D.E. Edmunds and C. Goldie – for most helpful discussions, Y. Ishikawa, N. Jacob – for access to his legendary library, Y. Mishura and N. Sandrić. We thank our colleagues and friends who suffered for quite a while from our destructive search for counterexamples (*Do you know an example of a measure which fails to ... ?*), strange functions and many outer-worldly excursions – and our families who have us back in real life.

User's Guide

This book is not intended for linear reading – although this might well be possible – but invites the reader to browse, to read selectively and to look things up. We have, therefore, organized the material in self-contained chapters which treat different aspects of measure and integration theory. We assume that the reader has a basic knowledge of abstract measure and integration; the outline given in the 'panorama' (Chapter 1) is intended to refresh the reader's memory, to fix notation and to give a first non-technical introduction to the subject. The cross-reference [☞ $n.m$] appearing in the margin points towards essential counterexamples to the (positive) result at hand. Some supplementary material which is not always part of the mathematical curriculum is collected in Chapter 2; look it up once you need it.

☞ $n.m$

Cross-referencing. Throughout the text, [☞ $n.m$] and Example $n.m$ refers to counterexample m in Chapter n. Theorem $n.m$, Definition $n.m$, etc. point to the respective theorem, definition, etc. in the 'panorama' (Chapter 1) or the 'refresher' (Chapter 2). Equation m in Chapter n is denoted by ($n.m$). At the beginning of each chapter, we recall more specialized results and definitions which are particular to that chapter; these are numbered locally as 5A, 5B, 5C, ... (for Chapter 5, say) and they are mostly used within that chapter. Theorems, lemmas and corollaries may also appear in a counterexample; if needed, we use again local numbering $1, 2, 3, \dots$.

Finding stuff. Following Gelbaum & Olmsted [65] we have organized the examples by theme and all counterexamples are listed in the list of contents by (hopefully) meaningful names. We begin with examples on Riemann integration (Chapter 3) and move on to various aspects of the (abstract) Lebesgue integral (Chapters 4–19). The chapters on Lebesgue integration follow 'The way of integration' (alluding to Fig. 1.3 in Chapter 1), i.e. beginning with measurable sets and σ-algebras to set functions, measurable functions, to integrals and theorems on integration. The subject index helps to find definitions, theorems and concepts, but it does not refer to specific counterexamples.

Notation. We tried to avoid specialized notation and we use commonly accepted standard notation, e.g. as in [MIMS]. The following list is intended to aid cross-referencing, so notation that is specific to a single section is generally not listed; numbers following entries are page numbers.

Unless otherwise stated, binary operations between functions such as $f \pm g$, $f \cdot g$, $f \wedge g$, $f \vee g$, comparisons $f \leqslant g$, $f < g$ or limiting relations $f_n \xrightarrow[n\to\infty]{} f$, $\lim_n f_n$, $\liminf_n f_n$, $\limsup f_n$, $\sup_i f_i$ or $\inf_i f_i$ are always understood pointwise.

General notation

positive	always in the sense $\geqslant 0$
negative	always in the sense $\leqslant 0$
increasing	$x \leqslant y \Rightarrow f(x) \leqslant f(y)$
decreasing	$x \leqslant y \Rightarrow f(x) \geqslant f(y)$
countable	finite or countably infinite
$\mathbb{N}$	natural numbers: 1, 2, 3, ...
$\mathbb{N}_0$	positive integers: 0, 1, 2, ...
$\mathbb{Z}, \mathbb{Q}, \mathbb{R}, \mathbb{C}$	integer, rational, real, complex numbers
$\overline{\mathbb{R}}$	$[-\infty, +\infty]$ (two-point compactification), 11, 38
$\inf\emptyset$, $\sup\emptyset$	$\inf\emptyset = +\infty$, $\sup\emptyset = -\infty$
$a \vee b$, $a \wedge b$	$\max\{a, b\}$, $\min\{a, b\}$
$\gcd(\cdot,\cdot)$	greatest common divisor
$\aleph_0$	cardinality of $\mathbb{N}$, 44
$\mathfrak{c}$	cardinality of $\mathbb{R}$, 44
ω_0	first infinite ordinal, ordinal number of $\mathbb{N}$, 44
ω_1	first uncountable ordinal, 45, 46
$\Omega = [0, \omega_1]$	ordinal space, 45, 46
$\Omega_0 = [0, \omega_1)$	countable ordinals, 45, 46

Sets and set operations

$A \uplus B$	union of disjoint sets
$A \vartriangle B$	$(A \setminus B) \uplus (B \setminus A)$
A^c	complement of A
$\overline{A}$	closure of A, 37
A°	open interior of A, 37
$A_n \uparrow A$	$A_n \subseteq A_{n+1}$, $A = \bigcup_n A_n$
$A_n \downarrow A$	$A_n \supseteq A_{n+1}$, $A = \bigcap_n A_n$
$\#A$	cardinality of A
$B_r(x)$	open (metric) ball $\{y\,;\ d(x,y) < r\}$

Families of sets

$\mathscr{A}, \mathscr{B}, \mathscr{C}$	generic families of sets
$\mathscr{A}^*$	μ^* measurable sets, 28 completion, 9
$\mathscr{A} \otimes \mathscr{B}$	product σ-algebra, 10, 21
$\mathscr{B}(X)$	Borel sets in X, 9
$\mathscr{L}(X)$	Lebesgue sets in X, 10
$\mathscr{O}(X)$	open sets in X, 36
$\mathscr{P}(X)$	all subsets of X
$\sigma(\mathscr{F})$	σ-algebra generated by $\mathscr{F}$, 9
$\sigma(\phi)$, $\sigma(\phi_i, i \in I)$	σ-algebra generated by the map(s) ϕ, resp. ϕ_i, 9

Measures and integrals

μ, ν	generic (positive) measures
μ_*, μ^*	inner and outer measure, 182, 100
δ_x	Dirac measure in x, 8
λ, λ^d	Lebesgue measure, 9
$\zeta, \zeta_X, \#(\cdot)$	counting measure on X, 8
$\mu\circ f^{-1}, f_*\mu$	image or push-forward measure, 8, 24

$\mu \times \nu$	product of measures, 10, 21
$\mu * \nu$	convolution, 26
$\mu \ll \nu$	absolute continuity, 29
$\mu \perp \nu$	singular measures, 29
$\frac{d\nu}{d\mu}$	Radon–Nikodým derivative, 29
$\operatorname{supp} \mu$	support of a measure, 123
$\overline{\int}, \underline{\int}$	upper, lower R-integral, 2

Functions and spaces

$\mathbb{1}_A$	$\mathbb{1}_A(x) = \begin{cases} 1, & x \in A, \\ 0, & x \notin A, \end{cases}$
$\operatorname{sgn}(x)$	$\mathbb{1}_{(0,\infty)}(x) - \mathbb{1}_{(-\infty,0)}(x)$
$f(A)$	$\{f(x);\ x \in A\}$
$f^{-1}(\mathscr{B})$	$\{f^{-1}(B);\ B \in \mathscr{B}\}$
f^+	$\max\{f(x), 0\}$ positive part
f^-	$-\min\{f(x), 0\}$ negative part
$\{f \in B\}$	$\{x;\ f(x) \in B\}$
$\{f \geqslant \lambda\}$	$\{x;\ f(x) \geqslant \lambda\}$, etc.
$f * g$	convolution, 26
$\operatorname{supp} f$	support $\overline{\{f \neq 0\}}$
$C(X)$	continuous functions on X
$C_b(X)$	bounded — —
$C_c(X)$	— — with compact support
L^p, L^∞	Lebesgue spaces, 15
$\|f\|_p, \|f\|_{L^p}$	$(\int \|f\|^p \, d\mu)^{1/p}$, $p < \infty$
$\|f\|_\infty, \|f\|_{L^\infty}$	$\operatorname{esssup} f :=$ $\inf\{c;\ \mu\{\|f\| \geqslant c\} = 0\}$, 14

List of Topics and Phenomena

Topic	Possible consequence	Example
μ is not finite	▶ μ not continuous from above	[☞ 5.10]
	▶ range of μ not closed	[☞ 6.17]
	▶ Jensen's inequality does not hold	
	▶ $p \leqslant q \not\Rightarrow L^q \subseteq L^p$	[☞ 18.1]
	▶ no series test for integrability	[☞ 10.5]
	▶ Egorov's theorem fails	[☞ 11.11]
	▶ convergence in probability $\not\Rightarrow$ convergence in measure	[☞ 11.5]
μ is not σ-finite	▶ no unique product measure	[☞ 16.1]
	▶ Fubini's and Tonelli's theorem fail	[☞ 16.8–16.18]
	▶ Radon–Nikodým's theorem fails	[☞ 17.1–17.3]
	▶ Lebesgue's decomposition theorem fails	[☞ 17.11]
	▶ there is no positive integrable function	[☞ 10.20]
	▶ limits in probability not unique	[☞ 11.6]
	▶ $f_n \to f$ in probability $\not\Rightarrow f_{n_k} \to f$ a.e.	[☞ 11.7]
	▶ L^p, $1 \leqslant p < \infty$, not separable	[☞ 18.14]
	▶ $(X, \mathscr{A}^*, \mu^*\vert_{\mathscr{A}^*}) \neq$ completion of $(X, \mathscr{A}, \mu)$	[☞ 9.9]
	▶ trace of a regular measure not regular	[☞ 9.22]
μ does not have the finite subset property	▶ $\int_A f\,d\mu = \int_A g\,d\mu$ for all $A \not\Rightarrow f = g$ a.e.	[☞ 10.23]
	▶ $\exists f \in L^\infty$ s.t. $\Lambda_f(g) = \int fg\,d\mu$, $g \in L^1$, satisfies $\Vert\Lambda_f\Vert < \Vert f\Vert_{L^\infty}$	[☞ 18.21]
	▶ $\int \vert fg\vert\,d\mu \leqslant C\Vert g\Vert_{L^q} \not\Rightarrow f \in L^p$	[☞ 18.19]
μ is not locally finite	▶ $C_b(X) \cap L^p(\mu)$ not dense in $L^p(\mu)$, $1 \leqslant p < \infty$	[☞ 18.16]
μ is not regular	▶ Lusin's theorem fails	[☞ 13.16]
	▶ $C_b(X) \cap L^p(\mu)$ not dense in $L^p(\mu)$, $1 \leqslant p < \infty$	[☞ 18.16]
	▶ there exists $\nu \neq \mu$ s.t. $\int f\,d\mu = \int f\,d\nu$ for all $f \in C_c(X)$	[☞ 18.26, 18.27]

Topic	Possible consequence	Example
X is not separable	▶ $\mathscr{B}(X)$ not generated by open balls ▶ $\mathscr{B}(X)$ not countably generated ▶ $\operatorname{supp}\mu \neq$ smallest closed set F such that $\mu(X \setminus F) = 0$ ▶ $\mu(X) \neq \mu(\operatorname{supp}\mu)$ ▶ finite measures not tight	[☞4.14] [☞4.10] [☞6.2] [☞6.3] [☞5.26]
X is not a metric space	▶ compact sets not Borel ▶ fewer Baire sets than Borel sets ▶ pointwise limits of measurable functions not measurable ▶ finite measures not outer regular ▶ inner compact regular $\not\Rightarrow$ inner regular	[☞4.15] [☞4.24] [☞8.16] [☞9.18] [☞9.18]
X is not σ-compact	▶ locally finite $\not\Rightarrow$ σ-finite ▶ inner regular $\not\Rightarrow$ inner compact regular	[☞5.16] [☞9.20]
X is not locally convex	▶ only trivial dual space ▶ no Bochner integral	[☞18.11] [☞18.34]
X has cardinality $> \mathfrak{c}$	▶ the diagonal is not in $\mathscr{P}(X) \otimes \mathscr{P}(X)$ ▶ $\mathscr{B}(X) \otimes \mathscr{B}(X) \neq \mathscr{B}(X \times X)$ ▶ metric not jointly measurable with respect to $\mathscr{B}(X) \otimes \mathscr{B}(X)$	[☞15.9] [☞15.6] [☞15.10]
$\mathscr{A}$ is too small, e.g. discrete	▶ 'few' measurable functions $f : X \to \mathbb{R}$ ▶ factorization lemma fails	[☞8.2] [☞8.20]
$\mathscr{A}$ is too big, e.g. discrete	▶ 'many' measurable functions $f : X \to \mathbb{R}$ ▶ 'few' non-atomic measures	[☞8.1] [☞6.15]
$\mathscr{A}$ not countably generated	▶ two-valued measures which are not a point mass	[☞6.10]
role of 'small' sets	▶ Lebesgue null sets may be uncountable/of second category ▶ $B + B = \mathbb{R}$ for a Lebesgue null set B ▶ $2^{\mathfrak{c}}$ many Lebesgue sets but 'only' $\mathfrak{c}$ many Borel sets ▶ $f' = 0$ a.e. $\not\Rightarrow$ f constant ▶ f a.e. continuous $\not\Rightarrow$ $f = g$ a.e. for g continuous ▶ $\mu_n \to \mu$ weakly $\not\Rightarrow$ $\mu_n(B) \to \mu(B)$ for all B ▶ support of a probability measure may have measure 0	[☞7.4, 7.9] [☞7.27] [☞4.20] [☞2.6, 14.5] [☞13.1] [☞19.5] [☞6.3]
lack of countability	▶ many theorems fail for nets, e.g. classical convergence theorems, Egorov's and Lévy's continuity theorem ▶ uncountable supremum of measurable functions are not measurable ▶ $\mathscr{B}(X)^{\otimes I}$ is 'small' for I uncountable	[☞12.9] [☞11.12, 19.12] [☞8.18] [☞15.8, 4.17]

Topic	Possible consequence	Example
	▶ projective limit of consistent family may not exist	[☞16.21]
	▶ $t \longmapsto f(t,x)$ cts. $\forall x$ $\not\Rightarrow t \longmapsto \int f(t,x)\,\mu(dx)$ cts.	[☞14.12]
lack of uniform integrability	▶ $(f_n)_{n\in\mathbb{N}} \subseteq L^1$, $f_n \to 0$ a.e. $\not\Rightarrow \int f_n \to 0$	[☞12.1, 12.7]
	▶ $f_n \to f$, $f_n' \to g$ everywhere $\not\Rightarrow f' = g$ a.e.	[☞14.8]
	▶ $f_n \to f$ in probability $\not\Rightarrow f_{n_k} \to f$ in measure	[☞11.8]
	▶ sequential weak compactness fails	[☞18.29]
L^1, L^∞ are special	▶L^∞ separable if, and only if, $\dim L^\infty < \infty$	[☞18.15]
	▶$(L^1)^* \supsetneqq L^\infty$	[☞18.21, 18.22]
	▶$(L^\infty)^* \supsetneqq L^1$	[☞18.24]
	▶ not uniformly convex	[☞18.32]
⚠ atom	▶ comparison: different definitions of atom	[☞6.11]
⚠ absolute continuity	▶ comparison: different definitions of absolute continuity	[☞17.4]
⚠ convergence in measure	▶ comparison: convergence in measure vs. in probability	[☞p. 20 Fig. 1.4 pp. 221, 226–227]
⚠ weak convergence	▶ weak convergence of measures is not weak convergence in the sense of functional analysis	[☞p. 371, 19.7]
⚠ Baire σ-algebra	▶ comparison of different definitions of Baire sets	[☞4.23]

1

A Panorama of Lebesgue Integration

The idea of measuring area and volume by infinitesimal (exhaustion) methods was already known to the ancient Greeks. This may be seen as the first example of 'integration'. The precursor of our modern notion of integration begins with the creation of the infinitesimal calculus by Newton and Leibniz. For Newton, the derivative was the primary operation of calculus and the integral was just the primitive, i.e. the antiderivative. Leibniz followed a more geometric approach, defining the integral as a sum of infinitesimal quantities which represent the area below the graph of a curve, thus establishing the integral as an object in its own right. Of course, both Newton and Leibniz were describing essentially the same object, and the history of integration is, in some sense, the attempt to reconcile both definitions. A short overview of the early history of integration is given in Section 1.11 at the end of this chapter. For us, the modern theory of integration starts in the year 1854 with Riemann's Habilitationsschrift [143].

1.1 Modern Integration. 'Also zuerst: Was hat man unter $\int_a^b f(x)\,dx$ zu verstehen?'[1]

Riemann's answer to (t)his question is the following definition [143, Section 4]:

Definition 1.1 A bounded function $f : [a, b] \to \mathbb{R}$ defined on a compact interval $[a, b] \subseteq \mathbb{R}$ is **integrable** (in the sense of Riemann) if the limit

$$\int_a^b f(x)\,dx = \lim_{|\Pi|\to 0} \sum_{i=1}^{n} (x_i - x_{i-1}) f(\xi_i), \quad |\Pi| := \max_{1\leqslant i\leqslant n} |x_i - x_{i-1}|, \tag{1.1}$$

taken along all finite partitions $\Pi = \{a = x_0 < x_1 < \cdots < x_n = b\}$ and for any choice of intermediary points $\xi_i \in [x_{i-1}, x_i]$ exists and is finite.

[1] Riemann [143, p. 239] – To begin with: What is the meaning of $\int_a^b f(x)\,dx$?

Riemann immediately gives two necessary and sufficient conditions for the convergence of (1.1), cf. [MIMS, p. 443] for a modern proof. Denote by Π a finite partition of $[a, b]$ and write $D_i := \sup_{[x_{i-1},x_i]} f - \inf_{[x_{i-1},x_i]} f$ for the oscillation of a function f in the ith partition interval $[x_{i-1}, x_i]$.

(R1) The limit in (1.1) exists if, and only if, for all finite partitions Π of $[a, b]$,

$$\lim_{|\Pi|\to 0} \sum_{i=1}^{n} D_i \cdot |x_i - x_{i-1}| = 0.$$

(R2) The limit in (1.1) exists if, and only if,

$$\forall \epsilon > 0,\ \sigma > 0 \quad \exists \delta > 0 \quad \forall \Pi,\ |\Pi| \leqslant \delta \quad \sum_{i:D_i>\sigma} |x_{i-1} - x_i| < \epsilon.$$

In retrospect, Riemann's condition (R2) marks the beginning of the study of outer (Lebesgue) measure. We will see in Theorem 1.28 below that a bounded function f is Riemann integrable if, and only if, the set of its discontinuity points is a Lebesgue null set.

From (R2) it is clear that the Riemann integral is capable of dealing with functions which are discontinuous on a (countable) dense subset. This fact was already illustrated by Riemann in [143] using the function

$$f(x) = \sum_{n=1}^{\infty} \frac{h(nx)}{n^2}, \qquad h(x) = \begin{cases} x - k, & \text{if } x \in \left(k - \frac{1}{2}, k + \frac{1}{2}\right),\ k \in \mathbb{Z}, \\ 0, & \text{if } x = k \pm \frac{1}{2},\ k \in \mathbb{Z}, \end{cases}$$

which is discontinuous on the set $Q = \{p/(2n);\ \gcd(p, 2n) = 1\}$; see Fig. 3.3. Hankel [73, pp. 199–200] observed that f is an example of a function such that

$$F(x) := \int_0^x f(t)\,dt$$

is continuous, but $F'(x) = f(x)$ fails if $x \in Q$, i.e. F is not a primitive of f [☞ 3.7].

After the publication of Riemann's 1854 thesis in 1867, his definition of the integral was widely accepted, and it is still one of the most important and widely used notions of integration. The presentation was quickly streamlined, notably by the introduction of upper and lower sums and integrals which make Riemann's criterion (R1) more tractable.

Definition 1.2 (Thomae [183], Darboux [37], Volterra [195]) For a bounded function $f : [a, b] \to \mathbb{R}$ we call

$$S_\Pi[f] := \sum_{x_{i-1},x_i\in\Pi} m_i \cdot (x_i - x_{i-1}), \qquad m_i := \inf_{x\in[x_{i-1},x_i]} f(x),$$

$$S^{\Pi}[f] := \sum_{x_{i-1},x_i\in\Pi} M_i\cdot(x_i-x_{i-1}), \qquad M_i := \sup_{x\in[x_{i-1},x_i]} f(x),$$

the **lower** and **upper Darboux sums** and

$$\underline{\int_a^b} f(x)\,dx := \sup_{\Pi\subseteq[a,b]} S_{\Pi}[f] \quad\text{and}\quad \overline{\int_a^b} f(x)\,dx := \inf_{\Pi\subseteq[a,b]} S^{\Pi}[f]$$

(sup and inf range over all finite partitions Π of $[a,b]$) the **lower** and **upper Riemann–Darboux integrals**.

Using the lower and upper integrals we can show the following integrability criterion.

Theorem 1.3 ([MIMS, p. 443]) *A bounded function $f: [a,b]\to\mathbb{R}$ is Riemann integrable if, and only if,*

$$-\infty < \underline{\int_a^b} f(x)\,dx = \overline{\int_a^b} f(x)\,dx < \infty.$$

The common (finite) value is the Riemann integral $\int_a^b f(x)\,dx$.

The development of the Riemann integral and the concept of a function go hand in hand. Up to Cauchy, functions were (implicitly) thought to be smooth, after Cauchy to be continuous; from 1867, Riemann integrable functions were seen to be the most general and still reasonable functions. But soon there were first examples of non-Riemann integrable functions, and other shortcomings of the Riemann integral were discovered:

1° The rather limited scope of Riemann integrable functions. The (proper) Riemann integral makes sense only on bounded sets and for bounded functions [☞ 3.2, 3.3], it behaves badly under compositions [☞ 3.11] and there are rather natural and simple non-integrable functions [☞ 3.4, 3.5].

2° If the Riemann integral is extended to two dimensions, the familiar formula

$$\iint_{[a,b]\times[c,d]} f(x,y)\,dx\,dy = \int_a^b \left[\int_c^d f(x,y)\,dy\right] dx = \int_c^d \left[\int_a^b f(x,y)\,dx\right] dy$$

may become senseless since some, or all, of the one-dimensional integrals might not exist [☞ 3.20].

3° Riemann's theory does not fix the difference between integral and primitive. There are integrable functions f such that $F(x) = \int_a^x f(t)\,dt$ is not everywhere differentiable, i.e. not a proper primitive. Worse, there are everywhere differentiable functions F whose derivative F' is not integrable [☞ 14.2, 14.3].

4° The Riemann integral behaves rather badly if one wants to interchange limits and integrals. Among other pathologies, one can construct a uniformly bounded sequence of Riemann integrable functions $(f_n)_{n\in\mathbb{N}}$ on $[0,1]$ such that $\lim_{n\to\infty} f_n(x) = f(x)$ but f is not Riemann integrable [☞ 3.13, 3.14, 3.16].

1.2 The Idea Behind Lebesgue Integration

Part of the problem with Riemann's definition is that the approximation procedure used in (1.1) is based on given partitions of the domain $[a,b]$ of the function $f: [a,b] \to \mathbb{R}$, i.e. these partitions need not relate to the behaviour of f.

Lebesgue's idea in [100, 101] is to split the range $f([a,b])$ of a bounded function $f: [a,b] \to \mathbb{R}$ into equal intervals, say $J_1, \dots, J_k$, and to determine those sets $I_1, \dots, I_k \subseteq [a,b]$ such that $I_i = f^{-1}(J_i)$. The corresponding approximations of the integral would be

$$U = \sum_{i=1}^{k} |I_i| \cdot \sup J_i \quad \text{and} \quad L = \sum_{i=1}^{k} |I_i| \cdot \inf J_i, \tag{1.2}$$

where $|A|$ denotes the total length of the set A. If we choose an equidistant partitioning of mesh δ, the value of the upper approximation is $U = L + \delta \cdot |[a,b]| = L + \delta \cdot (b-a)$, i.e. it is enough to restrict one's attention to L. Notice that the resulting partition of the domain depends on f. Before we give proper definitions and discuss the implications of this approach, let us consider a simple example.

Example 1.4 Consider an oscillating periodic function, e.g. $f(x) = \sin^2(n\pi x)$ with $n \in \mathbb{N}$ and $x \in [0,1)$, cf. Fig. 1.1. Using the relation $\sin^2\alpha = \frac{1}{2}(1-\cos 2\alpha)$ it is easy to determine the integral of f,

$$\int_0^1 \sin^2(n\pi x)\,dx = \frac{1}{2}\int_0^1 (1-\cos(2n\pi x))\,dx = \frac{1}{2}\int_0^1 dx = \frac{1}{2},$$

but the upper and lower Darboux sums for an equidistant partition of $[0,1]$, $\Pi = \{0 = x_0 < \cdots < x_k = 1\}$, with mesh $|\Pi| = \frac{1}{k} \geqslant \frac{1}{n}$ $\left(\text{or a general partition such that } \min_i(x_i - x_{i-1}) \geqslant \frac{1}{n}\right)$ are easily seen to be 1 and 0, respectively.

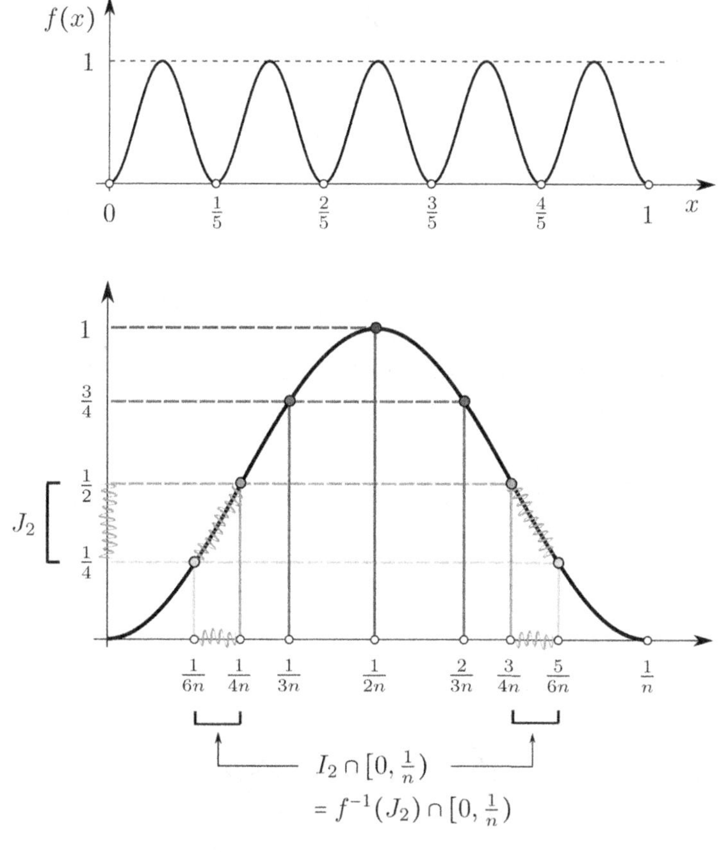

Figure 1.1 The oscillating periodic function $f(x) = \sin^2(n\pi x)$ for $n = 5$ (upper panel) and the choice of the domain partition I_k for an equidistant range partition J_k (lower panel).

However, Lebesgue's approach using $k = 4$ and $J_i = \left[\frac{i-1}{k}, \frac{i}{k}\right)$ gives, over the first period,

$$I_i \cap \left[0, \tfrac{1}{n}\right) = \left\{0 \leqslant x < \tfrac{1}{n};\ \tfrac{i-1}{k} \leqslant f(x) < \tfrac{i}{k}\right\} = [x_{i-1}, x_i) \cup (x_{8-i}, x_{8-(i-1)}]$$

with $x_0 = 0$, $x_1 = 1/6n$, $x_2 = 1/4n$, $x_3 = 1/3n$, $x_4 = 1/2n$, $x_5 = 2/3n$, $x_6 = 3/4n$, $x_7 = 5/6n$ and $x_8 = 1/n$; see Fig. 1.1 (lower panel). Thus, in $[0, 1/n)$ we get

$$\sum_{i=1}^{4} \frac{i-1}{4} \left[(x_i - x_{i-1}) + (x_{8-(i-1)} - x_{8-i})\right] = \left(\frac{1}{24} + \frac{1}{12} + \frac{1}{4}\right)\frac{1}{n} = \frac{3}{8n}.$$

Since there are n periods in $[0, 1]$, we have $L = \frac{3}{8}$ (and $U = \frac{3}{8} + \frac{1}{4} = \frac{5}{8}$). This is already a reasonable approximation of the true value $\frac{1}{2}$ and, if one uses $\frac{1}{2}(U+L)$, it even happens to be the exact value.

This example makes it clear that Lebesgue's approach is better suited to deal with (rapidly) oscillating integrands, in particular, when the oscillations approach a condensation point as is the case for $x \longmapsto \sin^2 \frac{1}{x}$ as $x \to 0$.

1.3 Lebesgue Essentials – Measures and σ-Algebras

Let us recast Lebesgue's approximation of a function $f \geqslant 0$ from below by slicing its range **horizontally** as shown in Fig. 1.2. The level sets

$$A_k^n := \begin{cases} \{k2^{-n} \leqslant f < (k+1)2^{-n}\} & \text{for } k = 0, 1, 2, \dots, n2^n - 1, \\ \{f \geqslant n\} & \text{for } k = n2^n, \end{cases}$$

can be used to define step functions

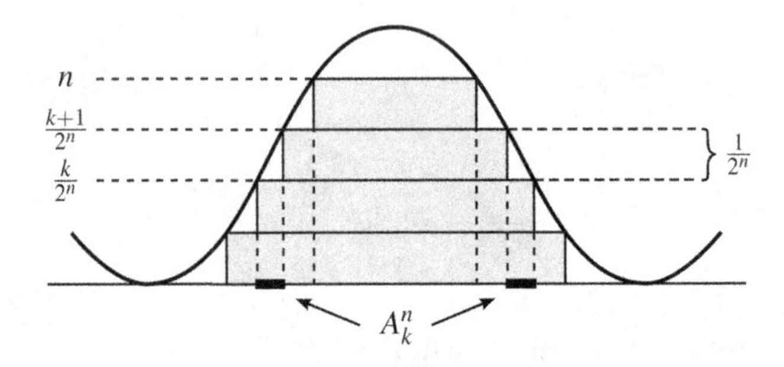

Figure 1.2 The function f sits like a 'Mexican hat' (a sombrero) over the approximating simple functions.

$$\phi_n(x) := \sum_{k=0}^{n2^n} k2^{-n}\mathbb{1}_{A_k^n}(x)$$

which approximate f from below. Coming from below has the advantage that f need not be bounded; instead, we use a moving cut-off level n which kicks in on the set $A_{n2^n}^n$. From Fig. 1.2 we see that

(i) $0 \leqslant \phi_n \leqslant \phi_{n+1} \leqslant f$ and $\phi_n \uparrow f$;
(ii) $|\phi_n(x) - f(x)| \leqslant 2^{-n}$ if $x \in \{f < n\}$; in particular, if f is bounded, the sequence ϕ_n approximates f uniformly.

We are interested in the nature of the level sets A_k^n. Property (i) requires that we are able to subdivide the level sets A_k^n finitely often. If we want to integrate the sum of two functions f, g, the level sets of $f + g$ will be expressed through finite unions and intersections of the level sets of f and g.

Therefore, the level sets form a family of sets which is closed if we repeat the usual set operations (intersection, union, taking complements and differences) finitely or – for limits – countably infinitely often. This requirement leads naturally to the notion of a σ-algebra.

Definition 1.5 Let $X \neq \emptyset$ be any set, and denote by $\mathscr{P}(X)$ its power set. A **σ-algebra** $\mathscr{A}$ on X is a family of subsets of X with the following properties:

$$X \in \mathscr{A}, \tag{Σ_1}$$

$$A \in \mathscr{A} \Rightarrow A^c \in \mathscr{A}, \tag{Σ_2}$$

$$(A_n)_{n\in\mathbb{N}} \subseteq \mathscr{A} \Rightarrow \bigcup_{n\in\mathbb{N}} A_n \in \mathscr{A}. \tag{Σ_3}$$

Because of (Σ_1) and (Σ_2) we have $\emptyset \in \mathscr{A}$, and using $A_1 \cup A_2 \cup \emptyset \cup \dots$ in (Σ_3) shows that $\mathscr{A}$ is stable under finite unions. With de Morgan's laws we get that $\mathscr{A}$ is also stable under finite and countably infinite intersections and this is also true for differences as $A \setminus B = A \cap B^c$ is a combination of complementation and intersection.

The second ingredient needed for the construction of the integral is a 'gauge' for the size of the level sets A_k^n. In Example 1.4 we naively took the 'length' of the interval and there was no problem since the level sets were relatively simple. In the general case we need a function defined on all possible level sets which is compatible with (countably often repeated) set operations. This is the rationale for the following definition.

Definition 1.6 Let $X \neq \emptyset$ be any set. A (positive) **measure** is a set function $\mu : \mathscr{A} \to [0, \infty]$ satisfying

$$\mathscr{A} \text{ is a } \sigma\text{-algebra on } X, \tag{M_0}$$

$$\mu(\emptyset) = 0, \tag{M_1}$$

$$(A_n)_{n\in\mathbb{N}} \subseteq \mathscr{A} \text{ pairwise disjoint} \Rightarrow \mu\left(\bigcup\nolimits^{\bullet}_{n\in\mathbb{N}} A_n\right) = \sum_{n\in\mathbb{N}} \mu(A_n). \tag{M_2}$$

The pair $(X, \mathscr{A})$ is called a **measurable space** and $(X, \mathscr{A}, \mu)$ is called a **measure space**. The measure space is called **finite**, if $\mu(X) < \infty$, and **σ-finite**, if there exists a sequence $(F_n)_{n\in\mathbb{N}} \subseteq \mathscr{A}$ such that $X = \bigcup_{n\in\mathbb{N}} F_n$ and $\mu(F_n) < \infty$. A set $A \in \mathscr{A}$ is often called a **measurable set**.

The requirements (M_0)–(M_2) lead to a rich family of set functions with many further properties; see [MIMS, pp. 24, 28]. For example, if $A, B, A_n, B_n \in \mathscr{A}$:

(a) $A \cap B = \emptyset \Rightarrow \mu(A \cup B) = \mu(A) + \mu(B)$ (additive)
(b) $A \subseteq B \Rightarrow \mu(A) \leqslant \mu(B)$ (monotone)
(c) $A \subseteq B,\ \mu(A) < \infty \Rightarrow \mu(B \setminus A) = \mu(B) - \mu(A)$
(d) $\mu(A \cup B) + \mu(A \cap B) = \mu(A) + \mu(B)$ (strongly additive)
(e) $\mu(A \cup B) \leqslant \mu(A) + \mu(B)$ (subadditive)
(f) $A_n \uparrow A \Rightarrow \mu(A) = \sup_n \mu(A_n) = \lim_n \mu(A_n)$ (continuous from below)
(g) $B_n \downarrow B,\ \mu(B_1) < \infty \Rightarrow \mu(B) = \inf_n \mu(B_n) = \lim_n \mu(B_n)$ (continuous from above)
(h) $\mu\Big(\bigcup_{n\in\mathbb{N}} A_n\Big) \leqslant \sum_{n\in\mathbb{N}} \mu(A_n)$. ($\sigma$-subadditive)

Example 1.7 Here are some of the most commonly used measures and σ-algebras. Unless otherwise indicated, $(X, \mathscr{A}, \mu)$ is an arbitrary measure space.

σ-**Algebra** $\mathscr{A}$	**Typical measure** on $(X, \mathscr{A})$
(a) The **indiscrete σ-algebra**: $\{\emptyset, X\}$ – this is the smallest possible σ-algebra on X.	▸ The **trivial measure** $\tau(\emptyset) = 0$ and $\tau(X) = \infty$.
(b) The **discrete σ-algebra**: $\mathscr{P}(X)$ – this is the largest possible σ-algebra on X.	▸ As a rule of thumb, rich σ-algebras admit only poor (i.e. simple) measures: $\mathscr{P}(X)$ can support the **trivial measure** from (a), the **counting measure** $\zeta(A) = \#A$, or **Dirac's delta function (point mass)** at $x \in X$ $$\delta_x(A) = \begin{cases} 0, & \text{if } x \notin A, \\ 1, & \text{if } x \in A. \end{cases}$$
(c) The **trivial σ-algebra**: $\mathscr{T}_\mu = \{A \in \mathscr{A};\ \mu(A) = 0 \text{ or } \mu(A^c) = 0\}$.	▸ This construction works for every measure space $(X, \mathscr{A}, \mu)$.
(d) The **co-countable σ-algebra**: $\{A \subseteq X;\ \#A \leqslant \#\mathbb{N} \text{ or } \#A^c \leqslant \#\mathbb{N}\}$ on an uncountable set X [☞ Example 4A].	▸ The **co-countable** (probability) measure $$\mu(A) = \begin{cases} 0, & \text{if } A \text{ is countable}, \\ 1, & \text{if } A^c \text{ is countable}, \end{cases}$$ [☞ Example 5A].
(e) The **trace σ-algebra**: Let $E \subseteq X$. $\mathscr{A}_E = E \cap \mathscr{A} := \{E \cap A;\ A \in \mathscr{A}\}$.	▸ If $E \in \mathscr{A}$, the restriction $\mu|_E(A) := \mu(E \cap A)$ is a measure on the trace measurable space $(E, \mathscr{A}_E)$; [☞ 5.9] if $E \notin \mathscr{A}$.
(f) The **pre-image σ-algebra**: Let $\phi : X \to X'$ be any map and $\mathscr{A}'$ a σ-algebra on X'. $$\phi^{-1}(\mathscr{A}') = \{\phi^{-1}(A');\ A' \in \mathscr{A}'\}.$$	▸ If μ is a measure on $(X, \mathscr{A})$, then $$\mu'(A') := \mu(\phi^{-1}(A'))$$ is called the **image measure** or **push-forward measure** of μ under ϕ. Notation: $\mu\circ\phi^{-1}$, $\phi_*\mu$ or $\phi(\mu)$.

(g) The **σ-algebra generated by a set** $F \subseteq X$: This is the smallest σ-algebra on X containing the set F: $\sigma(F) = \{\emptyset, F, F^c, X\}$.

(h) The **σ-algebra generated by a family of sets** $\mathscr{F}$: This is the smallest σ-algebra containing the family $\mathscr{F}$: $\sigma(\mathscr{F}) = \bigcap\{\mathscr{B}\,;\ \mathscr{F} \subseteq \mathscr{B}, \mathscr{B}\ \sigma\text{-algebra}\}$.

(i) Let $\phi_i : X \to X_i$, $i \in I$ be arbitrarily many mappings and assume that $\mathscr{A}_i$ is a σ-algebra in X_i. The **σ-algebra generated by the family of mappings** $(\phi_i)_{i\in I}$, $\sigma(\phi_i, i \in I) = \sigma\left(\bigcup_{i\in I} \phi_i^{-1}(\mathscr{A}_i)\right)$, is the smallest σ-algebra that makes all ϕ_i measurable (see Definition 1.8 further on).

(j) The **completed σ-algebra**: Let $\mathscr{F} \subseteq \mathscr{A}$ be a (not necessarily proper) sub-σ-algebra,

$$\mathscr{N}_\mu = \{N \in \mathscr{A}\,;\ \mu(N) = 0\}$$

the family of all measurable null sets, and

$$\mathscr{N}_\mu^* = \{N^* \subseteq X\,;\ \exists N \in \mathscr{N}_\mu,\ N^* \subseteq N\}$$

the family of all subsets of measurable null sets.
The **completion** of $\mathscr{F}$ is the σ-algebra $\mathscr{F}^* := \sigma(\mathscr{F}, \mathscr{N}^*)$. One can show that

$$\begin{aligned}\mathscr{F}^* &= \{F \vartriangle N^*\,;\ F \in \mathscr{F},\ N^* \in \mathscr{N}^*\}\\ &= \big\{F^*\,;\ \exists A, B \in \mathscr{F},\ A \subseteq F^* \subseteq B,\\ &\qquad\qquad \mu(B \setminus A) = 0\big\}.\end{aligned}$$

▶ The **completion** $\bar{\mu}$ of the measure μ (defined on $\mathscr{F}$) is the measure $\bar{\mu}$ on the measurable space $(X, \mathscr{F}^*)$ given by

$$\bar{\mu}(F^*) := \frac{1}{2}(\mu(A) + \mu(B)), \quad F^* \in \mathscr{F}^*,$$

where the sets $A, B \in \mathscr{F}$ are such that $\mu(B \setminus A) = 0$ and $A \subseteq F^* \subseteq B$. The former ensures that $\bar{\mu}$ is well-defined, i.e. independent of the choice of the sets A and B.
Since $\mathscr{F} \subseteq \mathscr{F}^*$, $\bar{\mu}$ is an extension of μ.

(k) Let X be a topological space and $\mathscr{O}$ the family of all open sets. The **Borel** or **topological σ-algebra** is the σ-algebra generated by the open sets $\mathscr{B}(X) = \sigma(\mathscr{O})$. Since a set is open if its complement is closed, $\mathscr{B}(X)$ is also generated by the closed sets. If X is a metric space which is the union of countably many compact sets $X = \bigcup_{n\in\mathbb{N}} K_n$ (e.g. if X is locally compact and separable), then $\mathscr{B}(X)$ is also generated by the compact sets [☞4.15, 4.16].

(l) The Borel sets in $\mathbb{R}^d$, $\mathscr{B}(\mathbb{R}^d)$, are generated by any of the following families: The open sets, the closed sets, the compact sets, the open balls $B_r(q)$ (radius $r \in \mathbb{Q}^+$, centre $q \in \mathbb{Q}^d$), the rectangles $\times_{i=1}^d [a_i, b_i)$ (with rational $a_i, b_i \in \mathbb{Q}$).

▶ Most measures used in analysis are defined on the Borel sets (or their completion, the Lebesgue sets, cf. Example (n)). The prime example of a measure on $\mathscr{B}(\mathbb{R}^d)$ is d-dimensional Lebesgue measure λ^d. Since the structure of the Borel sets is quite complicated, one defines λ^d on a sufficiently rich generator

$$\lambda^d\left(\mathop{\times}_{i=1}^{d} [a_i, b_i)\right) = \prod_{i=1}^{d} (b_i - a_i).$$

We will see in Theorem 1.40 that this characterizes λ^d uniquely.

(m) If $A \subseteq \mathbb{R}^d$, then $\mathscr{B}(A)$ is the Borel σ-algebra which is generated by the relatively open subsets of A. It is not hard to see that $\mathscr{B}(A)$ coincides with the trace σ-algebra $A \cap \mathscr{B}(\mathbb{R}^d)$.

▶ Use λ^d_A, the trace of Lebesgue measure λ^d on the trace-σ-algebra; see Example (e).

(n) The **Lebesgue σ-algebra** or **Lebesgue sets** $\mathscr{L}(\mathbb{R}^d)$ are the completion, see Example (j), of the Borel sets with respect to Lebesgue measure.

▶ Use the completion $\bar{\lambda}^d$ of λ^d; see Example (j).

(o) The **product σ-algebra** $\mathscr{A} \otimes \mathscr{B}$ is the σ-algebra $\sigma(\mathscr{A} \times \mathscr{B})$ generated by all generalized 'rectangles', i.e. sets of the form $A \times B \in \mathscr{A} \times \mathscr{B}$ with $A \in \mathscr{A}$ and $B \in \mathscr{B}$.

▶ Let $(X, \mathscr{A}, \mu)$ and $(Y, \mathscr{B}, \nu)$ be σ-finite measure spaces. Similar to the construction of Lebesgue measure, the product measure ρ is defined first on the sets $A \times B \in \mathscr{A} \times \mathscr{B}$ of a generator,

$$\rho(A \times B) := \mu(A)\nu(B),$$

and from the general theory it is known that this characterizes ρ on $\mathscr{A} \otimes \mathscr{B}$, cf. Theorem 1.33.

1.4 Lebesgue Essentials – Integrals and Measurable Functions

Let us return to the original problem of integrating a function. A real-valued function $f : X \to \mathbb{R}$ whose level sets $\{a \leqslant f < b\}$ are in a σ-algebra $\mathscr{A}$ on X is called measurable. The observation

$$\{a \leqslant f < b\} = \{f \geqslant a\} \cap \{f < b\} = f^{-1}([a, \infty)) \cap f^{-1}((-\infty, b))$$

explains the following slightly more general definition.

Definition 1.8 Let $(X, \mathscr{A})$ and $(Y, \mathscr{B})$ be two measurable spaces. A mapping $f : X \to Y$ is called $\mathscr{A}/\mathscr{B}$ **measurable**, if

$$\forall B \in \mathscr{B} \ : \ f^{-1}(B) \in \mathscr{A}. \tag{1.3}$$

If Y is a topological space equipped with its Borel sets, then measurable functions f are also called **Borel maps** or **Borel functions**.

Remark 1.9 (a) If $\mathscr{B}$ is generated by some family $\mathscr{H}$, then (1.3) is equivalent to the requirement that $f^{-1}(H) \in \mathscr{A}$ for all $H \in \mathscr{H}$. In particular, if we consider $\mathbb{R}$ equipped with the Borel σ-algebra $\mathscr{B}(\mathbb{R})$, then measurability of $f : X \to \mathbb{R}$ means that $\{f \leqslant a\} \in \mathscr{A}$ for all $a \in \mathbb{R}$ or $\{f > b\} \in \mathscr{A}$ for all $b \in \mathbb{R}$; see [MIMS, pp. 54, 60].

Since the pre-image of an open set under a continuous function is open, continuous functions are always Borel measurable.

(b) The pre-image σ-algebras from Example 1.7.(f), (i) are the smallest σ-algebras in X such that the map $\phi : X \to X'$, resp. all maps $\phi_i : X \to X_i$, become measurable [MIMS, p. 55].

(c) Let $(X, \mathscr{A})$, $(Y, \mathscr{B})$ and $(Z, \mathscr{C})$ be measurable spaces and $f : X \to Y$ and $g : Y \to Z$ be $\mathscr{A}/\mathscr{B}$, resp. $\mathscr{B}/\mathscr{C}$, measurable maps. The composition $f \circ g$ is $\mathscr{A}/\mathscr{C}$ measurable [MIMS, p. 54].

(d) Let $(Z, \mathscr{C})$ and $(X_i, \mathscr{A}_i)$, $i = 1, 2$, be measurable spaces. The product σ-algebra, see Example 1.7.(o), is constructed in such a way that a mapping $f = (f_1, f_2) : Z \to X_1 \times X_2$ is $\mathscr{C}/\mathscr{A}_1 \otimes \mathscr{A}_2$ measurable if, and only if, the coordinate maps $f_i : Z \to X_i$ are $\mathscr{C}/\mathscr{A}_i$ measurable [MIMS, p. 149]. This construction is analogous to the definition of an initial topology (also known as the weak or limit or projective topology).

(e) The family of Borel measurable real-valued functions is a vector space which is closed under countable pointwise infima and suprema. In particular, the lim inf, lim sup and lim of a sequence of Borel measurable functions is again a Borel measurable function – possibly with values in the extended real line $[-\infty, \infty]$; see [MIMS, pp. 66, 67].

☞8.15-8.18

A **simple function** is a function $\phi : X \to \mathbb{R}$ of the form

$$\phi(x) = \sum_{i=1}^{n} \alpha_i \mathbb{1}_{A_i}(x), \quad \alpha_i \in \mathbb{R},\ A_i \in \mathscr{A}. \tag{1.4}$$

Without loss we can assume that $A_i = \{\phi = \alpha_i\} = \{x \in X\,;\ \phi(x) = \alpha_i\}$. We write $\mathcal{E} = \mathcal{E}(X)$ for the family of all simple functions on a measurable space $(X, \mathscr{A})$. A simple function is obviously measurable and attains at most finitely many values. The following characterization of measurability of real functions is important for the development of the integral. It is, at the same time, an important structural result for the class of measurable functions. Its proof is, essentially, summed up in Fig. 1.2.

Theorem 1.10 (sombrero lemma [MIMS, pp. 64, 65]) *Let $(X, \mathscr{A})$ be any measurable space. A positive function $f : X \to [0, \infty]$ is $\mathscr{A}/\mathscr{B}(\overline{\mathbb{R}})$ measurable if, and only if, there exists an increasing sequence of simple functions $\phi_n \geqslant 0$ such that $f = \sup_{n \in \mathbb{N}} \phi_n$. If f is bounded, then ϕ_n approximates f uniformly.*

☞8.22

Remark 1.11 In measure theory one often works with the **extended real line** $\overline{\mathbb{R}} = [-\infty, \infty]$ which is the two-point compactification of $\mathbb{R}$ [☞Example 2.1(f)]; in addition to the usual neighbourhoods of points $x \in \mathbb{R}$, we have the neighbourhoods of $\pm\infty$ which are all sets containing $(x, +\infty]$, resp. $[-\infty, y)$, for some $x, y \in \mathbb{R}$. The Borel sets are $\mathscr{B}(\overline{\mathbb{R}}) = \sigma(\mathscr{B}(\mathbb{R}), \{\infty\}, \{-\infty\})$ and $\mathscr{B}(\mathbb{R})$ is the trace $\mathbb{R} \cap \mathscr{B}(\overline{\mathbb{R}})$.

It is clear how to add or multiply $x, y \in [-\infty, +\infty]$ unless both $x, y = \pm\infty$. For these cases, we agree that

$$\infty + \infty := \infty, \quad -\infty - \infty := -\infty, \quad \text{and} \quad 0 \cdot (\pm\infty) := 0,$$

while expressions of the type '$\infty - \infty$', '$-\infty + \infty$' or '$\pm\infty/\infty$' **are not defined**. Unless otherwise stated, we equip $\mathbb{R}$ and $\overline{\mathbb{R}}$ with the Borel σ-algebra $\mathscr{B}(\mathbb{R})$ and $\mathscr{B}(\overline{\mathbb{R}})$.

If we combine the idea that the integral represents the area below the graph of a positive function with the sombrero lemma, we naturally arrive at the following definition.

Definition 1.12 Let $(X, \mathscr{A}, \mu)$ be a measure space.

(a) The **integral** of a real-valued simple function $\phi \in \mathcal{E}(X)$ is

$$I_\mu(\phi) := \sum_{\alpha \in \phi(X)} \alpha \mu\{\phi = \alpha\} \in \mathbb{R}. \tag{1.5}$$

(b) The **integral** of a positive measurable function $f : X \to [0, \infty]$ is

$$\int f \, d\mu := \int f(x) \, \mu(dx) := \sup \{ I_\mu(\phi) ; \; \phi \in \mathcal{E}(X), \; 0 \leqslant \phi \leqslant f \} \in [0, \infty]. \tag{1.6}$$

It is obvious that both (1.5) and (1.6) are well-defined, i.e. independent of the representation, resp. approximation. Since the (finite) sum in (1.5) is linear and monotone, this is also true for the functional $\phi \longmapsto I_\mu(\phi)$. While it is easy to see that (1.6) extends (1.5) – i.e. $\int \phi \, d\mu = I_\mu(\phi)$ – it is not clear that the supremum preserves linearity. The following theorem comes to the rescue.

Theorem 1.13 (Beppo Levi [MIMS, p. 75]) *Let $(X, \mathscr{A}, \mu)$ be a measure space and $(f_n)_{n \in \mathbb{N}}$ an increasing sequence $0 \leqslant f_n \leqslant f_{n+1} \leqslant \ldots$ of positive measurable functions. The limit $f := \sup_{n \in \mathbb{N}} f_n : X \to [0, \infty]$ is measurable and*

$$\int \sup_{n \in \mathbb{N}} f_n \, d\mu = \sup_{n \in \mathbb{N}} \int f_n \, d\mu. \tag{1.7}$$

If $f : X \to [0, \infty]$ is measurable, then the sombrero lemma (Theorem 1.10) tells us that there is an increasing sequence of positive simple functions such that $\phi_n \uparrow f$. Therefore, the supremum $\int f \, d\mu$ is, in fact, an increasing limit $\lim_{n \to \infty} I_\mu(\phi_n)$. This means that on the positive measurable functions the functional $f \longmapsto \int f \, d\mu$ is monotone, additive and positively homogeneous.

The following equivalent formulation of (1.7) is often useful [MIMS, p. 77]:

☞12.5

$$\int \sum_{n=1}^{\infty} g_n \, d\mu = \sum_{n=1}^{\infty} \int g_n \, d\mu \quad \text{for positive, measurable } (g_n)_{n\in\mathbb{N}}. \tag{1.8}$$

In order to integrate measurable functions of an arbitrary sign, we use the decomposition $f = f^+ - f^-$ into the positive part $f^+ := \max\{f, 0\}$ and negative part $f^- := \max\{-f, 0\}$. Since $f^\pm$ are positive and measurable, $\int f^\pm \, d\mu$ exist and it is natural to **extend the integral by linearity**. As $\int f^\pm \, d\mu$ may be $+\infty$, we must avoid expressions of the form '$\infty - \infty$'; see Remark 1.11.

Definition 1.14 A function $f : X \to \overline{\mathbb{R}}$ on a measure space $(X, \mathscr{A}, \mu)$ is said to be (μ-)**integrable**, if it is $\mathscr{A}/\mathscr{B}(\overline{\mathbb{R}})$ measurable and if $\int f^+ \, d\mu, \int f^- \, d\mu < \infty$. In this case we call

$$\int f \, d\mu := \int f^+ \, d\mu - \int f^- \, d\mu \in (-\infty, \infty) \tag{1.9}$$

the (μ-)**integral** of f. The set of all real-valued, resp. $\overline{\mathbb{R}}$-valued, μ-integrable functions is denoted by $\mathcal{L}^1(\mu)$, resp. $\mathcal{L}^1_{\overline{\mathbb{R}}}(\mu)$.

By construction, the integral is a **linear** functional $f \longmapsto \int f \, d\mu$ which **preserves positivity**, i.e. $f \geqslant 0 \;\Rightarrow\; \int f \, d\mu \geqslant 0$. In particular, the integral is **monotone**, $f \leqslant g \Rightarrow \int f \, d\mu \leqslant \int g \, d\mu$, and satisfies the **triangle inequality** $|\int f \, d\mu| \leqslant \int |f| \, d\mu$. It is clear from the inequality $|\mathbb{1}_A f| \leqslant |f|$ that for every measurable set $A \in \mathscr{A}$ and integrable f, the function $\mathbb{1}_A f$ is again integrable. In this case we write

$$\int_A f \, d\mu := \int \mathbb{1}_A \cdot f \, d\mu, \quad A \in \mathscr{A}, \; f \in \mathcal{L}^1_{\overline{\mathbb{R}}}(\mu). \tag{1.10}$$

The construction of the integral is summed up in Fig. 1.3.

1.5 Spaces of Integrable Functions

We define on $\mathcal{L}^1_{\overline{\mathbb{R}}}(\mu)$ a functional $\|f\|_1 := \int |f| \, d\mu$. Clearly,

$$0 \leqslant \|f\|_1 < \infty, \quad \|\alpha f\|_1 = |\alpha| \|f\|_1, \quad \text{and} \quad \|f + g\|_1 \leqslant \|f\|_1 + \|g\|_1,$$

i.e. $f \longmapsto \|f\|_1$ has almost all properties needed for a norm. However,

$$\begin{aligned} \epsilon\mu\{|f| > \epsilon\} \leqslant \int_{\{|f|>\epsilon\}} |f| \, d\mu \leqslant \int |f| \, d\mu &= \sup_{n\in\mathbb{N}} \int_{\{f\neq 0\}} n \wedge |f| \, d\mu \\ &\leqslant \sup_{n\in\mathbb{N}} \left(n\mu\{f \neq 0\}\right), \end{aligned} \tag{1.11}$$

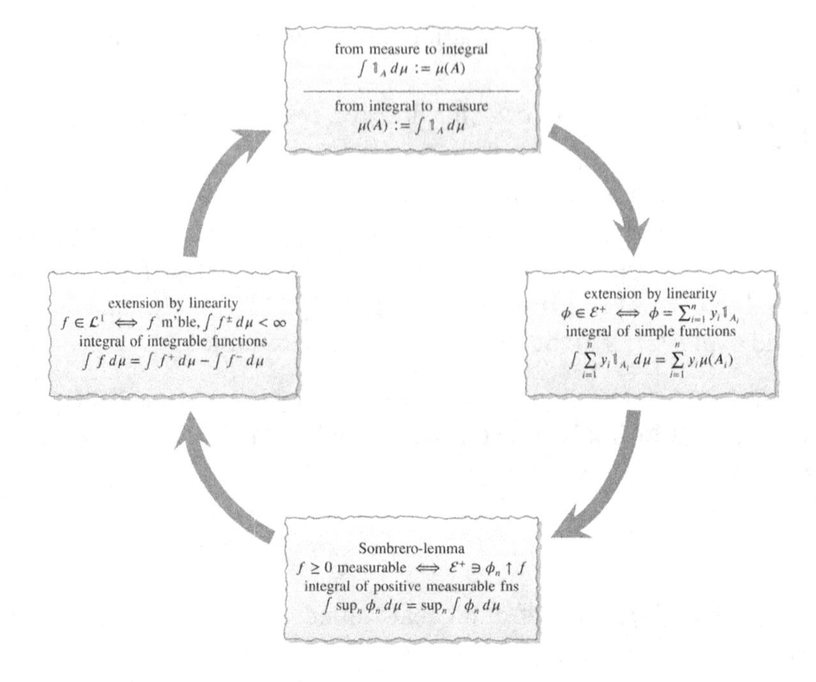

Figure 1.3 The way of integration: from measures to integrals – and back.

which shows that $\|f\|_1 = 0$ if, and only if, $\mu\{f \neq 0\} = 0$. In order to turn $(\mathcal{L}^1_{\overline{\mathbb{R}}}(\mu), \|\cdot\|_1)$ into a normed space, we have to identify functions which differ on sets of measure null.

Definition 1.15 Let $(X, \mathscr{A}, \mu)$ be any measure space. A property $\Pi = \Pi(x)$, $x \in X$, holds **almost everywhere** (a.e.) if the set $\{x \in X\,;\ \Pi(x) \text{ does not hold}\}$ is a subset of a measurable **null set**, i.e. $N \in \mathscr{A}$ such that $\mu(N) = 0$.

In particular, two functions $f, g : X \to \overline{\mathbb{R}}$ coincide a.e., if $\{f \neq g\}$ is a subset of a μ-null set. Note that $\{f \neq g\}$ need not be measurable if one of the functions f, g is not measurable.

Definition 1.16 Let $p \in (0, \infty]$. The space of ***p*th order integrable** functions is given by

$$\mathcal{L}^p_{\overline{\mathbb{R}}}(\mu) := \Big\{f : X \to \overline{\mathbb{R}}\,;\ f \text{ measurable}, \|f\|_p < \infty\Big\}, \tag{1.12}$$

and we set for $0 < p < \infty$, resp. $p = \infty$,

$$\|f\|_p = \left(\int |f|^p \, d\mu\right)^{\frac{1}{p}}, \quad \text{resp.} \quad \|f\|_\infty = \operatorname{esssup}|f| = \inf\{c\,;\ \mu\{|f| > c\} = 0\}. \tag{1.13}$$

Any two functions satisfying $f = g$ μ-a.e. are equivalent, $f \sim g$, and the family of equivalence classes of functions in $\mathcal{L}^p_{\overline{\mathbb{R}}}(\mu)$ is denoted by $L^p(\mu)$. The spaces $L^p(\mu)$ are called **Lebesgue spaces**.

Remark 1.17 (a) If $f \in \mathcal{L}^p_{\overline{\mathbb{R}}}(\mu)$, $0 < p < \infty$, then

$$n^p \mu\{|f|^p > n^p\} \leqslant \int_{\{|f|^p > n^p\}} |f|^p \, d\mu \leqslant \int |f|^p \, d\mu < \infty$$

implies that $\mu\{|f| = \infty\} = 0$. Thus, every $f \in \mathcal{L}^p_{\overline{\mathbb{R}}}(\mu)$ has an $\mathbb{R}$-valued representative, and there is no need to introduce the notation $L^p_{\overline{\mathbb{R}}}(\mu)$.

(b) It is not hard to check that the relation '$\sim$' from Definition 1.16 is an equivalence relation. Let us denote, for a moment, the equivalence classes by $[f]$. In view of (1.11) it is clear that

$$\|[f]\|_{L^p} := \inf\{\|g\|_p \,;\; g \in \mathcal{L}^p_{\overline{\mathbb{R}}}(\mu),\; g \sim f\} = \|f\|_p$$

and $\|[f]\|_{L^p} = 0 \iff [f] = [0]$. In particular, $\|\cdot\|_{L^1}$ is indeed a norm on $L^1(\mu)$. We will see below that this is also true for $(L^p(\mu), \|\cdot\|_{L^p})$ and all $p \in (1, \infty]$.

(c) Since $\|f\|_p^p = \||f|^p\|_1$, we conclude from (1.11) and the elementary inequality $|f + g|^p \leqslant 2^p \max(|f|^p, |g|^p) \leqslant 2^p (|f|^p + |g|^p)$ that the spaces $\mathcal{L}^p(\mu)$ and $L^p(\mu)$, $0 < p \leqslant \infty$ are linear spaces. A bit of care is needed when arguing with L^p-'functions': as usual, operations (addition, limits, suprema, etc.) on the equivalence classes $[f]$ are reduced to the corresponding operations on the representative f. Since each operation adds a null set, we must make sure that the union of these null sets is still a null set. Obviously, all operations involving **countably many representatives** are safe, and we follow the custom to identify L^p-'functions' with their representatives, and $\|\cdot\|_{L^p}$ with $\|\cdot\|_p$.

Hölder's inequality is one of the most versatile tools in the study of L^p-spaces.

Theorem 1.18 (Hölder's inequality [MIMS, p. 117]) *Let*[2] *$p, q \in [1, \infty]$ be conjugate indices, i.e. $p^{-1} + q^{-1} = 1$, and $f \in L^p(\mu)$ and $g \in L^q(\mu)$. Then $fg \in L^1(\mu)$ and*

$$\int |fg| \, d\mu \leqslant \left(\int |f|^p \, d\mu\right)^{1/p} \left(\int |g|^q \, d\mu\right)^{1/q} \tag{1.14}$$

(with the obvious modification if $p = \infty$ or $q = \infty$).

[2] If $p \in (0, 1)$, then $q < 0$, and Hölder's inequality (1.14) is reversed [☞ 18.9].

In particular, Hölder's inequality can be used to prove the triangle inequality for the functional $f \longmapsto \|f\|_p$, which is also known as **Minkowski's inequality**.

☞10.11 18.9

Corollary 1.19 (Minkowski's inequality [MIMS, p. 118]) *Let $p \in [1,\infty]$ and $f, g \in L^p(\mu)$. Then $f+g \in L^p(\mu)$ and*

$$\left(\int |f+g|^p \, d\mu\right)^{1/p} \leqslant \left(\int |f|^p \, d\mu\right)^{1/p} + \left(\int |g|^p \, d\mu\right)^{1/p} \tag{1.15}$$

(with the obvious modification if $p=\infty$).

One can recover from Theorem 1.18 and Corollary 1.19 the classical Hölder and Minkowski inequalities for real or complex sequences $(a_n)_{n\in\mathbb{N}} \subseteq \mathbb{C}$: just use the counting measure $\mu := \zeta_{\mathbb{N}}$ and interpret $(a_n)_{n\in\mathbb{N}}$ as a (trivially measurable) function $a : \mathbb{N} \to \mathbb{C}$ on the measurable space $(\mathbb{N}, \mathscr{P}(\mathbb{N}))$; see Example 1.7.(b).

Theorem 1.20 (Riesz–Fischer [MIMS, pp. 121, 123]) *The spaces $(L^p(\mu), \|\cdot\|_p)$, $p \in [1,\infty]$, are **Banach spaces**. In particular, every Cauchy sequence converges:*

$$\forall (f_n)_n \subseteq L^p(\mu), \quad \lim_{m,n\to\infty} \|f_n - f_m\|_p = 0, \ \exists f \in L^p(\mu) \ : \ \lim_{n\to\infty} \|f_n - f\|_p = 0.$$

Moreover, there is a subsequence $(f_{n(k)})_{k\in\mathbb{N}}$ which converges μ-a.e. to f, that is $f(x) = \lim_{k\to\infty} f_{n(k)}(x)$ for μ-a.e. $x \in X$.

The topological dual space $(L^p(\mu))^*$ of the Banach space $L^p(\mu)$, $p \in [1,\infty]$, consists of all continuous linear functionals $\Lambda : L^p(\mu) \to \mathbb{R}$. Hölder's inequality together with the following Riesz representation theorem allows us to identify $(L^p(\mu))^*$ with $L^q(\mu)$.

☞18.20-18.24

Theorem 1.21 ([MIMS, pp. 241, 243]) *Let $(X, \mathscr{A}, \mu)$ be a σ-finite measure space and let $1 \leqslant p < \infty$ and $1 < q \leqslant \infty$ be conjugate indices, i.e. $p^{-1} + q^{-1} = 1$. Every continuous linear functional $\Lambda : L^p(\mu) \to \mathbb{R}$ is of the form $\Lambda(u) = \int uf \, d\mu$ for a unique element $f \in L^q(\mu)$. In particular, one can identify $(L^p(\mu))^*$ and $L^q(\mu)$.*

If $1 < p, q < \infty$, σ-finiteness is actually not needed; see Dunford & Schwartz [51, Theorem IV.8.1]; the dual of $L^\infty(\mu)$ is the space of all finitely additive signed measures [51, Theorem IV.8.16]. As a consequence of Theorem 1.21 we can express, in the σ-finite setting, the L^q-norm using the L^p-norm

☞18.19

$$\|f\|_{L^q} = \sup_{u\neq 0} \frac{\int fu \, d\mu}{\|u\|_{L^p}}, \quad 1 \leqslant q \leqslant \infty, \ p^{-1} + q^{-1} = 1 \tag{1.16}$$

(the case $p=\infty$ can be found in [205, Lemma γ, p. 358]).

If X is a locally compact (metric) space, then L^∞ is often replaced by the space

$C_b(X)$ or $C_c(X)$ of continuous functions which are bounded or have compact support. In this case we have the following theorem; see [MIMS, p. 244] (locally compact metric spaces) or [150, Theorem 2.14] (locally compact setting).

Theorem 1.22 (Riesz representation [MIMS, p. 244], [150, Theorem 2.14]) *Let $(X, \mathscr{O})$ be a locally compact Hausdorff space and $\Lambda : C_c(X) \to \mathbb{R}$ be a positive linear functional. There is a uniquely determined inner-compact and outer-open regular measure μ on $(X, \mathscr{B}(X))$ such that*

☞ 18.26 18.27

$$\Lambda(u) = \int u \, d\mu, \quad u \in C_c(X). \tag{1.17}$$

The notions of inner and outer regularity are discussed in Chapter 9 and in the counterexamples 18.26, 18.27.

Both Hölder's and Minkowski's inequalities can be seen as particular cases of (higher-dimensional) convexity inequalities – see [MIMS, pp. 128–132, Problem 13.25]. Here we mention only the most basic version of a convexity inequality.

Theorem 1.23 (Jensen's inequality [MIMS, p. 126]) *Let $V : [0, \infty) \to [0, \infty)$ be a convex function which is extended to $[0, \infty]$ by $V(\infty) := \infty$. If $(X, \mathscr{A}, \mu)$ is a measure space with $\mu(X) \leqslant 1$, then*

$$V\left(\int f \, d\mu\right) \leqslant \int V(f) \, d\mu \in [0, \infty] \tag{1.18}$$

for every measurable $f : X \to [0, \infty]$. In particular, $f \in L^1(\mu)$ if $V(f) \in L^1(\mu)$.

1.6 Convergence Theorems

Among the most important features of Lebesgue integration are the convergence theorems. They address the problem of whether we can interchange limits of functions with the integral. Throughout we allow for improper limits, i.e. attaining values in $[-\infty, \infty]$; if the limiting function is integrable, then Remark 1.17.(a) tells us that $\pm\infty$ is only attained on a μ-null set. An overview over all convergence modes is given in Fig. 1.4 below.

We have already used a convergence theorem, in order to define the integral: Beppo Levi's theorem (Theorem 1.13). Here is the version for monotone sequences of functions with arbitrary sign; it shows, in particular, that the positivity $f_n \geqslant 0$ in Theorem 1.13 is all about the existence of an integrable minorant – to wit $g \equiv 0$.

Theorem 1.24 (monotone convergence [MIMS, pp. 96, 97]) *Assume that*

☞ 12.4 12.5

$(f_n)_{n\in\mathbb{N}} \subseteq L^1(\mu)$ is an increasing [resp. decreasing] sequence of integrable functions and set $f := \lim_{n\to\infty} f_n \in [-\infty, \infty]$. The limit f is in $L^1(\mu)$ if, and only if, $\sup_{n\in\mathbb{N}} \int f_n \, d\mu < \infty$ [resp. $\inf_{n\in\mathbb{N}} \int f_n \, d\mu > -\infty$]; in this case, we have

$$\lim_{n\to\infty} \int f_n \, d\mu = \int \lim_{n\to\infty} f_n \, d\mu \in \mathbb{R}. \tag{1.19}$$

In fact, Theorem 1.24 follows easily from Beppo Levi's theorem applied to the sequence $f_n - f_1 \geqslant 0$. If the sequence $(f_n)_{n\in\mathbb{N}}$ is not increasing – this means that $\lim_{n\to\infty} f_n$ need not exist – but integrably minorized, the technique used for Theorem 1.24 still gives the following result.

☞ 12.2 12.4

Theorem 1.25 (Fatou's lemma [MIMS, p. 78]) *Let $(f_n)_{n\in\mathbb{N}}$ be a sequence of measurable functions which has an integrable minorant, i.e. $g(x) \leqslant f_n(x)$ μ-a.e. for $g \in L^1(\mu)$. Then*

$$\int \liminf_{n\to\infty} (f_n - g) \, d\mu \leqslant \liminf_{n\to\infty} \int (f_n - g) \, d\mu \in [0, \infty]. \tag{1.20}$$

If $(f_n)_{n\in\mathbb{N}}$ converges a.e. and has both an integrable minorant and majorant, we obtain Lebesgue's dominated convergence theorem.

☞ 12.1 12.12

Theorem 1.26 (Lebesgue; dominated convergence [MIMS, p. 97]) *Let $(f_n)_{n\in\mathbb{N}}$ be a sequence of measurable functions such that*

(i) *$|f_n(x)| \leqslant g(x)$ for all $n \in \mathbb{N}$, μ-almost all $x \in X$ and some* ***integrable majorant*** *$g \in L^1(\mu)$;*
(ii) *$f(x) = \lim_{n\to\infty} f_n(x)$ for μ-almost all $x \in X$.*

Under these assumptions, $f_n \in L^1(\mu)$ and $f \in L^1(\mu)$ and

$$\lim_{n\to\infty} \|f_n - f\|_1 = 0 \quad \text{and} \quad \lim_{n\to\infty} \int f_n \, d\mu = \int \lim_{n\to\infty} f_n \, d\mu. \tag{1.21}$$

Among the most useful applications of the dominated convergence theorem are criteria for continuity and differentiability of parameter-dependent integrals.

☞ 14.9-14.11

Example 1.27 ([MIMS, pp. 99, 100]) Let $(X, \mathscr{A}, \mu)$ be a measure space, (a, b) a non-degenerate open interval of $\mathbb{R}$ and $u : (a, b) \times X \to \mathbb{R}$ a function such that $x \longmapsto u(t, x)$ is in $L^1(\mu)$ for every $t \in (a, b)$. Define the parameter-dependent integral $U(t) := \int u(t, x) \, \mu(dx)$. If

☞ 12.10

(a) (i) $t \longmapsto u(t, x)$ is continuous for μ-a.e. $x \in X$;
(ii) $|u(t, x)| \leqslant g(x)$ for all $t \in (a, b)$, μ-a.e. $x \in X$ and some $g \in L^1(\mu)$;
then $t \longmapsto U(t)$ is continuous for all $t \in (a, b)$.
(b) (i) $t \longmapsto u(t, x)$ is differentiable for μ-a.e. $x \in X$;

(ii) $|\partial_t u(t,x)| \leqslant g(x)$ for all $t \in (a,b)$, μ-a.e. $x \in X$ and some $g \in L^1(\mu)$;
then $t \longmapsto U(t)$ is differentiable for all $t \in (a,b)$ and

$$\frac{d}{dt} U(t) = \int \frac{\partial}{\partial t} u(t,x) \, \mu(dx).$$

In both cases the majorizing condition (ii) is a give-away for the proof: reduce the assertions to limits $t_n \to t \in (a,b)$ and

$$\lim_{n\to\infty} U(t_n), \quad \text{resp.} \quad \lim_{n\to\infty} \frac{U(t_n) - U(t)}{t_n - t},$$

and combine the continuity/differentiability of the integrand with the dominated convergence theorem.

Another consequence of Lebesgue's dominated convergence theorem is the characterization of Riemann integrability.

Theorem 1.28 ([MIMS, pp. 103, 104]) *A bounded function $f : [a,b] \to \mathbb{R}$ is Riemann integrable if, and only if, the discontinuity points of f are a (subset of a) Borel measurable null set for one-dimensional Lebesgue measure λ.*

☞ 3.4 3.17

If f is Riemann integrable and Borel measurable, then $f \in L^1(\lambda)$, and the Riemann and Lebesgue integrals coincide

$$\int_{[a,b]} f(x) \, \lambda(dx) = \int_a^b f(x) \, dx. \tag{1.22}$$

If f is not bounded or if the interval $[a,b]$ is replaced by an open or infinite interval, one has to use improper Riemann integrals which are defined as limits of Riemann integrals. Some, but not all, of these improper Riemann integrals can be understood using Lebesgue's theory. There is a kind of converse to Lebesgue's dominated convergence theorem which also explains the role of the majorizing condition in Theorem 1.26. For the statement we need two more definitions.

Definition 1.29 A sequence of measurable functions $f_n : X \to \mathbb{R}$ converges to a measurable function $f : X \to \mathbb{R}$ **in probability** if

$$\forall \epsilon > 0 \quad \forall F \in \mathscr{A}, \ \mu(F) < \infty \ : \ \lim_{n\to\infty} \mu\left(F \cap \{|f - f_n| > \epsilon\}\right) = 0. \tag{1.23}$$

If (1.23) holds without the restriction $\mu(F) < \infty$, then we say that $f_n \to f$ **in measure.**[3]

[3] Different conventions for the notion 'convergence in measure' are used in the literature. Some authors consider only what we call here 'convergence in probability' and refer to this as convergence in measure, e.g. [MIMS], or introduce convergence in measure only for finite measures, e.g. [19].

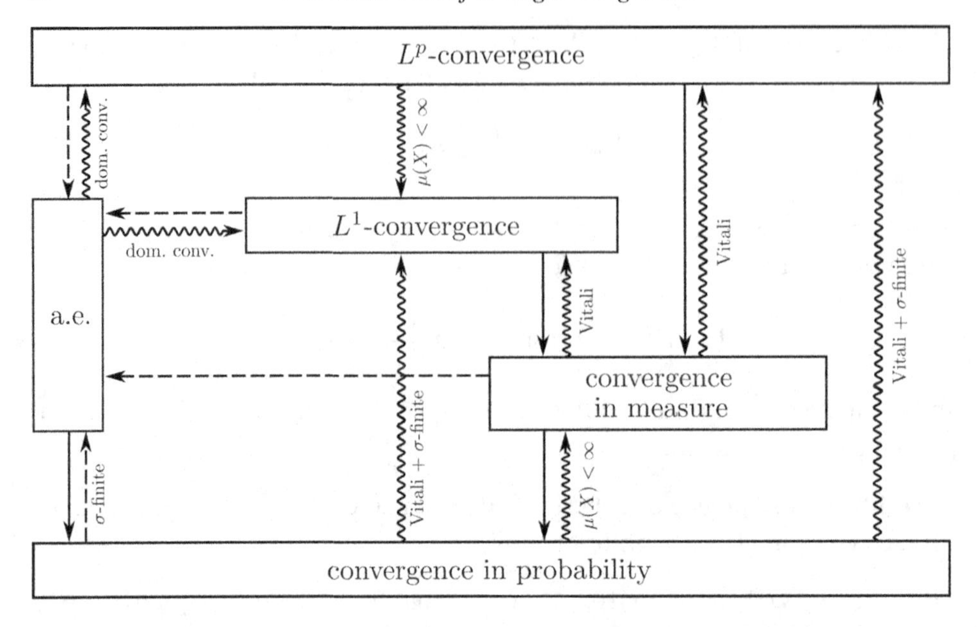

Figure 1.4 An overview of the modes of convergence for a sequence $(f_n)_{n\in\mathbb{N}}$. Solid black arrows indicate implications, dashed black arrows indicate that the implication holds for a subsequence (under the condition given) and wiggly black arrows show that the implication holds using the theorem or additional condition mentioned next to the arrow; see also the tables in Example 11.5 for further convergence modes.

There is no big difference between convergence in probability and convergence in measure if we are on a finite or σ-finite measure space. Things begin to fall apart if we consider more general measure spaces. Note that convergence in measure implies both convergence in probability and a.e. convergence of a subsequence $(f_{n(k)})_{k\in\mathbb{N}}$. Conversely, a.e. convergence implies convergence in probability, but not necessarily convergence in measure.

☞11.5

The second ingredient is, essentially, a compactness condition in the space $L^1(\mu)$.

Definition 1.30 ([MIMS, p. 258]) A family $(f_i)_{i\in I} \subseteq L^1(\mu)$ is **uniformly integrable**[4] if there exists an integrable function $g \in L^1(\mu)$ such that

$$\lim_{R\to\infty} \sup_{i\in I} \int_{\{|f_i|>Rg\}} |f_i| \, d\mu = 0. \tag{1.24}$$

In finite measure spaces we may take in (1.24) the integrable function $g \equiv 1$.

[4] 'Uniform integrability' is used under various names and with various definitions. A thorough discussion can be found in [MIMS, Chapter 22]. The variant given here is inspired by the usual definition in finite measure spaces and it is equivalent to the definition given in [MIMS].

If, as in Theorem 1.26, the family $(f_i)_{i\in I}$ is bounded by an integrable function, $|f_i| \leqslant g$, then $(f_i)_{i\in I}$ is uniformly integrable.

Theorem 1.31 (Vitali's convergence theorem [MIMS, p. 262]) *Let $(X, \mathscr{A}, \mu)$ be a measure space and $(f_n)_{n\in\mathbb{N}} \subseteq L^p(\mu)$, $p \in [1, \infty)$, be a sequence of functions which converges in measure to some measurable function f. The following assertions are equivalent:*

☞12.15

(a) $\lim_{n\to\infty} \|f_n - f\|_p = 0$;

(b) $(|f_n|^p)_{n\in\mathbb{N}}$ *is uniformly integrable;*

(c) $\lim_{n\to\infty} \int |f_n|^p \, d\mu = \int |f|^p \, d\mu.$

If the measure space is σ-finite, then it is enough to assume convergence in probability.

It is possible to state Theorem 1.31 for an arbitrary (non-σ-finite) measure space and with convergence in probability. In this case we set

$$S := \bigcup_{n\in\mathbb{N}} \bigcup_{k\in\mathbb{N}} \left\{|f_n| > \frac{1}{k}\right\}$$

and note that $\mu\left\{|f_n| > \frac{1}{k}\right\} < \infty$ since $f_n \in L^p$. Thus, $\mu|_S$ is σ-finite and the limit in probability $f_n \to f$ is a.e. unique on S; on S^c, we **define** $f := 0$ (this is possible due to the non-uniqueness of limits in probability on general measure spaces [☞11.6]).

1.7 Product Measure, Fubini and Tonelli

Let $(X, \mathscr{A}, \mu)$ and $(Y, \mathscr{B}, \nu)$ be two measure spaces. The construction of a measure on the product set $X \times Y$ is inspired by the way we calculate the area of a rectangle – the only difference being that we use generalized rectangles of the form $A \times B \in \mathscr{A} \times \mathscr{B}$ and replace 'width' by $\mu(A)$ and 'height' by $\nu(B)$.

Definition 1.32 A measure ρ on the product $(X \times Y, \mathscr{A} \otimes \mathscr{B} := \sigma(\mathscr{A} \times \mathscr{B}))$ satisfying

$$\rho(A \times B) := \mu(A)\nu(B), \quad A \in \mathscr{A},\ B \in \mathscr{B}, \tag{1.25}$$

is called a **product measure** (with marginal measures μ and ν). If the product measure is unique, it is denoted by $\mu \times \nu$. The space $(X, \mathscr{A}, \mu) \otimes (Y, \mathscr{B}, \nu) := (X \times Y, \mathscr{A} \otimes \mathscr{B}, \mu \times \nu)$ is called the **product measure space**.

As usually happens with definitions, existence and uniqueness need to be established. With a view towards Theorem 1.40, uniqueness is guaranteed if both measures μ and ν are σ-finite. Existence follows either from Carathéodory's extension procedure (Theorem 1.41 and Theorem 16A) or by the following direct construction, interpreting any axis-parallel cross-section of a measurable set $E \subseteq X \times Y$ as an 'infinitesimal' rectangle; see Fig. 1.5. This construction exchanges the existence problem for a measurability problem of the cross-sections. The following theorem addresses this point. For simplicity, we state it in the σ-finite version, which also guarantees uniqueness of the product measure.

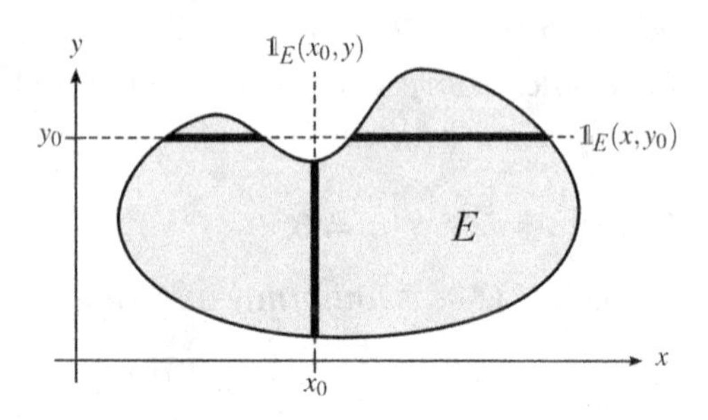

Figure 1.5 Cavalieri's principle.

☞16.1 **Theorem 1.33** (existence of product measures [MIMS, p. 139]) *Let $(X, \mathscr{A}, \mu)$ and $(Y, \mathscr{B}, \nu)$ be two σ-finite measure spaces. The product measure $\mu \times \nu$ exists, it is unique and it is given for all $E \in \mathscr{A} \otimes \mathscr{B}$ by the iterated integrals*

$$\mu \times \nu(E) = \int_X \left[\int_Y \mathbb{1}_E(x, y)\, \nu(dy) \right] \mu(dx) = \int_Y \left[\int_X \mathbb{1}_E(x, y)\, \mu(dx) \right] \nu(dy). \tag{1.26}$$

In particular, if $E \in \mathscr{A} \otimes \mathscr{B}$, then the functions

$$\begin{aligned} x &\longmapsto \mathbb{1}_E(x, y), & y &\longmapsto \mathbb{1}_E(x, y), \\ x &\longmapsto \int_Y \mathbb{1}_E(x, y)\, \nu(dy), & y &\longmapsto \int_X \mathbb{1}_E(x, y)\, \mu(dx), \end{aligned} \tag{1.27}$$

are $\mathscr{A}$ resp. $\mathscr{B}$ measurable for all $x \in X$ and $y \in Y$.

For finite measures μ, ν, the proof of Theorem 1.33 is straightforward; one has to verify that the family of sets satisfying (1.26) and (1.27) is a σ-algebra which contains the rectangles $\mathscr{A} \times \mathscr{B}$. This construction can be extended to σ-finite measures if one uses the restrictions $\mu|_{E_k}$ and $\nu|_{F_k}$ for sequences $E_k \uparrow X$ and $F_k \uparrow Y$ such that $\mu(E_k) < \infty$ and $\mu(F_k) < \infty$.

With the sombrero lemma (Theorem 1.10, see also Fig. 1.2) we can extend Theorem 1.33 to all measurable functions $f : X \times Y \to [0, \infty]$.

☞16.10-16.14 **Corollary 1.34** (Tonelli [MIMS, p. 142]) *Let $(X, \mathscr{A}, \mu)$ and $(Y, \mathscr{B}, \nu)$ be two σ-finite measure spaces and $f : X \times Y \to [0, \infty]$ a positive $\mathscr{A} \otimes \mathscr{B}$ measurable* *function. Then*

$$\int_{X\times Y} f\, d(\mu\times\nu) = \int_X \int_Y f(x,y)\,\nu(dy)\,\mu(dx) = \int_Y \int_X f(x,y)\,\mu(dx)\,\nu(dy); \tag{1.28}$$

if one of these expressions is finite, then all are finite. Moreover, the functions

$$\begin{aligned} x &\longmapsto f(x,y), & y &\longmapsto f(x,y),\\ x &\longmapsto \int_Y f(x,y)\,\nu(dy), & y &\longmapsto \int_X f(x,y)\,\mu(dx), \end{aligned} \tag{1.29}$$

are $\mathscr{A}$ resp. $\mathscr{B}$ measurable for all $x \in X$ and $y \in Y$.

Similar to the construction of the class $\mathcal{L}^1$ in Definition 1.14 we deduce, by linearity, Fubini's theorem.

Corollary 1.35 (Fubini [MIMS, pp. 142, 143]) *Let $(X, \mathscr{A}, \mu)$ and $(Y, \mathscr{B}, \nu)$ be two σ-finite measure spaces and $f : X \times Y \to \overline{\mathbb{R}}$ an $\mathscr{A} \otimes \mathscr{B}$ measurable function. If at least one of the integrals*

☞16.15-16.18

$$\int_{X\times Y} |f|\, d(\mu\times\nu), \quad \int_X \int_Y |f(x,y)|\,\nu(dy)\,\mu(dx), \quad \int_Y \int_X |f(x,y)|\,\mu(dx)\,\nu(dy) \tag{1.30}$$

is finite, then all are finite, and $f \in L^1(\mu\times\nu)$. Moreover, the functions

$$\begin{aligned} x &\longmapsto f(x,y), & y &\longmapsto f(x,y),\\ x &\longmapsto \int_Y f(x,y)\,\nu(dy), & y &\longmapsto \int_X f(x,y)\,\mu(dx), \end{aligned} \tag{1.31}$$

are in $L^1(\mu)$ resp. $L^1(\nu)$ for ν-almost all $y \in Y$ resp. μ-almost all $x \in X$.

Let us conclude this section with three of the most prominent applications of Fubini's and Tonelli's theorems.

Example 1.36 (a) (**integration by parts** [MIMS, p. 144]). Assume that the functions $f, g : \mathbb{R} \to \mathbb{R}$ are measurable and integrable over a compact interval $[a, b] \subseteq \mathbb{R}$, and denote by $F(t) = \int_a^t f(x)\,dx$ and $G(t) = \int_a^t g(x)\,dx$ their primitives. Then one has

$$F(b)G(b) - F(a)G(a) = \int_a^b f(t)G(t)\,dt + \int_a^b F(t)g(t)\,dt. \tag{1.32}$$

(b) (**layer-cake formula** [MIMS, p. 146]). Let $(X, \mathscr{A}, \mu)$ be a σ-finite measure space and $f : X \to [0, \infty]$ be a measurable function. Then we can express the Lebesgue integral as an (improper) Riemann integral

$$\int f\,d\mu = \int_0^\infty \mu\{f \geqslant t\}\,dt. \tag{1.33}$$

The function $t \longmapsto \mu^f(t) := \mu\{f \geqslant t\}$ is called the (left-continuous) **distribution function** of f under μ. If we happen to know that $f \in L^1(\mu)$ or $\mu^f \in L^1(\lambda)$, we can drop the assumption of σ-finiteness since we can express the set $\{f > 0\} = \bigcup_{n\in\mathbb{N}} \left\{f > \frac{1}{n}\right\}$ as a countable union of sets with finite μ-measure [☞ 10.5].

(c) (**Minkowski's inequality for double integrals** [MIMS, p. 148]). Let $(X, \mathscr{A}, \mu)$ and $(Y, \mathscr{B}, \nu)$ be σ-finite measure spaces and $f : X \times Y \to \overline{\mathbb{R}}$ be $\mathscr{A} \otimes \mathscr{B}$ measurable. For all $p \in [1, \infty)$ we have

$$\left[\int_X \left[\int_Y |f(x,y)|\,\nu(dy)\right]^p \mu(dx)\right]^{1/p} \leqslant \int_Y \left[\int_X |f(x,y)|^p\,\mu(dx)\right]^{1/p} \nu(dy); \tag{1.34}$$

equality holds if $p = 1$.

1.8 Transformation Theorems

The technique of constructing an integral starting from a measure, cf. Fig. 1.3, can be applied in many situations. Take, for instance, the image or push-forward measure from Example 1.7.(f). Let $(X, \mathscr{A}, \mu)$ be a measure space, $(X', \mathscr{A}')$ a measurable space and $\phi : X \to X'$ a measurable map. By $\phi_*\mu(A') := \mu(\phi^{-1}(A'))$ the measure μ is 'transported' to a measure on $(X', \mathscr{A}')$. We can rewrite this definition in terms of integrals as

$$\int_{X'} \mathbb{1}_{A'}\,d\phi_*\mu = \int_X \mathbb{1}_{\phi^{-1}(A')}\,d\mu = \int_X \mathbb{1}_{A'}\circ\phi\,d\mu, \quad A' \in \mathscr{A}'. \tag{1.35}$$

By the linearity of the integral this extends to simple functions and then, using monotone convergence (Beppo Levi) and again linearity, we get the following theorem.

Theorem 1.37 ([MIMS, p. 154]) *Let $\phi : X \to X'$ be a measurable map between the measure space $(X, \mathscr{A}, \mu)$ and a measurable space $(X', \mathscr{A}')$. The $\phi_*\mu$-integral of a measurable function $f : X' \to \overline{\mathbb{R}}$ is given by*

$$\int_{X'} f\,d\phi_*\mu = \int_X f\circ\phi\,d\mu. \tag{1.36}$$

Formula (1.36) *holds either in* $[0,\infty]$ *for positive* $f \geqslant 0$ *or in* $\mathbb{R}$ *for* $f \in L^1(\phi_*\mu)$; *in the latter case,* $f \in L^1(\phi_*\mu) \iff f\circ\phi \in L^1(\mu)$.

Theorem 1.37 is indispensable in probability theory – the image measure under a random variable ϕ is the distribution of the random variable – but we may understand it also as an abstract version of the substitution rule for Lebesgue measure.

Let $X = X' = \mathbb{R}^d$, take Lebesgue measure $\mu = \lambda^d$ on X and consider the linear map $\phi : \mathbb{R}^d \to \mathbb{R}^d, \phi(x) = Mx$ for an invertible matrix $M \in \mathbb{R}^{d\times d}$. In order to identify $\phi_*\lambda^d$, take a cube $Q_\epsilon = [0,\epsilon)^d$ with side length $\epsilon > 0$ and observe that

$$\phi_*\lambda^d(Q_\epsilon) = \lambda^d(M^{-1}Q_\epsilon) = |\det M^{-1}| \cdot \epsilon^d = |\det M^{-1}| \cdot \lambda^d(Q_\epsilon). \tag{1.37}$$

In the last two equalities we use that $M^{-1}Q_\epsilon$ is the parallelepiped spanned by the column vectors of the matrix ϵM^{-1} whose volume is $\epsilon^d |\det M^{-1}|$. Since we can represent any half-open rectangle $R \subseteq \mathbb{R}^d$ as a union of disjoint, suitably shifted cubes Q_ϵ, (1.37) extends first to all rectangles in $\mathbb{R}^d$ and then, by the uniqueness theorem of measures [☞1.40], to all Borel sets. Thus, $\phi_*\lambda^d = |\det M^{-1}| \cdot \lambda^d$, and Theorem 1.37 tells us that

$$\int f(Mx)\,dx = \int f(y)|\det M^{-1}|\,dy. \tag{1.38}$$

Since $\phi^{-1}(x) = M^{-1}x$ is linear, $D\phi^{-1}(x) = M^{-1}$ is the derivative and, therefore, $|\det M^{-1}| = |\det D\phi^{-1}(x)|$. The 'next best thing' to a linear map is a function ϕ which is bijective and can locally be approximated by linear maps, in other words ϕ is a C^1-diffeomorphism. Combining such a localized linearization with (1.37) eventually leads to the following **substitution rule** for Lebesgue integrals.

Theorem 1.38 (Jacobi's transformation theorem [MIMS, p. 166]) *Assume that* $U, V \subseteq \mathbb{R}^d$ *are open sets,* $\psi : U \to V$ *a* C^1*-diffeomorphism and* $\lambda_W := \lambda^d(\cdot \cap W)$ d*-dimensional Lebesgue measure on* $(W, \mathscr{B}(W))$, $W = U, V$. *Then*

$$\int_V f(y)\,dy = \int_U f(\psi(x))|\det D\psi(x)|\,dx \tag{1.39}$$

holds in $[0,\infty]$ *for all positive measurable* $f : V \to [0,\infty]$ *and in* $\mathbb{R}$ *for all* $f \in L^1(\lambda_V)$; *in the latter case* $f \in L^1(\lambda_V) \iff |\det D\psi| f\circ\psi \in L^1(\lambda_U)$.

An important application of the transformation theorem to product measures is the convolution[5] of measures and functions.

[5] The German 'Faltung' or 'Faltungsprodukt' is often used in the older literature.

Definition 1.39 Let $(X,+)$ be a topological group and μ, ν σ-finite measures on $(X,\mathscr{B}(X))$. The **convolution** of μ and ν is the image measure $(\mu\times\nu)\circ\alpha^{-1}$ under the continuous addition map $\alpha: X\times X\to X$, $\alpha(x,y)=x+y$.

Convolutions appear naturally in probability theory in connection with the distribution of the sum of independent random variables, but also in the Cauchy product in the summation of series in many other applications. The assumption of σ-finiteness ensures that the product measure $\mu\times\nu$ is well-defined – see however the footnote on p. 295 – and that we can freely use Tonelli's theorem. The transformation theorem [☞Theorem 1.37] in combination with Tonelli's theorem [☞Corollary 1.34] gives the following useful integral formulae for convolutions. For all $B\in\mathscr{B}(X)$,

$$\mu*\nu(B)=\iint_{X\times X}\mathbb{1}_B(x+y)\,\mu\times\nu(dx,dy) \tag{1.40}$$

$$=\int_X\mu(B-y)\,\nu(dy)=\int_X\nu(B-x)\,\mu(dx). \tag{1.41}$$

This shows immediately that the mapping $(\mu,\nu)\longmapsto\mu*\nu$ is symmetric and bilinear. Moreover, the integral with respect to a measure obtained by convolution is

$$\int_X f(z)\,\mu*\nu(dz)=\int_X\int_X f(x+y)\,\mu(dx)\,\nu(dy)$$

for all positive measurable functions $f: X\to\mathbb{R}$ or all integrable functions $f\in L^1(\mu*\nu)=L^1((\mu\times\nu)\circ\alpha^{-1})$.

If $X=\mathbb{R}^d$, we can also define the convolution of two positive integrable functions $f,g\in L^1(\lambda^d)$, $f,g\geqslant 0$, if we consider the measures $\mu=f\lambda^d$ and $\nu=g\lambda^d$. The convolution $\mu*\nu$ has again a density with respect to Lebesgue measure and the density is of the form

$$f*g(x):=\int_{\mathbb{R}^d}f(x-y)g(y)\,dy=\int_{\mathbb{R}^d}f(y)g(x-y)\,dy. \tag{1.42}$$

This definition clearly extends to all (not necessarily positive) $f,g\in L^1(\lambda^d)$ and the **convolution** product **of functions** $(f,g)\longmapsto f*g$ is symmetric, bilinear and produces an element $f*g\in L^1(\lambda^d)$ [MIMS, pp. 157–159]. Using Young's inequality, we can extend the convolution to functions $f\in L^p(\lambda^d)$ and $g\in L^q(\lambda^d)$ such that $f*g\in L^r(\lambda^d)$, where $p,q,r\in[1,\infty)$ satisfy $\frac{1}{p}+\frac{1}{q}=1+\frac{1}{r}$ [MIMS, p. 163, Problem 15.14].

1.9 Extension of Set Functions and Measures

We will now turn to the problem of how one can construct a measure. Consider, for example, Lebesgue measure. It is hopeless to tell immediately the volume $\lambda(B)$ of an arbitrary set $B \in \mathscr{B}(\mathbb{R}^d)$, but for some simple sets like d-dimensional 'rectangles' $\times_{i=1}^{d}[a_i, b_i)$ the formula 'volume = length × height × width ×...' gives the correct value. Since the d-dimensional rectangles generate the Borel σ-algebra $\mathscr{B}(\mathbb{R}^d)$, there is hope that knowledge of the volume of rectangles characterizes Lebesgue measure on all Borel sets. Note, however, that there is no constructive way to build an arbitrary Borel set from rectangles.

In order to ensure uniqueness of the extension, it is enough to know that intersections of rectangles are again rectangles, and that we can exhaust $\mathbb{R}^d$ by rectangles. More precisely

Theorem 1.40 (uniqueness of measures [MIMS, p. 34]) *Let $(X, \mathscr{A})$ be a measurable space and $\mathscr{S} \subseteq \mathscr{A}$ be a collection of measurable sets such that*

☞ 5.18

(i) *$\mathscr{S}$ is $\cap$-stable, i.e. $S \cap T \in \mathscr{S}$ for all $S, T \in \mathscr{S}$;*
(ii) *there is an exhausting sequence $(S_n)_{n\in\mathbb{N}} \subseteq \mathscr{S}$, $S_n \uparrow X$.*

If μ, ν are measures on $(X, \mathscr{A})$ such that $\mu|_{\mathscr{S}} = \nu|_{\mathscr{S}}$ and $\mu(S_n) = \nu(S_n) < \infty$, then the measures coincide on $\sigma(\mathscr{S})$.

Theorem 1.40 is most interesting if $\mathscr{A} = \sigma(\mathscr{S})$. If $\mu(X) = \nu(X) < \infty$, we can replace $\mathscr{S}$ by $\mathscr{S} \cup \{X\}$, and this family automatically satisfies condition (ii) with $S_n = X$.

If we want to extend a set function μ defined on a family $\mathscr{S}$ to a measure on $\sigma(\mathscr{S})$, we need at least that $\mu|_{\mathscr{S}}$ is a **measure relative to** $\mathscr{S}$, i.e.

$$\emptyset \in \mathscr{S}, \quad \mu(\emptyset) = 0,$$

$$\begin{gathered} \forall (S_n)_{n\in\mathbb{N}} \subseteq \mathscr{S} \text{ pairwise disjoint:} \\ S = \bigcupdot_{n\in\mathbb{N}} S_n \in \mathscr{S} \quad \text{and} \quad \mu(S) = \sum_{n=1}^{\infty} \mu(S_n). \end{gathered} \tag{1.43}$$

Such set functions are often called **pre-measures**. In view of Theorem 1.40, the family $\mathscr{S}$ should be $\cap$-stable (otherwise the extension may not be unique) but it turns out that we need a further structural property for $\mathscr{S}$ which is automatically satisfied for the rectangles: $\mathscr{S}$ must be a **semi-ring**, i.e. a family of sets with

$$\emptyset \in \mathscr{S}, \tag{S$_1$}$$

$$S, T \in \mathscr{S} \Rightarrow S \cap T \in \mathscr{S}, \tag{S$_2$}$$

$$\begin{gathered} \text{for } S, T \in \mathscr{S} \text{ there exist finitely many disjoint} \\ S_1, S_2, \dots, S_n \in \mathscr{S} \text{ such that } S \setminus T = \bigcupdot_{i=1}^{n} S_i. \end{gathered} \tag{S$_3$}$$

☞5.5 9.13

Theorem 1.41 (existence of measures; Carathéodory [MIMS, pp. 39, 40]) *Let $\mathcal{S} \subseteq \mathcal{P}(X)$ be a semi-ring and $\mu : \mathcal{S} \to [0, \infty]$ a measure relative to $\mathcal{S}$, i.e. a set function satisfying* (1.43)*. Then μ has an extension to a measure μ on $\sigma(\mathcal{S})$. If $\mathcal{S}$ contains an exhausting sequence $(S_n)_{n\in\mathbb{N}}$, $S_n \uparrow X$ such that $\mu(S_n) < \infty$ for all $n \in \mathbb{N}$, then the extension is unique.*

The idea behind Carathédory's construction is instructive (whereas the details are tiresome). First one defines a set function on all subsets $A \subseteq X$

$$\mu^*(A) = \inf\left\{\sum_{n=1}^{\infty} \mu(S_n);\ S_n \in \mathcal{S},\ \bigcup_{n\in\mathbb{N}} S_n \supseteq A\right\} \tag{1.44}$$

(as usual, $\inf \emptyset = \infty$). This set function is an **outer measure**, i.e. a set function satisfying

$$\mu^*(\emptyset) = 0, \tag{OM$_1$}$$

$$A \subseteq B \Rightarrow \mu^*(A) \leqslant \mu^*(B), \tag{OM$_2$}$$

$$\mu^*\Big(\bigcup_{n\in\mathbb{N}} A_n\Big) \leqslant \sum_{n\in\mathbb{N}} \mu^*(A_n); \tag{OM$_3$}$$

its restriction to the σ-algebra of μ^* **measurable sets**

$$\mathcal{A}^* := \{A \subseteq X;\ \mu^*(Q) = \mu^*(Q \cap A) + \mu^*(Q \setminus A) \quad \forall Q \subseteq X\} \tag{1.45}$$

(it is not obvious that $\mathcal{A}^*$ is indeed a σ-algebra!) is a measure. Since $\sigma(\mathcal{S}) \subseteq \mathcal{A}^*$ – equality is the exception [☞9.9] – $\mu := \mu^*|_{\sigma(\mathcal{S})}$ is the extension we have been looking for.

Note that $(X, \mathcal{A}^*, \mu^*|_{\mathcal{A}^*})$ is a **complete measure space** (see Example 1.7.(j)), i.e.

$$\forall N \subseteq N^* \in \mathcal{A}^* : N \in \mathcal{A}^* \quad \text{and} \quad \mu^*(N) = 0. \tag{1.46}$$

In other words: subsets of $\mathcal{A}^*$ measurable null sets are again in $\mathcal{A}^*$ and, therefore, null sets.

We have not touched on the problem of how to construct, in general, good set functions μ on $\mathcal{S}$, and there are various possibilities. A brief introduction is given in [MIMS, Chapter 18]; see also [123].

1.10 Signed Measures and Radon–Nikodým

A simple way to construct a new measure ν on a given measure space $(X, \mathscr{A}, \mu)$ is to consider densities. Let $f : X \to [0, \infty]$ be a measurable function and set

$$\nu(A) := \int_A f(x)\,\mu(dx) := \int \mathbb{1}_A(x) f(x)\,\mu(dx), \quad A \in \mathscr{A}. \tag{1.47}$$

Because of the linearity of the integral and the Beppo Levi theorem it is not difficult to check that ν is a measure. Moreover, every μ-null set is also a ν-null set. This is to say, we have

$$\forall N \in \mathscr{A},\ \mu(N) = 0 : \quad \nu(N) = 0. \tag{1.48}$$

Definition 1.42 Let μ, ν be two measures on the measurable space $(X, \mathscr{A})$. We call ν **absolutely continuous (with respect to μ)**, and write $\nu \ll \mu$, if (1.48) holds.

We say that μ **and** ν **are (mutually) singular**, and write $\mu \perp \nu$, if there exists a set $N \in \mathscr{A}$ such that $\mu(N) = 0$ and $\nu(X \setminus N) = 0$.

The following theorem gives a complete characterization of measures having a density.

Theorem 1.43 (Radon–Nikodým [MIMS, pp. 230, 301]) *Let μ, ν be two measures on a measurable space $(X, \mathscr{A})$ and assume that μ is σ-finite. The following assertions are equivalent:* ☞17.1

(a) $\nu(A) = \int_A f(x)\,\mu(dx)$, $A \in \mathscr{A}$, *for some μ-a.e. unique measurable function* $f : X \to [0, \infty]$.
(b) ν *is absolutely continuous with respect to μ, i.e. $\nu \ll \mu$.*

Moreover, $\nu(X) < \infty$ if, and only if, $f \in L^1(\mu)$, and ν is σ-finite if, and only if, $\mu\{f = \infty\} = 0$.

With the Radon–Nikodým theorem we can decompose any measure ν into an absolutely continuous and a singular part with respect to the underlying reference measure.

Corollary 1.44 (Lebesgue decomposition [MIMS, pp. 235, 306]) *Let $(X, \mathscr{A}, \mu)$ be a σ-finite measure space. Every σ-finite measure ν on $(X, \mathscr{A})$ can be written as $\nu = \nu^{\perp} + \nu^{\circ}$, where $\nu^{\circ} \ll \mu$ and $\nu^{\perp} \perp \mu$. This decomposition is, up to μ-null sets, unique.* ☞17.11

It is natural to consider in (1.47) not only positive densities f, but functions $f \in L^1(\mu)$ of arbitrary sign. The resulting set function $\nu(A)$ is still σ-additive but not positive. This is the first example of a so-called signed measure.

Definition 1.45 Let $(X, \mathscr{A})$ be a measurable space. A **signed measure** is a set function $m : \mathscr{A} \to [-\infty, \infty]$ which satisfies $m(\emptyset) = 0$ and is σ-additive, i.e.

$$(A_n)_{n\in\mathbb{N}} \subseteq \mathscr{A} \text{ pairwise disjoint } \Rightarrow m\Big(\biguplus_{n\in\mathbb{N}} A_n\Big) = \sum_{n\in\mathbb{N}} m(A_n) \in [-\infty, \infty]. \tag{M$_2$}$$

Typical examples of signed measures are

- $m = \mu - \nu$ where at least one of the measures μ, ν is finite;
- $m(A) = \int_A f(x)\,\mu(dx)$ for any (positive) measure μ and $f \in L^1(\mu)$.

Before we discuss the general structure of signed measures, let us briefly explain the meaning of σ-additivity (M_2) for a signed measure.

Remark 1.46 (a) Setting $A_1 = A$, $A_2 = B$ and $A_n = \emptyset$ for $n \geqslant 3$, we deduce from (M_2) that a signed measure is finitely additive. If A and B are disjoint, the well-definedness of $m(A \uplus B) = m(A) + m(B)$ requires that m takes values either in $(-\infty, \infty]$ or in $[-\infty, \infty)$. Indeed, if there is a set $A \in \mathscr{A}$ with $m(A) = \pm\infty$, we see that $m(X) = \pm\infty$, which means that $m(A) = \infty$ and $m(B) = -\infty$ is impossible. In particular, the partial sums $\sum_{n=1}^{N} m(A_n)$, $N \in \mathbb{N}$, appearing in (M_2) exist either in $(-\infty, \infty]$ or $[-\infty, \infty)$.

(b) Let $\pi : \mathbb{N} \to \mathbb{N}$ be a reordering, i.e. any bijective map. Clearly, we have $\bigcup_{n\in\mathbb{N}} A_n = \bigcup_{n\in\mathbb{N}} A_{\pi(n)}$, and therefore the condition (M_2) implies that the series $\sum_{n\in\mathbb{N}} m(A_n) \in [-\infty, +\infty]$ is invariant under reordering. In view of (a), this can occur only if one of the following three mutually exclusive alternatives holds:

- all $m(A_n) \in (-\infty, +\infty]$ and at least one $m(A_{n_0}) = +\infty$;
- all $m(A_n) \in [-\infty, +\infty)$ and at least one $m(A_{n_0}) = -\infty$;
- all $m(A_n) \in (-\infty, +\infty)$ and $\sum_{n\in\mathbb{N}} |m(A_n)| < \infty$.

In particular, the set function $\mu(B) := \sum_{n\in B\cap\mathbb{N}} (-1)^n \frac{1}{n}$, is not a signed measure since the conditionally convergent series $\sum_{n=1}^{\infty} (-1)^n \frac{1}{n}$ can be rearranged so as to attain any value in $[-\infty, \infty]$.

The following theorem shows that any signed measure can be written as a difference of two (positive) measures. In particular, we can reduce many results for signed measures to assertions on (positive) measures.

Theorem 1.47 (Hahn–Jordan decomposition [19, pp. 175, 176]) *Assume that $m : \mathscr{A} \to [-\infty, \infty]$ is a signed measure. There exist disjoint sets $A^\pm \in \mathscr{A}$ such* *that* $X = A^+ \uplus A^-$ *and*

$$\begin{aligned} m(A^+) \geqslant 0 \quad &and \quad \forall B \subseteq A^+,\ B \in \mathscr{A}\ :\ m(B) \geqslant 0, \\ m(A^-) \leqslant 0 \quad &and \quad \forall B \subseteq A^-,\ B \in \mathscr{A}\ :\ m(B) \leqslant 0. \end{aligned} \tag{1.49}$$

Moreover, $\mu(B) := m^+(B) := m(B \cap A^+)$ *and* $\nu(B) := m^-(B) := -m(B \cap A^-)$ *are positive measures satisfying*

$$m = \mu - \nu \quad and \quad \mu \perp \nu. \tag{1.50}$$

The decomposition of $X = A^+ \uplus A^-$ is called the **Hahn decomposition** and the representation $m = m^+ - m^-$ is called the **Jordan decomposition**.

From Remark 1.46.(a) we know that at least one of the measures $m^\pm$ is finite. Moreover, the Hahn decomposition, hence the Jordan decomposition, is 'essentially' unique in the sense that for any two decompositions $X = A^+ \uplus A^-$ and $X = C^+ \uplus C^-$ satisfying (1.49) the sets $A^+ \cap C^-$ and $A^- \cap C^+$ satisfy both conditions mentioned in (1.49).

1.11 A Historical Aperçu From the Beginnings Until 1854 [6]

The technique of 'integration' may well be traced back to ancient times when Greek scientists determined the volume of regular bodies by (infinitesimal) exhaustion methods, e.g. Eudoxos of Knidos (c. 390–c. 337 BC) or Archimedes (c. 287–c. 212 BC). Exhaustion was also used in pre-calculus times, e.g. Kepler (1571–1630) and Cavalieri (1598–1647), but the history of integration properly starts with the discovery of calculus by Leibniz (1646–1716) and Newton (1643–1727). Starting from the concept of a derivative, Newton understood integration as 'anti-derivative' and the value of a definite integral $\int_a^b f(x)\,dx$ was for him the difference $F(b)-F(a)$ of a primitive, i.e. a function F with derivative $F' = f$. Leibniz followed a more geometric approach, defining the integral as a sum of infinitely small quantities (*summa omnium*) – the notation $\int y$ and $\int y\,dx$ reminiscent of a sum is due to Leibniz, the name integration was proposed by J. Bernoulli [28, § 620 *et seq.*] – and he discovered in this way that the integral is the inverse operation to differentiation, arriving, like Newton, at the relation $f(b) - f(a) = \int_a^b f'(x)\,dx$. Since Leibniz's infinitesimals were quite vague and controversial, it was the Newtonian point of view of the integral as anti-derivative that persisted. Euler (1707–1783) was one of the main supporters of this stance, propagating differentiation as the primary operation of calculus.

But there were problems. Fourier (1768–1830) tried to represent an 'arbitrary'

[6] Our historical comments are mainly based on Cantor [31], Hawkins [77], Medvedev [116, Chapter 4] and Pesin [135].

function f on $(-\pi, \pi)$ in the form

$$f(x) = \frac{1}{2}a_0 + \sum_{n=1}^{\infty} (a_n \cos nx + b_n \sin nx).$$

He thought that it was enough to determine the coefficients a_n, b_n and he even found the recipe to do so:

$$a_n = \frac{1}{\pi}\int_{-\pi}^{\pi} f(x)\cos nx\,dx \quad \text{and} \quad b_n = \frac{1}{\pi}\int_{-\pi}^{\pi} f(x)\sin nx\,dx.$$

But what would the primitives of $f(x)\cos nx$ and $f(x)\sin nx$ be? Fourier assumed that every function could be represented as

$$f(x) = \frac{1}{2\pi}\int_{-\infty}^{\infty} f(\alpha)\,d\alpha \int_{-\infty}^{\infty} \cos p(x-\alpha)\,dp$$

without caring about the (non-)convergence of the cosine integral; in his reasoning, $x \longmapsto \cos p(x-\alpha)$ is differentiable and this property is inherited by the integral, hence there is a primitive, after all. Nowadays we know that not all functions can be represented in terms of Fourier series or -integrals, nevertheless Fourier's idea was to be a major driving force in the development of the concept of functions and integrals.

Cauchy (1789–1857) was one of the first mathematicians to separate integration from differentiation. Motivated by the questions raised by Fourier, he started from the notion of a (uniformly) continuous[7] function and defined the integral of a function f as a limit of the Cauchy sums

$$\int_a^b f(x)\,dx = \lim_{|\Pi|\to 0}\sum_{i=1}^{n}(x_i - x_{i-1})f(x_{i-1}), \quad |\Pi| := \max_{1\leqslant i\leqslant n}(x_i - x_{i-1}), \tag{1.51}$$

along all possible finite partitions $\Pi = \{a = x_0 < x_1 < \cdots < x_n = b\}$. Cauchy showed that this limit always exists for bounded continuous functions. He used the continuity of f to prove that $\frac{d}{dx}\int_a^x f(t)\,dt = f(x)$ and from this he deduced the fundamental theorem $\int_a^b F'(x)\,dx = F(b) - F(a)$ for all F such that F' exists and is continuous. Thus, for continuous functions f, all primitives (in the sense of Newton) are of the form $F(x) = \int_a^x f(t)\,dt + C$. In order to deal with (some) discontinuous functions, Cauchy proposed using the principal value integral, that is $\lim_{\epsilon\to 0}\left(\int_a^{c-\epsilon} f(x)\,dx + \int_{c+\epsilon}^b f(x)\,dx\right)$ if f fails to be continuous at c, but this idea did not catch on until Hardy's time and it also did not solve the problem of integrating more general discontinuous functions. Although Cauchy identified all possible primitives of continuous functions, this required the notion of

[7] Cauchy did not distinguish between continuity and uniform continuity.

an integral; thus, Newton's idea to **define** the integral in terms of a primitive is circular if based on Cauchy's writings. Only in 1905 did Lebesgue [104] discover an 'integral-free' construction of primitives of continuous functions.

1.12 Appendix: H. Lebesgue's Seminal Paper

Before his thesis [101] was published, Lebesgue explained his new approach to integration in a note in the *Comptes Rendus Hebdomadaires de l'Academie des Sciences* **132** (1901) 1–3. This text contains all the important ideas and it is worth reading even after more than 100 years.[8]

¶1 H. Lebesgue: On a generalization of the definite integral ¶1

For continuous functions there is no difference between the concepts of integral and primitive. Riemann defined an integral of certain discontinuous functions, but not all derivatives of functions are integrable in the sense of Riemann. Therefore, the problem of finding primitives cannot be solved by integration, and one would like to have a definition of an integral which contains, as a particular case, Riemann's integral and which allows one to solve the problem of finding primitives *.

In order to define the integral of a continuous and increasing function

$$y(x)\ (a \leqslant x \leqslant b),$$

one divides the interval (a, b) in subintervals and adds the quantities obtained by multiplying the length of the subinterval by one of the values y if x is in this subinterval. If x is in the interval (a_i, a_{i+1}), y takes values between certain bounds m_i, m_{i+1}, and, conversely, if y is between m_i and m_{i+1}, then x is between a_i and a_{i+1}. Instead of considering the division obtained from varying x, that is instead of fixing the numbers a_i, one could have chosen a division from varying y, i.e. from the numbers m_i. One knows that the first method (fixing the a_i) leads to the definition of Riemann and to the definition of the upper and lower integrals introduced by M. Darboux. Let us consider the second method.

¶2 Assume that the function y takes values between m and M. We fix ¶2

$$m = m_0 < m_1 < m_2 < \cdots < m_{p-1} < M = m_p.$$

We have $y = m$ if x is contained in a set E_0 and $m_{i-1} < y \leqslant m_i$ if x is contained in a set E_i.

[8] Translation from the French by the authors. The original page breaks and page numbers n are indicated by ¶n in the running text and in the margins.

* These two *a priori* requirements for any generalization of the integral are indeed compatible, since every, in the sense of Riemann integrable, derivative function has one of its primitives as its integral.

From now on we will denote by λ_0, λ_i the measures of these sets. Let us consider the following two sums:

$$m_0\lambda_0 + \sum m_i\lambda_i, \quad m_0\lambda_0 + \sum m_{i-1}\lambda_i.$$

If, given that the maximal distance between two consecutive m_i tends to zero, the above sums converge to the same limit independently of the choice of the values m_i, then this limit is, by definition, the integral of y and y will be called *integrable*.

Let us consider a set of points from (a, b); one can cover these points in infinitely many ways by countably infinitely many intervals; the lower limit of the sum of the lengths of these intervals is the measure of the set. A set E is said to be *measurable* if the sum of its measure and the measure of the set of points which are not contained in E yields the measure of (a, b) **. Here are two properties of these sets: given infinitely many measurable sets E_i, the set of points which are contained in at least one of them is measurable; if any two of the E_i have no points in common, the measure of the set is the sum of the measures of the E_i. The set of points which are contained in each of the E_i is measurable.

It is natural to consider first those functions such that the sets which appear in the definition of the integral are measurable. One finds: *if a function, whose absolute value is bounded from above, is such that for any A and B the set of the points x for which one has $A < y \leqslant B$ is measurable, then it is integrable* using the procedure laid out above. Such a function will be called *summable*. The integral of a summable function is between the lower and the upper integral. Consequently, *if a function, which is integrable in the sense of Riemann, is summable, then the integrals from both definitions coincide.* That is, *every function integrable in Riemann's sense is summable* since the set of its discontinuity points is a set of measure zero; and one can show that a function is summable if, after neglecting a set of points x of measure zero, ¶3 it is continuous at each of the remaining points. This property allows one immediately to construct functions which are not integrable in the sense of Riemann but still summable. Let $f(x)$ and $\phi(x)$ be two continuous functions, and let $\phi(x)$ not be identically zero; any function which differs from $f(x)$ only at the points of an everywhere dense set of measure zero and which has at these points the value $f(x) + \phi(x)$ is summable without being integrable in the sense of Riemann. *Example:* The function which is equal to 0 if x is irrational and equal to 1 if x is rational. This construction shows that the set of summable functions has a cardinality which is larger than that of the continuum. Here are two properties of this set:

1° *If f and ϕ are summable, so are $f + \phi$ and $f\phi$* and the integral of $f + \phi$ is the sum of the integrals of f and of ϕ.

¶3

** If one adds to these sets certain suitably chosen sets of measure zero, one obtains the measurable sets in the sense of Mr. Borel (*Leçons sur la théorie des fonctions*).

2° *If a sequence of summable functions has a limit, this is a summable function.*

The set of summable functions obviously contains $y = k$ and $y = x$; thus, because of 1°, it contains all polynomials and since it contains all limits, because of 2°, it contains all continuous functions and all limits of continuous functions, that is the functions of the first class (see Baire, *Annali di Matematica*, 1899), it contains all functions of the second class, etc.

In particular, since *every derivative of a function whose absolute value is bounded from above* is of the first class, *it is summable* and one can show that *its integral, considered as a function of its upper limit, is among its primitives.*

Here is a geometric application: if $|f'|, |\phi'|, |\psi'|$ are bounded from above, the length of the curve

$$x = f(t), \quad y = \phi(t), \quad z = \psi(t)$$

is given by the integral of $\sqrt{f'^2 + \phi'^2 + \psi'^2}$. If $\phi = \psi = 0$, one obtains the total variation of the function f having bounded variation. Should f', ϕ', ψ' not exist, one can obtain an almost identical theorem if one replaces the derivatives by Dini's numbers [Dini's derivatives].

(29 April 1901)

2

A Refresher of Topology and Ordinal Numbers

In this chapter we collect some basic constructions (the Cantor set, the Cantor function) and topics (basic point-set topology, cardinal and ordinal numbers, the ordinal space) which are not always covered in elementary courses on analysis and measure and integration. Since we will frequently use these concepts, we include them here for further reference.

2.1 A Modicum of Point-Set Topology

Our standard references on topology are the monographs by Willard [200] and Engelking [53]; the interplay between topology and measure with a focus on 'small' sets is discussed in Oxtoby [133]. The material below (and much more) can be found in these books.

Topologies. Let $X \neq \emptyset$ be any set. A **topology** on X is a family of sets denoted by $\mathscr{O} = \mathscr{O}(X) \subseteq \mathscr{P}(X)$ and satisfying

$$\emptyset, X \in \mathscr{O}, \tag{$\mathscr{O}_1$}$$

$$\text{intersections of finitely many sets in } \mathscr{O} \text{ belong to } \mathscr{O}, \tag{$\mathscr{O}_2$}$$

$$\text{unions of arbitrarily many sets in } \mathscr{O} \text{ belong to } \mathscr{O}. \tag{$\mathscr{O}_3$}$$

The sets in $\mathscr{O}$ are called **open sets** and the pair $(X, \mathscr{O})$ is said to be a **topological space**.

We can describe a topology $\mathscr{O}$ with the help of neighbourhoods of all points $x \in X$. A set N such that $N \supseteq U(x)$ for some open set $U(x) \in \mathscr{O}$ containing x is called a **neighbourhood** of x. If N is itself open, it is called an **open neighbourhood**. A **neighbourhood base** at $x \in X$ is a family of sets $\mathscr{U}_x \subseteq \mathscr{O}$ such that

$$\forall U \in \mathscr{O},\ x \in U \quad \exists V \in \mathscr{U}_x \ :\ x \in V \subseteq U.$$

Conversely, the families $\mathcal{U}_x$, $x \in X$, uniquely determine $\mathcal{O}$:

$$U \in \mathcal{O} \iff \forall x \in U \quad \exists V \in \mathcal{U}_x \; : \; x \in V \subseteq U.$$

If every $x \in X$ has a countable neighbourhood base $\mathcal{U}_x$, then the space X is said to be **first countable**.

There are various notions connected with openness, for example

closed sets: A set $F \subseteq X$ is closed if $F^c = X \setminus F$ is open.

F_σ**-sets**: A set $F \subseteq X$ is an F_σ-set if it is the countable union of closed sets: $F = \bigcup_{n\in\mathbb{N}} F_n$. An F_σ-set need not be closed.

G_δ**-sets**: A set $U \subseteq X$ is a G_δ-set if it is the countable intersection of open sets: $U = \bigcap_{n\in\mathbb{N}} U_n$. A G_δ-set need not be open.

compact set: A set $K \subseteq X$ is compact if, for every open cover $\bigcup_i U_i \supseteq K$, there is a finite subcover $U_{i(1)} \cup \cdots \cup U_{i(n)} \supseteq K$. Note that some authors define compactness only in Hausdorff spaces. This condition ensures that compact sets are closed; see [200, Theorem 17.5] or [53, Theorem 3.1.8].

locally compact: This means that for every point $x \in X$ there is a compact set $K = K(x)$ containing an open neighbourhood of x.

basis of the topology $\mathcal{O}$ is a family $\mathcal{B} \subseteq \mathcal{O}$ such that every $U \in \mathcal{O}$ can be obtained as a union of sets from $\mathcal{B}$. If there is a countable basis $\mathcal{B}$, then the space X is said to be **second countable**.

Hausdorff space or T_2**-space**: we call $(X, \mathcal{O})$ Hausdorff if the topology is rich enough to distinguish points in X, i.e. if any two points $x \neq y$ have disjoint neighbourhoods $U(x), V(y) \in \mathcal{O}$, $U(x) \cap V(y) = \emptyset$. This is an example of a so-called **separation axiom**, usually denoted by T_i, $i = 1, 2, 3, 3a, 4$, which essentially expresses how rich $\mathcal{O}$ is. For example, under T_3 we can separate closed sets from points, and under T_4 we can separate any two closed disjoint sets.

open interior of $A \subseteq X$: $A^\circ = \bigcup\{U\,;\; U \subseteq A,\; U \in \mathcal{O}\}$ is the largest open set contained in A.

closure of $A \subseteq X$: $\overline{A} = \bigcap\{F\,;\; A \subseteq F,\; F^c \in \mathcal{O}\}$ is the smallest closed set containing A.

The notion of topology resembles that of a σ-algebra, see (Σ_1)–(Σ_3) on p. 7, but a topology is, in general, not stable under countable intersections and complementation, while a σ-algebra is, in general, not stable under arbitrary unions. There is often a close interplay between topology and measurability. Recall that the Borel σ-algebra $\mathcal{B}(X)$ is the smallest σ-algebra in X containing $\mathcal{O}$; therefore, the Borel σ-algebra is also called the **topological σ-algebra**. This is the reason why a change of topology can be used to produce counterexamples in measure and integration. Let us briefly list some of the most important topologies and the corresponding Borel σ-algebras.

Example 2.1 Throughout $X \neq \emptyset$ is some non-empty set.

(a) **Normed and metric spaces**. Let $d : X \times X \to [0, \infty]$ be a metric, i.e. a map which is definite: $d(x, y) = 0 \iff x = y$, symmetric: $d(x, y) = d(y, x)$, and satisfies the triangle inequality: $d(x, y) \leqslant d(x, z) + d(z, y)$. The family of **open (metric) balls** $B_r(x) = \{y \in X\,;\, d(x, y) < r\}$, $x \in X$, $r > 0$, is a neighbourhood base for the topology induced by the metric d.

   Since any norm $x \longmapsto |x|$ defines a metric $d(x, y) = |x - y|$, norm topologies are special cases of metric topologies.

(b) **Discrete topology**: $\mathscr{O} = \mathscr{P}(X)$. This is the largest (or finest) possible topology on X. It is induced by the **discrete metric** $d(x, y) = 0$ if $x = y$ and, if $x \neq y$, $d(x, y) = c$ for some $c \in (0, \infty]$ (usually $c = 1$ or $c = \infty$). Note that $B_c(x) = \{x\}$. The corresponding Borel σ-algebra is again the power set $\mathscr{P}(X)$.

(c) **Indiscrete topology**: $\mathscr{O} = \{\emptyset, X\}$. This is the smallest (or coarsest) possible topology on X. The corresponding Borel σ-algebra is $\{\emptyset, X\}$.

(d) **Co-finite topology**: $\mathscr{O} = \{U \subseteq X\,;\, \#U^c < \infty\} \cup \{\emptyset, X\}$. If X is finite, $\mathscr{O}$ is the discrete topology. The co-finite topology is the smallest (coarsest) topology such that singletons $\{x\}$ are closed sets. Moreover, every subset $A \subseteq X$ is compact. The corresponding Borel σ-algebra is the co-countable σ-algebra [☞ Example 1.7.(d)].

(e) **Relative (or trace) topology**: Let $A \subseteq X$ by any subset and $\mathscr{O} = \mathscr{O}(X)$ any topology on X. The relative topology in A is given by

$$\mathscr{O}(A) := A \cap \mathscr{O}(X) := \{A \cap U\,;\, U \in \mathscr{O}\}.$$

   Note that a set $K \subseteq A$ is compact in $\mathscr{O}(A)$ if, and only if, it is compact in $\mathscr{O}(X)$. Moreover, the set A is both open and closed in $(A, \mathscr{O}(A))$, no matter which properties A had in the original space $(X, \mathscr{O})$. The corresponding Borel σ-algebra is the trace σ-algebra $A \cap \mathscr{B}(X)$ [☞ Example 1.7.(e)].

(f) **Two-point compactification of** $\mathbb{R}$: Denote by (a, b), $[-\infty, a)$ and $(b, \infty]$ intervals in $\overline{\mathbb{R}}$. A set $N = N(x) \subseteq \overline{\mathbb{R}}$ is a neighbourhood of $x \in \overline{\mathbb{R}}$, if for some $h > 0$

$$N \supseteq \begin{cases} (x - h, x + h), & \text{if } x \in \mathbb{R}, \\ [-\infty, -h), & \text{if } x = -\infty, \\ (h, \infty], & \text{if } x = \infty. \end{cases}$$

   In particular, the intervals $[-\infty, a)$ and $(b, \infty]$ are open intervals in this topology. Note that the usual topology in $\mathbb{R}$ appears as the trace topology: $\mathcal{O}(\mathbb{R}) = \mathbb{R} \cap \mathcal{O}(\overline{\mathbb{R}})$. Every set $B^* \in \mathcal{B}(\overline{\mathbb{R}})$ is of the form $B^* = B \cup S$ for $B \in \mathcal{B}(\mathbb{R})$ and $S \in \{\emptyset, \{-\infty\}, \{+\infty\}, \{-\infty, +\infty\}\}$.

(g) **Order topology**: Assume that there is a linear order '$<$' [☞ 2.2] on X. The open intervals in this topology are the sets $(a, b) = \{x \in X\,;\ a < x < b\}$. An example of an order topology is the ordinal space [☞ Section 2.4].

(h) **Sorgenfrey topology**: Let $X = \mathbb{R}$ or any linearly ordered [☞ 2.2] space. The Sorgenfrey topology on $\mathbb{R}$ – also known as the Sorgenfrey line or the right half-open interval topology – is the topology where the basic neighbourhoods of a point $x \in \mathbb{R}$ are of the form $[x, y) = \{z \in \mathbb{R}\,;\ x \leqslant z < y\}$. In this topology, the sets $(-\infty, x)$, $[x, y)$, $[x, +\infty)$ are both open and closed; the sets (x, y) and $(x, +\infty)$ are open but not closed. The Sorgenfrey topology is first countable (i.e. each point has a countable neighbourhood base) but not second countable (i.e. it has no countable base), hence not metrizable. A set is compact in the Sorgenfrey topology if, and only if, it is countable and nowhere dense (in the Euclidean topology of $\mathbb{R}$). The corresponding Borel σ-algebra coincides with the Borel σ-algebra generated by $\mathbb{R}$ equipped with the usual Euclidean metric.

(i) **Product topology**: Let $(X, \mathcal{O})$ be any topological space and I an index set. The (finite or infinite) product of X is the set

$$X^I := \bigtimes_{i \in I} X = \{f\,;\ f : I \to X\};$$

by $\pi_i : X^I \to X$, $f \longmapsto \pi_i(f) := f(i)$ we denote the ith coordinate projection. The canonical product topology on X^I is the topology whose basis consists of sets of the form

$$\bigcap_{k=1}^{n} \pi_{i(k)}^{-1}(U_{i(k)}), \quad n \in \mathbb{N},\ U_{i(k)} \in \mathcal{O},$$

i.e. products $\bigtimes_{i \in I} U_i$ where only finitely many of the $U_i \neq X$.

In this topology, a sequence $(f_n)_{n \in \mathbb{N}} \subseteq X^I$ converges to $f \in X^I$ if, and only if, $f_n(i) \to f(i)$ in X for each $i \in I$. Therefore, the product topology is often called the topology of pointwise convergence (of the coordinates). By **Tychonoff's theorem**, products of the form $\bigtimes_{i \in I} K_i$ where $K_i \subseteq X$ is compact, are compact sets. If I is at most countable and the topology of X has a countable base, then the corresponding Borel σ-algebra is the product σ-algebra $\mathcal{B}(X)^{\otimes I}$.

Polish spaces. Among the most important topological spaces in analysis and measure theory are Polish spaces. A Hausdorff topological space $(X, \mathcal{O})$ is called a **Polish space** if it can be equipped with a metric d such that (X, d) is a complete metric space; the metric d generates the topology $\mathfrak{O}$ and $\mathfrak{O}$ has a countable basis (second countable).[1] Every Polish space is first countable (since it is a metric space) and it is separable; the latter property is, in metric spaces, equivalent to the existence of a countable basis of the topology [53, Corollary 1.3.8, Theorem 4.1.15] or [200, Theorem 16.9, 16.11].

Example 2.2 (Bourbaki [21, Chapter IX.6.1]) The following spaces are Polish spaces.

(a) $\mathbb{R}^d$ with the Euclidean metric $d(x, y) = |x - y|$ is a Polish space.
(b) The Hilbert cube $[0, 1]^{\mathbb{N}}$ with the metric $d(x, y) := \sum_{i=1}^{\infty} 2^{-i}|x_i - y_i|$, where $x = (x_i)_{i\in\mathbb{N}}$, $y = (y_i)_{i\in\mathbb{N}}$ is a Polish space.
(c) If (X_i, d_i), $i = 1, \dots, n$, are Polish spaces, then $(X_1 \times \cdots \times X_n, d_1 + \cdots + d_n)$ is Polish.
(d) If (X, d) is a Polish space and $F \subseteq X$ a closed subset, then $(F, d|_{F\times F})$ is Polish.
(e) If (X, d) is a Polish space and $G \subseteq X$ an open subset, then G is Polish if equipped with a suitable metric (which is, in general, not the restriction of d). More generally, Mazurkiewicz showed that G is Polish if, and only if, G is a G_δ-set of the Polish space (X, d).
(f) If we combine Example 2.2.(e) and (b), we see that a topological space X is Polish if, and only if, it is homeomorphic to a G_δ-set of the Hilbert cube $[0, 1]^{\mathbb{N}}$.

Category. The Baire category theorem is one of those abstract theorems in functional analysis which is used for existence proofs. Baire's theorem involves a topological notion of 'smallness' of sets.

Definition 2.3 Let $(X, \mathfrak{O})$ be a topological space. A set $A \subseteq X$ is **dense** if $\overline{A} = X$ and **nowhere dense** if $(\overline{A})^\circ = \emptyset$.

A subset $B \subseteq X$ is of **first category** (in the sense of Baire) or **meagre** if $B = \bigcup_{n\in\mathbb{N}} A_n$ with countable many nowhere dense sets A_n. All other subsets of X are said to be of **second category**.

Baire's category theorem now reads as follows.

Theorem 2.4 (Baire) *Let X be a locally compact Hausdorff space or a completely metrizable space. If $D_n \subseteq X$, $n \in \mathbb{N}$, are countably many dense open sets, then $D := \bigcap_{n\in\mathbb{N}} D_n$ is dense.*

[1] The name 'Polish space' was proposed by Bourbaki [21], but it is not always used in topology books and one has to collect the material from various sources.

Any space where the intersection of countably many open dense sets is again dense is a **Baire space**. The following theorem shows that every Baire space is of second category.

Theorem 2.5 *A topological space $(X, \mathfrak{O})$ is of second category if, and only if, the intersection of countably many dense open sets $D_n \subseteq X$ is non-empty.*

Let us list a few consequences of these theorems and definitions.

Example 2.6 (a) Let A, A' be nowhere dense sets. Then any $B \subseteq A$, $\overline{A}$ and $A \cup A'$ are nowhere dense.

(b) Let A, A_n, $n \in \mathbb{N}$, be sets of first category. The sets $B \subseteq A$ and $\bigcup_{n\in\mathbb{N}} A_n$ are of first category.

(c) Any (bounded or unbounded) interval $I \subseteq \mathbb{R}$ is of second category.

(d) If $A \subseteq \mathbb{R}$ is of first category, then A^c is dense. Indeed, if $A = \bigcup_{n\in\mathbb{N}} A_n$ for nowhere dense sets A_n, then $A \subseteq \bigcup_{n\in\mathbb{N}} \overline{A}_n$ and $A^c \supseteq \bigcap_{n\in\mathbb{N}} \overline{A}_n^{\,c}$. Since the sets $\overline{A}_n^{\,c}$ are open and dense, it follows from Baire's theorem that A^c is dense.

2.2 The Axiom of Choice and Its Relatives

Throughout the text we will need some basic set theory. For most purposes, the 'naive approach' to set theory is more than enough, see Halmos [71], and we do not want to delve into the details of the Zermelo–Fraenkel axiomatics. In case you ever need (or want) to look up details, the books by Kechris [89] and Jech [85] would be our first choice; the survey paper by Bingham & Ostaszewski [17] is another good resource.

Let X_i, $i \in I$, be arbitrarily many non-empty sets and consider the infinite product

$$X := \bigtimes_{i\in I} X_i := \Big\{ f\,;\ f : I \to \bigcup_{i\in I} X_i,\ f(i) \in X_i \Big\}.$$

If I is finite, say $\#I = n$, then we can identify X with the family of all n-tuples $\{(x_1, \dots, x_n)\,;\ x_i \in X_i, i = 1, \dots, n\}$ and it is clear that $X \neq \emptyset$. If I is countably infinite, we have to use the induction principle to construct some $(x_k)_{k\in\mathbb{N}}$ to make sure that $X \neq \emptyset$. If I is not countable, then it is not clear at all whether the product $\bigtimes_{i\in I} X_i$ is empty. At this point, the axiom of choice comes in.

Axiom of Choice (AC) Let $\{X_i\,;\ i \in I\}$ be a collection of non-empty sets. There exists a set $C \subseteq \bigcup_{i\in I}\{i\} \times X_i$ which contains exactly one element from each set $\{i\} \times X_i$, $i \in I$.

It is known that AC is independent of all other Zermelo–Fraenkel (ZF) axioms, i.e. it cannot be proved within ZF; if we extend ZF by AC, we usually speak of ZFC.

The axiom of choice can be stated in different ways and comes in several guises. For this we need the concept of an order.

Well-ordering. A set X is **ordered** if there is an order relation '$\preceq$' which is reflexive, antisymmetric and transitive. We speak of a **partial order** if we want to stress that X contains elements which cannot be compared '$\preceq$'. By contrast, in a **linearly** (or **totally**) **ordered** space, any two elements of X can be compared. We write $x \prec x'$ if $x \preceq x'$ and $x \neq x'$. If X is a linearly ordered set, then we call $[x] = \{y \in X\,;\ y \prec x\}$ the **initial segment** determined by x.

If we want to compare linearly ordered sets X and Y, we should not just use a bijection (as for cardinality) but a bijection f that preserves the ordering on X and Y, i.e. $x \preceq_X x' \implies f(x) \preceq_Y f(x')$; f is called an **order isomorphism**.

An ordering is called a **well-ordering** if every $\emptyset \neq A \subseteq X$ has a smallest element. A deep result of Zermelo tells us that every set can be well-ordered.

Theorem 2.7 (well-ordering theorem) *Every set can be equipped with an order-relation such that $(X, \preceq)$ is well-ordered.*

Note that the order in the well-ordering theorem may be quite different from the natural order of the set X – think about the sets $\mathbb{Q}$ or $\mathbb{C}$. The proof of the well-ordering theorem relies on the axiom of choice, in fact: it is equivalent to the axiom of choice. But there are further equivalences.

Theorem 2.8 *The axiom of choice is equivalent to each of the following properties.*

Hausdorff's maximality principle / Kuratowski–Zorn lemma *Every partially ordered set $(X, \preceq)$ contains a linearly ordered subset $Y \subseteq X$ which is maximal, i.e. any other linearly ordered subset $Z \subseteq X$ satisfies $Z \subseteq Y$.*

Zorn's lemma *Every partially ordered set $(X, \preceq)$, with the property that every linearly ordered subset $Y \subseteq X$ has an upper bound, contains a maximal element.*

An accessible proof that AC is equivalent to the statements in Theorems 2.7 and 2.8 is [80, Theorem 3.12].

The working analyst or probabilist does not encounter the AC too often, but let us remind you of the proofs of the Hahn–Banach theorem or the existence of a (Hamel) basis in a vector space or Tychonoff's theorem (existence of ultrafilters). In the present text we will use AC in the construction of the ordinal space [☞Section 2.3, 2.4] and of non-measurable sets [☞7.22].

Transfinite induction. When dealing with well-ordered sets we often need to extend the principle of mathematical induction to an uncountable setting. The induction principle asserts that: *If S is a subset of a well-ordered set X and if for all $x \in X$ the initial segment $[x] \subseteq S$, then $S = X$.* We will use it often in the following form:

Theorem 2.9 (transfinite induction) *Let X be a well-ordered set with smallest element $1 \in X$ and $S(x)$ be a statement which may or may not be true. If*

(a) *$S(1)$ is true,*
(b) *$S(y)$ is true for all $y \prec x$ implies that $S(x)$ is true,*

then $S(x)$ is true for all $x \in X$.

The continuum hypothesis. We will see below [☞2.3] that there are several 'degrees' of infinity. Of these, countable sets – those which can be bijectively mapped to $\mathbb{N}$, where the cardinality is called 'aleph nought' $\aleph_0$ – and uncountable sets should be familiar. The typical example of an uncountable set is $\mathbb{N}^{\mathbb{N}}$ or $\{0, 1\}^{\mathbb{N}}$ and the proof of non-countability uses Cantor's diagonal procedure; see [MIMS, p. 12]. Since $\{0, 1\}^{\mathbb{N}}$ can be identified with $[0, 1]$ or $\mathbb{R}$, the cardinality of $\{0, 1\}^{\mathbb{N}}$ is called the **continuum** $\mathfrak{c}$. While it is clear that $\aleph_0 < \mathfrak{c}$, one cannot prove that there is a set $\mathbb{N} \subseteq S \subseteq \mathbb{R}$ such that $\aleph_0 < \#S < \mathfrak{c}$. Since, by AC, there is a smallest uncountable cardinal $\aleph_1 > \aleph_0$, the question is whether $\mathfrak{c} = \aleph_1$. Again, on the basis of ZFC one cannot decide this, and so we need the **continuum hypothesis**.

Continuum Hypothesis (CH) $\aleph_1 = \mathfrak{c}$ or, equivalently, $\aleph_1 = 2^{\aleph_0}$.

Note that the equality $\mathfrak{c} = 2^{\aleph_0}$ follows directly from the bijections between $\mathbb{R}$ and $[0, 1]$ and $\{0, 1\}^{\mathbb{N}}$.

The generalized continuum hypothesis (GCH), which we will not need here, asserts that one has $\aleph_{\alpha+1} = 2^{\aleph_\alpha}$ if we set up a hierarchy of alephs tagged by an ordinal number α and its immediate successor $\alpha + 1$; see p. 46 further on.

2.3 Cardinal and Ordinal Numbers

Cardinal numbers. Let X, Y be any two sets. If there is a bijection $f : X \to Y$ we say that X **and** Y **are of the same cardinal type** and we write $\#X = \#Y$; $\#X$ is the **cardinal number** of the set X. If $X = \{1, 2, \dots, n\}$, then we set $\#X := n$, i.e. the cardinal number of a finite set is the number of its elements. Things are different for infinite sets: here we encounter various degrees of infinity. The simplest one is **countable infinity**: a set X is countably infinite if there is a bijection $f : X \to \mathbb{N}$. The cardinality of $\mathbb{N}$ is usually denoted by $\#\mathbb{N} = \aleph_0$. Using Cantor's diagonal procedure and the base-2 (dyadic) representation of all numbers $x \in [0,1]$, one can show that

$$\#[0,1] = \#\{0,1\}^{\mathbb{N}} = 2^{\aleph_0} > \aleph_0.$$

Usually, one writes $\mathfrak{c}$ for the cardinality of the continuum $[0,1]$. It is straightforward to see that $\mathbb{N}, \mathbb{Z}, \mathbb{Q}, \mathbb{N} \times \mathbb{N}$ are of type $\aleph_0$, while $[0,1], \mathbb{R}, \mathbb{R}^2, \mathbb{C}$ are of type $\mathfrak{c}$. These considerations also show the first principles of cardinal arithmetic: let $\mathfrak{a}, \mathfrak{b}$ be cardinal numbers of the sets A and B, respectively, and assume that $A \cap B = \emptyset$. We define

(a) $\mathfrak{a} + \mathfrak{b}$ as the cardinality of $A \cup B$;
(b) $\mathfrak{a}\mathfrak{b}$ as the cardinality of $A \times B$;
(c) $\mathfrak{a}^{\mathfrak{b}}$ as the cardinality of $A^B = \{f \,;\, f : B \to A\}$.

We write $\#A \leqslant \#B$ if there is an injection $i : A \to B$. A slightly deeper result, the **Cantor–Bernstein theorem** tells us that if both $\#A \leqslant \#B$ and $\#B \leqslant \#A$ hold, then $\#A = \#B$. We write $\#A < \#B$ if $\#A \leqslant \#B$ but $\#A \neq \#B$. A typical example is $\#A < \#\mathscr{P}(A)$. This result is due to Cantor. If $\mathfrak{a}$ is an *infinite* cardinal number, then

(a) $\mathfrak{a} + \mathfrak{a} = \mathfrak{a}$ – adapt the argument that is used to enumerate $\mathbb{Z}$;
(b) $\mathfrak{a}^2 = \mathfrak{a}$ – adapt the argument that is used to enumerate $\mathbb{N} \times \mathbb{N}$;
(c) $\mathfrak{b} \leqslant \mathfrak{a} \implies \mathfrak{b} + \mathfrak{a} = \mathfrak{a}$ – observe that $\mathfrak{a} \leqslant \mathfrak{a} + \mathfrak{b} \leqslant \mathfrak{a} + \mathfrak{a} = \mathfrak{a}$;
(d) $0 < \mathfrak{b} \leqslant \mathfrak{a} \implies \mathfrak{b} \cdot \mathfrak{a} = \mathfrak{a}$ – observe that $\mathfrak{a} \leqslant \mathfrak{b} \cdot \mathfrak{a} \leqslant \mathfrak{a} \cdot \mathfrak{a} = \mathfrak{a}$;
(e) $\mathfrak{a} < 2^{\mathfrak{a}}$ – let A be a set such that $\#A = \mathfrak{a}$. By Cantor's theorem, $\mathfrak{a} < \#\mathscr{P}(A)$ and we may identify $\mathscr{P}(A)$ with $\{0,1\}^A$ using the bijection $A \leftrightarrow \mathbb{1}_A \in \{0,1\}$.

These considerations show that there is a hierarchy of cardinals $\aleph_0, 2^{\aleph_0}, 2^{(2^{\aleph_0})}$, etc. which we can describe only by using the concept of **ordinal numbers**.

Ordinal numbers. We can now define ordinal numbers. By analogy to cardinal numbers we use order isomorphisms to assign an ordinal number to each well-ordered set. We say that two linearly ordered sets X and Y are of the same **order type** if there exists some order isomorphism $f : X \to Y$. If X is, in addition, well-ordered, we attach to X an **ordinal number** $\omega(X)$. [2]

We write $\omega(\emptyset) = 0$, $\omega(\{1, \dots, n\}) = n$ and $\omega(\mathbb{N}) = \omega_0$, but $\mathbb{Q}$ does not have an

[2] From a logical point of view this is a 'naive definition'. We use implicitly that the equivalence classes of well-ordered sets are themselves a set, and this will eventually lead to problems of the type 'does the set of all sets contain itself as a set?' In order to avoid this, the proper definition of ordinal numbers is the following: we call a set E **transitive** if it has the property that $x \in E \implies x \subseteq E$. For example, the sets $\emptyset, \{\emptyset, \{\emptyset\}\}, \{\emptyset, \{\emptyset\}, \{\emptyset, \{\emptyset\}\}\}$, etc. are transitive (and can be used to define $\mathbb{N}_0$). An ordinal number is a transitive set which is well-ordered by the relation $\in$; see Jech [85, Chapter I.2].

ordinal number (if we use the standard ordering '$\leqslant$') as it is not well-ordered. We can use the initial segment $[x]$ to compare ordinal numbers. For sets A, B with $\alpha = \omega(A)$ and $\beta = \omega(B)$, we define

$$\alpha \prec \beta \overset{\text{def}}{\iff} \exists b \in B \ : \ A \text{ and } [b] \text{ are order isomorphic}$$

and we write $\alpha \preceq \beta$ if either $\alpha \prec \beta$ or $\alpha = \beta$. This definition does not depend on the actual sets but only on their order type. With some effort one can show that

(a) $\forall a \in A \,:\, [a]$ and A can never be order isomorphic; this excludes, in particular, that $\alpha \prec \alpha$.
(b) $\forall a, a' \in A \,:\,$ if $[a]$ and $[a']$ are order isomorphic, then $a = a'$.
(c) For any two ordinal numbers α, β exactly one of the following alternatives holds: $\alpha \prec \beta$, $\alpha = \beta$, $\beta \prec \alpha$.

Notice that the ordinal numbers are themselves well-ordered. The above results allow us to identify α with the set $[\alpha]$.

Using Zermelo's well-ordering theorem, we can construct an uncountable well-ordered set Ω with the property that there is a largest element $\omega_1 \in \Omega$ such that the initial segments $[\alpha]$ for $\alpha \prec \omega_1$ are countable sets. We call Ω the **ordinal space**, ω_1 the **first uncountable ordinal** and $\Omega_0 = \Omega \setminus \{\omega_1\}$ are the **countable ordinal numbers**. Because of the well-ordering, each ordinal has an immediate successor which is usually denoted by $\omega_1^+ = \omega_1 + 1$. Note that some ordinals, e.g. ω_1, are limit ordinal numbers: they do not have an immediate predecessor.

Let us denote by 0 the smallest element of Ω (it exists since Ω is well-ordered). Thus we get a hierarchy of ordinal numbers

$$0, 1, 2, 3, \dots, \omega_0, \omega_0 + 1, \omega_0 + 2, \dots, 2\omega_0, 2\omega_0 + 1, 2\omega_0 + 2, \dots$$

where $n \cdot \omega_0$ means the ordinal of the union of n 'clones' of Ω. The smallest ordinal larger than these is ω_0^2 and proceeding as above we get ω_0^3, ω_0^4, etc. All these ordinals are countable ordinals and there is no way to reach ω_1 in this way. In other words: if $A \subseteq \Omega \setminus \{\omega_1\}$ is countable, then $\sup A \prec \omega_1$.

Cardinal numbers (reprise). We can now return to the discussion of cardinal numbers. For each infinite cardinal $\mathfrak{a}$ consider the set

$$c(\mathfrak{a}) = \{\mathfrak{b} \text{ infinite cardinal number}\,;\ \mathfrak{b} < \mathfrak{a}\}$$

($c(\mathfrak{a})$ is the initial segment $[\mathfrak{a}]$ for the order on the cardinal numbers). Since $c(\mathfrak{a})$ is a well-ordered set, it has an ordinal number, say α. We define $\aleph_\alpha := \mathfrak{a}$.

An equivalent way to define the $\aleph$s is via **transfinite induction**:

$$\aleph_\alpha \text{ is the smallest cardinal such that } \aleph_\beta < \aleph_\alpha \text{ for all } \beta \prec \alpha.$$

Since $c(\aleph_0) = \emptyset$ and $\omega(\emptyset) = 0$ this definition is consistent with the previous notation. $\aleph_0$ is the smallest infinite cardinal number; $\aleph_1$ is then the next strictly larger cardinal number (use the ordering introduced for cardinal numbers), i.e. the first cardinal number of **some** uncountable set. This means, in particular, that the set $\{\mathfrak{a}\,;\ \mathfrak{a} < \aleph_1\}$ is countable. Recall that the continuum $[0,1]$ is uncountable, but it is not clear how $\aleph_1$ and $\mathfrak{c} = 2^{\aleph_0}$ compare. The **continuum hypothesis** states that $\aleph_1 = 2^{\aleph_0}$.

The above construction gives a hierarchy of $\aleph$s and – writing $\alpha + 1$ for the immediate successor of the ordinal number α – the **generalized continuum hypothesis** states that $\aleph_{\alpha+1} = 2^{\aleph_\alpha}$.

Comment Proofs of the results on counting are in [MIMS, Chapter 2]; the monograph by Hewitt & Stromberg [80, Sections I.3, I.4] has an introduction to ordinal numbers written for analysts.

2.4 The Ordinal Space

Let $\Omega = [0,\omega_1]$ be the **ordinal space** which has the first uncountable ordinal ω_1 as its maximal element. If we embed Ω into the space of all ordinal numbers, $\Omega = [\omega_1]$ is the initial segment in the notation of Section 2.3. We will now introduce a topology on Ω. In order to do so, we specify a basis of the topology; all open sets can be obtained as arbitrary unions of the following basis sets:

$$[0,\beta) := \{\gamma \in \Omega\,;\ \gamma \prec \beta\},$$
$$(\alpha,\omega_1] := \{\gamma \in \Omega\,;\ \alpha \prec \gamma\},$$
$$(\alpha,\beta) := [0,\beta) \cap (\alpha,\omega_1],$$

where $\alpha \prec \beta \prec \omega_1$. Since $\beta + 1$ is the immediate successor of $\beta \prec \omega_1$, we have $(\alpha, \beta+1) = (\alpha,\beta]$. In this topology, the point ω_1 does not have a countable neighbourhood base: each open neighbourhood would contain some set $(\alpha_n,\omega_1]$, but $\bigcap_{n\in\mathbb{N}}(\alpha_n,\omega_1] \neq \{\omega_1\}$ since ω_1 is a limit ordinal, i.e. it cannot be reached by a sequence of ordinals $\alpha_n \prec \omega_1$.

A similar argument shows that $\Omega_0 := \Omega \setminus \{\omega_1\}$, hence Ω, is not separable: if $D \subseteq \Omega_0$ were a dense countable subset, then $\sup D \prec \omega_1$ since $\sup D$ is a countable ordinal; this contradicts the assumption that D is dense.

Given the above topology, Ω is a compact space. Assume that $\mathscr{U}$ is an open cover of Ω, i.e. $\Omega = \bigcup_{U\in\mathscr{U}} U$. Without loss of generality we can assume that the sets $U \in \mathscr{U}$ are basis sets. Let $\alpha_1 \in \Omega$ be the least element such that $(\alpha_1,\omega_1]$

is contained in some $U_1 \in \mathcal{U}$. If $\alpha_1 \neq 1$, let $\alpha_2 \in \Omega$ be the least element such that $(\alpha_2, \alpha_1]$ is in some $U_2 \in \mathcal{U}$ and continue this procedure. If we could find infinitely many different α_n, we would have a sequence $\alpha_1 \succ \alpha_2 \succ \dots$ without a smallest element; this contradicts the fact that Ω, hence $\{\alpha_n\,;\ n \in \mathbb{N}\}$ is well-ordered. Therefore, $\{U_1, \dots, U_n\} \subseteq \mathcal{U}$ is a finite sub-cover, i.e. Ω and all sets $[0, \alpha]$, $\alpha \in \Omega$, are compact.

The set $\Omega_0 := \Omega \setminus \{\omega_1\}$ of countable ordinals is sequentially compact, i.e. every sequence $(\alpha_n)_{n\in\mathbb{N}} \subseteq \Omega_0$ has an accumulation point $\alpha \in \Omega_0$. This follows easily from the compactness of Ω. We get first that an accumulation point $\alpha \in \Omega$ exists. But since α is a subsequential limit of some $(\alpha'_n)_{n\in\mathbb{N}} \subseteq (\alpha_n)_{n\in\mathbb{N}}$ and ω_1 is a limit ordinal, it is clear that $\alpha \neq \omega_1$, so $\alpha \in \Omega_0$.

Every continuous function $f : \Omega_0 \to \mathbb{R}$ is **eventually constant**, i.e. there is some α_f such that $f(\alpha) = f(\beta)$ for all $\alpha_f \prec \alpha, \beta$. In order to see this, we show first that there is a sequence $(\alpha_n)_{n\in\mathbb{N}}$ such that

$$\alpha_n \prec \beta \Rightarrow |f(\beta) - f(\alpha_n)| < \frac{1}{n}.$$

If such a sequence did not exist, there would be some $n_0 \in \mathbb{N}$ and, by recursion, an increasing sequence $(\beta_i)_{i\in\mathbb{N}}$ such that $|f(\beta_i) - f(\beta_{i+1})| \geqslant 1/n_0$ for every $i \in \mathbb{N}$. Since $\beta_i \to \beta := \sup_{i\in\mathbb{N}} \beta_i \prec \omega_1$, we would get by continuity that $f(\beta_i) \to f(\beta) \in \{+\infty, -\infty\}$, which is impossible.

Let $\alpha_f := \sup_{n\in\mathbb{N}} \alpha_n$. Then, by definition, $f(\alpha) = f(\beta)$ for all $\alpha, \beta \succ \alpha_f$. Note that this proof shows that we can extend any continuous f defined on Ω_0 uniquely to Ω.

2.5 The Cantor Set: A Nowhere Dense, Perfect Set

The classical Cantor ternary set is an example of a **perfect** and **nowhere dense** set with zero Lebesgue measure. Recall that a (non-empty) perfect set is a closed set such that each of its points is a limit point; such sets are necessarily uncountable. A nowhere dense set is a set whose closure has empty interior. A thorough discussion of (the topological properties of) the Cantor set can be found in [200].

We will follow Cantor's construction [30]; see also [MIMS, p. 3]. Let $(a_n)_{n\in\mathbb{N}_0}$ be the sequence in $(0, 1)$ whose general term is given by $a_n = \frac{1}{2}\left(\frac{2}{3}\right)^{n+1}$. Observe that $\sum_{n=0}^{\infty} a_n = 1$. Set $C_0 := [0, 1]$. Recursively, we obtain the following closed sets:

C_1 by removing from the middle of C_0 an open interval I_2 of length $a_0 = \frac{1}{3}$;
C_2 by removing from the middles of the two parts of C_1 the open intervals I_{02} and I_{22}, each of length $\frac{1}{2}a_1 = \frac{1}{9}$;

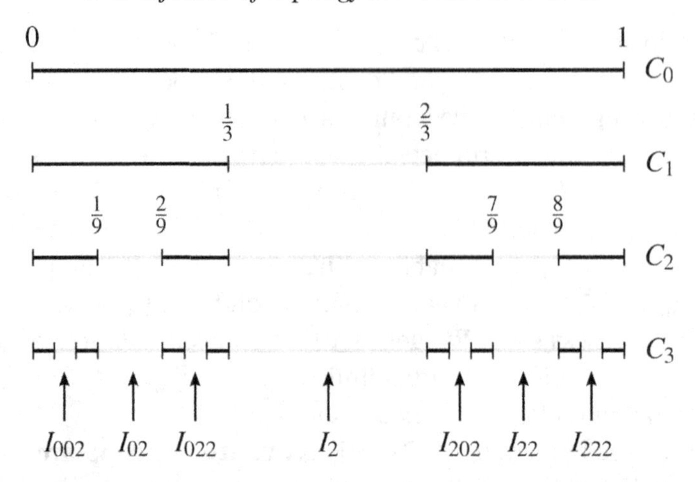

Figure 2.1 The figure shows the first three steps in the construction of Cantor's middle-third set.

C_3 by removing from the middles of the four parts of C_2 the open intervals $I_{002}, I_{022}, I_{202}$ and I_{222}, each of length $\frac{1}{4}a_2 = \frac{1}{27}$;

...

C_{n+1} by removing from the middles of the 2^n parts of C_n the open intervals $I_{t_1 t_2 \dots t_{n-1} t_n 2}$, $t_i \in \{0, 2\}$, each of length $2^{-n}a_n = \frac{1}{3^{n+1}}$;

see Fig. 2.1, and we define $C := \bigcap_{n \in \mathbb{N}_0} C_n$.

The 0–2–sequence appearing as label of the open intervals $I_{t_1 t_2 \dots t_{n-1} t_n 2}$ which have been removed in the $(n+1)$st step encodes the ternary (i.e. base-three, 3-adic) expansion of the right end point of the interval, i.e.

$$\sup I_{t_1 t_2 \dots t_{n-1} t_n 2} = \sum_{i=1}^{n} \frac{t_i}{3^i} + \frac{2}{3^{n+1}} = 0.t_1 t_2 \dots t_{n-1} t_n 2.$$

This shows that C consists of the points $\{0.t_1 t_2 t_3 \dots\ ;\ t_i \in \{0, 2\}\}$ and, using Cantor's diagonal method, we conclude that C is not countable. In fact, the map

$$\Phi : 0.t_1 t_2 t_3 \dots t_n \dots (\text{ternary}) \longleftrightarrow 0.\frac{t_1}{2}\frac{t_2}{2}\frac{t_3}{2} \dots \frac{t_n}{2} \dots (\text{binary})$$

establishes a bijection between C and $[0, 1]$ showing that the cardinality of C is that of the continuum: $\mathfrak{c}$. As usual, we identify the numbers

$$0.t_1 \dots t_{n-1} t_n \overline{(p-1)} \dots \quad \text{and} \quad 0.t_1 \dots t_{n-1}(t_n + 1)\overline{0} \dots$$

($t_n \neq 1$ if $p = 2$ and $t_n \neq 2$ if $p = 3$) to enforce uniqueness of the representation, but this identification is obviously respected by the map Φ.

By construction, every $x \in C$ is approached by a sequence of endpoints of removed intervals, and since these endpoints are themselves in C, the set C is perfect. Moreover, there is no open interval in $[0,1]$ which does not intersect with some of the removed open intervals, so C is nowhere dense. Since the total length of the 2^n intervals removed in the $(n+1)$st step equals $2^n \cdot 3^{-n-1} = a_n$, it follows that

$$\lambda(C) = 1 - \sum_{n=0}^{\infty} a_n = 0,$$

i.e. C has Lebesgue measure zero.

2.6 The Cantor Function and Its Inverse

The ***Cantor function*** or **devil's staircase** is a function $F : [0,1] \to [0,1]$ which is increasing, continuous, Lebesgue a.e. differentiable with derivative $F' = 0$ – but it is not absolutely continuous. This means, in particular, that

$$F(x) - F(0) \neq \int_0^x F'(t)\,dt,$$

no matter whether we use the Riemann or the Lebesgue integral.

The idea of the construction of F is to use the classical Cantor middle-thirds set $C \subseteq [0,1]$ with zero Lebesgue measure [☞ Section 2.5] as the set of points where F is strictly increasing.

To achieve this, recall that we have removed in the $(n+1)$st step of the construction of C exactly 2^n intervals $I_{t_1 t_2 \dots t_n 2}$, $t_i \in \{0,2\}$, of length 3^{-n-1}; the string $t_1 \dots t_n 2$ stands for the triadic representation of the right endpoint of the removed interval.

We construct a sequence of functions $F_n : [0,1] \to [0,1]$ by

$$F_{n+1}(x) := \begin{cases} 0, & \text{if } x = 0, \\ \frac{1}{2}\sum_{i=1}^{k} t_i 2^{-i} + 2^{-(k+1)}, & \text{if } x \in I_{t_1 t_2 \dots t_k 2} \text{ for } k \in \{0, \dots, n\}, \\ 1, & \text{if } x = 1, \end{cases} \tag{2.1}$$

and interpolate linearly between these values to define $F_n(x)$ for all other x; see Fig. 2.2. We are going to show that $F(x) = \lim_{n\to\infty} F_n(x)$ exists and defines an increasing and continuous function.

The definition of the F_ns shows that

$$|F_n(x) - F_{n+1}(x)| \leqslant \frac{1}{2^{n+1}}.$$

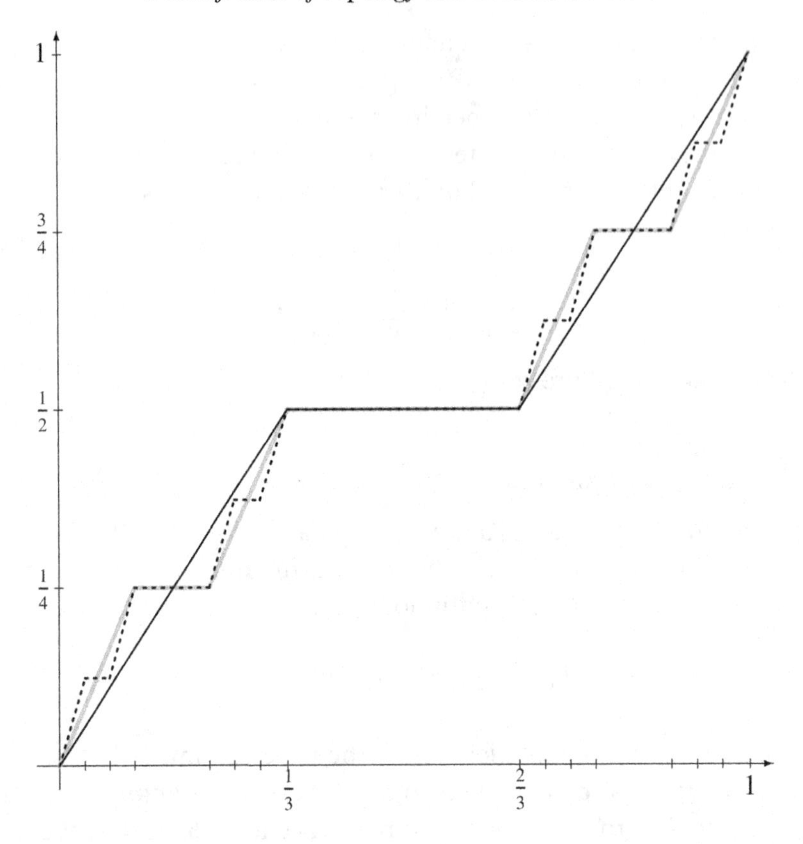

Figure 2.2 The first stages of the construction of the Cantor function: F_1 is represented by a black line, F_2 by a grey line and F_3 is the dashed black line.

This follows from the observation that F_{n+1} is obtained from F_n by modifying the graph only on the sets $I_{t_1 t_2 \dots t_n 2}$ by replacing the diagonal line by a diagonal–flat–diagonal line; this happens only within a range of 2^{-n} units.

We conclude that the convergence of $F_n \to F$ is uniform, i.e. F inherits the continuity and monotonicity from the F_ns.

For $x \in [0,1] \setminus C$ there is some open interval $I_n = I_{t_1 t_2 \dots t_{n-1} 2}$ containing x. Since $F|_{I_n} = F_n|_{I_n}$ we conclude that $F'(x) = F'_n(x) = 0$. Since $\lambda(C) = 0$ we see that $F' = 0$ Lebesgue a.e. Thus,

$$F(x) - F(0) \neq \int_0^x F'(t)\,dt, \quad x > 0,$$

which shows that F is not absolutely continuous.

The inverse of the Cantor function. Since the Cantor function F is increasing, we may interpret it as the distribution function of a (singular) measure on the

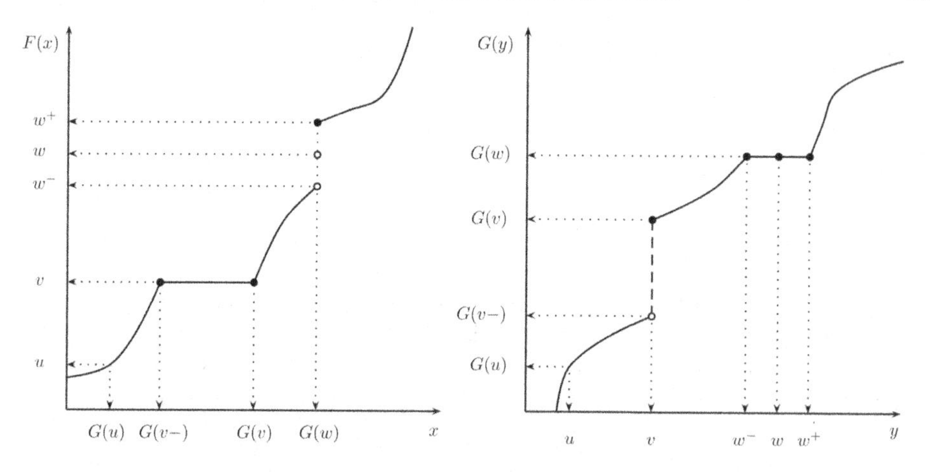

Figure 2.3 An increasing function F and its generalized inverse $G = F^{-1}$.

space $([0,1], \mathscr{B}[0,1])$ given by $\mu_F(a,b] = F(b) - F(a)$, $0 \leqslant a < b \leqslant 1$. In order to work with μ_F it is useful to consider the **generalized (right-continuous) inverse** of F:

$$G(y) := F^{-1}(y) := \inf\{x \in [0,1]\,;\ F(x) > y\},$$

cf. Fig. 2.3. By definition, $G : [0,1] \to [0,1]$ is an increasing and right-continuous function satisfying $F(G(y)) = y$ and $G(F(x)) \geqslant x$ and $G(F(x)) = x$ if $F(x)$ is a continuity point of G; a discussion on generalized inverses can be found in [BM, Lemma 18.15]. G has jump discontinuities if F has a level stretch, and vice versa.

The last remark is a strong hint how one can construct the function G explicitly. Recall the definition of the **Rademacher functions** $R_k : [0,1] \to [-1,1]$, $k \in \mathbb{N}_0$,

$$R_0 = \mathbb{1}_{[0,1]},\quad R_1 = \mathbb{1}_{\left[0,\frac{1}{2}\right)} - \mathbb{1}_{\left[\frac{1}{2},0\right]},\quad R_2 = \mathbb{1}_{\left[0,\frac{1}{4}\right)} - \mathbb{1}_{\left[\frac{1}{4},\frac{1}{2}\right)} + \mathbb{1}_{\left[\frac{1}{2},\frac{3}{4}\right)} - \mathbb{1}_{\left[\frac{3}{4},1\right]},\dots$$

The graphs of the first four Rademacher functions are shown in Fig. 2.4.

Lemma 2.10 *The generalized inverse of the Cantor function F is given by*

$$G(y) = 2\sum_{k=1}^{\infty} 3^{-k}\frac{1}{2}(1 - R_k(y)) = \frac{1}{2} - \sum_{k=1}^{\infty} 3^{-k} R_k(y), \quad y \in [0,1], \tag{2.2}$$

cf. Fig. 2.5.

Proof Since $0 \leqslant 1 - R_k \leqslant 2$, it is easy to see that the sum defining the function G converges uniformly. In particular, G inherits the right-continuity from its summands $1 - R_k$; it is also obvious that G is monotone increasing.

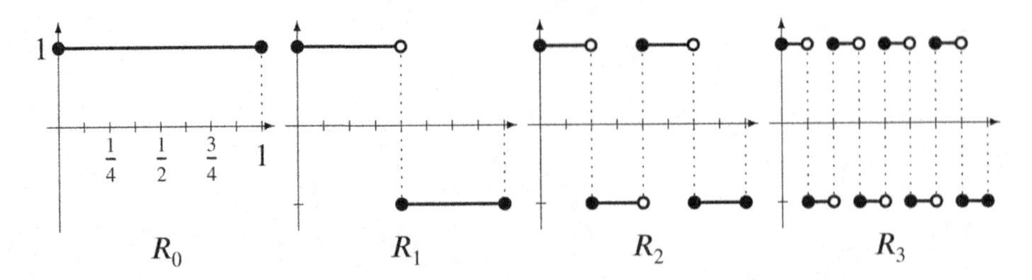

Figure 2.4 The first four Rademacher functions. In each generation, all intervals of constancy are divided into two equal halves and the sign of the function flips to the opposite sign on the right half. Thus, the newly added jumps have negative height, while the already existing jumps become positive.

Take $y \in [0,1)$ with binary representation $y = \sum_{i=1}^{\infty} s_i 2^{-i}$, $s_i \in \{0,1\}$. By the definition of the Rademacher functions, it holds that

$$\forall k \in \mathbb{N} \; : \; \frac{1}{2}(1 - R_k(y)) = s_k.$$

In particular, if $y = \sum_{i=1}^{n} s_i 2^{-i} + 2^{-n-1}$ for some $n \in \mathbb{N}$, then

$$G(y) = \sum_{k=1}^{n} 3^{-k}(2s_k) + 2 \cdot 3^{-(n+1)} =: x.$$

The same is true for the generalized inverse F^{-1}, i.e. $F^{-1}(y) = x$. Indeed, if we set $t_i := 2s_i$, then x is the right end point of the interval $I_{t_1 \dots t_n 2}$ which appears in the construction of the Cantor function F [☞(2.1)], and it follows from the definition of F that

$$F(x) = \sum_{i=1}^{n} 2^{-i}\frac{t_i}{2} + 2^{-(n+1)} = \sum_{i=1}^{n} 2^{-i}s_i + 2^{-(n+1)} = y.$$

Since x is the right end point of the constancy interval $I_{t_1 \dots t_n 2}$, we know from the construction of F that $F(x')$ is strictly larger than $F(x) = y$ for any $x' > x$. Hence,

$$F^{-1}(y) = \inf\{x' \in [0,1]\, ; \; F(x') > y\} = x,$$

and we see that $F^{-1}(y) = G(y)$ for every $y \in [0,1)$, which can be written in the form $y = \sum_{i=1}^{n} s_i 2^{-i} + 2^{-n-1}$. As

$$D = \left\{ \sum_{i=1}^{n} s_i 2^{-i} + 2^{-n-1} \, ; \; n \in \mathbb{N}, \; s_i \in \{0,1\} \right\} = \{k2^{-n} \, ; \; k = 0, \dots, 2^{n-1}, \; n \in \mathbb{N}\}$$

is dense in $[0,1]$, the right-continuity of G and F^{-1} gives $F^{-1}(y) = G(y)$ for all $y \in [0,1)$. For $y = 1$ equality is immediate from the definition of F and G. □

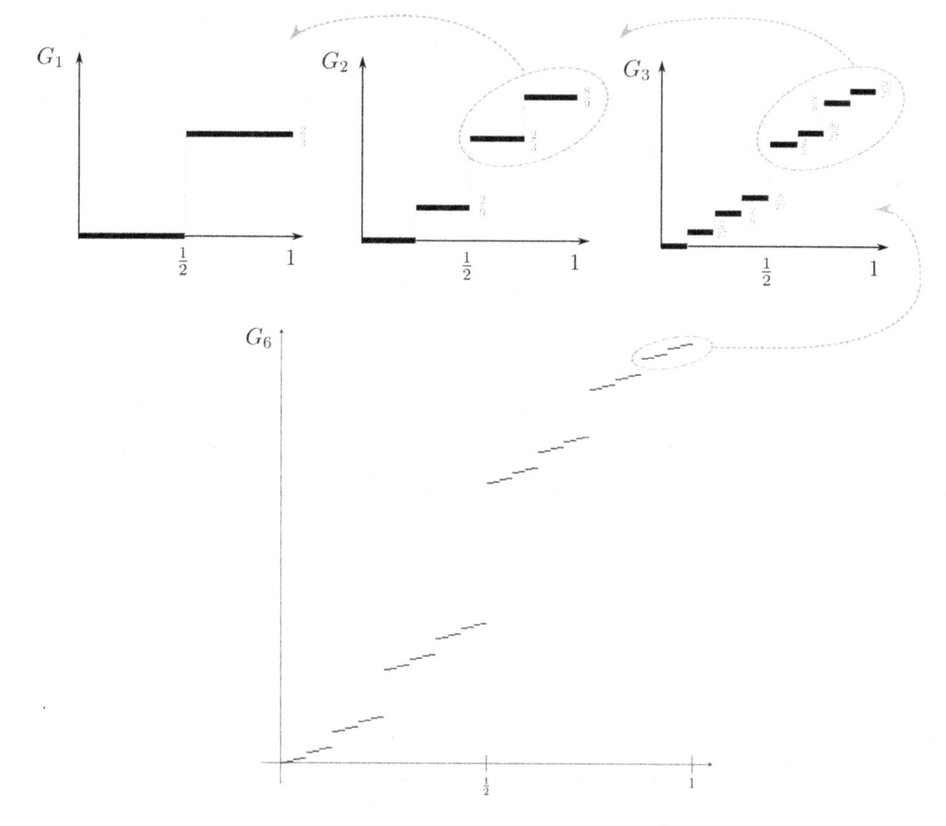

Figure 2.5 Plots of the partial sums $G_n = 2\sum_{k=1}^{n} 3^{-k}\frac{1}{2}(1 - R_k)$ for the values $n \in \{1, 2, 3, 6\}$. For $n \to \infty$ the sequence G_n converges to the generalized inverse of the Cantor function.

The above proof that $G = F^{-1}$ can also be understood intuitively. Because of the right-continuity, it is enough to identify G and F^{-1} on a dense subset of $(0, 1]$, say, the dyadic numbers $D \subsetneq (0, 1)$ where the function G jumps. If we arrange the dyadic numbers by 'generations', $D^k := \{i/2^k\,;\ i = 1, \dots, 2^k - 1\}$, then $D^k \setminus D^{k-1}$ are the 'new' numbers in generation k. Clearly,

$$\forall n \in \mathbb{N} \quad \forall i2^{-n} \in D^n \setminus D^{n-1} \ :\ \Delta(1 - R_k)(i2^{-n}) = \begin{cases} 2, & \text{if } k = n, \\ 0, & \text{if } k < n, \\ -2, & \text{if } k > n. \end{cases}$$

This shows that

$$\Delta G\,(i2^{-n}) = 2 \cdot 3^{-n} - 2 \sum_{k=n+1}^{\infty} 3^{-k} = 3^{-n} \quad \text{for all} \quad i2^{-n} \in D^n \setminus D^{n-1}.$$

Thus, there is a bijection between the jumps of the function G and the intervals of constancy of F. If we can show that the jumps of G and the intervals of constancy of F can be transformed into each other by a reflection at the 45-degree line in the (x, y)-plane, we are obviously done. Using the symmetry of the Rademacher functions w.r.t. the point $\binom{1/2}{1/2}$ it is clear that the jump of height $\frac{1}{3}$ is located in the middle of the line $\left\{\frac{1}{2}\right\}\times[0,1]$ in the (x, y)-plane. Again by symmetry and scaling, the jump of height $\frac{1}{9}$ is in the middle of the line $\left\{\frac{1}{4}\right\}\times\left[0,\frac{1}{3}\right]$ and $\left\{\frac{3}{4}\right\}\times\left[\frac{1}{3},1\right]$ and so on.

Using the fact that $F\circ G(y) = y$, $F(0) = G(0) = 0$ and $F(1) = G(1) = 1$, we can now evaluate integrals w.r.t. the Lebesgue–Stieltjes measure μ_F induced by the Cantor function F: for every bounded Borel function $f : [0, 1] \to \mathbb{R}$ we have

$$\begin{aligned}\int_{[0,1]} f(x)\,\mu(dx) = \int_0^1 f(x)\,dF(x) &\overset{x=G(y)}{=} \int_0^1 f(G(y))\,dF\circ G(y) \\ &= \int_0^1 f(G(y))\,dy.\end{aligned} \tag{2.3}$$

With a bit more effort, we can show that $f \in L^1([0, 1], \mathscr{B}[0, 1], \mu)$ if, and only if, $f\circ G \in L^1([0, 1], \mathscr{B}[0, 1], dy)$; see Theorem 1.37.

3

Riemann Is Not Enough

The Riemann integral [☞ 1.1] is still one of the most important and useful notions of integration. If $f : [a, b] \to \mathbb{R}$ is a bounded function defined on a compact interval $[a, b] \subseteq \mathbb{R}$, $\Pi = \{a = x_0 < x_1 < \cdots < x_{n(\Pi)} = b\}$ a generic finite partition and $m_i = \inf_{[x_{i-1}, x_i]} f$, resp. $M_i = \sup_{[x_{i-1}, x_i]} f$, the infimum, resp. supremum, of f in the ith partition interval, then one defines the lower and upper Riemann–Darboux integrals as

$$\underline{\int_a^b} f(x)\, dx := \sup_{\Pi} \underbrace{\sum_{i=1}^{n(\Pi)} m_i (x_i - x_{i-1})}_{\text{lower Darboux sum}},$$

$$\overline{\int_a^b} f(x)\, dx := \inf_{\Pi} \underbrace{\sum_{i=1}^{n(\Pi)} M_i (x_i - x_{i-1})}_{\text{upper Darboux sum}} = -\underline{\int_a^b} (-f)(x)\, dx,$$

and their common finite value, if it exists, is the **Riemann integral**

$$\int_a^b f(x)\, dx = \underline{\int_a^b} f(x)\, dx = \overline{\int_a^b} f(x)\, dx.$$

Alternatively, the Riemann integral is the limit along all finite partitions Π with $|\Pi| = \max_i (x_i - x_{i-1}) \to 0$ and for any choice of points $\xi_i \in [x_{i-1}, x_i]$:

$$\int_a^b f(x)\, dx = \lim_{|\Pi| \to 0} \underbrace{\sum_{i=1}^{n(\Pi)} f(\xi_i)(x_i - x_{i-1})}_{\text{Riemann sum}}.$$

The Riemann integral has quite a few shortcomings – see the discussion at the end of Section 1.1 and the counterexamples in this chapter – but probably the most unsatisfactory feature is that one needs Lebesgue's theory in or-

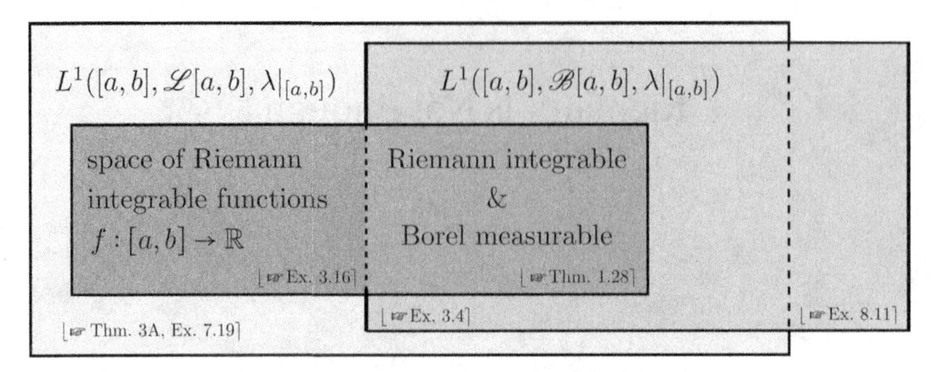

Figure 3.1 The figure shows the relations between the space of Riemann integrable functions, the space of Borel measurable functions with finite Lebesgue integral and the space of Lebesgue measurable functions with finite Lebesgue integral.

der to characterize all Riemann integrable functions. To wit *a bounded function* $f : [a, b] \to \mathbb{R}$ *is Riemann integrable if, and only if, the set of discontinuity points* $D^f := \{x \in (a, b);\ f \text{ is discontinuous at } x\}$ *is a Lebesgue null set* [☞Theorem 1.28]. Observe that the set D^f is for **any** function $f : [a, b] \to \mathbb{R}$ a Borel set [☞8.9], and so its Lebesgue measure $\lambda(D^f)$ is well-defined.

Moreover, any *Borel measurable and Riemann integrable function is Lebesgue integrable* [☞Theorem 1.28], i.e. the Lebesgue integral extends the (proper) Riemann integral. A summary of what may happen is given in Fig. 3.1.

Theorem 3A *Every Riemann integrable function* $f : [a, b] \to \mathbb{R}$ *is Lebesgue measurable, hence Lebesgue integrable, and the values of the Riemann and the Lebesgue integral coincide.*

Proof Take sequences $(u_n)_{n\in\mathbb{N}}, (v_n)_{n\in\mathbb{N}}$ of Borel measurable step functions such that $u_n \uparrow f$ and $v_n \downarrow f$. The step functions are Borel measurable, hence Lebesgue measurable, and so $U := \sup_{n\in\mathbb{N}} u_n$ and $V := \inf_{n\in\mathbb{N}} v_n$ are Lebesgue measurable. By the Riemann integrability of f and the monotone convergence theorem,

$$\int_{[a,b]} V(x)\,\lambda(dx) = \sup_{n\in\mathbb{N}} \int_{[a,b]} v_n(x)\,\lambda(dx) = \sup_{n\in\mathbb{N}} \int_a^b v_n(x)\,dx = \int_a^b f(x)\,dx$$

and, analogously,

$$\int_{[a,b]} U(x)\,\lambda(dx) = \inf_{n\in\mathbb{N}} \int_{[a,b]} u_n(x)\,\lambda(dx) = \inf_{n\in\mathbb{N}} \int_a^b u_n(x)\,dx = \int_a^b f(x)\,dx.$$

Thus, $\int_{[a,b]}(V(x) - U(x))\,\lambda(dx) = 0$. As $V \geqslant f \geqslant U$, this gives $U = V = f$

Lebesgue almost everywhere, and so f is Lebesgue measurable. Moreover, the Riemann and the Lebesgue integrals coincide. □

Having said this, let us point out that the improper Riemann integral has features which cannot (or only very clumsily) be captured by Lebesgue integration [☞ 10.8, 10.9], and it would be a gross misunderstanding to claim that Lebesgue integration makes the Riemann integral obsolete. A typical example is the sine integral

$$\operatorname{Si}(x) = \int_0^x \frac{\sin t}{t}\,dt, \qquad \lim_{x\to\infty} \operatorname{Si}(x) = \frac{\pi}{2}.$$

While $\operatorname{Si}(x)$ exists as both Lebesgue and Riemann integral, the limit as $x \to \infty$, i.e. the integral $\int_0^\infty t^{-1} \sin t\,dt$, exists as an improper Riemann integral but not as a Lebesgue integral since $\int_0^\infty t^{-1}|\sin t|\,dt = \infty$; [☞ 10.8] for a full discussion. Let us point out the importance of the sine integral in harmonic analysis, e.g. for the (inversion of the) Fourier transform [MIMS, pp. 217–220], [185, Chapter I], [207, Chapter XVI, vol. 2, pp. 242–258] or as a prototype of a singular integral [67, Chapter 4].

3.1 The Riemann–Darboux upper integral is not additive

Consider the interval $[0,1]$ and define $f := \mathbb{1}_{[0,1]\cap\mathbb{Q}}$ and $g := \mathbb{1}_{[0,1]\setminus\mathbb{Q}}$. Every step function f_1, resp. g_1, with $f_1 \geqslant f$, resp. $g_1 \geqslant g$, satisfies $f_1 \geqslant \mathbb{1}_{[0,1]}$ and $g_1 \geqslant \mathbb{1}_{[0,1]}$; this follows easily from the fact that both $[0,1]\cap\mathbb{Q}$ and $[0,1]\setminus\mathbb{Q}$ have non-empty intersections with any non-empty interval $(a,b) \subseteq [0,1]$. Therefore,

$$1 + 1 = \overline{\int_0^1} f(x)\,dx + \overline{\int_0^1} g(x)\,dx,$$

and

$$\overline{\int_0^1} (f(x) + g(x))\,dx = \overline{\int_0^1} \mathbb{1}_{[0,1]}(x)\,dx = \int_0^1 \mathbb{1}_{[0,1]}(x)\,dx = 1.$$

Comment In general, the upper integral is subadditive, i.e.

$$\overline{\int_a^b} (f(x) + g(x))\,dx \leqslant \overline{\int_a^b} f(x)\,dx + \overline{\int_a^b} g(x)\,dx,$$

and the lower integral is superadditive as $\underline{\int_a^b} f(x)\,dx = -\overline{\int_a^b} (-f(x))\,dx$.

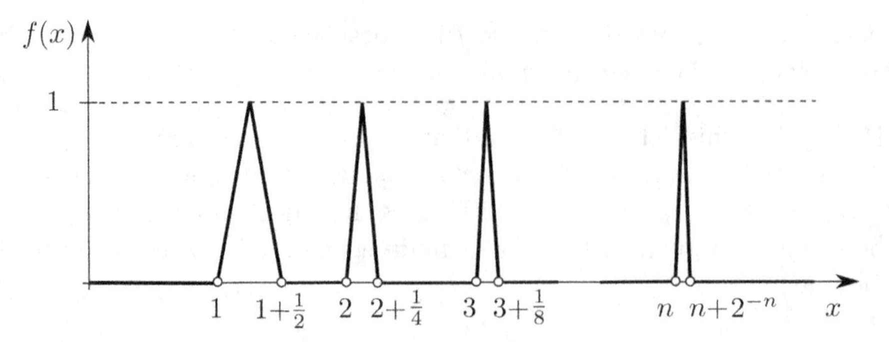

Figure 3.2 The function considered in Example 3.2.

3.2 Why one should define Riemann integrals on bounded intervals

Consider the half-line $[0, \infty)$ and a continuous function $f(x)$ as shown in Fig. 3.2. Clearly, the Lebesgue integral of f exists and has the value

$$\int_{[0,\infty)} f(x)\,\lambda(dx) = \frac{1}{2}\sum_{n=1}^{\infty}\frac{1}{2^n} = \frac{1}{2}.$$

If we try to set up Riemann sums, we encounter the following problem: fix some $m \in \mathbb{N}$ and partition $[0, \infty)$ into non-overlapping intervals $I_n := [n2^{-m}, (n+1)2^{-m})$, $n \in \mathbb{N}_0$. In each interval we select some supporting point $\xi_n \in I_n$. If possible, we take ξ_n so that $f(\xi_n) = 1$, otherwise we take any $\xi_n \in I_n$. The corresponding Riemann sum is

$$\sum_{n\in\mathbb{N}_0} f(\xi_n)\cdot((n+1)2^{-m} - n2^{-m}) = \sum_{n\in\mathbb{N}} 2^{-m} = \infty,$$

since we have infinitely many occurrences of $f(\xi_n) = 1$.

This shows that we can always pick an infinite Riemann sum, i.e. the Riemann integral is not well-defined in this case.

3.3 There are no unbounded Riemann integrable functions

Let $f : [a, b] \to \mathbb{R}$ be an unbounded function, say, $\sup_{x\in[a,b]} f(x) = \infty$. There exists a sequence $(x_n)_{n\in\mathbb{N}} \subseteq [a, b]$ such that $f(x_n) \geqslant n$, and by compactness there is a convergent subsequence $(x_{n(k)})_{k\in\mathbb{N}}$ with $n(k) \to \infty$ and $x_{n(k)} \to x_\infty$ for some $x_\infty \in [a, b]$. By construction, f is unbounded in every neighbourhood of x_∞. In particular, the Darboux upper sum

$$S^{\Pi}[f] := \sum_{i=1}^{n} M_i(t_i - t_{i-1}), \qquad M_i := \sup_{x\in[t_{i-1},t_i]} f(x)$$

is infinite for any partition $\Pi = \{a = t_0 < \cdots < t_n = b\}$; hence, f is not Riemann integrable.

Comment The situation is much different for the Lebesgue integral; for instance, $f(x) := \frac{1}{\sqrt{x}}\mathbb{1}_{(0,1)}(x)$ is unbounded and $\int_{[0,1]} f(x)\,\lambda(dx) < \infty$, cf. [MIMS, p. 108]. As we have seen above, f fails to be (properly) Riemann integrable.

3.4 A function which is not Riemann integrable

Consider the interval $[0,1]$ and Dirichlet's jump function $\mathbb{1}_{[0,1]\cap\mathbb{Q}}$. Let $\sum_{i=1}^n s_i \mathbb{1}_{J_i}$ be a step function, i.e. $s_i \in \mathbb{R}$ and $J_i \subseteq [0,1]$ are finitely many disjoint intervals covering $[0,1]$. If

$$\sum_{i=1}^n s_i \mathbb{1}_{J_i} \leqslant \mathbb{1}_{[0,1]\cap\mathbb{Q}} \quad \text{or} \quad \sum_{i=1}^n s_i \mathbb{1}_{J_i} \geqslant \mathbb{1}_{[0,1]\cap\mathbb{Q}}$$

then we have $s_i \leqslant 0$ or $s_i \geqslant 1$, respectively. In particular, the upper Riemann–Darboux integral of Dirichlet's jump function is one and the lower Riemann–Darboux integral is zero.

Comment This is the typical example of a non-Riemann but Lebesgue integrable function. The Lebesgue integral has the value $\int \mathbb{1}_{[0,1]\cap\mathbb{Q}}\,d\lambda = 0$ since $\mathbb{Q}$ and $\mathbb{Q} \cap [0,1]$ are countable sets, hence, Borel sets.

We may change the above example slightly: consider Thomae's function

$$f(x) = \begin{cases} 0, & \text{if } x \in [0,1] \setminus \mathbb{Q}, \\ q^{-1}, & \text{if } x = \frac{p}{q} \in \mathbb{Q} \cap [0,1],\ \gcd(p,q) = 1, \end{cases}$$

see Fig. 3.4. In contrast to Dirichlet's jump function $\mathbb{1}_{[0,1]\cap\mathbb{Q}}$, the function f **is** Riemann integrable [☞ 3.8].

3.5 Yet another function which is not Riemann integrable

The following example of a non-Riemann integrable function is due to Smith [167] from 1875. Smith used this construction in order to disprove a claim by Hankel [73, § 7, p. 87] that perfect sets (Hankel calls them 'zerstreut') always have measure zero. Note the similarity to Cantor's construction [☞ Section 2.5] which appeared some years later in Cantor [30]; at about the same time, other mathematicians came up independently with similar ideas, for example Volterra [194] and du Bois-Reymond [48, p. 128, footnote].

Split the interval $[0,1)$ into m equally long half-open intervals $I_1, \dots, I_m$ and remove the last interval I_m. Iterate this procedure with each remaining interval I_k, $k = 1, \dots, m-1$. After n steps, all remaining intervals have a total length of $(1 - 1/m)^n$ and this tends to zero as $n \to \infty$. By construction, the limiting set does not contain any open interval, hence it is nowhere dense.

If one removes, in each step, less than the mth part of the remaining lengths, one arrives at a 'fat' Cantor set in the spirit of Example 7.3: split $[0,1)$ into m half-open sub-intervals $I_1, \dots, I_m$ and remove I_m. In step 2, split each of the remaining intervals into m^2 intervals and remove in each case the last interval. In the kth step, split the still remaining intervals into m^k intervals and remove the last of those, and so on. The sum of the remaining 'lengths' after n steps is $\prod_{k=1}^{n}(1 - 1/m^k)$. Since this product converges to $\prod_{k=1}^{\infty}(1 - 1/m^k) > 0$, the limiting set, say S, has positive 'length'.

In view of Riemann's condition (R2) [☞ p. 2], the function $f(x) = \mathbb{1}_S(x)$ is not Riemann integrable.

3.6 A non-Riemann integrable function where a sequence of Riemann sums is convergent

We have seen in Example 3.4 that Dirichlet's jump function $\mathbb{1}_{\mathbb{Q}\cap[0,1]}$ is not Riemann integrable. Consider any partition $\Pi = \{t_0 = 0 < t_1 < \cdots < t_n = 1\}$, $n \in \mathbb{N}$, of $[0,1]$ and pick in each interval of the partition some irrational number $\xi_i \in (t_{i-1}, t_i)$, $i = 1, \dots, n$. The corresponding Riemann sum satisfies

$$S(\Pi) = \sum_{i=1}^{n} \mathbb{1}_{\mathbb{Q}\cap[0,1]}(\xi_i)(t_i - t_{i-1}) = 0,$$

and we have, trivially, $S(\Pi) \to 0$ as the mesh $|\Pi| \to 0$.

A similar phenomenon occurs if we pick equidistant partitions with mesh $1/n$, $\Pi_n = \{t_i = i/n\,;\ i = 0, \dots, n\}$, and the left end point as supporting point $\xi_i = t_{i-1}$. The corresponding Riemann sum is

$$S(\Pi_n) = \sum_{i=1}^{n} \mathbb{1}_{\mathbb{Q}\cap[0,1]}(t_{i-1})(t_i - t_{i-1}) = \sum_{i=1}^{n}(t_i - t_{i-1}) = 1,$$

and we have, trivially, $S(\Pi_n) \to 1$ as $n \to \infty$.

Comment In order to verify Riemann integrability using Riemann sums, one has to make sure that **all possible partitions** and **all possible intermediate points** lead to the same limit. Checking just a particular sequence of Riemann sums is not enough to ensure integrability.

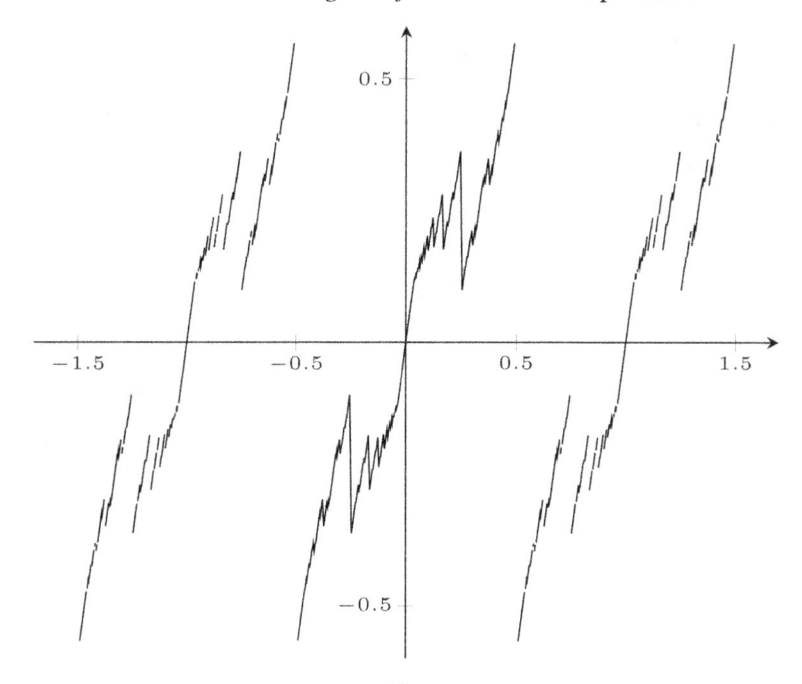

Figure 3.3 Plot of the partial sum $\sum_{n=1}^{N} h(nx)n^{-2}$ on the interval $[-1.5, 1.5]$ for $N = 15$. The picture shows three periods of the function; in the left and right panels the discontinuities are shown as jumps, in the middle panel the jumps are replaced by vertical lines to give a better impression of the shape of the function.

3.7 A Riemann integrable function without a primitive

Riemann's integrability criterion (R2) [☞p. 2] can be used to show that the Riemann integral is capable of dealing with functions which are discontinuous on a (countable) dense subset. The series defining the function (see Fig. 3.3)

$$f(x) = \sum_{n=1}^{\infty} \frac{h(nx)}{n^2}, \qquad h(x) = \begin{cases} x - k, & \text{if } x \in \left(k - \frac{1}{2}, k + \frac{1}{2}\right),\ k \in \mathbb{Z}, \\ 0, & \text{if } x = k \pm \frac{1}{2},\ k \in \mathbb{Z} \end{cases}$$

(i.e. h is a 1-periodic sawtooth-like function) converges for every x and, if x can be written as $x = p/(2n)$ for $p \in \mathbb{Z}$ and $2n \in \mathbb{N}$ being relatively prime, one has (see [143, p. 242])

$$f(x\pm) = f(x) \mp \frac{1}{2n^2} \sum_{i=1}^{\infty} \frac{1}{(2i-1)^2} = f(x) \mp \frac{\pi^2}{16n^2}.$$

Thus, f is discontinuous on all rational numbers of the form $p/(2n)$ and given in their simplest (completely cancelled) form; this is a dense subset of $\mathbb{R}$. Clearly, on every compact interval there are only finitely many points of the form $x = p/(2n)$ such that $f(x-) - f(x+) = \pi^2/(8n^2)$ exceeds a given $\sigma > 0$, i.e. Riemann's integrability criterion (R2) is satisfied and $\int_a^b f(x)\,dx$ exists for all $a \leqslant b$.

If $h > 0$, we have

$$\frac{F(x+h) - F(x)}{h} = \frac{1}{h}\int_x^{x+h} f(t)\,dt \xrightarrow[h\downarrow 0]{} f(x+),$$
$$\frac{F(x) - F(x-h)}{h} = \frac{1}{h}\int_{x-h}^{x} f(t)\,dt \xrightarrow[h\downarrow 0]{} f(x-),$$

which means that F' does not exist if $f(x+) \neq f(x-)$, e.g. if $x = p/(2n)$ for relatively prime $p, 2n \in \mathbb{N}$.

Comment The above construction is due to Riemann [143]. Hankel [74, § 12, pp. 197–200] observed that Riemann's example is also an example of a function f such that

$$F(x) := \int_0^x f(t)\,dt$$

is a continuous, but not (everywhere) differentiable function. In particular, it is not true that $F'(x) = f(x)$ for all x, i.e. F is not a primitive of f.

Further examples of this type, also involving the Lebesgue integral, can be found in Chapter 14.

3.8 A Riemann integrable function whose discontinuity points are dense

Consider **Thomae's function**, named after the mathematician Carl Johannes Thomae,

$$f(x) = \begin{cases} 0, & \text{if } x \in [0,1] \setminus \mathbb{Q}, \\ q^{-1}, & \text{if } x = \frac{p}{q} \in \mathbb{Q} \cap [0,1],\ \gcd(p,q) = 1, \end{cases}$$

cf. Fig. 3.4.

We claim that f is Riemann integrable. Since f is bounded, it suffices to prove that f is Lebesgue a.e. continuous [☞ Theorem 1.28]. We show that f is sequentially continuous at any $x \in [0,1] \setminus \mathbb{Q}$. For $x \in [0,1] \setminus \mathbb{Q}$ let $(x_n)_{n\in\mathbb{N}} \subseteq [0,1]$ be such that $x_n \to x$. As $f(x) = f(y)$ for $y \in [0,1] \setminus \mathbb{Q}$, we may assume that $(x_n)_{n\in\mathbb{N}} \subseteq [0,1] \cap \mathbb{Q}$. Write $x_n = p_n/q_n$ where we assume that $\gcd(p_n, q_n) = 1$.

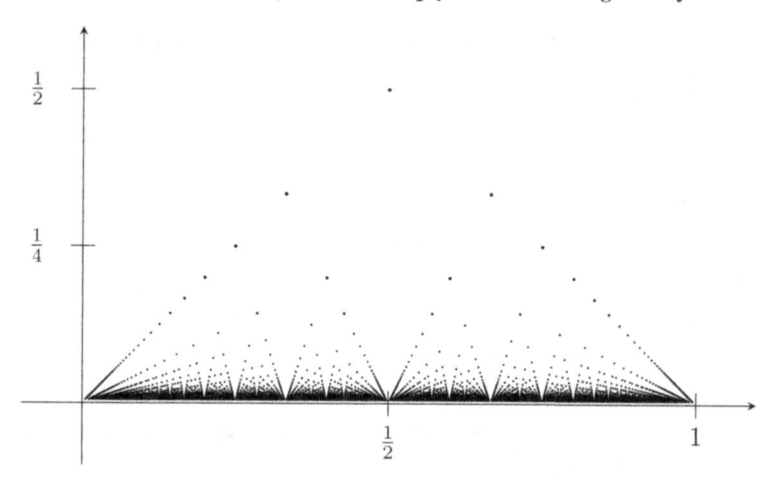

Figure 3.4 Plot of Thomae's function on the interval $(0, 1)$. Because of its resemblance to popping corn, it is also called 'popcorn function'.

Then $q_n \to \infty$ as $n \to \infty$. Indeed: if $(q_n)_{n\in\mathbb{N}}$ is bounded, then $p_n \leqslant q_n$ shows that also $(p_n)_{n\in\mathbb{N}}$ is bounded, and therefore there exist subsequences $(q_{n(k)})_{k\in\mathbb{N}}$ and $(p_{n(k)})_{k\in\mathbb{N}}$ such that $q_{n(k)} \to q$ and $p_{n(k)} \to p$ for some $p, q \in \mathbb{N}$. Hence, $x_{n(k)} = p_{n(k)}/q_{n(k)} \to p/q$ but also $x_{n(k)} \to x$, contradicting our assumption $x \notin \mathbb{Q}$. Thus, we conclude that $q_n \to \infty$ as $n \to \infty$, which implies

$$\lim_{n\to\infty} f(x_n) = \lim_{n\to\infty} \frac{1}{q_n} = 0 = f(x).$$

This finishes the proof that f is Lebesgue a.e. continuous, hence Riemann integrable. Consequently, the Riemann integral $\int_0^1 f(x)\,dx$ exists, and it coincides with the Lebesgue integral $\int_{[0,1]} f\,d\lambda$ [☞Theorem 1.28]. The Lebesgue integral equals zero since the set $\{f \neq 0\}$ is countable, i.e. a Lebesgue null set. Moreover, f is discontinuous at each $x \in \mathbb{Q} \cap [0, 1]$, which implies that the set of discontinuity points of f is dense in $[0, 1]$.

3.9 Semicontinuity does not imply Riemann integrability

Construct a fat Cantor set $C \subseteq [0, 1]$ [☞7.3 or 3.5] with Lebesgue measure $\lambda(C) = \alpha \in (0, 1)$. Since C does not contain any open ball, the function $\mathbb{1}_C(x)$ is discontinuous for each $x \in C$. As $\lambda(C) > 0$, it follows that f is not Riemann integrable [☞Theorem 1.28]. But, C being a compact set, $\mathbb{1}_C$ is upper semicontinuous.

3.10 A function which has the intermediate value property but is not Riemann integrable

The continuity of a function $f: [a,b] \to \mathbb{R}$ implies Riemann integrability of f. Moreover, continuity of f implies that f has the intermediate value property, i.e. $f([c,d]) \subseteq \mathbb{R}$ is an interval for any subinterval $[c,d] \subseteq [a,b]$. At first glance, the intermediate value property seems to be close to continuity – it is not so easy to construct a discontinuous function which has the intermediate value property – and one might suspect that the intermediate value property is a sufficient condition for Riemann integrability. This, however, is false. The following construction yields a function $f: [0,1] \to \mathbb{R}$ which has the intermediate value property but is not Riemann integrable; further examples of this kind can be found in [72].

There are continuum many triplets (x,y,c) such that $0 \leqslant x < y \leqslant 1$ and $c \in \mathbb{R}$. Let ω be the smallest ordinal number [☞ Section 2.3] of cardinality $\mathfrak{c}$ – it exists because of Zorn's lemma [☞ p. 2.8] – and let $\{(x_\alpha, y_\alpha, c_\alpha);\ \alpha \prec \omega\}$ be a well-ordering [☞ p. 42] of all triplets (x,y,c), $0 \leqslant x < y \leqslant 1$, $c \in \mathbb{R}$. Using transfinite induction, we assign recursively a value of the function f at each point:

- ▶ Pick any $z_0 \in (x_0, y_0)$ and set $f(z_0) := c_0$.
- ▶ Assume that $z_\gamma \in (x_\gamma, y_\gamma)$ has been chosen for all $\gamma \prec \alpha$ and some $\alpha \prec \omega$. Since $A := \{z_\gamma;\ \gamma \prec \alpha\}$ has cardinality strictly less than $\mathfrak{c}$, it follows that $(x_\alpha, y_\alpha) \setminus A$ is non-empty, and we can pick $z_\alpha \in (x_\alpha, y_\alpha) \setminus A$. Set $f(z_\alpha) := c_\alpha$.

We extend the function f to the interval $[0,1]$ by setting $f(z) := 0$ for all z for which f has not yet been defined.

If $(x,y) \subseteq [0,1]$ is an open interval and $c \in \mathbb{R}$ a real number, then there exists by construction some $z \in (x,y)$ such that $f(z) = c$. This means that f maps any open interval $I \neq \emptyset$ onto $\mathbb{R}$. In particular, f has the intermediate value property, but the unboundedness of f implies that f is not Riemann integrable [☞ 3.3].

Comment As we have seen, the intermediate value property does not imply Riemann integrability. Conversely, Riemann integrability does not imply the intermediate value property; take e.g. $[a,b] = [0,2]$ and $f(x) = \mathbb{1}_{[0,1]}(x)$, then f is Riemann integrable but does not have the intermediate value property.

3.11 A Lipschitz continuous function g and a Riemann integrable function f such that $f \circ g$ is not Riemann integrable

Construct a fat Cantor set $C \subseteq [0,1]$ [☞7.3] whose Lebesgue measure is $\lambda(C) = \alpha \in (0,1)$. Recall that C is constructed by removing recursively open intervals [☞7.3, Section 2.5]; below, (a,b) stands for any such interval.

We define on $[0,1]$ the following functions: $f(x) := \mathbb{1}_{\{1\}}(x)$ is clearly Riemann integrable and

$$g(x) := \begin{cases} 1, & \text{if } x \in C, \\ 1 - \frac{1}{2}(b-a) + \left|x - \frac{1}{2}(b+a)\right|, & \text{if } x \in (a,b) \end{cases}$$

is easily seen to be Lipschitz continuous: $|g(x) - g(y)| \leqslant |x - y|$.

Nevertheless, $\mathbb{1}_C(x) = f(g(x))$ is not Riemann integrable [☞3.9].

Comment The composition $g \circ f$ of a Lipschitz continuous function g and a Riemann integrable function f is always Riemann integrable; see [MIMS, p. 448]. With a little more effort, one can show that it is enough to assume that g is continuous; see [52, Lemma 1.2.12, Theorem 1.2.13].

3.12 The composition of Riemann integrable functions need not be Riemann integrable

[☞3.11]

3.13 An increasing sequence of Riemann integrable functions $0 \leqslant f_n \leqslant 1$ such that $\sup_n f_n$ is not Riemann integrable

Let $(q_n)_{n\in\mathbb{N}}$ denote an enumeration of the set $[0,1] \cap \mathbb{Q}$ and set $f_n = \mathbb{1}_{\{q_1,\dots,q_n\}}$. The functions f_n are Riemann integrable, satisfy $0 \leqslant f_n \leqslant 1$, and increase towards Dirichlet's jump function $\mathbb{1}_{[0,1]\cap\mathbb{Q}}$, which is not Riemann integrable [☞3.4].

3.14 A decreasing sequence of Riemann integrable functions $0 \leqslant f_n \leqslant 1$ such that $\inf_n f_n$ is not Riemann integrable

Let C be a geometric fat Cantor set [☞7.3] with $\lambda(C) = \alpha > 0$. To wit, we could remove in each step the middle quarter, so that $\alpha = \frac{1}{2}$. The construction shows that $C = \bigcap_{n\in\mathbb{N}_0} C_n$, where each C_n is a closed set which is made up of 2^n disjoint intervals; moreover, the distance between two adjacent intervals is at least 4^{-n}.

This shows that we can construct continuous functions $f_n : [0,1] \to [0,1]$ such that

$$\{f_n = 1\} = C_n + \left[-4^{-(n+1)}, 4^{-(n+1)}\right] = \left\{x + y;\ x \in C_n,\ |y| \leqslant 4^{-(n+1)}\right\}$$

and

$$\operatorname{supp} f_n \subseteq C_n + (-\epsilon_n, \epsilon_n) = \{x + y;\ x \in C_n,\ |y| < \epsilon_n\}$$

for some $\frac{1}{4}4^{-n} < \epsilon_n < \frac{1}{2}4^{-n}$.

By construction, $f_n \downarrow \mathbb{1}_C$, each f_n is continuous, hence Riemann integrable, but $\mathbb{1}_C$ is not [☞ 3.9].

3.15 Limit theorems for Riemann integrals are sub-optimal

In order to interchange Riemann integrals and limits, one usually requires uniform convergence. The following example illustrates that uniform convergence is often too strong. Let $(a_n)_{n\in\mathbb{N}}$ be a sequence of positive numbers and define a sequence of tent functions $f_n : [0,1] \to [0,\infty)$ by

$$f_n(x) = \begin{cases} 2a_n nx, & \text{if } 0 \leqslant x \leqslant \frac{1}{2n}, \\ 2a_n(1 - nx), & \text{if } \frac{1}{2n} < x \leqslant \frac{1}{n}, \\ 0, & \text{if } \frac{1}{n} < x \leqslant 1. \end{cases}$$

We have $\lim_{n\to\infty} f_n(x) = 0$ for all $x \in [0,1]$. The convergence $f_n \to f$ is uniform if, and only if, $\lim_{n\to\infty} a_n = 0$.

Moreover,

$$\int_0^1 f_n(x)\,dx = \frac{a_n}{2n},$$

which means that the integrals converge to 0 if, and only if, $\lim_{n\to\infty} a_n/n = 0$; hence

$$\lim_{n\to\infty} \int_0^1 f_n(x)\,dx = \int_0^1 \lim_{n\to\infty} f_n(x)\,dx = 0.$$

This condition is much weaker than $\lim_{n\to\infty} a_n = 0$, which is equivalent to the uniform convergence of the integrands.

Comment There is a relatively unknown result by Arzelà [6], which was independently rediscovered (for continuous functions) by Osgood [131], giving the best possible result for a 'Riemannian' convergence theorem.

Theorem (Arzelà–Osgood) *Let $f_n: [a,b] \to \mathbb{R}$ be a sequence of Riemann integrable functions which is uniformly bounded, i.e.* $\sup_{n\in\mathbb{N}} \sup_{x\in[a,b]} |f_n(x)| < \infty$. *If* $\lim_{n\to\infty} f_n(x) = f(x)$, *and if the pointwise limit f is Riemann integrable, then*

$$\lim_{n\to\infty} \int_a^b |f_n(x) - f(x)|\,dx = 0 \quad \textit{and} \quad \lim_{n\to\infty} \int_a^b f_n(x)\,dx = \int_a^b f(x)\,dx.$$

One should compare this result with the dominated convergence theorem for Lebesgue integrals. In fact, Lebesgue refers in his doctoral dissertation [101, footnote p. 29] to Osgood [131] (but he gives the wrong year of publication: 1894). A modern presentation with various proofs (in a Riemannian spirit) and historical details is in Luxemburg [108]. A very accessible proof can be found in Dyer & Edmunds [52, Theorem 1.7.16]. Note that the additional integrability assumption on the limit f cannot be avoided [☞ 3.13,3.14].

3.16 The space of Riemann integrable functions is not complete

We will construct a sequence of Riemann integrable functions $f_n: [0,1] \to \mathbb{R}$ such that

$$\lim_{m,n\to\infty} \int_0^1 |f_n(x) - f_m(x)|\,dx = 0$$

holds, but there exists no Riemann integrable function $f: [0,1] \to \mathbb{R}$ such that

$$\lim_{n\to\infty} \int_0^1 |f_n(x) - f(x)|\,dx = 0. \tag{3.1}$$

In order to distinguish between Lebesgue and Riemann integrals, we denote them by $\int_{[0,1]} \dots d\lambda$ and $\int_0^1 \dots\, dx$, respectively.

Let $C = C^{(\alpha)}$, $\alpha = \frac{1}{2}$, be a fat Cantor set in $[0,1]$ [☞ 7.3], and denote the different stages of its construction by $C_n \downarrow C$. Set $f_n(x) := \mathbb{1}_{C_n}(x)$. The sequence f_n decreases monotonically to $\mathbb{1}_C$, so we get by dominated convergence

$$\lim_{n\to\infty} \int_{[0,1]} |\mathbb{1}_C - f_n|\,d\lambda = 0.$$

The Riemann integral of a Borel measurable Riemann integrable function coincides with the Lebesgue integral [☞ Theorem 1.28], and therefore the Riemann integrability of f_n gives

$$\lim_{m,n\to\infty} \int_0^1 |f_n(x) - f_m(x)|\,dx = \lim_{m,n\to\infty} \int_{[0,1]} |f_n - f_m|\,d\lambda$$

$$\leqslant \lim_{m\to\infty} \int_{[0,1]} |f_m - \mathbb{1}_C|\, d\lambda + \lim_{n\to\infty} \int_{[0,1]} |\mathbb{1}_C - f_n|\, d\lambda = 0.$$

Assume there were some Riemann integrable f such that (3.1) holds. Since Riemann integrable functions are Lebesgue integrable [☞Theorem 3A], the above calculation shows that $f = \mathbb{1}_C$ Lebesgue a.e. Define

$$N := \{x \in [0,1]\,;\ f(x) \neq \mathbb{1}_C(x) \text{ or } f \text{ discontinuous at } x\}.$$

The set of discontinuity points of the Riemann integrable function f has Lebesgue measure zero [☞Theorem 1.28] and so $\lambda(N) = 0$. By the construction of the set C, $\lambda(C \setminus N) = \lambda(C) = \frac{1}{2}$, i.e. there is some $x \in C \cap (0,1) \setminus N$ such that $f(x) = \mathbb{1}_C(x) = 1$, and f is continuous at x.

Take any open neighbourhood I of x. Since $[0,1]\setminus C$ is dense, the intersection $U := I \cap ((0,1) \setminus C)$ is non-empty and satisfies $\lambda(U) = \lambda(U \setminus N) > 0$. Thus, we can find some $x_0 \in I$ such that $f(x_0) = \mathbb{1}_C(x_0) = 0$ and f is continuous at x_0. If we take smaller and smaller neighbourhoods I, we can construct a sequence $x_n \to x$ such that $\lim_{n\to\infty} f(x_n) = 0 \neq 1 = f(x)$. Since x is a continuity point of f, we have reached a contradiction.

3.17 An example where integration by substitution goes wrong

Consider the (Riemann) integral $\int_{-1}^{1} 2|t|\, dt$. The function $\phi(t) := t^2$ satisfies $|\phi'(t)| = 2|t|$, and therefore a change of the integration variable $t \rightsquigarrow x = \phi(t)$ gives

$$2 = \int_{-1}^{1} 2|t|\, dt \neq \int_{0}^{1} dx = 1.$$

This shows that a coordinate change must be bijective.

Comment [☞10.17] for a further example.

3.18 A Riemann integrable function which is not Borel measurable

Take $A \in \mathcal{L}(\mathbb{R})$ such that A is not Borel measurable and $A \subseteq C$ for the classical middle-thirds Cantor set C; for the existence of such a non-Borel measurable set [☞7.20]. Consider

$$f(x) := \mathbb{1}_A(x), \quad x \in \mathbb{R}.$$

As $\{f = 1\} = A \notin \mathscr{B}(\mathbb{R})$, the function f is not Borel measurable. It follows from the closedness of the Cantor set C that f is continuous at every point $x \in \mathbb{R} \setminus C$. This means that the set of discontinuities of f is contained in the

Borel null set C. Since f is bounded, this implies that f is Riemann integrable [☞Theorem 1.28].

Comment Riemann integrable functions need not be Borel measurable, but they are Lebesgue measurable [☞Theorem 3A]. If $f : [a, b] \to \mathbb{R}$ is Riemann integrable and Borel measurable, then f is Lebesgue integrable [☞Theorem 1.28]. Our counterexample shows that the assumption of Borel measurability cannot be dropped.

3.19 A non-Riemann integrable function f which coincides a.e. with a continuous function

Consider the space $X = [0, 1]$, and let $f = 1 - \mathbb{1}_{[0,1]\cap\mathbb{Q}} = \mathbb{1}_{[0,1]\setminus\mathbb{Q}}$. As $f = 1$ a.e., it is obvious that f coincides a.e. with a continuous function. Since $1 - f$ is Dirichlet's discontinuous function which is not Riemann integrable [☞3.4], f is not Riemann integrable.

This example also shows that 'f coincides a.e. with a continuous function' and 'f is a.e. continuous' are different notions. In the latter case, f would be Riemann integrable, since a bounded function on a compact interval is Riemann integrable if, and only if, the set of its discontinuities is a Lebesgue null set [☞Theorem 1.28].

3.20 A Riemann integrable function on $\mathbb{R}^2$ whose iterated integrals are not Riemann integrable

Using the same idea as for functions on $\mathbb{R}$, one can extend the Riemann integral to closed rectangles $[a, b] \times [c, d]$ and, by a further limiting procedure, to all compact subsets of $\mathbb{R}^2$. Here we describe the construction only on a rectangle. Denote by Π a finite partition of $[a, b]\times[c, d]$ into finitely many non-overlapping rectangles. Refining the partition, if necessary, we may assume that the set of the lower left corners of all rectangles in the partition is of the form $\Pi_x \times \Pi_y$ where $\Pi_x = \{a = x_0 < x_1 < \cdots < x_m = b\}$ and $\Pi_y = \{c = y_0 < y_1 < \cdots < y_n = d\}$ are finite partitions of $[a, b]$ and $[c, d]$, respectively, with mesh $|\Pi_x| = \max_{1\leqslant i\leqslant m}(x_i - x_{i-1})$ and $|\Pi_y| = \max_{1\leqslant j\leqslant n}(y_j - y_{j-1})$; see Fig. 3.5. The Riemann integral is given by the following limit (if it exists):

$$\iint_{[a,b]\times[c,d]} f(x, y)\, d(x, y) := \lim_{|\Pi|\to 0} \sum_{i,j=1}^{m,n} f(\xi_i, \eta_j)(x_i - x_{i-1})(y_j - y_{j-1}),$$

where $(\xi_i, \eta_j) \in [x_{i-1}, x_i] \times [y_{j-1}, y_j]$ is arbitrary and where the limit is taken along all partitions as described above with mesh $|\Pi| := \max\{|\Pi_x|, |\Pi_y|\}$. It is

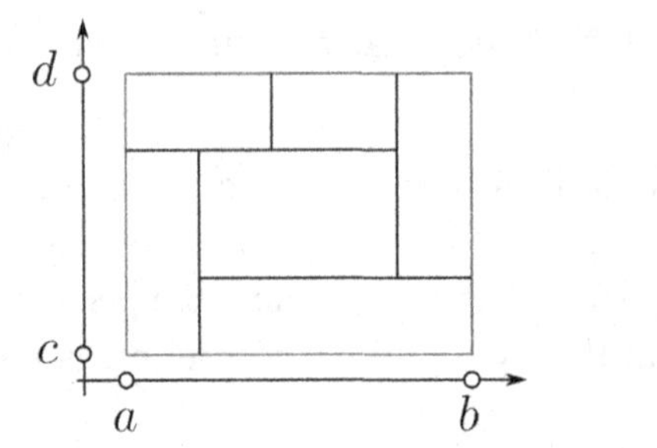

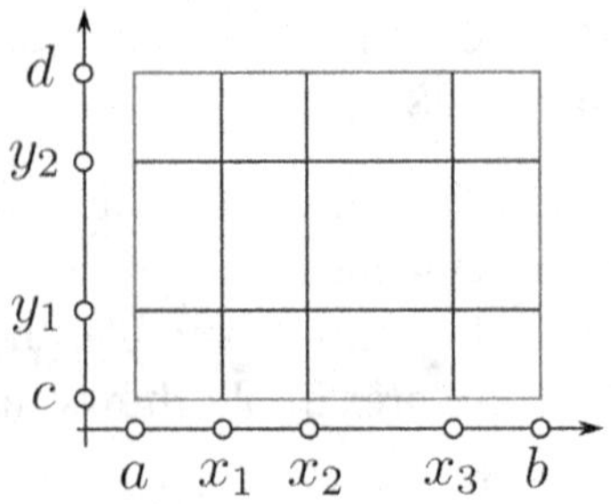

Figure 3.5 Bird's-eye view: original partition (left panel) and refined partition $\Pi_x \times \Pi_y$ (right panel).

obvious how to define the upper and lower Riemann–Darboux sums and integrals and how to extend Riemann's criterion (R2), cf. p. 2, to the two-dimensional case. In particular, every bounded function on a rectangle whose discontinuity points are a (two-dimensional) Lebesgue null set, is integrable in the sense of Riemann.

Consider the functions $f, g\colon [0,1] \times [0,1] \to [0,1]$ which are defined for a fixed $N \in \mathbb{N}$ by

$$g(x,y) = \begin{cases} 1, & (x,y) \in \left\{\frac{1}{N}, \frac{2}{N}, \dots, \frac{N-1}{N}\right\} \times (\mathbb{Q} \cap [0,1]), \\ 0, & \text{otherwise}, \end{cases}$$

$$f(x,y) = \max\{(g(x,y), g(y,x))\}.$$

By construction, $f(x, i/N) = \mathbb{1}_{\mathbb{Q}\cap[0,1]}(x)$ and $f(i/N, y) = \mathbb{1}_{\mathbb{Q}\cap[0,1]}(y)$ for every $i = 1, 2, \dots, N-1$, i.e. the Riemann integrals

$$\int_0^1 f(x, \tfrac{i}{N})\, dx \quad \text{and} \quad \int_0^1 f(\tfrac{i}{N}, y)\, dy, \quad i = 1, 2, \dots, N-1,$$

do not exist. The function f, however, is bounded with countably many discontinuity points $\{f = 1\}$, thus f is Riemann integrable. This may also be seen directly by constructing a sequence of lower and upper step functions. Let $k > N$ and set $\ell_k(x,y) := 0$ and

$$u_k(x,y) := \sum_{i=1}^{N-1} \left(\mathbb{1}_{[0,1]\times[\frac{i}{N}-\frac{1}{k}, \frac{i}{N}+\frac{1}{k}]}(x,y) + \mathbb{1}_{[\frac{i}{N}-\frac{1}{k}, \frac{i}{N}+\frac{1}{k}]\times[0,1]}(x,y) \right).$$

Then we have $\ell_k(x,y) \leqslant f(x,y) \leqslant u_k(x,y)$ and from this it is clear that f is integrable with integral

$$0 \leqslant \iint_{[0,1]\times[0,1]} f(x,y)\,d(x,y) \leqslant \iint_{[0,1]\times[0,1]} u_k(x,y)\,d(x,y) \leqslant 4\frac{N-1}{k} \xrightarrow[k\to\infty]{} 0.$$

This shows that the double integral

$$\iint_{[0,1]\times[0,1]} f(x,y)\,d(x,y)$$

exists, but the iterated integrals

$$\int_0^1 \left[\int_0^1 f(x,y)\,dx\right] dy \quad \text{and} \quad \int_0^1 \left[\int_0^1 f(x,y)\,dy\right] dx$$

are not defined.

3.21 Upper and lower integrals do not work for the Riemann–Stieltjes integral

Let $f, \phi : [a,b] \to \mathbb{R}$ be two functions which are left- or right-continuous. A Riemann–Stieltjes sum is an expression of the form

$$S_\Pi(f,\phi) = \sum_{i=1}^{n} f(\xi_i)(\phi(x_i) - \phi(x_{i-1})),$$

where $\Pi = \{a = x_0 < \cdots < x_n = b\}$ is a partition of $[a,b]$ and $\xi_i \in [x_{i-1}, x_i]$ is any intermediate point. If the Riemann–Stieltjes sums converge for any choice of partitions such that $\max_{1\leqslant i\leqslant n}(x_i - x_{i-1}) \to 0$ and for any choice of intermediate points, then the limit is denoted by $\int_a^b f\,d\phi$ and it is called the **Riemann–Stieltjes** integral.

If $\phi(x) = x$, this yields the classical Riemann integral. The following properties of the Riemann–Stieltjes integral are well known; see e.g. Kestelman [90, Chapter XI]:

(a) $\int_a^b f\,d\phi$ exists for all $f \in C[a,b]$ if, and only if, ϕ is a function of bounded variation;

(b) $f \longmapsto \int_a^b f\,d\phi$ and $\phi \longmapsto \int_a^b f\,d\phi$ are linear (if all integrals exist);

(c) if $\int_a^b f\,d\phi$ exists, then $\int_a^c f\,d\phi$ and $\int_c^b f\,d\phi$ exist for all $c \in (a,b)$ and
$$\int_a^b f\,d\phi = \int_a^c f\,d\phi + \int_c^b f\,d\phi;$$

(d) if $\int_a^b f\,d\phi$ exists, so does $\int_a^b \phi\,df$ and we have
$$\int_a^b f\,d\phi + \int_a^b \phi\,df = f(b)\phi(b) - f(a)\phi(a);$$

(e) if $\int_a^b f\,d\phi$ exists, then f and ϕ cannot have common discontinuities.

Even if f is bounded and ϕ is monotone, it is not possible to define the Riemann–Stieltjes integral using upper and lower Darboux–Stieltjes-type sums. Consider $[a, b] = [-1, 1]$ and the functions

$$f(x) = \begin{cases} 0, & \text{if } x \in [-1, 0), \\ 1, & \text{if } x \in [0, 1], \end{cases} \quad \text{and} \quad \phi(x) = \begin{cases} 0, & \text{if } x \in [-1, 0], \\ 1, & \text{if } x \in (0, 1]. \end{cases}$$

Since f and ϕ have a common discontinuity, $\int_{-1}^{1} f\, d\phi$ does not exist. Consider a partition $\Pi = \{-1 = x_0 < \cdots < x_n = 1\}$ such that there is some i_0 with $x_{i_0-1} < 0 < x_{i_0}$. The value of $S_\Pi(f, \phi)$ is $f(\xi_{i_0})$, which may be 0 or 1, depending on the choice of the intermediate point $\xi_{i_0} \in [x_{i_0-1}, x_{i_0}]$, and so there cannot be a limit.

Let us consider the upper and lower Darboux–Stieltjes sums for a partition $\Pi = \{-1 = x_0 < \cdots < x_n = 1\}$:

$$U_\Pi(f, \phi) = \sum_{i=1}^{n} \sup_{[x_{i-1}, x_i]} f \cdot (\phi(x_i) - \phi(x_{i-1})) = 1,$$

while

$$L_\Pi(f, \phi) = \sum_{i=1}^{n} \inf_{[x_{i-1}, x_i]} f \cdot (\phi(x_i) - \phi(x_{i-1})) = \begin{cases} 0, & \text{if } 0 \notin \Pi, \\ 1, & \text{if } 0 \in \Pi. \end{cases}$$

Thus,

$$\sup_\Pi L_\Pi(f, \phi) = 1 = \inf_\Pi U_\Pi(f, \phi)$$

but the limits $\lim_{|\Pi| \to 0} S_\Pi(f, \phi)$ and $\lim_{|\Pi| \to 0} L_\Pi(f, \phi)$ do not exist.

Comment If f is bounded, ϕ is monotone and if $\int_a^b f\, d\phi$ exists, then

$$\lim_{|\Pi| \to 0} S_\Pi(f, \phi) = \lim_{|\Pi| \to 0} L_\Pi(f, \phi) = \lim_{|\Pi| \to 0} U_\Pi(f, \phi) = \int_a^b f\, d\phi.$$

If f is bounded, ϕ is monotone and continuous, then

$$\lim_{|\Pi| \to 0} L_\Pi(f, \phi) = \sup_\Pi L_\Pi(f, \phi) \quad \text{and} \quad \lim_{|\Pi| \to 0} U_\Pi(f, \phi) = \inf_\Pi U_\Pi(f, \phi).$$

3.22 The Riemann–Stieltjes integral does not exist if integrand and integrator have a common discontinuity

[☞ 3.21]

4

Families of Sets

Let $X \neq \emptyset$ be a set, and denote by $\mathscr{P}(X) = \{A;\ A \subseteq X\}$ its power set. A family $\mathscr{A} \subseteq \mathscr{P}(X)$ with $\emptyset \in \mathscr{A}$ is called a

- **σ-algebra** [☞Definition 1.5] if $\mathscr{A}$ is closed under complements and countable unions,
- **algebra** if $\mathscr{A}$ is closed under complements and finite unions,
- **ring** if $\mathscr{A}$ is closed under set-theoretic differences and finite unions,
- **semi-ring** [☞p. 27] if $\mathscr{A}$ is closed under finite intersections and for every $A, B \in \mathscr{A}$, $A \subseteq B$, there exist pairwise disjoint sets $A_1, \dots, A_n \in \mathscr{A}$ such that $A \setminus B = \bigcup_{j=1}^{n} A_j$.

It is straightforward to see that every σ-algebra is an algebra, every algebra is a ring and every ring is a semi-ring. Moreover, σ-algebras have the property that all set operations – unions, intersections, complements and differences – can be repeated countably often without leaving the σ-algebra; the same is true for algebras and finitely many set operations. Using the identity $A \cap B = A \setminus (A \setminus B)$ we see that a ring is also closed under finite intersections.

If $\mathscr{A}$ is a σ-algebra on X, then the pair $(X, \mathscr{A})$ is a **measurable space**. There are the following three fundamental ways to generate a σ-algebra on a set X. Firstly, given any family $\mathscr{G} \subseteq \mathscr{P}(X)$, there exists a smallest σ-algebra $\mathscr{A}$ on X which contains $\mathscr{G}$. We write $\mathscr{A} = \sigma(\mathscr{G})$ and call $\mathscr{G}$ a **generator** of $\mathscr{A}$. Secondly, if $(Y_i, \mathscr{B}_i)$, $i \in I$, is a family of measurable spaces and $f_i : X \to Y_i$ a family of mappings, then there is a smallest σ-algebra $\mathscr{A}$ on X which makes all f_i $\mathscr{A}/\mathscr{B}_i$ measurable, and we denote this σ-algebra by

$$\mathscr{A} = \sigma(f_i, i \in I).$$

Thirdly, if X is a topological space, then the topology $\mathscr{O}$ induces a σ-algebra on X: the **Borel σ-algebra** $\mathscr{B}(X)$ is the smallest σ-algebra containing all open sets, i.e. $\mathscr{B} = \sigma(\mathscr{O})$. Equivalently, $\mathscr{B}(X)$ is the smallest σ-algebra containing all closed sets. The σ-algebra generated by the compact sets may be smaller or larger than the Borel σ-algebra [☞4.15, 4.16].

A σ-algebra $\mathscr{A}$ is **countably generated** if there exists a countable generator, i.e. a sequence $(G_n)_{n\in\mathbb{N}} \subseteq \mathscr{A}$ such that $\mathscr{A} = \sigma(G_n;\ n \in \mathbb{N})$. A set $A \in \mathscr{A}$, $A \neq \emptyset$, is an **atom of** $\mathscr{A}$ if A does not contain any proper measurable subset $B \in \mathscr{A}$, i.e.

$$B \in \mathscr{A},\ B \subseteq A \Rightarrow B = \emptyset \text{ or } B = A. \tag{4.1}$$

Examples of measurable spaces are listed in Example 1.7. There are some σ-algebras which appear in many of our counterexamples but which are not always discussed in classical text books on measure theory. For the reader's convenience we present some additional material on these σ-algebras, namely, the co-countable σ-algebra [☞Example 4A], a prominent σ-algebra on the ordinal space [☞Example 4B] and the product σ-algebra [☞Example 4C].

Example 4A (Co-countable σ-algebra) For an uncountable set X, the σ-algebra of all co-countable sets is defined by

$$\mathscr{A} := \{A \subseteq X;\ \text{either } A \text{ or } A^c \text{ is at most countable}\}.$$

One can quickly verify that $\mathscr{A}$ is indeed a σ-algebra, i.e. (Σ_1) $X \in \mathscr{A}$, (Σ_2) $\mathscr{A}$ is closed under complements and (Σ_3) closed under countable unions.

(Σ_1): $X^c = \emptyset$ is countable, so $X \in \mathscr{A}$.

(Σ_2): Since the condition defining $\mathscr{A}$ treats A and A^c in a symmetric way, it is clear that $A \in \mathscr{A} \iff A^c \in \mathscr{A}$.

(Σ_3): Let $(A_n)_{n\in\mathbb{N}} \subseteq \mathscr{A}$. If all sets A_n are at most countable, then so is their union $A = \bigcup_{n\in\mathbb{N}} A_n$ and thus $A \in \mathscr{A}$. If, say, A_{n_0} is not countable, then $A_{n_0}^c$ is at most countable and

$$\left(\bigcup_{n\in\mathbb{N}} A_n\right)^c = \bigcap_{n\in\mathbb{N}} A_n^c \subseteq A_{n_0}^c$$

shows that $\left(\bigcup_{n\in\mathbb{N}} A_n\right)^c$ is at most countable, so $\bigcup_{n\in\mathbb{N}} A_n \in \mathscr{A}$.

By definition, each singleton $\{x\}$ is in $\mathscr{A}$. If $A \in \mathscr{A}$, then either A or A^c is an at most countable union of singletons, and so

$$\mathscr{A} = \sigma(\{x\};\ x \in X).$$

The co-countable σ-algebra $\mathscr{A}$ is generated by a topology. Indeed, if we consider X as a topological space with the co-finite topology [☞Example 2.1.(d)], then $\mathscr{B}(X) = \mathscr{A}$.

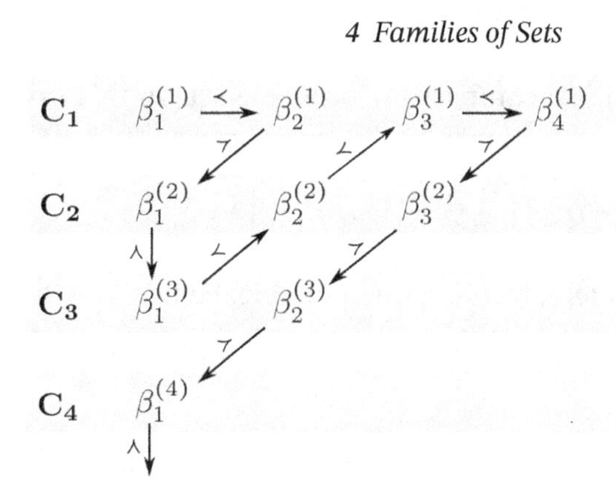

Figure 4.1 Counting scheme for the construction of the nested sequences $(\beta_k^{(n)})_{k\in\mathbb{N}} \subseteq C_n$.

Example 4B Let $\Omega = [0, \omega_1]$ be the ordinal space [☞Section 2.4], and consider the following σ-algebra:

$$\mathscr{A} := \{A \subseteq \Omega;\ A \cup \{\omega_1\} \text{ or } A^c \cup \{\omega_1\} \text{ contains an uncountable compact set}\}.$$

Since every countable set in $\Omega_0 = [0, \omega_1)$ must be contained in an interval $[0, \alpha]$ for some $\alpha < \omega_1$ – take an enumeration $(\alpha_n)_{n\in\mathbb{N}}$ and set $\alpha := \sup_n \alpha_n \prec \omega_1$ – it is clear that

$$A \text{ is uncountable} \iff \forall \alpha \in \Omega_0 \quad \exists \beta \in A : \alpha \prec \beta.$$

This entails

$$C_n \subseteq \Omega \text{ compact and uncountable} \Rightarrow \bigcap_{n=1}^{\infty} C_n \text{ uncountable.} \tag{4.2}$$

Indeed: take $C_n \subseteq \Omega$, $n \in \mathbb{N}$, compact and uncountable, and fix $\alpha \in \Omega_0$. Since each C_n is uncountable, we can construct recursively sequences $(\beta_k^{(n)})_{k\in\mathbb{N}} \subseteq C_n$ such that $\alpha \prec \beta_1^{(1)} \prec \beta_2^{(1)} \prec \beta_1^{(2)} \prec \beta_1^{(3)} \prec \beta_2^{(2)} \prec \beta_3^{(1)} \prec \beta_4^{(1)} \prec \ldots$; see Fig. 4.1. The supremum $\beta := \sup_{k\in\mathbb{N}} \beta_k^{(n)}$ does not depend on $n \in \mathbb{N}$, and it follows by compactness that $\beta \in C_n$ for all $n \in \mathbb{N}$. Hence, $\alpha \prec \beta \in \bigcap_{n\in\mathbb{N}} C_n$, which proves that $\bigcap_{n\in\mathbb{N}} C_n$ is uncountable.

We can now check that $\mathscr{A}$ is indeed a σ-algebra. The conditions (Σ_1) $(\emptyset \in \mathscr{A})$ and (Σ_2) $(A \in \mathscr{A} \iff A^c \in \mathscr{A})$ are clear. For (Σ_3) assume that $(A_n)_{n\in\mathbb{N}} \subseteq \mathscr{A}$ is a sequence and set $A := \bigcup_{n\in\mathbb{N}} A_n$. Two scenarios are possible:

Case 1. At least one $A_n \cup \{\omega_1\}$ contains an uncountable compact set C. Then so does $A \cup \{\omega_1\} := \bigcup_{n\in\mathbb{N}} A_n \cup \{\omega_1\}$, i.e. $A \in \mathscr{A}$.

Case 2. No $A_n \cup \{\omega_1\}$ contains a compact uncountable set. In this case, each $A_n^c \cup \{\omega_1\} \supseteq C_n$ for suitable uncountable compact sets C_n and

$$A^c \cup \{\omega_1\} = \bigcap_{n\in\mathbb{N}} A_n^c \cup \{\omega_1\} \supseteq \bigcap_{n\in\mathbb{N}} C_n =: C.$$

The set C is compact and, by (4.2), uncountable. Hence, $A \in \mathscr{A}$.

Since $\mathscr{A}$ contains all countable sets and all uncountable compact sets, it follows that $\mathscr{B}(\Omega_0) \subseteq \mathscr{A}$.

Example 4C (product σ-algebra) Let $(X_i, \mathscr{A}_i)$ be a family of measurable spaces for some index set I, e.g. $I = [0,1]$. The **product space** $\bigtimes_{i\in I} X_i$ is defined by

$$X^I := \bigtimes_{i\in I} X_i := \left\{ f : I \to \bigcup_{i\in I} X_i \,;\ \forall i \in I \ :\ f(i) \in X_i \right\}.$$

For $K \subseteq I$ denote by $\pi_K : X^I \to X^K, f \longmapsto f|_K$ the coordinate projection. A **cylinder set** or **finitely based set** with basis $K \subseteq I$ is a set of the form $\pi_K^{-1}(A)$ where $\#K < \infty$ and $A \in \bigtimes_{i\in K} \mathscr{A}_i$. The **product σ-algebra** $\mathscr{A}^{\otimes I}$ is the smallest σ-algebra containing all cylinder sets. Equivalently, $\mathscr{A}^{\otimes I}$ is the smallest σ-algebra which makes all projections $\pi_{\{i\}} : X^I \to X_i$, $i \in I$, measurable, i.e.

$$\mathscr{A}^{\otimes I} = \sigma(\pi_{\{i\}} \,;\ i \in I) = \sigma\left(\left\{\pi_{\{i\}}^{-1}(A) \,;\ A \in \mathscr{A}_i, i \in I\right\}\right).$$

4.1 A Dynkin system which is not a σ-algebra

A family $\mathscr{D} \subseteq \mathscr{P}(X)$ with $X \in \mathscr{D}$ is a **Dynkin system** if $\mathscr{D}$ is closed under the formation of complements and under the formation of countable unions of pairwise disjoint sets.

The following example shows that there are Dynkin systems which are not σ-algebras, i.e. $(D_i)_{i\in\mathbb{N}} \subseteq \mathscr{D}$ need not imply $\bigcup_{i\in\mathbb{N}} D_i \in \mathscr{D}$ if the sets D_i are not pairwise disjoint. Let $X = \{1, 2, 3, \dots, 2k-1, 2k\}$ for some fixed $k \in \mathbb{N}$. The family $\mathscr{D} = \{A \subseteq X \,;\ \#A \text{ is even}\}$ is a Dynkin system, but not a σ-algebra.

Since X contains an even number of elements, $\mathscr{D}$ is closed under the formation of complements. Moreover, $\emptyset \in \mathscr{D}$. If the intersection of $A, B \in \mathscr{D}$ contains an odd number of elements, then $A \cup B$ contains an odd number of elements. For example,

$$A = \{1,2\} \in \mathscr{D}, \quad B = \{2,3,4,5\} \in \mathscr{D},$$

whereas

$$A \cup B = \{1,2,3,4,5\} \notin \mathscr{D}.$$

This means that $\mathscr{D}$ is not closed under finite unions, and so $\mathscr{D}$ is not a σ-algebra. If $A, B \in \mathscr{D}$ are disjoint, it is clear that A, B and also $A \uplus B$ contain an even number of elements; hence, $A \cup B \in \mathscr{D}$. Since there are at most finitely many disjoint subsets of X, this proves $\mathscr{D}$ is closed under countably infinite unions of pairwise disjoint sets.

4.2 A monotone class which is not a σ-algebra

Let $(X, \mathscr{A})$ be a measurable space. A **monotone class** is a family $\mathscr{M} \subseteq \mathscr{A}$ which contains $\emptyset$ and X and which is stable under the formation of countable unions (resp. intersections) of increasing (resp. decreasing) sequences.

Consider the real line $X = \mathbb{R}$ with its Borel sets $\mathscr{A} = \mathscr{B}(\mathbb{R})$. The family of sets $\mathscr{M} := \{(-\infty, n);\ n \in \mathbb{Z}\} \cup \{\emptyset, \mathbb{R}\}$ is closed under countable unions and intersections. Take $I \subseteq \mathbb{Z}$ and set $m := \inf I$ and $M := \sup I$, then

$$\bigcup_{i \in I}(-\infty, i) = (-\infty, M) \in \mathscr{M} \quad \text{and} \quad \bigcap_{i \in I}(-\infty, i) = (-\infty, m) \in \mathscr{M},$$

using the conventions $\emptyset = (-\infty, -\infty)$ and $\mathbb{R} = (-\infty, +\infty)$. Since $\mathscr{M}$ is not stable under the formation of complements, it is not a σ-algebra.

Comment This example even shows that there are monotone classes which are stable under countable intersections without being a σ-algebra. The following variant shows that there are monotone classes which are stable under the formation of complements without being a σ-algebra. Let $\mathscr{M}$ be as before and set $\mathscr{M}' := \mathscr{M} \cup \{M^c;\ M \in \mathscr{M}\}$. Using the above argument, it is clear that $\mathscr{M}'$ is a monotone class and that $M \in \mathscr{M}'$ if, and only if, $M^c \in \mathscr{M}'$. But we have $(-1, 1) = (-\infty, 1) \cap (-1, \infty) \notin \mathscr{M}'$. Note, however, that a monotone class which is stable both under finite intersections and the formation of complements, must be a σ-algebra. Indeed, $A \cup B = (A^c \cap B^c)^c$ shows that $\mathscr{M}$ is closed under finite unions, and this property allows us to treat countable unions of arbitrary sets as countable unions of increasing sets $\bigcup_{n \in \mathbb{N}} A_n = \bigcup_{n \in \mathbb{N}}(A_1 \cup \cdots \cup A_n)$.

4.3 A σ-algebra which contains all singletons but no non-trivial interval

Consider the co-countable σ-algebra on $\mathbb{R}$ [☞ Example 4A]

$$\mathscr{A} := \{A \subseteq \mathbb{R};\ A \text{ or } A^c \text{ is at most countable}\}.$$

Obviously, each singleton $\{x\}$ is in $\mathscr{A}$, while $\mathscr{A}$ does not contain any non-trivial interval.

4.4 There is no σ-algebra with $\#\mathscr{A} = \#\mathbb{N}$

Let $(X, \mathscr{A})$ be a measurable space, and suppose that $\#\mathscr{A} = \#\mathbb{N}$. Then the sets

$$A(x) := \bigcap_{A \in \mathscr{A}, A \ni x} A, \quad x \in X, \tag{4.3}$$

are in $\mathscr{A}$, since $A(x)$ is a countable intersection of measurable sets. Write $\mathscr{A}_0$ for the family of atoms of $\mathscr{A}$ [☞(4.1)]. We claim that $\#\mathscr{A}_0 = \#\mathbb{N}$.

1° $A(x) \in \mathscr{A}_0$ is an atom which contains x. *Indeed:* if $A(x)$ were not an atom, there would be $B \subseteq A(x)$ such that $B \in \mathscr{A}$, $B \neq \emptyset$ and $B \neq A(x)$. We may assume that $x \in B$, otherwise we would take $B' := A(x) \setminus B$ instead of B. Thus, B is part of the intersection appearing in (4.3) so that $B \supseteq A(x)$. Hence, $B = A(x)$ which is impossible.

2° $\mathscr{A}$ has countably infinitely many atoms. *Indeed:* since $\#\mathscr{A} = \#\mathbb{N}$, there are countably infinitely many disjoint sets in $\mathscr{A}$, and therefore the construction (4.3) yields at least $\#\mathbb{N}$ many atoms. The claim follows as there cannot be more atoms than members of $\mathscr{A}$.

Since $\mathscr{A}$ contains all countable unions of sets from $\mathscr{A}_0$, and since there are more than countably many such unions, it is clear that $\#\mathscr{A} > \#\mathbb{N}$.

4.5 A σ-algebra which has no non-empty atoms

Let I be an uncountable set, e.g. $I = [0, 1]$, and consider $X = \{0, 1\}^I$ with the product σ-algebra $\mathscr{A} := \mathscr{P}(\{0, 1\})^{\otimes I}$, cf. Example 4C. For $K \subseteq I$ denote by $\pi_K : \{0, 1\}^I \to \{0, 1\}^K$ the coordinate projection. The following lemma shows that every $A \in \mathscr{A}$ is of the form $\pi_S^{-1}(B)$ for some countable set $S \subseteq I$.

Lemma *Let $I = [0, 1]$. For every $A \in \mathscr{A}$ there is a countable set $S = S_A \subseteq I$ such that*

$$f \in X,\ w \in A\ :\ f|_S = w|_S \ \Rightarrow\ f \in A. \tag{4.4}$$

Proof Set $\Sigma := \{A \subseteq X\,;\ (4.4) \text{ holds for some countable set } S = S_A\}$. We claim that Σ is a σ-algebra.

1° Clearly, $\emptyset \in \Sigma$.

2° Let $A \in \Sigma$ with $S = S_A$. Then we find for A^c and S that

$$f \in X,\ w \in A^c,\ f|_S = w|_S \ \Rightarrow\ f \notin A;$$

otherwise, we would have $f \in A$ and then (4.4) would imply that $w \in A$. This means that (4.4) holds for A^c with $S = S_A = S_{A^c}$, i.e. $A^c \in \Sigma$.

3° Let $(A_n)_{n\in\mathbb{N}} \subseteq \Sigma$ and set $S := \bigcup_{n=1}^{\infty} S_{A_n}$. Obviously, S is countable and

$$f \in X,\ w \in \bigcup_{n=1}^{\infty} A_n,\ f|_S = w|_S \Rightarrow \exists n_0 : w \in A_{n_0},\ f|_{S_{A_{n_0}}} = w|_{S_{A_{n_0}}}$$
$$\overset{(4.4)}{\Rightarrow} f \in A_{n_0}, \text{ hence, } f \in \bigcup_{n=1}^{\infty} A_n.$$

This proves that $\bigcup_{n=1}^{\infty} A_n \in \Sigma$.

Since all sets of the form $\{\pi_{\{t\}} \in C\}$, $t \in I$, $C \in \mathscr{P}(\{0,1\})$, are contained in Σ, we find

$$\mathscr{A} = \sigma(\pi_{\{t\}};\ t \in I) \subseteq \sigma(\Sigma) = \Sigma. \qquad \square$$

Now assume that $A_0 \in \mathscr{A}$, $A_0 \neq \emptyset$, is an atom, i.e.

$$B \in \mathscr{A},\ B \subseteq A_0 \Rightarrow B = \emptyset \text{ or } B = A_0.$$

By the above lemma, A_0 has basis S for a countable set $S \subseteq I$, i.e. $A_0 = \pi_S^{-1}(B)$ for some set B. Take $i \in I \setminus S$, define $S' = S \cup \{i\}$ and consider the set $\pi_{S'}^{-1}(B')$ where $B' = B \times \{0\}$, say. Then $\pi_{S'}^{-1}(B') \subsetneqq A_0$ and $\emptyset \neq \pi_{S'}^{-1}(B') \in \mathscr{A}$, contradicting our assumption that A_0 is an atom.

4.6 An increasing family of σ-algebras whose union fails to be a σ-algebra

Consider on $\mathbb{N} = \{1, 2, \dots\}$ the families $\mathscr{G}_n := \{\{1\}, \dots, \{n\}\}$ and define the sets $A_n := \{1, 2, \dots, n\}$ and $A_n^c = \mathbb{N} \setminus A_n = \{n+1, n+2, \dots\}$. For each $n \in \mathbb{N}$ the family

$$\mathscr{A}_n := \{A \subseteq \mathbb{N};\ \text{either } A \subseteq A_n \text{ or } A_n^c \subseteq A\}$$

is easily seen to be a σ-algebra. Moreover, $\mathscr{A}_n = \sigma(\mathscr{G}_n)$. The inclusion '$\supseteq$' follows immediately from the observation that $\mathscr{G}_n \subseteq \mathscr{A}_n$. Conversely, if $A \in \mathscr{A}_n$ we have to distinguish between two cases:

Case 1. $A \subseteq A_n$. Then A is a finite union of sets from $\mathscr{G}_n$, hence $A \in \sigma(\mathscr{G}_n)$.
Case 2. $A_n^c \subseteq A$. Then A^c satisfies $A^c \subseteq A_n$ and we conclude that $A^c \in \sigma(\mathscr{G}_n)$, hence $A \in \sigma(\mathscr{G}_n)$.

From $\mathscr{G}_1 \subseteq \mathscr{G}_2 \subseteq \dots$ we get that $\mathscr{A}_1 \subseteq \mathscr{A}_2 \subseteq \dots$. In order to see that $\bigcup_{n\in\mathbb{N}} \mathscr{A}_n$ fails to be a σ-algebra, we take the sets $B_n := A_n \cap 2\mathbb{N}$ and observe that $B_n \in \mathscr{A}_n$; however, $\bigcup_{n\in\mathbb{N}} B_n = 2\mathbb{N}$ is in none of the $\mathscr{A}_n$, hence not in $\bigcup_{n\in\mathbb{N}} \mathscr{A}_n$.

Comment If $\mathscr{A}_n$, $n \in \mathbb{N}$, is a sequence of nested σ-algebras, i.e. $\mathscr{A}_n \subseteq \mathscr{A}_{n+1}$ for all $n \in \mathbb{N}$, then the union $\bigcup_{n\in\mathbb{N}} \mathscr{A}_n$ is a σ-algebra if, and only if, there is an index $N \in \mathbb{N}$ such that $\mathscr{A}_n = \mathscr{A}_{n+1}$ for all $n \geqslant N$ ⌊☞ 4.7⌋.

4.7 The union of countably many strictly increasing σ-algebras is never a σ-algebra

Let $(\mathscr{A}_n)_{n\in\mathbb{N}}$ be a strictly increasing sequence of σ-algebras, i.e. $\mathscr{A}_n \subsetneqq \mathscr{A}_{n+1}$ for all $n \in \mathbb{N}$, and set $\mathscr{A}_\infty := \bigcup_{n\in\mathbb{N}} \mathscr{A}_n$. The following proof, essentially due to Broughton & Huff [27], shows that $\mathscr{A}_\infty$ cannot be a σ-algebra.

Since the sequence $(\mathscr{A}_n)_{n\in\mathbb{N}}$ is strictly increasing, we may assume that $\mathscr{A}_1 \neq \{\emptyset, X\}$. Recall also the notion of a trace σ-algebra ⌊☞ Example 1.7.(e)⌋

$$B \cap \mathscr{A}_n := \{B \cap A;\ A \in \mathscr{A}_n\}.$$

1° *Claim:* There exists a set $E \in \mathscr{A}_1$ such that $(E \cap \mathscr{A}_{n+1}) \setminus (E \cap \mathscr{A}_n) \neq \emptyset$ for infinitely many $n \in \mathbb{N}$.

To see this, assume – to the contrary – that for some n and some $B \in \mathscr{A}_1$ we have

$$B \cap \mathscr{A}_n = B \cap \mathscr{A}_{n+1} \quad \text{and} \quad B^c \cap \mathscr{A}_n = B^c \cap \mathscr{A}_{n+1}.$$

If $U \in \mathscr{A}_{n+1} \setminus \mathscr{A}_n$, then

$$U = \underbrace{(B \cap U)}_{\in B\cap\mathscr{A}_{n+1} = B\cap\mathscr{A}_n \subseteq \mathscr{A}_n} \cup \underbrace{(B^c \cap U)}_{\in B^c\cap\mathscr{A}_{n+1} = B^c\cap\mathscr{A}_n \subseteq \mathscr{A}_n},$$

leading to the contradiction $U \in \mathscr{A}_n$. Thus the claim holds with either $E = B$ or $E = B^c$.

2° Let E be the set from Step 1 and denote by $n_1, n_2, \ldots$ an increasing sequence for which the assertion of Step 1 holds. Then

$$\mathscr{F}_k := E \cap \mathscr{A}_{n_k}, \quad k \in \mathbb{N},$$

is a strictly increasing sequence of σ-algebras over the set E. Again we may assume that $\mathscr{F}_1 \neq \{\emptyset, E\}$. As in Step 1, we find some $E_1 \in \mathscr{F}_1$ such that E_1 is not trivial (i.e. $E_1 \neq \emptyset$ and $E_1 \neq E$) and $(E_1 \cap \mathscr{F}_{k+1}) \setminus (E_1 \cap \mathscr{F}_k) \neq \emptyset$ holds for infinitely many k.

3° Now we repeat Step 2 and construct recursively a sequence of σ-algebras $\mathscr{A}_{i_1} \subseteq \mathscr{A}_{i_2} \subseteq \mathscr{A}_{i_3} \ldots$ and a sequence of sets $E_1 \supseteq E_2 \supseteq E_3 \ldots$ such that

$$E_k \in \mathscr{A}_{i_k} \quad \text{and} \quad E_{k+1} \in (E_k \cap \mathscr{A}_{i_{k+1}}) \setminus (E_k \cap \mathscr{A}_{i_k}).$$

4° The sets $F_k := E_k \setminus E_{k+1}$ have the property that they are disjoint and $F_k \in \mathscr{A}_{i_{k+1}} \setminus \mathscr{A}_{i_k}$. Since the σ-algebras are increasing, we have

$$\bigcup_{n\in\mathbb{N}} \mathscr{A}_n = \bigcup_{k\in\mathbb{N}} \mathscr{A}_{i_k},$$

which means that we can restrict ourselves to a subsequence. This means that we can assume that $i_k = k$.

5° Without loss of generality, we can identify F_k with $\{k\}$ and assume that the $\mathscr{A}_n$ are σ-algebras on $\mathbb{N}$ such that $\{k\} \in \mathscr{A}_{k+1} \setminus \mathscr{A}_k$. Let B_n be the smallest set in $\mathscr{A}_n$ such that $n \in B_n$. Then $n \in B_n \subseteq \{n, n+1, n+2, ...\}$ and $B_n \neq \{n\}$. Moreover

$$m \in B_n \Rightarrow B_m \subseteq B_n \quad \text{since} \quad m \in B_n \cap B_m \in \mathscr{A}_m.$$

Now define $n_1 = 1$ and pick n_{k+1} recursively: $n_{k+1} \in B_{n_k}$ such that $n_{k+1} \neq n_k$. Then $B_{n_1} \supseteq B_{n_2} \supseteq ....$ Set $E = \{n_2, n_4, n_6, ...\}$. If $\mathscr{A}_\infty$ were a σ-algebra, then $E \in \mathscr{A}_n$ for some n, thus $E \in \mathscr{A}_{n_{2k}}$ for some k. Then $\{n_{2k}, n_{2k+2}, ...\} \in \mathscr{A}_{n_{2k}}$ and thus $B_{n_{2k}} \subseteq \{n_{2k}, n_{2k+2}, ...\}$. This contradicts the fact $n_{2k+1} \in B_{n_{2k}}$.

Comment Let $(\mathscr{R}_n)_{n\in\mathbb{N}}$ be a sequence of σ-rings over a set X, i.e. $\mathscr{R}_n \subseteq \mathscr{P}(X)$ is a family which is closed under set-theoretic differences and countable unions. If $(\mathscr{R}_n)_{n\in\mathbb{N}}$ is an increasing sequence, i.e. $\mathscr{R}_n \subseteq \mathscr{R}_{n+1}$ for all $n \in \mathbb{N}$, then the union $\mathscr{R} = \bigcup_n \mathscr{R}_n$ is a σ-ring if, and only if, there is some index $N \in \mathbb{N}$ such that $\mathscr{R}_N = \mathscr{R}_{N+1} = ...$; see Overdijk et al. [132]. In particular, if $(\mathscr{R}_n)_{n\in\mathbb{N}}$ is a sequence of **strictly** increasing σ-algebras, i.e. $\mathscr{R}_n \subsetneqq \mathscr{R}_{n+1}$ for all $n \in \mathbb{N}$, then the union $\mathscr{R} = \bigcup_n \mathscr{R}_n$ is never a σ-algebra.

4.8 A countably generated σ-algebra containing a sub-σ-algebra which is not countably generated

Let $(X, \mathscr{A})$ be a measurable space. A σ-algebra is **countably generated** if there is a sequence $(A_n)_{n\in\mathbb{N}} \subseteq \mathscr{A}$ such that $\mathscr{A} = \sigma(A_n, n \in \mathbb{N})$.

The Borel σ-algebra $\mathscr{B}(\mathbb{R})$ is countably generated – consider for example $[q, \infty)$, $q \in \mathbb{Q}$ – but its sub-σ-algebra

$$\mathscr{A} := \{A \subseteq \mathbb{R};\ A \text{ or } A^c \text{ is countable}\},$$

cf. Example 4A, is not countably generated. Suppose that $\mathscr{A}$ is countably generated, i.e. $\sigma(A_n, n \in \mathbb{N}) = \mathscr{A}$ for some $(A_n)_{n\in\mathbb{N}} \subseteq \mathscr{A}$. Without loss, we may assume that each A_n is countable, otherwise replace it by its complement. Since A_n is a countable subset of $C := \bigcup_{n\in\mathbb{N}} A_n$ we have $A_n \in \sigma(\{c\}, c \in C)$, and so

$$\mathscr{A} = \sigma(A_n, n \in \mathbb{N}) \subseteq \sigma(\{c\}, c \in C).$$

The reverse inclusion '$\supseteq$' holds as $\{c\} \in \mathscr{A}$ for any $c \in C$ by the very definition of $\mathscr{A}$; thus

$$\mathscr{A} = \sigma(\{c\}, c \in C).$$

It can easily be verified that

$$\mathscr{F} := \{A;\ A \subseteq C\} \cup \{A \cup C^c;\ A \subseteq C\}$$

is a σ-algebra on $\mathbb{R}$; moreover, $\{c\} \in \mathscr{F}$ for any $c \in C$. Hence, $\mathscr{A} \subseteq \mathscr{F}$. This is a contradiction to the fact that $\{x\} \in \mathscr{A}$, $\{x\} \notin \mathscr{F}$ for any $x \in \mathbb{R} \setminus C$ (which is non-empty as C is countable).

Comment For a measurable space $(X, \mathscr{A})$ denote by $\sigma(\mathscr{A}, A)$ the smallest σ-algebra containing $\mathscr{A}$ and the set $A \subseteq X$. It can be shown that there are measurable spaces $(X, \mathscr{A})$ for which there exists a set $A \notin \mathscr{A}$ such that $\sigma(\mathscr{A}, A)$ is countably generated but $\mathscr{A}$ fails to be countably generated, cf. [140, p. 15]. For yet another possibility to construct a countably generated σ-algebra containing a sub-σ-algebra which is not countably generated [☞4.9].

4.9 Two countably generated σ-algebras whose intersection is not countably generated

We will show that there exists a countably generated σ-algebra $\mathscr{A}$ on $(0,1)$ such that its intersection with the Borel σ-algebra $\mathscr{B}(0,1)$ is the co-countable σ-algebra [☞Example 4A], i.e.

$$\mathscr{A} \cap \mathscr{B}(0,1) = \{A \subseteq (0,1);\ \#A \leqslant \#\mathbb{N} \text{ or } \#A^c \leqslant \#\mathbb{N}\}.$$

This means that the co-countable σ-algebra – which is not countably generated [☞4.8] – is the intersection of two countably generated σ-algebras. The construction of $\mathscr{A}$ is due to Rao [140, p. 16]. We set

$$\mathscr{U}(0,1) := \{A \subseteq (0,1);\ A \text{ and } (0,1) \setminus A \text{ are uncountable}\}.$$

Lemma *There exists a bijective function $f : (0,1) \to (0,1)$ such that the pre-image $f^{-1}(B) \notin \mathscr{B}(0,1)$ for any $B \in \mathscr{B}(0,1) \cap \mathscr{U}(0,1)$.*

Proof There exist continuum many Borel sets. Let ω_1 be the first uncountable ordinal and $(A_0, B_0), \dots, (A_\alpha, B_\alpha), \dots$, $\alpha < \omega_1$, be a well-ordering of all pairs of disjoint uncountable Borel sets in $(0,1)$. Construct inductively a sequence of pairs by choosing $x_\alpha \in A_\alpha$, $y_\alpha \in B_\alpha$ such that $\langle x_\alpha; y_\alpha\rangle \notin \{\langle x_\beta; y_\beta\rangle;\ \beta < \alpha\}$. Note that this is possible since any uncountable Borel set has continuum many elements. Define a bijective mapping

$$f : (0,1) \to (0,1), \quad \begin{cases} f(x_\alpha) = y_\alpha, \\ f(y_\alpha) = x_\alpha, \\ f(z) = z, \quad \text{for all other } z. \end{cases}$$

It remains to check that $f^{-1}(B) \notin \mathscr{B}(0,1)$ for all $B \in \mathscr{B}(0,1) \cap \mathscr{U}(0,1)$. To this end, we first notice that $f^{-1}(B) \cap B \in \mathscr{U}(0,1)$ for any $B \in \mathscr{B}(0,1) \cap \mathscr{U}(0,1)$.

Indeed: any uncountable Borel set $B \in \mathscr{B}(0,1)$ contains uncountably many disjoint Borel sets which are uncountable, say B_α. Since each pair (B_α, B_β) appears in the well-ordering, there is a pair $\langle x; y\rangle$ in the chosen sequence such that $x \in B_\alpha$ and $y \in B_\beta$. As $x \in B_\alpha \subseteq B$ and $x = f^{-1}(y) \in f^{-1}(B_\beta) \subseteq f^{-1}(B)$ we have $x \in B \cap f^{-1}(B)$. As there are uncountably many such pairs (B_α, B_β), it follows from our construction of the sequence that $f^{-1}(B) \cap B$ is uncountable. The complement of $f^{-1}(B) \cap B$ contains $(0,1) \setminus B$, which is uncountable for $B \in \mathscr{U}(0,1)$. Therefore, $f^{-1}(B) \cap B \in \mathscr{U}(0,1)$ for all $B \in \mathscr{B}(0,1) \cap \mathscr{U}(0,1)$.

Now let $B \in \mathscr{B}(0,1) \cap \mathscr{U}(0,1)$. It follows from the bijectivity of f and our previous consideration that $A := f^{-1}(B) \cap B \in \mathscr{U}(0,1)$ and $f^{-1}(A) = A$. If $f^{-1}(B)$ were a Borel set, then $A \in \mathscr{U}(0,1) \cap \mathscr{B}(0,1)$, and therefore the pair (A, A^c) would appear in the well-ordering; therefore at least one element in A would be mapped to A^c, which means that $f^{-1}(A) = A$ cannot hold. Hence, $f^{-1}(B)$ cannot be Borel. □

Let f be the function constructed in the lemma. We claim that the σ-algebra $\mathscr{F} := f^{-1}(\mathscr{B}(0,1)) \cap \mathscr{B}(0,1)$ coincides with the co-countable σ-algebra $\mathscr{C}$. Since $\{x\} = f^{-1}(\{f(y)\}) \cap \{x\}$ it is clear that $\mathscr{F}$ contains all singletons, and so $\mathscr{F} \supseteq \mathscr{C}$. If $A \in \mathscr{U}(0,1)$ has a representation of the form $A = f^{-1}(B)$ for $B \in \mathscr{B}(0,1)$, then $B \in \mathscr{U}(0,1)$, and thus, by the construction of f, $A = f^{-1}(B) \notin \mathscr{B}(0,1)$. This shows that $\mathscr{F} \cap \mathscr{U}(0,1) = \emptyset$, i.e. $\mathscr{F}$ does not contain any sets which are uncountable and have uncountable complement, hence $\mathscr{F} \subseteq \mathscr{C}$.

4.10 A Borel σ-algebra which is not countably generated

Let X be a topological space. If the topology has a countable basis – e.g. if X is a separable metric space – then the Borel σ-algebra is countably generated. This property of the Borel σ-algebra breaks down in general topological spaces.

Let X be any uncountable set. Consider X as topological space with the co-finite topology on X [☞Example 2.1.(d)], then $\mathscr{B}(X)$ is the co-countable σ-algebra [☞Example 4A], which is not countably generated [☞4.8].

4.11 $\sigma(\mathscr{G})$ can only separate points if $\mathscr{G}$ does

A family $\mathscr{G} \subseteq \mathscr{P}(X)$ **separates points** in X if for any $x, y \in X$, $x \neq y$, there exists a set $G \in \mathscr{G}$ such that $x \in G$ and $y \notin G$. Since $\mathscr{G} \subseteq \sigma(\mathscr{G})$ it is trivial that

$$\mathscr{G} \text{ separates points } \Rightarrow \sigma(\mathscr{G}) \text{ separates points.}$$

The converse is also true: if $\sigma(\mathscr{G})$ separates points, then $\mathscr{G}$ separates points. In order to prove this, fix $x, y \in X$, $x \neq y$, and define

$$\Sigma := \{A \subseteq X\,;\; x, y \in A \text{ or } x, y \notin A\}.$$

It is easy to verify that Σ is a σ-algebra. If $\mathscr{G}$ does not separate points, then $\mathscr{G} \subseteq \Sigma$ and, hence, $\sigma(\mathscr{G}) \subseteq \sigma(\Sigma) = \Sigma$ which means that $\sigma(\mathscr{G})$ does not separate points.

Comment As an immediate consequence, we get the following result: *If a σ-algebra $\mathscr{A}$ separates points, then* ***any*** *generator $\mathscr{G}$ of $\mathscr{A}$ must separate points.*

4.12 A family $\mathscr{G}$ of intervals whose endpoints form a dense subset of $\mathbb{R}$ but $\sigma(\mathscr{G}) \subsetneqq \mathscr{B}(\mathbb{R})$

Let $\mathscr{G} = \{[-q, q]\,;\; q \in \mathbb{Q}\}$. By construction, all rationals are endpoints of the intervals in $\mathscr{G}$. While every set $A \in \sigma(\mathscr{G})$ is a Borel set, it is also symmetric with respect to the origin. Thus, any non-symmetric $B \in \mathscr{B}(\mathbb{R})$ cannot be in $\sigma(\mathscr{G})$.

4.13 Intersection and the σ-operation do not commute: $\sigma\left(\bigcap_{n\in\mathbb{N}} \mathscr{G}_n\right) \subsetneqq \bigcap_{n\in\mathbb{N}} \sigma\left(\mathscr{G}_n\right)$

Let $(X, \mathscr{A})$ be a measurable space and $\mathscr{G}_n$, $n \in \mathbb{N}$, families of subsets such that $\mathscr{G}_{n+1} \supseteq \mathscr{G}_n$. While we always have $\sigma\left(\bigcap_{n\in\mathbb{N}} \mathscr{G}_n\right) \subseteq \bigcap_{n\in\mathbb{N}} \sigma\left(\mathscr{G}_n\right)$, we cannot expect equality.

Consider the real line with its Borel sets and denote by $\mathscr{J} = \{(r, q)\,;\; r, q \in \mathbb{Q}\}$ the family of all open intervals with rational endpoints; if $r > q$ we agree that $(r, q) = \emptyset$. Let $(I_n)_{n\in\mathbb{N}_0}$ be any enumeration of $\mathscr{J}$ such that $I_0 = \emptyset$, and let $\mathscr{J}_n := \{I_k\,;\; k > n\} \cup \{\emptyset\}$ be the family $\mathscr{J}$ without the first n non-trivial intervals.

Since $\mathbb{Q}$ is dense in $\mathbb{R}$, we can represent $I_1, \dots, I_n$ with countably many unions and intersections of intervals from $\mathscr{J}_n$, so that $\mathscr{B}(\mathbb{R}) = \sigma(\mathscr{J}) = \sigma(\mathscr{J}_n)$ for all $n \in \mathbb{N}$. As $\bigcap_{n\in\mathbb{N}} \mathscr{J}_n = \{\emptyset\}$, we see that

$$\{\emptyset, \mathbb{R}\} = \sigma(\{\emptyset\}) = \sigma\left(\bigcap_{n\in\mathbb{N}} \mathscr{J}_n\right) \subsetneqq \bigcap_{n\in\mathbb{N}} \sigma\left(\mathscr{J}_n\right) = \mathscr{B}(\mathbb{R}).$$

4.14 A metric space such that the σ-algebra generated by the open balls is smaller than the Borel σ-algebra

Consider $X = \mathbb{R}$ with the discrete metric d [☞ Example 2.1.(b)], i.e. $d(x, y) = 0$ if $x = y$ and, if $x \neq y$, $d(x, y) = c$ for some $c \in (0, \infty]$ (usually one uses $c = 1$ or $c = \infty$). In the metric space (X, d) the open balls are singletons or the whole space X,

$$B_r(x) = \{y \in X\,;\ d(x,y) < r\} = \begin{cases} \{x\}, & \text{if } r \in (0, c], \\ X, & \text{if } r > c, \end{cases}$$

and therefore the σ-algebra generated by the open balls is the co-countable σ-algebra [☞ Example 4A], i.e.

$$\sigma(B_r(x),\ x \in X,\ r > 0) = \sigma(\{x\},\ x \in X).$$

By construction, every set $U \subseteq X$ is open in the metric space (X, d). This means that the Borel σ-algebra $\mathscr{B}(X)$ coincides with the power set $\mathscr{P}(X)$, which is strictly larger than the co-countable σ-algebra.

Comment 1 In view of the above example, it is not surprising that there exist measures on the σ-algebra generated by the open balls which cannot be extended to the Borel σ-algebra $\mathscr{B}(X)$ [☞ 5.7]. Moreover, the example shows that continuous functions are not necessarily measurable with respect to the σ-algebra generated by the open balls; since in the discrete topology all functions are continuous, we can take $f = \mathbb{1}_A$ for A such that A and A^c are uncountable.

Comment 2 For separable metric spaces (X, d) the above pathology cannot happen: if there exists a countable dense subset, then any open set can be written as a countable union of open balls, and this gives

$$\mathscr{B}(X) = \sigma(U\,;\ U \text{ open}) \subseteq \sigma(B_r(x),\ x \in X,\ r > 0);$$

the converse inclusion is obvious.

4.15 The σ-algebra generated by the compact sets can be larger than the Borel σ-algebra (compact sets need not be Borel sets)

Let X be a topological space. By definition, the Borel σ-algebra $\mathscr{B}(X)$ is the smallest σ-algebra containing all open sets. Since every closed set is the complement of an open set, $\mathscr{B}(X)$ can be equivalently defined as the smallest σ-algebra containing all closed sets. In general, the Borel σ-algebra does not contain all compact sets. The reason is that – in general topological spaces – compact sets need not be closed.

Consider $X = \mathbb{R}$ with the co-finite topology [☞ Example 2.1.(d)], i.e. a set $U \subseteq X$ is open if, and only if, $U = \emptyset$ or $X \setminus U$ is finite. The associated Borel σ-algebra $\mathscr{B}(X)$ coincides with the co-countable σ-algebra [☞ Example 4A],

$$A \in \mathscr{B}(X) \iff \#A \leqslant \#\mathbb{N} \text{ or } \#A^c \leqslant \#\mathbb{N}.$$

It is not hard to see that **every** set $A \subseteq X$ is compact in the co-finite topology, see e.g. [172, Example 18], which means that $\mathscr{P}(X)$ is the smallest σ-algebra containing all compact sets. For instance, the compact set $A = [0, 1]$ is not in $\mathscr{B}(X)$.

Comment If (X, d) is a metric space, then the σ-algebra $\mathscr{K}$ generated by the compact sets cannot be larger than the Borel σ-algebra:

$$\mathscr{K} := \sigma(\{C \subseteq X\,;\ C \text{ compact}\}) \subseteq \mathscr{B}(X).$$

This follows from the fact that compact sets in metric spaces are closed, i.e. their complements are open sets, cf. [200, Theorem 17.5]. Under the additional assumption that (X, d) is σ-compact, it follows that $\mathscr{K} = \mathscr{B}(X)$; see the comment in Example 4.16.

4.16 The σ-algebra generated by the compact sets can be smaller than the Borel σ-algebra

Let X be a topological space with Borel σ-algebra $\mathscr{B}(X)$, i.e. the smallest σ-algebra containing all open sets. If there are 'many' open sets, then there exist only 'few' compact sets, and therefore the Borel σ-algebra can be strictly larger than the σ-algebra generated by the compact sets.

Consider on $X = \mathbb{R}$ the discrete metric d [☞ Example 2.1.(b)], i.e. $d(x, y) = 0$ if $x = y$ and $d(x, y) = 1$ if $x \neq y$. As $B_r(x) = \{x\}$ for every $r \in (0, 1]$, each singleton $\{x\}$ is open, and so $\mathscr{B}(X) = \mathscr{P}(X)$. Note that the only compact sets in X are finite sets. Consequently, the smallest σ-algebra $\mathscr{K}$ containing all compact sets is the co-countable σ-algebra [☞ Example 4A],

$$\mathscr{K} = \{A \subseteq X\,;\ \#A \leqslant \#\mathbb{N} \text{ or } \#A^c \leqslant \#\mathbb{N}\}.$$

In particular, $\mathscr{K}$ is strictly smaller than $\mathscr{P}(X) = \mathscr{B}(X)$, e.g. $[0, 1] \in \mathscr{B}(X) \setminus \mathscr{K}$.

Comment A metric space (X, d) is said to be **σ-compact** if there exists a sequence $(K_n)_{n\in\mathbb{N}} \subseteq X$ of compact sets such that $K_n \uparrow X$. If (X, d) is a σ-compact metric space, then the Borel σ-algebra is the smallest σ-algebra containing all compact sets, i.e.

$$\mathscr{B}(X) = \sigma(\{C \subseteq X\,;\ C \text{ compact}\}) =: \mathscr{K}.$$

Every compact set in a metric space is closed, i.e. it can be written as the complement of an open set, cf. [200, Theorem 17.5], and this gives $\mathscr{K} \subseteq \mathscr{B}(X)$. Now fix a sequence $(K_n)_{n\in\mathbb{N}} \subseteq X$ of compact sets with $K_n \uparrow X$, and let $F \subseteq X$ be closed. As $F \cap K_n$ is compact, it follows that

$$F = \bigcup_{n\in\mathbb{N}} (F \cap K_n) \in \mathscr{K}.$$

Since $\mathscr{B}(X)$ is the smallest σ-algebra containing all closed sets, we conclude that $\mathscr{B}(X) \subseteq \mathscr{K}$.

4.17 A topology such that every non-empty Borel set has uncountably many elements

For $I = [0, 1]$ equip $X = \{0, 1\}^I$ with the product topology [☞ Example 2.1.(i)]. The corresponding Borel σ-algebra is the product σ-algebra $\mathscr{B}(X) = \mathscr{P}(\{0, 1\})^{\otimes I}$ [☞ Example 4C]. Consequently, there exists for every $B \in \mathscr{B}(X)$ a countable set $S \subseteq I$ such that

$$y \in X,\ x \in B,\ x|_S = y|_S \Rightarrow y \in B \tag{4.5}$$

[☞ 4.5]. This implies that every non-empty Borel set is uncountable. Indeed: for given $B \in \mathscr{B}(X)$, $B \neq \emptyset$, take $S \subseteq I$ countable as in (4.5) and pick any $x \in B$. For $s \in I \setminus S$ set

$$y_s(t) := \begin{cases} x(t), & \text{if } t \in S, \\ 1, & \text{if } t = s, \\ 0, & \text{otherwise.} \end{cases}$$

Because of (4.5), we see that $y_s \in B$. As $y_s \neq y_{s'}$ for $s \neq s'$, this gives

$$\#B \geqslant \#\{y_s\,;\ s \in I \setminus S\} = \mathfrak{c}.$$

4.18 A metrizable and a non-metrizable topology having the same Borel sets

Denote by E the real line $\mathbb{R}$ equipped with the usual Euclidean topology $\mathscr{O}_E$ and by S the Sorgenfrey line [☞ Example 2.1.(h)], i.e. the real line equipped with the topology $\mathscr{O}_S$ generated by the half-open sets $\{[x, z), z > x\}$. The Sorgenfrey line is not metrizable. To prove this, we use the fact that any separable metrizable space has a countable base. Since $\mathbb{Q}$ is a countable dense subset, S is separable, and therefore it suffices to show that there is no countable base.

Fix some basis $\mathscr{U}$ for the topology of the Sorgenfrey line. Since $[x, x + 1)$ is open for any $x \in S$ there exists some $U_x \in \mathscr{U}$ such that $x \in U_x \subseteq [x, x+1)$. If $x < y$, then $x \notin U_y$, and this gives $U_x \neq U_y$. Hence, $U_x \neq U_y$ for $x \neq y$. This means that any basis $\mathscr{U}$ has at least $\#\mathbb{R} = \mathfrak{c}$ elements, hence S is not metrizable. However, the Borel σ-algebra $\mathscr{B}(S)$ coincides with the Borel σ-algebra $\mathscr{B}(E)$ generated by the Euclidean topology $\mathscr{O}_E$, i.e. $\sigma(\mathscr{O}_S) = \sigma(\mathscr{O}_E)$. In order to prove this, we need the following lemma.

Lemma *The Sorgenfrey line S is hereditarily Lindelöf, i.e. the open cover of any subset of S has a countable subcover. In other words, if $A \subseteq S$ is any set and $(U_i)_{i\in I} \subseteq \mathscr{O}_S$ such that $A \subseteq \bigcup_{i\in I} U_i$, then there exists a countable set $J \subseteq I$ such that $A \subseteq \bigcup_{j\in J} U_j$.*

Proof Take $A \subseteq S$ and an open cover $(U_i)_{i\in I} \subseteq \mathscr{O}_S$ of A. Denote by U_i° the open interior of U_i with respect to the Euclidean topology and set $U := \bigcup_{i\in I} U_i^\circ$. Since the Euclidean topology $\mathscr{O}_E$ has a countable base, there exists a countable set $J \subseteq I$ such that $U = \bigcup_{j\in J} U_j^\circ$. We claim that $N := A \setminus U$ is countable. Since $(U_i)_{i\in I}$ is an open cover of A, there exists for every $x \in N \subseteq A$ some $i \in I$ such that $x \in U_i$. The half-open intervals $[x, x+r)$, $r > 0$, are a neighbourhood base at x, and so there is some $r_x > 0$ such that $[x, x+r_x) \subseteq U_i$. In particular, $U \supseteq U_i^\circ \supseteq (x, x+r_x)$, which implies $N \cap (x, x+r_x) = \emptyset$. Consequently,

$$\{(x, x+r_x);\ x \in N\}$$

is a family of pairwise disjoint sets which are open in the Euclidean topology, and therefore the family, hence N, is countable. Pick for each $x \in N$ a set U_x from the open cover such that $x \in U_x$, then $(U_x)_{x\in N} \cup \{U_j\}_{j\in J}$ is a countable subcover of A. □

Corollary $\mathscr{B}(E) = \mathscr{B}(S)$.

Proof Every half-open interval $[x, z)$, $x < z$, is open in S, and so

$$(a,b) = \bigcup_{n\in\mathbb{N}} \left[a + \frac{1}{n}, b\right) \in \mathscr{B}(S)$$

for all $a < b$. Since the open intervals (a,b), $a, b \in \mathbb{Q}$, are a countable basis for the Euclidean topology, i.e. any open set in E can be written as a countable union of such intervals, it follows that the open sets of the Euclidean topology are contained in $\mathscr{B}(S)$; thus, $\mathscr{B}(E) \subseteq \mathscr{B}(S)$. Now take a set U which is open in the Sorgenfrey topology. By definition, there exists for each $x \in U$ some $N(x) \in \mathbb{N}$ such that $[x, x + \frac{1}{n}) \subseteq U$ for all $n \geqslant N(x)$. The sets

$$\left[x, x + \frac{1}{n}\right), \quad x \in U,\ n \geqslant N(x),$$

are an open cover of U, and so there exists, by the previous lemma, a countable subcover, i.e. we can pick countably many of the intervals, say $(I_n)_{n\in\mathbb{N}}$, such that $U = \bigcup_{n\in\mathbb{N}} I_n$. As $I_n \in \mathscr{B}(E)$ for each $n \in \mathbb{N}$, this gives $U \in \mathscr{B}(E)$ and thus $\mathscr{B}(S) \subseteq \mathscr{B}(E)$. □

Comment The choice of the intervals $(I_n)_{n\in\mathbb{N}}$ depends on the open set U. As we have seen, the Sorgenfrey topology does not have a countable base, and therefore it is **not** possible to find countably many intervals I_n, $n \in \mathbb{N}$, such that every open set $U \subseteq S$ can be written as

$$U = \bigcup_{n:I_n\subseteq U} I_n.$$

4.19 A σ-algebra which is not generated by any topology

We give an example of a σ-algebra $\mathscr{A}$ on a set X for which there does not exist a topology on X such that the associated Borel σ-algebra $\mathscr{B}(X)$ equals $\mathscr{A}$. The example is due to Lang [96] and relies on cardinality arguments.

We begin with a general result which can, for instance, be found in [50, II.9.4].

Lemma *Let Y be a set and $\mathscr{F}$ a family of maps $Y^{\mathbb{N}} \to Y$. For $A \subseteq Y$ define*

$$\mathscr{F}(A) := \bigcup_{f\in\mathscr{F}} \{f((y_i)_{i\in\mathbb{N}})\ ;\ (y_i)_{i\in\mathbb{N}} \subseteq A\}.$$

For every $G \subseteq Y$ with $G \subseteq \mathscr{F}(G)$ there exists a set $A \supseteq G$ with the following properties:

(a) *$A = \mathscr{F}(A)$.*
(b) *$A \supseteq G$ is the smallest set satisfying $A = \mathscr{F}(A)$, i.e. if $B \supseteq G$ is such that $B = \mathscr{F}(B)$ then $B \supseteq A$.*
(c) *If $\max\{\#\mathscr{F}, \#G\} \geqslant 2$ then $\#A \leqslant (\#\mathscr{F} \cdot \#G)^{\aleph_0}$* ($\aleph_0 = \#\mathbb{N}$ [☞ Section 2.3]).

As an immediate consequence we obtain an upper bound for the cardinality of σ-algebras.

Corollary *Let X be a set and $\mathscr{G} \subseteq \mathscr{P}(X)$ with $\#\mathscr{G} \geqslant 2$. Then $\#\sigma(\mathscr{G}) \leqslant (\#\mathscr{G})^{\aleph_0}$.*

Proof Denote by $\mathscr{F}$ the following family of maps:

$$\bigcap : \bigtimes_{i=1}^{\infty} \mathscr{P}(X) \to \mathscr{P}(X), \qquad (A_i)_{i\in\mathbb{N}} \longmapsto \bigcap_{i=1}^{\infty} A_i,$$

$$\bigcup : \bigtimes_{i=1}^{\infty} \mathscr{P}(X) \to \mathscr{P}(X), \qquad (A_i)_{i\in\mathbb{N}} \longmapsto \bigcup_{i=1}^{\infty} A_i,$$

$$\complement : \bigtimes_{i=1}^{\infty} \mathscr{P}(X) \to \mathscr{P}(X), \qquad (A_i)_{i\in\mathbb{N}} \longmapsto X \setminus A_1.$$

Fix $\mathscr{G} \subseteq \mathscr{P}(X)$ with $\#\mathscr{G} \geqslant 2$; without loss of generality, $\emptyset \in \mathscr{G}$. As $\mathscr{G} \subseteq \mathcal{F}(\mathscr{G})$, the above lemma shows that there exists a smallest family $\mathscr{A} \supseteq \mathscr{G}$ which is closed under $\mathcal{F}$, i.e. $\mathscr{A} = \mathcal{F}(\mathscr{A})$, and which satisfies $\#\mathscr{A} \leqslant (\#\mathscr{G})^{\aleph_0}$. By definition, $\sigma(\mathscr{G})$ is the smallest family which is closed under $\mathcal{F}$ and which contains $\mathscr{G}$; hence, $\sigma(\mathscr{G}) = \mathscr{A}$. In particular, $\#\sigma(\mathscr{G}) = \#\mathscr{A} \leqslant (\#\mathscr{G})^{\aleph_0}$. □

These preparations enable us to construct a σ-algebra which is not generated by any topology. Consider $X = \{0,1\}^{\mathbb{R}}$ which we equip with the infinite product σ-algebra $\mathscr{A} := \mathscr{P}(\{0,1\})^{\mathbb{R}}$, i.e. the σ-algebra generated by

$$\mathscr{G} := \{\pi_t^{-1}(A);\ A \subseteq \{0,1\},\ t \in \mathbb{R}\},$$

where $\pi_t : \{0,1\}^{\mathbb{R}} \to \{0,1\}, f \longmapsto f(t)$ denotes the projection onto the tth coordinate [☞ Example 4C]. Note that $\mathscr{G}$ separates points in X: if $f, g \in X$ are such that $f \neq g$, then there is a $t \in \mathbb{R}$ with $x := f(t) \neq g(t)$ and so $G := \pi_t^{-1}(\{x\}) \in \mathscr{G}$ satisfies $f \in G, g \notin G$. Clearly, $\#\mathscr{G} = \#\mathbb{R}$, and therefore it follows from the corollary that

$$\#\mathscr{A} \leqslant (\#\mathbb{R})^{\aleph_0} = \#\mathbb{R}. \tag{4.6}$$

Now suppose that there is a topology $\mathscr{O}$ on X with $\sigma(\mathscr{O}) = \mathscr{A}$. Since $\mathscr{G}$ separates points in X, so does $\mathscr{A} = \sigma(\mathscr{G})$, and this, in turn, implies that $\mathscr{O}$ separates points [☞ 4.11]. Consequently, for any $f, g \in X, f \neq g$, we can find an open set $U \in \mathscr{O}$ with $g \in U, f \notin U$. In other words, there exists a closed set F with $f \in F, g \notin F$. This gives

$$g \notin \overline{\{f\}} := \bigcap\{F \supseteq \{f\};\ F \subseteq X \text{ closed}\},$$

and so $\overline{\{f\}} \neq \overline{\{g\}}$ for $f \neq g$. Since the closed set $\overline{\{f\}}$ belongs to $\sigma(\mathscr{O})$ for every $f \in X$, we find that

$$\#\mathscr{A} = \#\sigma(\mathscr{O}) \geqslant \#\{\overline{\{f\}};\ f \in X\} = \#X = 2^{\#\mathbb{R}}.$$

This, however, contradicts (4.6) since, by Cantor's theorem, $2^{\#\mathbb{R}} > \#\mathbb{R}$.

Comment There exists a topology on $\mathbb{R}$ which generates the Lebesgue σ-algebra $\mathscr{L}(\mathbb{R})$; see [156]. More generally, if $(X, \mathscr{A}, \mu)$ is a σ-finite measure space and $\mathscr{A}$ is complete, i.e.

$$\mathscr{A} \supseteq \{N \subseteq X;\ \exists A \in \mathscr{A}, A \supseteq N\ :\ \mu(A) = 0\},$$

then $\mathscr{A}$ is generated by a topology. This can be proved using so-called liftings which induce a topology $\mathscr{O}$ on X, cf. [83, Chapter 5]. For σ-finite complete measure spaces the existence of a lifting is shown in [111], for the particular case of complete probability spaces see [20, Theorem 10.5.4].

4.20 A σ-algebra which is strictly between the Borel and the Lebesgue sets

Denote by $\mathscr{B}(\mathbb{R})$ and $\mathscr{L}(\mathbb{R})$ the Borel and Lebesgue σ-algebras on $\mathbb{R}$, respectively. We will show that the σ-algebra generated by the Souslin sets,

$$\Sigma := \sigma(\{S\,;\; S \subseteq \mathbb{R} \text{ is a Souslin set}\}),$$

satisfies $\mathscr{B}(\mathbb{R}) \subsetneqq \Sigma \subsetneqq \mathscr{L}(\mathbb{R})$, [☞introduction to Chapter 7 for the definition of Souslin sets]. Since there are Souslin sets which are not Borel [☞7.19], it is immediate that Σ is strictly larger than $\mathscr{B}(\mathbb{R})$. Moreover, every Souslin set is Lebesgue measurable, cf. [171, Theorem 4.3.1], and so $\Sigma \subseteq \mathscr{L}(\mathbb{R})$. In order to prove that Σ is strictly smaller than $\mathscr{L}(\mathbb{R})$, we use a cardinality argument based on the following general estimate:

$$\#\sigma(\mathscr{G}) \leqslant (\#\mathscr{G})^{\aleph_0}, \tag{4.7}$$

where $\aleph_0 = \#\mathbb{N}$ [☞Section 2.3] and the corollary in Example 4.19. Since the Borel σ-algebra $\mathscr{B}(\mathbb{R}^2)$ is generated by a countable family of sets, e.g. all open balls $B_r(x)$ with $r \in \mathbb{Q} \cap (0, \infty)$ and $x \in \mathbb{Q}^2$, it follows that

$$\#\mathscr{B}(\mathbb{R}^2) \leqslant (\#\mathbb{N})^{\aleph_0} = \mathfrak{c}.$$

Since every Souslin set $S \subseteq \mathbb{R}$ can be written as the projection of a Borel set in $\mathbb{R}^2$, cf. Theorem 1 in Example 7.19, it follows that there are at most continuum many Souslin sets in $\mathbb{R}$; for a more explicit proof of this fact see [MIMS, pp. 432-435]. Another application of (4.7) shows that $\#\Sigma \leqslant \mathfrak{c}$. There are, however, $2^{\mathfrak{c}}$ many Lebesgue sets, cf. [MIMS, p. 429], and so Σ is strictly smaller than $\mathscr{L}(\mathbb{R})$.

4.21 The Borel sets cannot be constructed by induction

Let $\mathscr{O}$ be the open sets in $\mathbb{R}$ and set

$$\mathscr{O}^c := \{U, U^c\,;\; U \in \mathscr{O}\} \quad \text{and} \quad \mathscr{O}^{c\sigma} := \left\{\bigcup_{n\in\mathbb{N}} U_n\,;\; U_n \in \mathscr{O}^c\right\}.$$

The Borel sets cannot be obtained by repeating this procedure countably often, i.e.

$$\bigcup_{n\in\mathbb{N}} \mathscr{F}_n \subsetneqq \mathscr{B}(\mathbb{R}) \quad \text{where} \quad \mathscr{F}_n := \underbrace{(\dots((\mathscr{O}^{c\sigma})^{c\sigma})\dots)^{c\sigma}}_{n \text{ times}}. \tag{4.8}$$

It is clear from the definition that $\mathscr{F}_n \subseteq \mathscr{B}(\mathbb{R})$ for all $n \in \mathbb{N}$; therefore, we have $\bigcup_{n\in\mathbb{N}} \mathscr{F}_n \subseteq \mathscr{B}(\mathbb{R})$. In order to prove that the inclusion is strict, we construct so-called universal sets[1] and use a diagonal argument.

We need some preparations. For $d \geqslant 1$ denote by $\mathscr{O}(\mathbb{R}^d)$ (resp. $\mathscr{C}(\mathbb{R}^d)$) the open (resp. closed) sets in $\mathbb{R}^d$, and define iteratively

$$\mathscr{G}_n(\mathbb{R}^d) := \left\{\bigcap_{i=1}^{\infty} A_i \,;\; A_i \in \mathscr{H}_{n-1}(\mathbb{R}^d)\right\}, \qquad \mathscr{G}_0(\mathbb{R}^d) := \mathscr{C}(\mathbb{R}^d),$$

$$\mathscr{H}_n(\mathbb{R}^d) := \left\{\bigcup_{i=1}^{\infty} A_i \,;\; A_i \in \mathscr{G}_{n-1}(\mathbb{R}^d)\right\}, \qquad \mathscr{H}_0(\mathbb{R}^d) := \mathscr{O}(\mathbb{R}^d).$$

Since $\emptyset$ and $\mathbb{R}^d$ are both open and closed, we have $\mathscr{G}_n(\mathbb{R}^d) \subseteq \mathscr{H}_{n+1}(\mathbb{R}^d)$ as well as $\mathscr{H}_n(\mathbb{R}^d) \subseteq \mathscr{G}_{n+1}(\mathbb{R}^d)$ for all $n \in \mathbb{N}_0$. The following properties are proved by induction on $n \in \mathbb{N}_0$:

1° $A \in \mathscr{G}_n(\mathbb{R}^d) \iff A^c \in \mathscr{H}_n(\mathbb{R}^d)$.

2° If $f : \mathbb{R}^d \to \mathbb{R}^k$ is a continuous function, then $f^{-1}(\mathscr{G}_n(\mathbb{R}^k)) \subseteq \mathscr{G}_n(\mathbb{R}^d)$ and $f^{-1}(\mathscr{H}_n(\mathbb{R}^k)) \subseteq \mathscr{H}_n(\mathbb{R}^d)$. For $n = 0$ this follows from the fact that pre-images of open (resp. closed) sets under continuous functions are open (resp. closed). For the induction use

$$f^{-1}(\mathbb{R}^k \setminus A) = \mathbb{R}^d \setminus f^{-1}(A) \quad \text{and} \quad f^{-1}\left(\bigcup_{i=1}^{\infty} A_i\right) = \bigcup_{i=1}^{\infty} f^{-1}(A_i).$$

3° $A \in \mathscr{G}_n(\mathbb{R}^d) \Rightarrow A^c \in \mathscr{H}_{n+1}(\mathbb{R}^d)$ and $A \in \mathscr{H}_n(\mathbb{R}^d) \Rightarrow A^c \in \mathscr{G}_{n+1}(\mathbb{R}^d)$. Because of **1°**, it is enough to verify the first implication. If $n = 0$, i.e. $A \subseteq \mathbb{R}^d$ is closed, then A^c is open and can be written as a union of the closed sets

$$A_i := \left\{x \in \mathbb{R}^d \,;\; d(x, A) := \inf_{y\in A} |y - x| \geqslant \frac{1}{i}\right\}, \qquad i \in \mathbb{N}.$$

Hence, $A \in \mathscr{H}_1(\mathbb{R}^d)$. The claim follows now by induction.

Let us return to the original problem in dimension $d = 1$. Using **3°**, it follows by

[1] A **universal set** is not a **universally measurable** set as is used in the theory of stochastic processes.

induction that $\mathscr{F}_n$ from (4.8) satisfies $\mathscr{F}_n \subseteq \mathscr{H}_{2n}(\mathbb{R})$ for all $n \in \mathbb{N}$. In particular,

$$\bigcup_{n\in\mathbb{N}} \mathscr{F}_n \subseteq \bigcup_{n\in\mathbb{N}} \mathscr{H}_n(\mathbb{R}) =: \mathscr{H}(\mathbb{R}).$$

In order to prove $\bigcup_{n\in\mathbb{N}} \mathscr{F}_n \subsetneqq \mathscr{B}(\mathbb{R})$, it suffices to show that $\mathscr{H}(\mathbb{R})$ is strictly smaller than $\mathscr{B}(\mathbb{R})$. We begin with an auxiliary result on universal sets.

Lemma *Let $n \in \mathbb{N}_0$.*

(a) *There is an $\mathscr{H}_n(\mathbb{R})$-universal set, i.e. a set $U \in \mathscr{H}_n(\mathbb{R}^2)$ with $U \subseteq [0,1] \times \mathbb{R}$ such that every $H \in \mathscr{H}_n(\mathbb{R})$ is a slice of U, i.e. there exists some $x \in [0,1]$ such that*

$$H = \{y \in \mathbb{R};\ (x,y) \in U\} =: U^x.$$

(b) *There exists a $\mathscr{G}_n(\mathbb{R})$-universal set, i.e. a set $V \in \mathscr{G}_n(\mathbb{R}^2)$ with $V \subseteq [0,1] \times \mathbb{R}$ such that every $G \in \mathscr{G}_n(\mathbb{R})$ can be written as $G = V^x$ for some $x \in [0,1]$.*

Proof Pick a sequence of continuous functions $f_i: [0,1] \to [0,1]$, $i \in \mathbb{N}$, with $f_i(0) = 0$, $f_i(1) = 1$ such that for every $(y_i)_{i\in\mathbb{N}} \subseteq [0,1]$ there exists an $x \in [0,1]$ with $f_i(x) = y_i$ for all $i \in \mathbb{N}$, [☞ 13.14] for the existence. Extend f_i continuously to $\mathbb{R}$ by setting $f_i(x) = x$ for $x \in \mathbb{R} \setminus [0,1]$.

We prove the assertion by induction and start with $n = 0$. Let $(I_k)_{k\in\mathbb{N}}$ be an enumeration of all open intervals with rational endpoints, and take a sequence $(a_k)_{k\in\mathbb{N}}$ such that $0 < a_1 < a_2 < \cdots < 1$ and $\lim_{k\to\infty} a_k = 1$. Define

$$W := \bigcup_{k\in\mathbb{N}} (a_k, a_{k+1}) \times I_k \subseteq [0,1] \times \mathbb{R}$$

and

$$U := \bigcup_{i\in\mathbb{N}} U_i \quad \text{with} \quad U_i := \{(x,y) \in [0,1] \times \mathbb{R};\ (f_i(x), y) \in W\}.$$

Clearly, W is open, i.e. $W \in \mathscr{H}_0(\mathbb{R}^2)$. From **2°** and the continuity of the mapping $(x,y) \longmapsto (f_i(x), y)$, it follows that $U \in \mathscr{H}_0(\mathbb{R}^2)$. Take an open set $H \subseteq \mathbb{R}$. If $H = \emptyset$, then $H = U^0$. For $H \neq \emptyset$ write $H = \bigcup_{i\in\mathbb{N}} I_{k_i}$ for suitable $k_i \in \mathbb{N}$. Pick any $x_i \in (a_{k_i}, a_{k_{i+1}}) \subseteq [0,1]$; then $I_{k_i} = W^{x_i}$. By construction of the sequence $(f_i)_{i\in\mathbb{N}}$, there exists an $x \in [0,1]$ such that $f_i(x) = x_i$ for all $i \in \mathbb{N}$. Then

$$H = \bigcup_{i\in\mathbb{N}} I_{k_i} = \bigcup_{i\in\mathbb{N}} W^{x_i} = \bigcup_{i\in\mathbb{N}} \{y \in \mathbb{R};\ (f_i(x), y) \in W\} = \bigcup_{i\in\mathbb{N}} (U_i)^x = U^x,$$

which shows that U is universal for the open sets.

For (b) consider $V := ([0,1] \times \mathbb{R}) \setminus U$. If $G \subseteq \mathbb{R}$ is closed, then there is some $x \in [0,1]$ such that $U^x = \mathbb{R} \setminus G$, i.e. $V^x = G$.

Suppose that (a) and (b) hold for some $n \in \mathbb{N}_0$, and denote by $U \in \mathscr{H}_n(\mathbb{R}^2)$, $V \in \mathscr{G}_n(\mathbb{R}^2)$ the corresponding universal sets. Set

$$\tilde{U} := \bigcup_{i=1}^{\infty} W_i \quad \text{with} \quad W_i := \{(x, y) \in [0,1] \times \mathbb{R}\,;\ (f_i(x), y) \in V\}.$$

As in the previous part, we find that $\tilde{U} \in \mathscr{H}_{n+1}(\mathbb{R}^2)$. If $H \in \mathscr{H}_{n+1}(\mathbb{R})$, i.e. $H = \bigcup_{i\in\mathbb{N}} G_i$ for $G_i \in \mathscr{G}_n(\mathbb{R})$, then there exist $x_i \in [0,1]$ such that $G_i = V^{x_i}$. By our choice of f_i, there is some $x \in [0,1]$ satisfying $f_i(x) = x_i$ for all $i \in \mathbb{N}$. Thus,

$$\tilde{U}^x = \bigcup_{i=1}^{\infty} (W_i)^x = \bigcup_{i=1}^{\infty} V^{f_i(x)} = \bigcup_{i=1}^{\infty} V^{x_i} = H,$$

i.e. (a) holds for $n+1$. By **1°**, $\tilde{V} := ([0,1]\times\mathbb{R})\setminus\tilde{U}$ does the job in (b) for $n+1$. □

Corollary 1 *$\mathscr{H}(\mathbb{R}) := \bigcup_{n\in\mathbb{N}} \mathscr{H}_n(\mathbb{R})$ is not a σ-algebra.*

Proof First, we construct an $\mathscr{H}(\mathbb{R})$-universal set. Take the sequence of continuous functions f_i, $i \in \mathbb{N}$, from the proof of the lemma, and let $U_n \in \mathscr{H}_n(\mathbb{R}^2)$ be $\mathscr{H}_n(\mathbb{R})$-universal. Define

$$U := \bigcup_{i=1}^{\infty} W_i \quad \text{with} \quad W_i := \{(x, y) \in [0,1] \times \mathbb{R}\,;\ (f_i(x), y) \in U_i\}.$$

Because of the continuity of $(x, y) \longmapsto (f_i(x), y)$, we have $W_i \in \mathscr{H}_i(\mathbb{R}^2)$, and so $U \in \bigcup_{i=1}^{\infty} \mathscr{H}_i(\mathbb{R}^2)$. Take $H \in \mathscr{H}(\mathbb{R})$. From the definition we know $H = \bigcup_{i=1}^{\infty} H_i$ for $H_i \in \mathscr{H}_i(\mathbb{R})$. Consequently, we can find $x_i \in [0,1]$ such that $U_i^{x_i} = H_i$. By the construction of f_i, we have $f_i(x) = x_i$, $i \in \mathbb{N}$, for some $x \in [0,1]$. Hence,

$$U^x = \bigcup_{i=1}^{\infty} W_i^x = \bigcup_{i=1}^{\infty} U_i^{f_i(x)} = \bigcup_{i=1}^{\infty} U_i^{x_i} = \bigcup_{i=1}^{\infty} H_i = H.$$

Consequently, U is $\mathscr{H}(\mathbb{R})$-universal. Now set

$$D := \{x \in \mathbb{R}\,;\ (x, x) \in U\} \subseteq [0,1].$$

We claim that $D \in \mathscr{H}(\mathbb{R})$ but $D^c = \mathbb{R}\setminus D \notin \mathscr{H}(\mathbb{R})$. Since the map $x \longmapsto (x, x)$ is continuous, **2°** gives

$$D_i := \{x \in \mathbb{R}\,;\ (x, x) \in W_i\} \in \mathscr{H}_i(\mathbb{R}), \quad i \in \mathbb{N},$$

and so

$$D = \bigcup_{i=1}^{\infty} D_i \in \bigcup_{i=1}^{\infty} \mathscr{H}_i(\mathbb{R}) = \mathscr{H}(\mathbb{R}).$$

It remains to prove that $D^c \notin \mathscr{H}(\mathbb{R})$. Suppose on the contrary that $D^c \in \mathscr{H}(\mathbb{R})$. Since U is $\mathscr{H}(\mathbb{R})$-universal, there is some $x \in [0,1]$ such that $U^x = D^c$. Hence,

$$x \in D \overset{\text{def. } D}{\iff} (x,x) \in U \overset{U^x = D^c}{\iff} x \in D^c,$$

which is a contradiction. We conclude that $D \in \mathscr{H}(\mathbb{R})$ and $D^c \notin \mathscr{H}(\mathbb{R})$, which means that $\mathscr{H}(\mathbb{R})$ is not a σ-algebra. □

Corollary 2 $\mathscr{H}(\mathbb{R}) = \bigcup_{n\in\mathbb{N}} \mathscr{H}_n(\mathbb{R}) \subsetneqq \mathscr{B}(\mathbb{R})$.

Proof By definition, $\mathscr{H}_n(\mathbb{R}) \subseteq \mathscr{B}(\mathbb{R})$ for all $n \in \mathbb{N}$ and thus $\mathscr{H}(\mathbb{R}) \subseteq \mathscr{B}(\mathbb{R})$. The inclusion is strict; otherwise $\mathscr{H}(\mathbb{R})$ would be a σ-algebra, in contradiction to the previous corollary. □

4.22 The Borel sets can be constructed by transfinite induction

We have seen in Example 4.21 that $\mathscr{B}(\mathbb{R})$ cannot be constructed from the open sets by taking unions and complements countably many times. The following construction shows that $\mathscr{B}(\mathbb{R})$ can be obtained by transfinite induction.

Let X be any non-empty set and $\mathscr{G} \subseteq \mathscr{P}(X)$. Let $\Omega = [0, \omega_1]$ be the ordinal space [☞ Section 2.4]. Define, by transfinite induction, the families

$$\mathscr{F}_0 := \mathscr{G}, \quad \mathscr{F}_1 := \mathscr{G}^{c\sigma}, \quad \dots, \quad \mathscr{F}_{\alpha+1} := (\mathscr{F}_\alpha)^{c\sigma}$$

and, if α is a limit ordinal, we set

$$\mathscr{F}_\alpha := \bigcup_{\beta<\alpha} \mathscr{F}_\beta; \quad \text{in particular, we have} \quad \mathscr{F}_\Omega := \mathscr{F}_{\omega_1} = \bigcup_{\alpha<\omega_1} \mathscr{F}_\alpha.$$

We claim that $\mathscr{F}_\Omega = \sigma(\mathscr{G})$. Clearly, $\mathscr{F}_\Omega$ is stable under the formation of complements. If $(F_n)_{n\in\mathbb{N}}$ is a sequence of sets from $\mathscr{F}_\Omega$, there are ordinals $\alpha_n < \omega_1$ such that $F_n \in \mathscr{F}_{\alpha_n}$ for each $n \in \mathbb{N}$. Set

$$F := \bigcup_{n\in\mathbb{N}} F_n \quad \text{and} \quad \alpha := \sup_{n\in\mathbb{N}} \alpha_n. \tag{4.9}$$

Since $\alpha < \omega_1$ – as a countable supremum of countable ordinals it is countable – we see that $F \in \mathscr{F}_{\alpha+1} \subseteq \mathscr{F}_\Omega$, thus $\mathscr{F}_\Omega$ is a σ-algebra.

By construction, $\mathscr{G} \subseteq \mathscr{F}_\Omega$, thus $\sigma(\mathscr{G}) \subseteq \mathscr{F}_\Omega$. We also have $\mathscr{F}_\Omega \subseteq \sigma(\mathscr{G})$ since each $F \in \mathscr{F}_\Omega$ is in some $\mathscr{F}_\alpha$ with $\alpha < \omega_1$, i.e. F can be obtained by countably many operations of complementation and countable unions of sets from $\mathscr{G}$.

Comment If we combine the above argument with the argument from Example 4.4, we can show that for any generator $\mathscr{G}$ such that $\#\mathscr{G} = \#\mathbb{N}$ the generated σ-algebra $\sigma(\mathscr{G})$ has the power of the continuum $\#\sigma(\mathscr{G}) = \#\mathscr{P}(\mathbb{N}) = \mathfrak{c}$.

Indeed, as in Example 4.4 we see that $\sigma(\mathscr{G})$ has countably infinitely many atoms

$$\mathscr{A}_0 := \Big\{A(x) := \bigcap_{G_n\in\mathscr{G}, G_n\ni x} G_n\,;\ x\in X\Big\} \subseteq \sigma(\mathscr{G}).$$

Since the elements in $\mathscr{A}_0$ are mutually disjoint, we see that $\#\sigma(\mathscr{A}_0) = \#\mathscr{P}(\mathbb{N})$, and we conclude that $\#\sigma(\mathscr{G}) \geqslant \mathfrak{c}$.

Conversely, we can count $\sigma(\mathscr{G}) = \bigcup_{\alpha<\omega_1} \mathscr{F}_\alpha$ using the families $\mathscr{F}_\alpha$. Clearly, $\#\mathscr{F}_1 = \#\mathbb{N}^\mathbb{N} = \mathfrak{c}$ and, by construction, $\#\mathscr{F}_2 \leqslant \mathfrak{c}^{\#\mathbb{N}} = \mathfrak{c}$. If $\alpha < \omega_1$, then $\mathscr{F}_\alpha$ has at most the cardinality of a countable sequence of families of cardinality $\mathfrak{c}$, hence $\#\mathscr{F}_\alpha \leqslant \mathfrak{c}$. Finally, since $\#\Omega \leqslant \mathfrak{c}$ we see that $\#\mathscr{F}_\Omega \leqslant \mathfrak{c}^2 = \mathfrak{c}$. For an alternative argument that $\#\sigma(\mathscr{G}) \leqslant \mathfrak{c}$ see the Corollary in Example 4.19.

In particular, there are continuum many Borel sets $\mathscr{B}(\mathbb{R})$, but there are $2^\mathfrak{c}$ many Lebesgue sets, see [MIMS, p. 429]. This is intuitively clear since the Lebesgue sets of, say, the real line contain **all** subsets of the Cantor set, and there are $\#\mathscr{P}(\mathbb{R}) = 2^\mathfrak{c}$ many of them.

4.23 (Non-)equivalent characterizations of the Baire σ-algebra

In a topological space $(X, \mathscr{O})$ the **Baire σ-algebra** $\mathscr{B}^\alpha(X)$ is the smallest σ-algebra $\mathscr{A}$ on X such that all continuous functions $f : (X, \mathscr{A}) \to (\mathbb{R}, \mathscr{B}(\mathbb{R}))$ are measurable. We call $B \in \mathscr{B}^\alpha(X)$ a **Baire (measurable) set.**[2] Since for $f \in C(X)$ the shifted function $f(\cdot) + a$, $a \in \mathbb{R}$, is again continuous, we have

$$\mathscr{B}^\alpha(X) = \sigma(\{f^{-1}(0, \infty)\,;\ f \in C(X)\}).$$

Moreover,

$$\mathscr{B}^\alpha(X) = \sigma(\{f^{-1}(\{0\})\,;\ f \in C(X)\}).$$

The inclusion $\supseteq$ is obvious, and the other inclusion follows from the fact that the continuous function $g(x) := \max\{0, f(x) - a\}$, $f \in C(X)$, $a \in \mathbb{R}$, satisfies $\{g = 0\} = \{f \leqslant a\}$. In the literature there are various alternative definitions of Baire sets, for example:

(a) $\mathscr{B}^\alpha_1(X)$: the smallest σ-algebra such that all continuous functions $f : X \to \mathbb{R}$ with compact support are measurable.
(b) $\mathscr{B}^\alpha_2(X)$: the smallest σ-algebra which contains all compact G_δ-sets.
(c) $\mathscr{B}^\alpha_3(X)$: the smallest σ-algebra which contains all closed G_δ-sets.

If X is a 'nice' topological space, then the families $\mathscr{B}^\alpha_i(X)$ coincide with $\mathscr{B}^\alpha(X)$.

[2] Baire sets are not to be confused with Baire spaces as they are used in category theory; see p. 40.

Theorem *Let $(X, \mathfrak{O})$ be a locally compact and σ-compact Hausdorff space, e.g. a Polish space. Then $\mathscr{B}_i^\alpha(X) = \mathscr{B}^\alpha(X)$ for $i = 1, 2, 3$.*

Proof For locally compact spaces there is the following version of Urysohn's lemma: if $K \subseteq X$ is compact and $U \subseteq X$ an open set containing K, then there exists a continuous function $\phi : X \to [0,1]$ such that $\phi|_K = 1$ and $\phi|_{X \setminus U} = 0$, cf. [53, Corollary 3.3.3]. Since X is σ-compact, we can find a sequence of compact sets $(K_n)_{n \in \mathbb{N}} \subseteq X$ with $K_n \uparrow X$. Compact subsets of Hausdorff spaces are closed, cf. [200, Theorem 17.5], and therefore we may apply Urysohn's lemma with $K = K_n$ and $U = X \setminus K_{n+1}$ to obtain $\phi_n \in C_c(X)$ such that $\phi_n|_{K_n} = 1$.

1⁰ $\mathscr{B}^\alpha(X) \subseteq \mathscr{B}_1^\alpha(X)$: If $f \in C(X)$, then $f_n := f\phi_n \in C_c(X)$ and $f_n \to f$ pointwise. Consequently, f is $\mathscr{B}_1^\alpha(X)$ measurable as pointwise limit of $\mathscr{B}_1^\alpha(X)$ measurable functions.

2⁰ $\mathscr{B}_1^\alpha(X) \subseteq \mathscr{B}_2^\alpha(X)$: For fixed $a \in \mathbb{R}$ and $f \in C_c(X)$ we define continuous functions $f_n(x) := \min\{f(x) - a, \phi_n(x) - 1\}$. Then

$$\{f_n \geqslant 0\} = \underbrace{\{f \geqslant a\}}_{\text{closed}} \cap \underbrace{\{\phi_n \geqslant 1\}}_{\text{compact}} \quad \text{and} \quad \{f_n \geqslant 0\} = \bigcap_{k \in \mathbb{N}} \underbrace{\left\{f_n > -\frac{1}{k}\right\}}_{\text{open}},$$

showing that $\{f_n \geqslant 0\}$ is a compact G_δ-set. Hence, $\{f_n \geqslant 0\} \in \mathscr{B}_2^\alpha(X)$, and $\{f \geqslant a\} = \bigcup_{n \in \mathbb{N}} \{f_n \geqslant 0\} \in \mathscr{B}_2^\alpha(X)$.

3⁰ $\mathscr{B}_2^\alpha(X) \subseteq \mathscr{B}_3^\alpha(X)$: This is obvious since compact subsets of the Hausdorff space X are closed, cf. [200, Theorem 17.5].

4⁰ $\mathscr{B}_3^\alpha(X) \subseteq \mathscr{B}^\alpha(X)$: let $F = \bigcap_{n \in \mathbb{N}} U_n$ be a closed G_δ-set. By Urysohn's lemma, there exists for each $n \in \mathbb{N}$ a continuous function $f_n : X \to [0,1]$ such that $f_n|_{F \cap K_n} = 1$ and $f_n|_{X \setminus U_n} = 0$. Thus,

$$F = \bigcap_{n \in \mathbb{N}} \{f_n = 0\}^c \in \mathscr{B}^\alpha(X).$$ □

The following examples show that the equivalence of these definitions breaks down in a general topological space.

$\mathscr{B}^\alpha(X) \neq \mathscr{B}_1^\alpha(X)$: Consider $X = \mathbb{R}$ with the discrete metric $d(x,y) = 0$ if $x = y$ and $d(x,y) = 1$ if $x \neq y$ [☞ Example 2.1.(b)]. Since every set $A \subseteq X$ is open, it follows that a set $K \subseteq X$ is compact if, and only if, it has at most finitely many elements. This means that every compactly supported function $f : X \to \mathbb{R}$ is zero except for finitely many points $x \in X$, and therefore $\mathscr{B}_1^\alpha(X)$ is contained in the co-countable σ-algebra [☞ Example 4A],

$$\Sigma := \{A \subseteq X;\ A \text{ or } X \setminus A \text{ is at most countable}\}.$$

On the other hand, $B := [0,1]$ is a Baire set since $B = X \setminus f^{-1}(0,\infty)$ for the continuous function

$$f(x) := d(x,[0,1]) := \inf\{|x-y|\,;\ y \in [0,1]\}, \quad x \in \mathbb{R}.$$

Since both B and B^c are uncountable, we have $B \notin \Sigma$ which implies that $\mathscr{B}^\alpha(X) \supsetneqq \mathscr{B}_1^\alpha(X)$.

$\mathscr{B}^\alpha(X) \neq \mathscr{B}_2^\alpha(X)$: Take $X = \mathbb{N}$ with the co-finite topology [☞ Example 2.1.(d)]. In this topology, every set $A \subseteq X$ is compact and can be written as a countable intersection of open sets; consequently, $\mathscr{P}(X)$ is the smallest σ-algebra containing all compact G_δ-sets. Since the only continuous functions $f : X \to \mathbb{R}$ are the constant functions, we conclude that $\mathscr{B}^\alpha(X) = \{\emptyset, X\}$.

$\mathscr{B}^\alpha(X) \neq \mathscr{B}_3^\alpha(X)$: There exists a Hausdorff space X which contains two points $x \neq y$ such that $\{x\}, \{y\}$ are G_δ-sets and

$$\forall f \in C(X) \ : \ f(x) = f(y), \tag{4.10}$$

cf. Thomas [184]. As $\{x\}$ is closed, it follows that $\{x\}$ is a closed G_δ-set, hence, $\{x\} \in \mathscr{B}_3^\alpha(X)$. We claim that $\{x\}$ is not a Baire set. In order to prove this, define

$$\Sigma := \{A \subseteq X\,;\ x \in A,\ y \in A \text{ or } x \notin A, y \notin A\},$$

and note that Σ is a σ-algebra. Moreover, (4.10) gives

$$x \in f^{-1}(0,\infty) \Rightarrow y \in f^{-1}(0,\infty)$$

for $f \in C(X)$, so $f^{-1}(0,\infty) \in \Sigma$ for $f \in C(X)$ implying that

$$\mathscr{B}^\alpha(X) = \sigma(\{f^{-1}(0,\infty)\,;\ f \in C(X)\}) \subseteq \sigma(\Sigma) = \Sigma.$$

In particular, $\{x\} \notin \mathscr{B}^\alpha(X)$.

4.24 The Baire σ-algebra can be strictly smaller than the Borel σ-algebra

Let $(X, \mathscr{O})$ be a topological space with Borel σ-algebra $\mathscr{B}(X)$ and Baire σ-algebra $\mathscr{B}^\alpha(X)$, i.e.

$$\mathscr{B}^\alpha(X) = \sigma(\{f^{-1}(0,\infty)\,;\ f \in C(X)\}).$$

Since $f^{-1}(0,\infty) \in \mathscr{B}(X)$ for every $f \in C(X)$, it is clear that $\mathscr{B}^\alpha(X) \subseteq \mathscr{B}(X)$. Moreover, if X is a metric space, every closed set $F \subseteq X$ can be written as $F = X \setminus f^{-1}(0,\infty)$ for $f \in C(X)$ – take $f(x) := d(x,F)$ – and it follows that $\mathscr{B}(X) = \mathscr{B}^\alpha(X)$. For general topological spaces the Baire σ-algebra may be strictly smaller than the Borel σ-algebra.

Consider the product space $X = \mathbb{R}^{[0,1]}$ with the product topology. By definition, we can identify each element $x \in X$ with a mapping $x : [0,1] \to \mathbb{R}$. If $f : X \to \mathbb{R}$ is a continuous function, then f depends on at most countably many indices, that is, there exists an at most countable set $S \subseteq [0,1]$ such that

$$x, y \in X,\ x|_S = y|_S \Rightarrow f(x) = f(y), \tag{4.11}$$

cf. [53, Problem 2.7.12(d)] or [146, Theorem 4]. This property carries over to all functions f which are of the form $f(x) = \mathbb{1}_A(x)$ for some $A \in \mathscr{B}^\alpha(X)$. In order to prove this, define

$$\Sigma := \left\{ A \subseteq X\,;\ \begin{array}{l} x \longmapsto f(x) := \mathbb{1}_A(x) \text{ satisfies (4.11)} \\ \text{for some countable set } S \subseteq [0,1] \end{array} \right\}.$$

Let us verify that Σ is a σ-algebra.

(Σ_1) $A = \emptyset \in \Sigma$ and $X \in \Sigma$ is trivial.

(Σ_2) Let $A \in \Sigma$ and choose $S \subseteq [0,1]$ such that (4.11) holds for $f = \mathbb{1}_A$ and $f = \mathbb{1}_X$. If $x, y \in X$ are such that $x|_S = y|_S$, then

$$\mathbb{1}_{A^c}(x) = \mathbb{1}_X(x) - \mathbb{1}_A(x) = \mathbb{1}_X(y) - \mathbb{1}_A(y) = \mathbb{1}_{A^c}(y),$$

i.e. (4.11) holds for $f = \mathbb{1}_{A^c}$. Thus, $A^c \in \Sigma$.

(Σ_3) For $(A_n)_{n\in\mathbb{N}} \subseteq \Sigma$ denote by $(S_n)_{n\in\mathbb{N}}$ the respective countable index sets. Set $A := \bigcup_{n\in\mathbb{N}} A_n$ and $S := \bigcup_{n\in\mathbb{N}} S_n$. By definition of S_n we have

$$\mathbb{1}_A(x) = \min\left\{1, \sum_{n=1}^{\infty} \mathbb{1}_{A_n}(x)\right\} = \min\left\{1, \sum_{n=1}^{\infty} \mathbb{1}_{A_n}(y)\right\} = \mathbb{1}_A(y)$$

for all $x, y \in X$ with $x|_S = y|_S$, and we conclude that $A \in \Sigma$.

Since (4.11) holds for $f \in C(X)$, it follows that $f^{-1}(0,\infty) \in \Sigma$. Hence,

$$\mathscr{B}^\alpha(X) = \sigma(\{f^{-1}(0,\infty)\,;\ f \in C(X)\}) \subseteq \sigma(\Sigma) = \Sigma.$$

The Borel sets are, in general, **not** determined by countably many indices, i.e. $\mathscr{B}(X) \subsetneqq \Sigma$. For instance, $B = \{\mathbb{1}_{[0,1]}\}$ is a closed set and hence $B \in \mathscr{B}(X)$. However, there is no countable set $S \subseteq [0,1]$ such that (4.11) holds for $f = \mathbb{1}_B$, i.e. $B \notin \mathscr{B}^\alpha(X)$. In order to see this, take any countable set $S \subseteq [0,1]$ and consider $x := \mathbb{1}_{[0,1]}$ and $y := \mathbb{1}_S$. Clearly, $x|_S = y|_S$, but $\mathbb{1}_B(x) = 1 \neq 0 = \mathbb{1}_B(y)$. Thus, $B \in \mathscr{B}(X) \setminus \Sigma \subseteq \mathscr{B}(X) \setminus \mathscr{B}^\alpha(X)$.

5

Set Functions and Measures

Let $X \neq \emptyset$ be any set and $\mathscr{G} \subseteq \mathscr{P}(X)$ a family of subsets of X. Throughout we assume that $\emptyset \in \mathscr{G}$. A **(positive) set function** is a map $\mu : \mathscr{G} \to [0, \infty]$. We assume that $\mu(\emptyset) = 0$. The property

$$\left.\begin{array}{r}(G_n)_{n\in\mathbb{N}} \subseteq \mathscr{G} \ \text{ pairwise disjoint} \\ \text{and } G = \bigcup\!\!\!\cdot_{n\in\mathbb{N}} G_n \in \mathscr{G}\end{array}\right\} \Rightarrow \mu(G) = \sum_{n\in\mathbb{N}} \mu(G_n) \tag{5.1}$$

is called **σ-additivity (relative to $\mathscr{G}$)**. A σ-additive set function is said to be a **pre-measure** or a **measure relative to** $\mathscr{G}$. A **measure** [☞ Definition 1.6] is a σ-additive set function defined on a σ-algebra [☞ Definition 1.5]. If $\mu(X) = 1$, then μ is said to be a **probability measure**. A **Borel measure** is a measure defined on the Borel σ-algebra $\mathscr{B}(X)$ [☞ Example 1.7.(k)] of a topological space X. Every σ-additive set function is also **(finitely) additive**, i.e. $\mu(G \mathbin{\dot\cup} H) = \mu(G) + \mu(H)$ for all $G, H \in \mathscr{G}$ with $G \mathbin{\dot\cup} H \in \mathscr{G}$. Some authors call a (finitely) additive set function a **content**. An additive set function is **finite** if $\mu(X) < \infty$ and σ-**finite** if there exists an increasing sequence of sets $G_n \in \mathscr{G}$, $G_n \uparrow X$ such that $\mu(G_n) < \infty$ for all $n \in \mathbb{N}$. If there exists for every $G \in \mathscr{G}$ a set $H \in \mathscr{G}$ such that $H \subseteq G$ and $0 < \mu(H) < \infty$, then μ has the **finite subset property**. For further properties and examples of measures we refer to the discussion in Section 1.3 of Chapter 1.

Sometimes it is useful to extend a measure μ to all subsets of X. This can be achieved by considering the **inner** and **outer measures** induced by the measure μ:

$$\begin{aligned} &\mu^*(E) := \inf\{\mu(A);\ A \supseteq E,\ A \in \mathscr{A}\} && \text{outer measure,} \\ &\mu_*(E) := \sup\{\mu(A);\ A \subseteq E,\ A \in \mathscr{A},\ \mu(A) < \infty\} && \text{inner measure.} \end{aligned}$$

Clearly, $\mu_* \leqslant \mu^*$, $\mu^*|_{\mathscr{A}} = \mu_*|_{\mathscr{A}} = \mu$; a full discussion of (general) inner and outer measures can be found at the beginning of Chapter 9.

The following two examples of 'exotic' measures play a central role in the counterexamples of this section.

Example 5A Let X be any uncountable set and $\mathscr{A}$ the co-countable σ-algebra [☞Example 4A]. The set function

$$\mu(A) := \begin{cases} 0, & \text{if } A \text{ is at most countable,} \\ 1, & \text{if } A^c \text{ is at most countable,} \end{cases}$$

is a measure on $(X, \mathscr{A})$. Since X is uncountable, there is no $A \subseteq X$ such that both A and A^c are countable, and so μ is well-defined. Moreover, $\mu(\emptyset) = 0$ is obvious. To prove σ-additivity, fix a sequence $(A_n)_{n\in\mathbb{N}} \subseteq \mathscr{A}$ of pairwise disjoint sets and set $A := \dot{\bigcup}_{n\in\mathbb{N}} A_n$. The following two cases can occur:

Case 1. A_n is countable for all $n \in \mathbb{N}$. In this case we have $\mu(A_n) = 0$ for all $n \in \mathbb{N}$ and $A = \bigcup_{n\in\mathbb{N}} A_n$ is countable. Thus,

$$\mu(A) = 0 = \sum_{n\in\mathbb{N}} \mu(A_n).$$

Case 2. A_n is not countable for some $n \in \mathbb{N}$. Since the sets are pairwise disjoint, we have $A_i \subseteq A_n^c$ for all $i \neq n$, and so $A_i \subseteq A_n^c$ is countable for $i \neq n$, i.e. $\mu(A_i) = 0$. Moreover, $A^c \subseteq A_n^c$ implies that $\mu(A) = 1$. Hence,

$$\mu(A) = 1 = \mu(A_n) = \sum_{k\in\mathbb{N}} \mu(A_k).$$

Example 5B (Dieudonné measure [46]) Let $\Omega = [0, \omega_1]$ be the ordinal space [☞Section 2.4] and consider the following σ-algebra [☞Example 4B]:

$$\mathscr{A} := \{A \subseteq \Omega;\ A \cup \{\omega_1\} \text{ or } A^c \cup \{\omega_1\} \text{ contains an uncountable compact set}\}.$$

On $(\Omega, \mathscr{A})$ the **Dieudonné measure** μ is defined as follows:

$$\mu(A) := \begin{cases} 0, & \text{if } A^c \cup \{\omega_1\} \text{ contains an uncountable compact set } C, \\ 1, & \text{if } A \cup \{\omega_1\} \text{ contains an uncountable compact set } C. \end{cases}$$

Since it cannot happen that both $A \cup \{\omega_1\}$ and $A^c \cup \{\omega_1\}$ contain an uncountable compact set, the set function μ is well-defined: otherwise there would be two disjoint uncountable compact sets, which is not possible; see (4.2) in Example 4B. Let us check that μ is a measure. Since the ordinal space Ω is compact, cf. Section 2.4, it is immediate that $\mu(\emptyset) = 0$. If $(A_n)_{n\in\mathbb{N}} \subseteq \mathscr{A}$ is a sequence of pairwise disjoint sets, then one of the following alternatives holds:

Case 1. There exists some $n \in \mathbb{N}$ such that $A_n \cup \{\omega_1\}$ contains an uncountable compact set C. Set $A := \bigcup_{n\in\mathbb{N}} A_n$, then $A \cup \{\omega_1\} \supseteq C$, and so $\mu(A) = 1$. Since there do not exist two uncountable compact subsets of Ω which are pairwise disjoint, cf. (4.2) in Example 4B, it follows from $A_i \subseteq \Omega \setminus C$ that $\mu(A_i) = 0$ for all $i \neq n$. Hence,

$$\mu(A) = \mu(A_n) = \sum_{i\geqslant 1} \mu(A_i).$$

Case 2. There does not exist an $n \in \mathbb{N}$ such that $A_n \cup \{\omega_1\}$ contains an uncountable compact set, i.e. $\mu(A_n) = 0$ for all $n \in \mathbb{N}$. Then $A_n^c \cup \{\omega_1\} \supseteq C_n$ for an uncountable compact set C_n, and we have for $A = \bigcup_{n\geqslant 1} A_n$,

$$A^c \cup \{\omega_1\} = \bigcap_{n\geqslant 1} (A_n^c \cup \{\omega_1\}) \supseteq \bigcap_{n\geqslant 1} C_n =: C.$$

Being an intersection of compact sets, C is compact; moreover it is uncountable; see (4.2) in Example 4B. Thus,

$$\mu(A) = 0 = \sum_{n\geqslant 1} \mu(A_n).$$

5.1 A class of measures where the $\mu(\emptyset) = 0$ is not needed in the definition

Assume that $\mu\colon \mathscr{A} \to [0,\infty]$ is a σ-additive set function on a σ-algebra; if $\mu(\emptyset) = 0$, then μ is a measure. We claim that μ is a measure if, and only if, there is some $A \in \mathscr{A}$ with $\mu(A) < \infty$.

If there is no such set A, then $\mu(A) = \infty$ for all $A \in \mathscr{A}$; in particular, $\mu(\emptyset) = \infty$ and μ cannot be a measure.

Otherwise, if there is some $A \in \mathscr{A}$ with $\mu(A) < \infty$, we get

$$\infty > \mu(A) = \mu(A \cup \emptyset \cup \emptyset \cup \dots) = \mu(A) + \sum_{i=2}^{\infty} \mu(\emptyset)$$

and this shows that we have necessarily $\mu(\emptyset) = 0$.

5.2 A set function which is additive but not σ-additive

Consider the natural numbers $\mathbb{N}$ and the power set $\mathscr{P}(\mathbb{N})$. The set function

$$\mu(A) := \begin{cases} \sum_{n\in A} 2^{-n}, & \text{if } A \in \mathscr{P}(\mathbb{N}),\ \#A < \infty, \\ +\infty, & \text{if } A \in \mathscr{P}(\mathbb{N}),\ \#A = \infty, \end{cases}$$

is monotone, additive but not σ-additive. In order to see additivity, take disjoint sets $A, B \in \mathscr{P}(\mathbb{N})$. If $\#A + \#B = \infty$, we have $\mu(A) + \mu(B) = \infty = \mu(A \cupdot B)$. If A and B are finite, we have

$$\mu(A \cupdot B) = \sum_{n \in A \cupdot B} 2^{-n} = \sum_{n \in A} 2^{-n} + \sum_{n \in B} 2^{-n} = \mu(A) + \mu(B).$$

If we take $A = \mathbb{N} = \bigcupdot_{n \in \mathbb{N}} \{n\}$ then

$$\infty = \mu(\mathbb{N}) \neq \sum_{n \in \mathbb{N}} \mu(\{n\}) = \sum_{n \in \mathbb{N}} 2^{-n} = 1,$$

which shows that μ is not σ-additive. Thus, μ is only finitely additive.

5.3 A finite set function which is additive but not σ-additive

Consider on $\mathbb{N}$ the co-finite algebra $\mathscr{A} := \{A \subseteq \mathbb{N}\,;\ \#A < \infty \text{ or } \#A^c < \infty\}$ and define a set function

$$\mu(A) := \begin{cases} 0, & \text{if } A \in \mathscr{A},\ \#A < \infty, \\ 1, & \text{if } A \in \mathscr{A},\ \#A = \infty. \end{cases}$$

Let $A, B \in \mathscr{A}$ be disjoint. Since $A \subseteq B^c$ we cannot have both sets infinite, i.e. additivity can be reduced to the following two cases:

Case 1. $\#A < \infty, \#B < \infty$. Then $\mu(A \cupdot B) = 0 = \mu(A) + \mu(B)$.
Case 2. $\#A < \infty, \#B = \infty$. Then $\mu(A \cupdot B) = 1 = \mu(A) + \mu(B)$.

If we take $A_n := \{n\}$, then σ-additivity fails:

$$1 = \mu(\mathbb{N}) \neq \sum_{n \in \mathbb{N}} \mu(\{n\}) = 0.$$

Comment It is possible to characterize **all** finitely additive set functions τ on the measure space $(X, \mathscr{P}(X))$ taking values 0 and 1 only. Let τ be a finitely additive zero–one set function and set $\mathscr{U} := \{A \subseteq X\,;\ \tau(A) = 1\}$. The family $\mathscr{U}$ is a **filter**:

(F1) $\emptyset \notin \mathscr{U}$;
(F2) if $A \in \mathscr{U}$ and $A \subseteq B$, then $B \in \mathscr{U}$;
(F3) if $A, B \in \mathscr{U}$, then $A \cap B \in \mathscr{U}$;

and even an **ultrafilter**, i.e. we have in addition to (F1)–(F3) the axiom

(UF) either A or A^c is in $\mathscr{U}$.

Only the third property (F3) needs a minute's thought. It follows from finite additivity, as $\tau(A\cap B) = \tau(A)+\tau(B)-\tau(A\cup B)$ and $\tau(A) = \tau(B) = \tau(A\cup B) = 1$.

Conversely, if $\mathcal{U}$ is an ultrafilter, satisfying the above axioms (F1)–(F3) and (UF), then the set function

$$\tau(A) := \begin{cases} 1, & \text{if } A \in \mathcal{U}, \\ 0, & \text{if } A \notin \mathcal{U}, \end{cases}$$

is a finitely additive zero–one set function. The proof is immediate.

Consequently, there is a one-to-one correspondence between ultrafilters and finitely additive set functions which have values zero and one.

Ultrafilters are the maximal elements in the family of filters; thus, it is not surprising that the existence of ultrafilters needs Zorn's lemma [☞ Theorem 2.8]. In topological spaces filters are used to define convergence. For a discussion of (ultra-)filters in topology we refer to [200, Chapter 12] and [53, Chapter 1.6].

5.4 Another finite set function which is additive but not σ-additive

Let λ be Lebesgue measure on $((0,\infty), \mathcal{B}(0,\infty))$ and define the function

$$\nu(A) := \operatorname*{Lim}_{n\to\infty} \frac{1}{n}\lambda(A \cap (0,n)), \quad A \in \mathcal{B}(0,\infty). \tag{5.2}$$

The limit Lim_n appearing in the above definition is a Banach limit, i.e. a shift-invariant, continuous linear functional on the space of bounded sequences ℓ^∞; see e.g. Rudin [151, pp. 85–86]. Its existence relies on the Hahn–Banach theorem, hence the axiom of choice.

Clearly, ν is a monotone set function which is additive since the limit is linear. Moreover, $\nu(A) \leqslant 1$. If we take $A_k = (k, k+1]$, we get

$$\nu(0,\infty) = \operatorname*{Lim}_{n\to\infty} \frac{1}{n}\lambda(0,n) = 1 \quad \text{and} \quad \nu(A_k) = \operatorname*{Lim}_{n\to\infty} \frac{1}{n}\lambda(A_k \cap (0,n)) = 0,$$

which means that

$$1 = \nu(0,\infty) = \nu\left(\dot{\bigcup_{k\geqslant 0}} A_k\right) \neq \sum_{k=0}^{\infty} \nu(A_k) = 0.$$

Comment We may replace in the above construction (5.2) the Banach limit Lim_n by the ordinary limit $\lim_n$ if we restrict ourselves to the class

$$\mathcal{C} := \left\{ A \in \mathcal{B}(0,\infty);\ \exists \lim_{n\to\infty} \frac{1}{n}\lambda(A\cap(0,n)) \right\}.$$

However, ν is not σ-additive on $\mathscr{C}$; worse, $\mathscr{C}$ is not a ring – consider, say, the sets

$$A = \bigcup_{k\in\mathbb{N}} [2k, 2k+1), \quad B = \bigcup_{k\in\mathbb{N}} [2k-1, 2k)$$

$$\text{and} \quad A' = \bigcup_{n\in\mathbb{N}} \{A \cap [2^{2n}, 2^{2n+1})\} \cup \{B \cap [2^{2n-1}, 2^{2n})\},$$

and check whether $A, A', A \cap A'$ appear in $\mathscr{C}$.

5.5 A set function with infinitely many extensions

The following example shows that a set function which is defined on too small a family of sets may have more than one extension to a measure. Let X be any set and take $A, B \subseteq X$ such that $A \cap B \neq \emptyset$, $A \setminus B \neq \emptyset$ and $B \setminus A \neq \emptyset$. We can define a set function on the family $\{\emptyset, A, B\}$ by setting

$$\nu(\emptyset) = 0, \ \nu(A) = \nu(B) = 1.$$

The smallest ring $\mathscr{R}$ containing the sets A and B, i.e. the ring generated by $\{A, B\}$, comprises the sets

$$\emptyset, \ A, \ B, \ A \cup B, \ A \cap B, \ A \setminus B, \ B \setminus A, \ (A \setminus B) \cup (B \setminus A).$$

Let $\alpha \in [0,1]$ be arbitrary and define a set function $\mu : \mathscr{R} \to [0, \infty]$ by

$$\mu(R) := \begin{cases} 0, & \text{if } R = \emptyset, \\ 1, & \text{if } R = A \text{ or } R = B, \\ \alpha, & \text{if } R = A \cap B, \\ 2-\alpha, & \text{if } R = A \cup B, \\ 1-\alpha, & \text{if } R = A \setminus B \text{ or } R = B \setminus A, \\ 2-2\alpha, & \text{if } R = (A \setminus B) \cup (B \setminus A). \end{cases}$$

This shows that ν has infinitely many extensions to the ring generated by the sets in the domain of ν. Since each μ can be extended to a measure on the σ-algebra $\sigma(\mathscr{R}) = \sigma(\{A, B\})$, this shows that ν has infinitely many extensions. Consequently, the σ-finiteness assumption in the uniqueness of measures, resp. extension of measures, theorem [☞Theorems 1.40, 1.41] is really needed.

5.6 A measure that cannot be further extended

Let X be an infinite set and pick a decreasing sequence $(A_n)_{n\in\mathbb{N}}$ of non-empty subsets $A_n \subseteq X$ such that $\bigcap_{n\in\mathbb{N}} A_n = \emptyset$. Write $\mathscr{A}_n := \sigma(A_1, \dots, A_n)$ and define a set function on $\mathscr{G} := \bigcup_{n\in\mathbb{N}} \mathscr{A}_n$ in the following way:

$$\mu(G) = \begin{cases} 1, & \text{if there is some } n \in \mathbb{N} \text{ such that } A_n \subseteq G, \\ 0, & \text{otherwise.} \end{cases}$$

Since each $\mathscr{A}_n$ is a σ-algebra (it is even a finite σ-algebra) it is not hard to check that $\mu|_{\mathscr{A}_n}$ is a measure. Notice that σ-additivity and additivity coincide since $\mathscr{A}_n$ has only finitely many elements.

In particular, μ is finitely additive on $\mathscr{G}$, but it is not σ-additive on $\mathscr{G}$ nor on $\sigma(\mathscr{G})$ since, by construction,

$$\mu\left(\bigcap_{n\in\mathbb{N}} A_n\right) = \mu(\emptyset) = 0 \neq 1 = \inf_{n\in\mathbb{N}} \mu(A_n).$$

5.7 A measure defined on the open balls which cannot be extended to the Borel sets

Let $X = \mathbb{R}$, $\mathscr{A}$ the co-countable σ-algebra [☞ Example 4A]

$$\mathscr{A} := \{A \subseteq X;\ A \text{ or } A^c \text{ is at most countable}\}$$

and μ the (probability) measure [☞ Example 5A]

$$\mu(A) = \begin{cases} 0, & \text{if } A \in \mathscr{A} \text{ is at most countable,} \\ 1, & \text{otherwise.} \end{cases}$$

We will now make X into a metric space by introducing the discrete metric d: $d(x,y) = 0$ if $x = y$ and $d(x,y) = \infty$ if $x \neq y$. Under this metric, the open balls (with finite radii) are exactly the singletons $\{x\}$, $x \in X$. On the other hand, every set $B \subseteq X$ is open (and closed). Thus, the σ-algebra generated by the open balls is the co-countable σ-algebra $\mathscr{A}$, while $\mathscr{B}(X)$ coincides with the power set $\mathscr{P}(X)$ [☞ 4.14].

This means that μ is a measure on the σ-algebra generated by the open balls. Assuming the continuum hypothesis, i.e. $\mathfrak{c} = \aleph_1$ [☞ p. 43], it can be shown that μ cannot be extended to a measure on the Borel σ-algebra $\mathscr{B}(X) = \mathscr{P}(X)$ [☞ 6.15].

5.8 A signed pre-measure on an algebra $\mathscr{R}$ which cannot be extended to a signed measure on $\sigma(\mathscr{R})$

Let $X = \mathbb{R}$ and consider the algebra

$$\mathscr{R} := \{F \subseteq \mathbb{R};\ \#F < \infty \text{ or } \#F^c < \infty\}.$$

Recall that a (signed) pre-measure is a (signed) σ-additive set function defined on a family of sets which is not necessarily a σ-algebra. Define for $F \in \mathscr{R}$

$$\mu(F) := \begin{cases} \#F \cap (-\infty, 0] - \#F \cap (0, \infty), & \text{if } \#F < \infty, \\ -\mu(F^c), & \text{if } \#F^c < \infty. \end{cases}$$

Clearly, $\mu(\emptyset) = 0$. Let $(F_n)_{n\in\mathbb{N}} \subseteq \mathscr{R}$ be a sequence of disjoint sets such that $F := \bigcupdot_{n\in\mathbb{N}} F_n \in \mathscr{R}$. If $\#F < \infty$, then at most finitely many of the F_n are non-empty, and σ-additivity is trivial. If $\#F^c < \infty$, then the disjointness implies that $\#F_{n_0}^c < \infty$ for exactly one index n_0. This means that $\#F_n < \infty$ for all $n \neq n_0$ and at most finitely of them are non-empty. Since

$$F_{n_0}^c = \Big(F \setminus \bigcupdot_{n\neq n_0} F_n\Big)^c = F^c \cupdot \bigcupdot_{n\neq n_0} F_n$$

is a union of disjoint sets and $F_{n_0}^c$ is finite, we get

$$-\mu(F_{n_0}) = \mu(F_{n_0}^c) = -\mu(F) + \sum_{n\neq n_0} \mu(F_n) \quad \text{therefore} \quad \mu(F) = \sum_{n\in\mathbb{N}} \mu(F_n).$$

This shows that μ is σ-additive relative to $\mathscr{R}$, i.e. a (signed) pre-measure on $\mathscr{R}$.

Since μ is unbounded from above and below, there is no σ-additive extension to $\sigma(\mathscr{R})$. This follows from the Hahn–Jordan decomposition of signed measures [☞ Rem. 1.46, Theorem 1.47]: $\mu = \mu^+ - \mu^-$ and at least one of the measures $\mu^\pm$ must be a finite measure.

5.9 A measure defined on a non-measurable set

Let $(X, \mathscr{A}, \mu)$ be a measure space with $\mathscr{A} \neq \mathscr{P}(X)$. Take $N \notin \mathscr{A}$ such that the inner measure [☞ p. 184] satisfies $\mu_*(N^c) = 0$. Define

$$\mathscr{A}_N := N \cap \mathscr{A} = \{N \cap A;\ A \in \mathscr{A}\} \quad \text{and} \quad \mu_N(A_N) := \mu(A), \quad A_N \in \mathscr{A}_N.$$

The family $\mathscr{A}_N$ is the trace σ-algebra [☞ Example 1.7.(e)] and it is not difficult to see that μ_N is a measure, once we know that μ_N is well-defined. In order to see this, assume that $N \cap A = N \cap B \neq \emptyset$ for $A, B \in \mathscr{A}$. As $A \setminus B \subseteq N^c$, it follows from the monotonicity of the inner measure that

$$\mu(A \setminus B) = \mu_*(A \setminus B) \leqslant \mu_*(N^c) = 0.$$

Analogously, $\mu(B \setminus A) = 0$. Consequently, $\mu(A) = \mu(B)$, and so $\mu_N(A_N)$ is independent of the representation of $A_N \in \mathscr{A}_N$.

Comment If μ is a finite measure, then the condition $\mu_*(N^c) = 0$ is usually stated in the form $\mu^*(N) = \mu(X)$, i.e. N is a set with 'full' outer measure. This follows easily from the identity $\mu(X) = \mu^*(N) + \mu_*(X \setminus N)$ [☞Theorem 9E].

5.10 A measure which is not continuous from above

Let λ be Lebesgue measure on $(\mathbb{R}, \mathscr{B}(\mathbb{R}))$ and consider the sets $A_n := [n, \infty)$. Then $A_n \downarrow \bigcap_{i=1}^{\infty} A_i = \emptyset$, but the sequence $\lambda(A_n) = \infty$ does not converge to 0.

5.11 A σ-finite measure which is not σ-finite on a smaller σ-algebra

Consider on $\mathbb{R}$ the Borel sets $\mathscr{B}(\mathbb{R})$ and the co-countable σ-algebra [☞Example 4A]

$$\mathscr{A} = \{A \subseteq \mathbb{R};\ A \text{ or } A^c \text{ is countable}\}.$$

It is clear that $\mathscr{A} \subseteq \mathscr{B}(\mathbb{R})$. Thus, we can consider Lebesgue measure on either of these σ-algebras. While on $\mathscr{B}(\mathbb{R})$ Lebesgue measure is σ-finite, this is not the case on $\mathscr{A}$, since $\lambda|_{\mathscr{A}}$ can take only the values 0 and ∞.

Comment If $(X, \mathscr{A}, \mu)$ is a σ-finite space with $\mu(X) = \infty$, then the restriction of μ to the trivial σ-algebra $\{\emptyset, X\}$ fails to be σ-finite.

5.12 A σ-finite measure μ on $\mathscr{B}(\mathbb{R})$ such that $\mu(I) = \infty$ for every non-trivial interval

Consider the measure $\mu(B) := \zeta_{\mathbb{Q}}(B) := \#(B \cap \mathbb{Q})$, $B \in \mathscr{B}(\mathbb{R})$, which counts the number of rational points in B. Clearly, $\zeta_{\mathbb{Q}}(a, b) = \infty$ for every non-empty open interval $(a, b) \neq \emptyset$, hence for all non-empty intervals. If $(q_n)_{n\in\mathbb{N}}$ is an enumeration of $\mathbb{Q}$, then the sets $B_n := (\mathbb{R} \setminus \mathbb{Q}) \cup \{q_1, \dots, q_n\}$ satisfy

$$B_n \uparrow \mathbb{R} \quad \text{and} \quad \zeta_{\mathbb{Q}}(B_n) = n,$$

i.e. $\zeta_{\mathbb{Q}}$ is σ-finite.

5.13 A σ-finite measure μ on $\mathscr{B}(\mathbb{R})$ which is not a Lebesgue–Stieltjes measure

The measure μ from the previous example [☞5.12] cannot be written in the form $\mu(a, b] = m(b) - m(a)$, i.e. it is not a Lebesgue–Stieltjes measure.

5.14 Infinite sums of finite measures need not be σ-finite

A measure μ on a measurable space $(X, \mathscr{A})$ is called **s-finite** if there are finite measures μ_n, $n \in \mathbb{N}$, on $(X, \mathscr{A})$ such that $\mu = \sum_{n\in\mathbb{N}} \mu_n$. S-finite measures need not be σ-finite. Consider, for instance, $X = \{0\}$ with the σ-algebra $\mathscr{A} := \{\emptyset, X\}$ and $\mu_n := \delta_0$ for $n \in \mathbb{N}$.

Comment The converse is always true: any σ-finite measure is s-finite. For a given σ-finite measure μ on $(X, \mathscr{A})$ choose $(A_n)_{n\in\mathbb{N}} \subseteq \mathscr{A}$ with $A_n \uparrow X$ and $\mu(A_n) < \infty$. Then the sets

$$F_1 := A_1, \qquad F_n := A_n \setminus \bigcup_{i=1}^{n-1} A_i, \quad n \geqslant 2,$$

are pairwise disjoint, have finite measure and $X = \dot{\bigcup}_{n\in\mathbb{N}} F_n$. Consequently, $\mu_n := \mu(\cdot \cap F_n)$ are finite measures and $\mu = \sum_{n\in\mathbb{N}} \mu_n$.

5.15 The image measure of a σ-finite measure is not necessarily σ-finite

The counting measure $\zeta := \sum_{(x,y)\in\mathbb{Z}^2} \delta_{(x,y)}$ on $(\mathbb{Z}^2, \mathscr{P}(\mathbb{Z}^2))$ is σ-finite, but its image measure under the canonical projection

$$p : \mathbb{Z}^2 \to \mathbb{Z}, \quad p(x, y) := x$$

is not σ-finite since $p_*\zeta(A) = \zeta(p^{-1}(A)) = \zeta(A \times \mathbb{Z}) = \infty$ for any non-empty $A \subseteq \mathbb{Z}$.

5.16 A locally finite measure need not be σ-finite

A measure on a topological space X with Borel σ-algebra $\mathscr{B}(X)$ is called **locally finite** if each point $x \in X$ has a neighbourhood of finite measure.

Equip the real line $\mathbb{R}$ with the discrete topology [☞ Example 2.1.(b)]; the corresponding Borel σ-algebra is the power set $\mathscr{B}(\mathbb{R}) = \mathscr{P}(\mathbb{R})$. The measure μ defined by

$$\mu(A) := \begin{cases} 0, & \text{if } A \text{ is countable,} \\ \infty, & \text{if } A \text{ is uncountable,} \end{cases}$$

is not σ-finite, but it is locally finite since for each $x \in \mathbb{R}$ the singleton $\{x\}$ is an open neighbourhood of x.

Comment In fact, $(\mathbb{R}, d)$ is locally compact but neither separable nor second countable, i.e. the topology has no countable base.

If X is a locally compact and second countable topological space, then it is σ-compact. The idea of the proof is simple: let $U(x)$ be a relatively compact neighbourhood of $x \in X$, i.e. the closure $K(x) = \overline{U}(x)$ is compact. Take a countable basis $(V_n)_{n\in\mathbb{N}}$ of the topology, then there is some $n(x) \in \mathbb{N}$ such that $x \in V_{n(x)} \subseteq U(x)$. By construction, $\overline{V}_{n(x)}$ is compact. Choose an enumeration $(n_i)_{i\in\mathbb{N}}$ of the countable set $\{n(x);\ x \in X\} \subseteq \mathbb{N}$, then $K_i = \overline{V}_{n_1} \cup \ldots \cup \overline{V}_{n_i}$ is compact and $K_i \uparrow X$. If μ is a locally finite measure on $(X, \mathscr{B}(X))$, then it is σ-finite.

In particular, locally compact, separable metric spaces are second countable, so locally finite measures on such spaces are always σ-finite.

5.17 Two measures on $\sigma(\mathscr{G})$ such that $\mu|_{\mathscr{G}} \leqslant \nu|_{\mathscr{G}}$ but $\mu \leqslant \nu$ fails

Consider Lebesgue measure λ on $((0,\infty), \mathscr{B}(0,\infty))$. The family of half-open intervals $\mathscr{G} = \{(a,\infty);\ a > 0\}$ generate $\mathscr{B}(0,\infty)$. Define two measures by

$$\nu(B) := \int_B \mathbb{1}_{[2,4]}\, d\lambda \quad \text{and} \quad \mu(B) := \nu(5 \cdot B), \quad B \in \mathscr{B}(0,\infty),$$

where $\alpha \cdot B := \{\alpha \cdot x;\ x \in B\}$ for $\alpha \in \mathbb{R}$. Since

$$\mu(a,\infty) = \nu(5a,\infty) \leqslant \nu(a,\infty),$$

it is immediate that $\mu|_{\mathscr{G}} \leqslant \nu|_{\mathscr{G}}$. However, we also have

$$\mu\left(\frac{3}{5}, \frac{4}{5}\right) = \nu(3,4) = 1 > 0 = \nu\left(\frac{3}{5}, \frac{4}{5}\right).$$

Comment This pathology cannot occur if $\mathscr{G}$ is a semi-ring, cf. [MIMS, p. 167].

5.18 Two measures on $\sigma(\mathscr{G})$ such that $\mu|_{\mathscr{G}} = \nu|_{\mathscr{G}}$ but $\mu \neq \nu$

Let $(X, \mathscr{A})$ be a measurable space and assume that the σ-algebra is generated by the family $\mathscr{G}$, i.e. $\mathscr{A} = \sigma(\mathscr{G})$. If $\mathscr{G}$ is stable under intersections and $\mathscr{G}$ contains an exhausting sequence $G_n \uparrow X$, then the implication

$$\mu|_{\mathscr{G}} = \nu|_{\mathscr{G}} \Rightarrow \mu|_{\mathscr{A}} = \nu|_{\mathscr{A}} \tag{5.3}$$

holds for any two measures μ, ν on $(X, \mathscr{A})$ with $\mu(G_n)+\nu(G_n) < \infty, n \in \mathbb{N}$. This is a classical version of the uniqueness of measures theorem [☞ Theorem 1.40]. The following examples show that the assumptions on the generator $\mathscr{G}$ are crucial. The implication (5.3) fails to hold in each of the following cases:

(a) $(X, \mathscr{A}) = (\mathbb{R}, \mathscr{B}(\mathbb{R}))$, $\mathscr{G} = \{(a,\infty);\ a \in \mathbb{Q}\}$, $\mu = \lambda$ Lebesgue measure and $\nu(A) := \#A$ counting measure. Clearly, $\lambda(a,\infty) = \infty = \nu(a,\infty)$ for all $a \in \mathbb{Q}$. The assumptions of the uniqueness of measures theorem are violated since there does not exist $(G_n)_{n\in\mathbb{N}} \subseteq \mathscr{G}$ with $\mu(G_n) + \nu(G_n) < \infty$.

(b) $X = \mathbb{R}$, $\mathscr{A}$ is the co-countable σ-algebra [☞ Example 4A], which is generated by $\mathscr{G} = \{A;\ \#A \leqslant \#\mathbb{N}\}$, $\mu = 0$ and [☞ Example 5A]

$$\nu(A) := \begin{cases} 0, & \text{if } A \text{ is at most countable,} \\ 1, & \text{if } A^c \text{ is at most countable.} \end{cases}$$

Thus, $\mu(A) = 0 = \nu(A)$. The assumptions of the uniqueness of measures theorem are not satisfied in this case since there does not exist a sequence $(G_n)_{n\in\mathbb{N}} \subseteq \mathscr{G}$ with $G_n \uparrow X$.

(c) $X = \{a, b, c, d\}$, $\mathscr{A} = \mathscr{P}(X) = \sigma(\mathscr{G})$ for $\mathscr{G} = \{X, \{a, b\}, \{b, c\}, \{c, d\}, \{d, a\}\}$ and

$$\mu = \frac{1}{4}\left(\delta_a + \delta_b + \delta_c + \delta_d\right), \qquad \nu = \frac{1}{2}\left(\delta_a + \delta_c\right).$$

It is obvious that μ and ν coincide on $\mathscr{G}$. The assumptions of the uniqueness of measures theorem are not satisfied since $\mathscr{G}$ is not stable under intersections.

Comment 1 Let (X, d) be a metric space and $\mathscr{A} = \mathscr{B}(X)$ the associated Borel σ-algebra. It is natural to ask whether every measure μ on $(X, \mathscr{A})$ is uniquely determined by its values on the balls. Davies [40] showed that this is not the case; he constructed a compact metric space (X, d) and two finite Borel measures $\mu \neq \nu$ such that $\mu(B) = \nu(B)$ for all closed balls B.

Comment 2 If μ and ν are probability measures (or finite measures with the same total mass), it is enough to assume that $\mathscr{G}$ is stable under intersections. Since $\mu(X) = \nu(X) = 1$, we may add X to the family $\mathscr{G}$ and use the constant sequence $G_n = X$: just note that $\sigma(\mathscr{G}) = \sigma(\mathscr{G} \cup \{X\})$.

5.19 Two measures $\mu \neq \nu$ such that $\int p\,d\mu = \int p\,d\nu$ for all polynomials

Consider Lebesgue measure λ on $(\mathbb{R}, \mathscr{B}(\mathbb{R}))$. For $u \in [0, 1]$ set

$$f_u(x) := \left(1 + u \sin x^{1/4}\right) \exp\left(-x^{1/4}\right) \mathbb{1}_{(0,\infty)}(x), \quad x \in \mathbb{R},$$

and denote by $\mu_u := f_u \lambda$ the measure with density f_u. By a change of variables, we have

$$\int_{(0,\infty)} x^k \sin(x^{1/4}) e^{-x^{1/4}}\,\lambda(dx) = 4 \int_{(0,\infty)} t^{4k+3} \sin(t) e^{-t}\,\lambda(dt).$$

Using integration by parts twice, we obtain

$$\int_{(0,\infty)} t^{n+2} \sin(t)e^{-t}\,\lambda(dt) = \frac{(n+2)(n+1)}{2} \int_{(0,\infty)} t^n \cos(t)e^{-t}\,\lambda(dt)$$

and

$$\int_{(0,\infty)} t^{n+2} \cos(t)e^{-t}\,\lambda(dt) = \frac{(n+2)(n+1)}{2} \int_{(0,\infty)} t^n \sin(t)e^{-t}\,\lambda(dt)$$

for all $n \geqslant 1$. By iteration, it follows that

$$\int_{(0,\infty)} x^k \sin(x^{1/4})e^{-x^{1/4}}\,\lambda(dx) = \frac{1}{2^{2k-1}}(4k+3)! \int_{(0,\infty)} t\cos(t)e^{-t}\,dt = 0$$

for all $k \geqslant 0$. This means that $\int_{\mathbb{R}} x^k\,\mu_u(dx)$ does not depend on $u \in [0,1]$, i.e. all measures μ_u, $u \in [0,1]$, have identical moments. Hence, $\int_{\mathbb{R}} p\,d\mu_u = \int_{\mathbb{R}} p\,d\mu_v$ for any polynomial p and all $u, v \in [0,1]$. But it is also obvious that $\mu_u \neq \mu_v$ if $u \neq v$.

Comment The example goes back to Stieltjes [179]. It shows that for a given sequence $(a_k)_{k\in\mathbb{N}_0} \subseteq \mathbb{R}$ there can exist infinitely many measures μ such that $a_k = \int_{\mathbb{R}} x^k\,\mu(dx)$, $k \geqslant 0$. This phenomenon is known as the indeterminate **moment problem**. A full discussion on the classical moment problem can be found in Akhiezer [4].

If μ, ν are measures on the space $(X, \mathscr{A})$, then a set $\mathcal{D} \subseteq L^1(\mu) \cap L^1(\nu)$ is called **(measure) determining** if

$$\int_X f\,d\mu = \int_X f\,d\nu \quad \text{for all } f \in \mathcal{D} \text{ implies that } \mu = \nu.$$

Our example shows that the polynomials are not determining for measures on $(\mathbb{R}, \mathscr{B}(\mathbb{R}))$. However, the polynomials **are** determining for measures on the space $([a,b], \mathscr{B}[a,b])$ with $-\infty < a < b < \infty$.

The key observation is that the polynomials are dense in $(C[a,b], \|\cdot\|_\infty)$, cf. [MIMS, p. 373], and therefore $\int_a^b x^k\,\mu(dx) = \int_a^b x^k\,\nu(dx)$, $k \in \mathbb{N}_0$, implies

$$\int_a^b f(x)\,\mu(dx) = \int_a^b f(x)\,\nu(dx), \qquad f \in C[a,b].$$

A standard approximation argument yields $\mu(B) = \nu(B)$ for any $B \in \mathscr{B}[a,b]$. See [MIMS, Chapter 17] for further details.

5.20 Two finite measures $\mu \neq \nu$ whose Fourier transforms coincide on an interval containing zero

Every finite measure μ on $(\mathbb{R}, \mathscr{B}(\mathbb{R}))$ is uniquely determined by its Fourier transform

$$\widehat{\mu}(\xi) := \frac{1}{2\pi} \int_{\mathbb{R}} e^{-ix\xi}\, \mu(dx), \quad \xi \in \mathbb{R},$$

i.e. if μ, ν are finite measures on $(\mathbb{R}, \mathscr{B}(\mathbb{R}))$ such that $\widehat{\mu} = \widehat{\nu}$ then $\mu = \nu$. If the Fourier transforms of two measures coincide only on some compact interval, then the measures need not be equal.

Consider the measure $\mu(dx) = f(x)\,\lambda(dx)$ with density

$$f(x) := 2\frac{1-\cos x}{x^2}\mathbb{1}_{\mathbb{R}\setminus\{0\}}(x) + \mathbb{1}_{\{0\}}(x), \quad x \in \mathbb{R}.$$

(The value 1 at $x = 0$ ensures that f is continuous, but it is otherwise irrelevant.) In order to prove that

$$\chi(\xi) := \max\{1 - |\xi|, 0\}, \quad \xi \in \mathbb{R},$$

is the Fourier transform of μ, we verify that the inverse Fourier transform $\check{\chi}$ equals f. Using that $\xi \longmapsto \chi(\xi)$ is an even function, we find that

$$\begin{aligned}\check{\chi}(x) = \int_{[-1,1]} \max\{1 - |\xi|, 0\} e^{ix\xi}\, d\xi &= \int_{[-1,1]} (1 - |\xi|) \cos(x\xi)\, d\xi \\ &= 2\int_0^1 (1 - \xi)\cos(x\xi)\, d\xi = f(x)\end{aligned}$$

for all $x \in \mathbb{R}$. Since χ and f are integrable with respect to Lebesgue measure, we may take the Fourier transform on both sides to conclude that $\widehat{\mu} = \widehat{f} = \chi$. Define another finite measure ν on $(\mathbb{R}, \mathscr{B}(\mathbb{R}))$ by

$$\nu(dx) = \frac{1}{4} f\left(\frac{x}{2}\right) \lambda(dx) + \pi\delta_0(dx).$$

Its Fourier transform is given by

$$\widehat{\nu}(\xi) = \frac{1}{2}\widehat{f}(2\xi) + \frac{1}{2} = \frac{1}{2}\max\{0, 1 - 2|\xi|\} + \frac{1}{2}$$

and it satisfies $\widehat{\mu}(\xi) = \widehat{\nu}(\xi)$ for all $|\xi| \leqslant \frac{1}{2}$.

However, $\mu \neq \nu$ since $\mu(\{0\}) = 0 \neq \pi = \nu(\{0\})$.

Comment An obvious modification of our example yields for fixed $R > 0$ two measures μ and ν whose Fourier transforms $\widehat{\mu}$ and $\widehat{\nu}$ coincide on the interval $[-R, R]$. Examples of this type can be found in Lukacs [107, pp. 29, 85 § 4.3 Examples 1, 2] and Ushakov [191, Appendix A, Examples 11, 12].

5.21 (Non)Equivalent definitions of the convolution of measures

If μ, ν are σ-finite measures on $(\mathbb{R}, \mathscr{B}(\mathbb{R}))$, then the convolution

$$(\mu * \nu)(B) = \int \mathbb{1}_B(x+y)\,(\mu \times \nu)(dx, dy), \quad B \in \mathscr{B}(\mathbb{R}), \tag{5.4}$$

is well-defined and, by Tonelli's theorem,

$$(\mu * \nu)(B) = \int \mu(B-y)\,\nu(dy) = \int \nu(B-x)\,\mu(dx), \tag{5.5}$$

i.e. we may define the convolution $\mu * \nu$ by any of the equivalent expressions in (5.4) and (5.5).

In particular, the convolution 'product' is commutative $\mu * \nu = \nu * \mu$. If one of the measures is not σ-finite, then there is, in general, no unique product measure $\mu \times \nu$ [☞ 16.1], and so we cannot define the convolution $\mu * \nu$ by (5.4).

The integrals in (5.5) are a bit more robust: if the measure ν is σ-finite, then $x \longmapsto \nu(B-x) = \int \mathbb{1}_B(x+y)\,\nu(dy)$ is measurable [☞ footnote on p. 295], and so the integral

$$\int \nu(B-x)\,\mu(dx)$$

exists; similarly, the integral

$$\int \mu(B-y)\,\nu(dy)$$

makes sense if μ is σ-finite. This gives a possibility to extend the convolution beyond σ-finite measures, but one needs to be careful: it can happen that both integrals exist yet do not give the same measure. Consider, for instance, Lebesgue measure λ and the counting measure ζ on $(\mathbb{R}, \mathscr{B}(\mathbb{R}))$. If $B \in \mathscr{B}(\mathbb{R})$ is a non-empty set, then $\zeta(B) \geqslant 1$, which implies

$$\int \zeta(B-x)\,\lambda(dx) = \int \zeta(B)\,\lambda(dx) = \infty.$$

From the translational invariance of Lebesgue measure it follows that

$$\int \lambda(B-y)\,\zeta(dy) = \begin{cases} 0, & \text{if } \lambda(B) = 0, \\ \infty, & \text{if } \lambda(B) > 0. \end{cases}$$

Consequently,

$$B \longmapsto \int \zeta(B-x)\,\lambda(dx) \quad \text{and} \quad B \longmapsto \int \lambda(B-y)\,\zeta(dy)$$

are both measures, but they are not equal. This means that there is no good way to define a commutative convolution product $\lambda * \zeta$.

Comment Both

$$\int \zeta(B - x)\,\lambda(dx) = \int \Big(\int \mathbb{1}_B(x + y)\,\zeta(dy) \Big)\,\lambda(dx)$$

and

$$\int \lambda(B - y)\,\zeta(dy) = \int \Big(\int \mathbb{1}_B(x + y)\,\lambda(dx) \Big)\,\zeta(dy)$$

are iterated integrals, i.e. our example shows that both iterated integrals exist but do not coincide. This pathology can happen only because ζ is not σ-finite, for further examples in this direction [☞ Chapter 16].

5.22 The convolution of σ-finite measures need not be σ-finite

If μ, ν are two finite measures on $(\mathbb{R}, \mathscr{B}(\mathbb{R}))$, then the convolution [☞ p. 26]

$$(\mu * \nu)(B) = \iint \mathbb{1}_B(x + y)\,\mu(dx)\,\nu(dy), \quad B \in \mathscr{B}(\mathbb{R}),$$

is again a finite measure. By contrast, σ-finiteness is not preserved under convolution. For example, Lebesgue measure λ on $(\mathbb{R}, \mathscr{B}(\mathbb{R}))$ is σ-finite, while the convolution $\lambda * \lambda$ is not σ-finite. Indeed, translation invariance of Lebesgue measure gives

$$(\lambda * \lambda)(B) = \iint \mathbb{1}_B(x + y)\,\lambda(dx)\,\lambda(dy) = \iint \mathbb{1}_B(x)\,\lambda(dx)\,\lambda(dy) = \infty$$

for every B with $\lambda(B) > 0$, i.e.

$$(\lambda * \lambda)(B) = \begin{cases} 0, & \text{if } \lambda(B) = 0, \\ \infty, & \text{if } \lambda(B) > 0. \end{cases}$$

Take any sequence $(B_n)_{n\in\mathbb{N}} \subseteq \mathscr{B}(\mathbb{R})$ with $B_n \uparrow \mathbb{R}$, then

$$(\lambda * \lambda)(B_n) \uparrow (\lambda * \lambda)(\mathbb{R}) = \infty,$$

and so $(\lambda * \lambda)(B_n) > 0$ for large $n \in \mathbb{N}$. Since $\lambda * \lambda$ takes only the values 0 and ∞, this means that $(\lambda * \lambda)(B_n) = \infty$ for large n. Consequently, $\lambda * \lambda$ is not σ-finite.

Comment The fact that convolution does not preserve σ-finiteness has interesting consequences for the iteration of convolutions, e.g. if μ, ν, ρ are σ-finite measures, then the convolution $\mu * (\nu * \rho)$ may not be well-defined. To make sense of the integral

$$\int_{\mathbb{R}} (\nu * \rho)(B - x)\,\mu(dx) = \int_{\mathbb{R}} \left(\int_{\mathbb{R}} \mathbb{1}_B(x+y)\,(\nu * \rho)(dy) \right) \mu(dx),$$

we need that $x \longmapsto (\nu * \rho)(B - x)$ is measurable; by Tonelli's theorem [☞ Corollary 1.34, footnote on p. 295], this is satisfied if $\nu * \rho$ is σ-finite but – as we have seen – σ-finiteness of $\nu * \rho$ may fail, and so the integrand may not be measurable. A possible way out is to change the order of integration and define

$$\begin{aligned} \mu * (\nu * \rho)(B) &:= \int_{\mathbb{R}} \mu(B - y)\,(\nu * \rho)(dy) \\ &= \int_{\mathbb{R}} \left(\int \mathbb{1}_B(x+y)\,\mu(dx) \right) (\nu * \rho)(dy). \end{aligned}$$

Since μ is σ-finite, the inner integral $y \longmapsto \mu(B - y) = \int \mathbb{1}_B(x+y)\,\mu(dx)$ is measurable, and so $\mu * (\nu * \rho)(B)$ is well-defined. This 'extended' convolution product is neither associative nor commutative; see also Example 5.21.

5.23 $\mu * \nu = \mu$ does not imply $\nu = \delta_0$

Let $(X, \mathscr{A})$ be a measurable space. For two measures μ, ν on $(X, \mathscr{A})$ define the convolution $\mu * \nu$ by

$$(\mu * \nu)(A) = \int_X \mu(A - x)\,\nu(dx), \quad A \in \mathscr{A},$$

whenever the integral is well-defined, i.e. $A - x \in \mathscr{A}$ and $x \longmapsto \mu(A - x)$ is measurable for all $x \in X$, $A \in \mathscr{A}$. If X is a metric space and μ, ν are inner compact regular [☞ p. 185] Borel measures with mass 1, $\mu(X) = \nu(X) = 1$, then

$$\mu * \nu = \mu \Rightarrow \nu = \delta_0,$$

cf. [192, Proposition I.4.7]. The following example shows that this implication does not hold, in general, if μ, ν are not inner compact regular Borel measures. Consider $X = \mathbb{R}$ with the co-countable σ-algebra $\mathscr{A}$ [☞ Example 4A] and the measure [☞ Example 5A]

$$\mu(A) := \begin{cases} 0, & \text{if } A \text{ is at most countable,} \\ 1, & \text{if } A^c \text{ is at most countable.} \end{cases}$$

By definition, μ is shift invariant, i.e. $\mu(A-x) = \mu(A)$ for all $x \in \mathbb{R}$ and $A \in \mathscr{A}$. This implies

$$(\mu * \mu)(A) = \int_{\mathbb{R}} \mu(A-x)\,\mu(dx) = \mu(A)\mu(X) = \mu(A), \quad A \in \mathscr{A},$$

but clearly $\mu \neq \delta_0$.

5.24 The push forward 'disaster' (image measures behaving badly)

Let $(X, \mathscr{A}, \mu)$ be a measure space, $(X', \mathscr{A}')$ a measurable space and $f : X \to X'$ an $\mathscr{A}/\mathscr{A}'$ measurable map. We do not assume that f is surjective, i.e. we do not necessarily have $f \circ f^{-1}(A') = A'$ for all $A' \subseteq X'$, cf. [MIMS, p. 14, Exercise 2.7].

The **image measure** or **push-forward** [☞ Example 1.7.(f), Theorem 1.37] of μ under f is the measure ν on $(Y, \mathscr{B})$ defined by

$$\nu(A') := \mu(f^{-1}(A')), \quad A' \in \mathscr{A}'.$$

By measurability, $f^{-1}(\mathscr{A}') \subseteq \mathscr{A}$ and since f^{-1} commutes with all set operations, it is easy to see that ν is indeed a measure. It is usually denoted by $\mu \circ f^{-1}$ or $f_*\mu$.

While $f^{-1}(A')$ is $\mathscr{A}$ measurable for all $A' \in \mathscr{A}'$, it is, in general, false that $f \circ f^{-1}(A') \in \mathscr{A}'$ (unless f is surjective). As an extreme case, it may happen that $f_*\mu$ is not even carried by $f(X)$ since $X' \setminus f(X)$ need not be measurable. Mind, however, that $f^{-1}(X' \setminus f(X)) = \emptyset$ is in $\mathscr{A}$, hence measurable.

More concretely, let λ be Lebesgue measure on $X' = [0,1]$ with its Borel σ-algebra $\mathscr{A}' = \mathscr{B}[0,1]$ and let $X \subseteq X'$ be a non-Lebesgue measurable set with inner measure $\lambda_*(X^c) = 0$ [☞ 7.22, 9.2].

Consider the trace σ-algebra $\mathscr{B}(X) := X \cap \mathscr{B}[0,1]$ and define μ as the restriction of λ, i.e.

$$\forall B \in \mathscr{B}(X) \ : \ \mu(B) := \lambda(B'),$$

where $B = X \cap B' \in X \cap \mathscr{B}[0,1] = \mathscr{B}(X)$ [☞ 5.9]. The map $i : X \to X' = [0,1]$, $x \longmapsto x$, satisfies $i_*\mu = \lambda$, since for any $B' \in \mathscr{B}[0,1]$

$$i_*\mu(B') = \mu(B) = \lambda(B') \quad \text{where } B = B' \cap X.$$

This shows that μ is carried by X, but the image measure $\lambda = i_*\mu$ is not carried by $i(X) = X$ since X is not even measurable.

Comment The name 'La catastrophe de la mesure image' and the above example are taken from Schwartz [161, p. 366].

5.25 The pull-back of a measure need not be a measure

Let $(X, \mathscr{A})$ and $(Y, \mathscr{B})$ be two measurable spaces. Assume that $f : X \to Y$ is a measurable map and μ is a measure on $(X, \mathscr{A})$. The push-forward of f under μ is the measure ν on $(Y, \mathscr{B})$ which is defined in the following way:

$$\nu(B) := \mu(f^{-1}(B)), \quad B \in \mathscr{B}$$

[☞ 5.24, Example 1.7.(f)]. The **pull-back** works the other way round: we start with a map $f : X \to Y$ and a measure ν on $(Y, \mathscr{B})$ and we want to construct a measure μ on $(X, \mathscr{A})$. A natural candidate would be

$$\mu(A) := \nu(f(A)), \quad A \in \mathscr{A},$$

which is, formally, the familiar pull-back used in functional analysis. However, $f^*\nu := \nu \circ f$ may be trivial or not a measure at all. There are several possibilities for this shortfall.

The map f is not injective. Consider $X = \mathbb{R}^2$ and $Y = \mathbb{R}$, both equipped with the Borel σ-algebra and let λ be Lebesgue measure on $Y = \mathbb{R}$. We use $f = \pi_1$, the canonical projection onto the first coordinate: $\pi_1 : \mathbb{R}^2 \to \mathbb{R}$, $(s, t) \longmapsto s$.

Decompose $A = [0, 2)^2$ into $A_1 = [0, 2) \times [0, 1)$ and $A_2 = [0, 2) \times [1, 2)$. Then

$$\mu(A) = \lambda(\pi_1(A)) = \lambda[0, 2) = 2;$$

in contrast, if μ were a measure, we would also have

$$\begin{aligned} 2 = \mu(A) = \mu(A_1 \uplus A_2) = \mu(A_1) + \mu(A_2) &= \lambda(\pi_1(A_1)) + \lambda(\pi_1(A_2)) \\ &= \lambda[0, 2) + \lambda[0, 2) = 4, \end{aligned}$$

which is impossible.

The map f is not surjective. Consider $X = \mathbb{R}$ and $Y = \mathbb{R}^2$, both equipped with the Borel σ-algebra and let λ^2 be Lebesgue measure on $Y = \mathbb{R}^2$. We use $f = \iota$, the canonical embedding: $\iota : \mathbb{R} \to \mathbb{R}^2$, $s \longmapsto (s, 0)$. In this case it is obvious that $\iota^*\lambda^2 = \lambda^2 \circ \iota \equiv 0$ is trivial.

The map f is bijective but does not preserve measurability. Consider $X = Y = \mathbb{R}$, both equipped with the Lebesgue σ-algebra $\mathscr{L}(\mathbb{R})$ [☞ Example 1.7.(n)], and let λ be Lebesgue measure on $Y = \mathbb{R}$. We use the bijective measurable function $f : \mathbb{R} \to \mathbb{R}$ from Example 8.13 whose inverse f^{-1} is not measurable. By construction, there are sets $A \in \mathscr{L}(\mathbb{R})$ such that $f(A) \notin \mathscr{L}(\mathbb{R})$, i.e. the very definition of the pull-back does not make sense.

Comment Pull-backs can easily be defined, if the map f is bi-measurable, i.e. f admits a pointwise defined inverse f^{-1} such that both $f : X \to Y$ and its inverse $f^{-1} : Y \to X$ are measurable. Clearly, the pull-back under f^{-1} is a push-forward under f, and vice versa. Although it is not regularly spelled out in the literature, this property is most important when we consider coordinate changes and Jacobi's general transformation theorem for integrals, cf. Theorem 1.38. General results on the measurability of direct images $f(A)$ are discussed in Examples 7.24 and 7.25.

5.26 A finite Borel measure which is not tight

Let (X, d) be a metric space with Borel σ-algebra $\mathscr{B}(X)$. A finite measure μ on $(X, \mathscr{B}(X))$ is said to be **tight** if for every $\epsilon > 0$ there is a compact set $K = K_\epsilon \subseteq X$ such that $\mu(X \setminus K) \leqslant \epsilon$. If (X, d) is separable and complete, then every finite Borel measure μ is tight, cf. [20, Theorem 7.1.7]. The situation is different if one of the assumptions on X fails, i.e. if X is not separable or not complete.

▶ X **is separable but not complete.** Finite measures on $(X, \mathscr{B}(X))$ need not be tight. Consider $[0, 1]$ equipped with the Euclidean metric and Lebesgue measure λ. Let $N \subseteq [0, 1]$ be such that N has inner measure zero, $\lambda_*(N) = 0$, and outer measure $\lambda^*(N) = 1$ [☞ 9.2]. In particular, N is not a Lebesgue measurable set. We consider N as a metric space with the induced metric and define a measure μ on the trace σ-algebra $\mathscr{L}_N := N \cap \mathscr{L}$ by

$$\mu(L_N) := \lambda(L)$$

where $L_N = L \cap N \in \mathscr{L}_N$ [☞ 5.9]. Clearly, $\mu(N) = \lambda[0, 1] = 1$, i.e. μ is a probability measure, but μ is not tight: every compact set $K \subseteq N$ is also compact in $[0, 1]$, and so $K \in \mathscr{L}$. As $\lambda_*(N) = 0$ it follows that $\lambda(K) = 0$ which implies $\mu(K) = 0$. Hence, $\mu(N \setminus K) = 1$ for any compact set K.

Note that N is separable as a subset of a separable metric space. However, N (equipped with the induced metric) is not complete. If it were a complete metric space, then it would follow from the completeness of the Euclidean space that N is closed, hence Lebesgue measurable, which contradicts our assumption on N.

▶ X **is complete but not separable.** Suppose that there is a finite measure μ on $(X, \mathscr{B}(X))$ which is not tight. Since X is complete, this entails that the support of μ [☞ Introduction to Chapter 6] has measure strictly smaller than $\mu(X)$, cf. [192, pp. 29, 31]. Denote by $\mathfrak{a}$ the minimal cardinality of any topology basis in X. As $\mu(\operatorname{supp}\mu) < \mu(X)$, it follows that $\mathfrak{a}$ is a **real measurable cardinal number**, i.e. there is a space Y of cardinality $\mathfrak{a}$ such that there exists a probability measure on $(Y, \mathscr{P}(Y))$ vanishing on all singletons, cf. [20,

Proposition 7.2.10]. Consequently, the existence of a non-tight finite measure on $(X, \mathscr{B}(X))$ implies the existence of a real measurable cardinal number. If we assume the axiom of choice, then it is possible to show that $\aleph_1$ is not real measurable [☞ 6.15].

Comment 1 Let (X, d) be a separable metric space. If μ is a finite measure on $(X, \mathscr{B}(X))$, then there exists for every $\epsilon > 0$ a closed, totally bounded[1] set $F \subseteq X$ such that $\mu(X \setminus F) \leqslant \epsilon$, cf. [192, Theorem 3.1]. Our counterexample shows that, in general, μ need not be tight, i.e. F cannot be chosen to be compact.

Comment 2 Tightness can also be defined for non-Borel measures. Let (X, d) be a metric space and $\mathscr{A}$ a σ-algebra on X containing all compact sets. A finite measure μ on $(X, \mathscr{A})$ is **tight** if there exists for every $\epsilon > 0$ a compact set $K \subseteq X$ such that $\mu(X \setminus K) \leqslant \epsilon$. As we have seen above, it is an open question whether there exists a finite Borel measure on a complete metric space which is not tight. However, it is possible to construct a complete metric space (X, d) and a finite measure μ on a measurable space $(X, \mathscr{A})$ such that μ fails to be tight.

Consider $[0, 1]$ equipped with the discrete metric d: $d(x, y) = 0$ if $x = y$ and $d(x, y) = 1$ if $x \neq y$ [☞ Example 2.1.(b)]. Obviously, (X, d) is a complete metric space. Let $\mathscr{A}$ be the co-countable σ-algebra on $[0, 1]$ [☞ Example 4A], and define a probability measure by

$$\mu(A) = \begin{cases} 0, & \text{if } A \text{ is at most countable,} \\ 1, & \text{if } A^c \text{ is at most countable.} \end{cases}$$

Every compact set is finite, and so $\mathscr{A}$ contains all compact sets. Moreover, we have $\mu(K) = 0$ for all compact sets K, i.e. μ is not tight. Note that $\mathscr{A}$ is strictly smaller than the Borel σ-algebra $\mathscr{B}(X) = \mathscr{P}(X)$ [☞ 4.14].

5.27 A translation-invariant Borel measure which is not a multiple of Lebesgue measure

Let μ be a measure on $(\mathbb{R}, \mathscr{B}(\mathbb{R}))$ which is invariant under translations, i.e. $\mu(x + B) = \mu(B)$ for all $x \in \mathbb{R}$ and $B \in \mathscr{B}(\mathbb{R})$. If the unit interval $[0, 1]$ has finite mass $\mu([0, 1]) < \infty$, then μ is a multiple of Lebesgue measure, i.e. $\mu = c\lambda$ for the constant $c = \mu([0, 1])$, cf. [MIMS, p. 35]. For measures μ which assign infinite mass to $[0, 1]$, this implication breaks down: for instance, the counting measure $\zeta_{\mathbb{R}}$ is a translation-invariant measure on $(\mathbb{R}, \mathscr{B}(\mathbb{R}))$ which is not a multiple of Lebesgue measure.

[1] A subset F of a metric space (X, d) is **totally bounded** if, for every $\epsilon > 0$, there are finitely many open metric balls $B_1, \dots, B_n$ of radius ϵ such that $F \subseteq \bigcup_{i=1}^n B_i$. If the metric space is complete, then F is totally bounded if, and only if, its closure $\overline{F}$ is compact, cf. [151, Theorem A4, p. 393].

Comment With some more effort, one can show that there are Borel measures which are invariant under translations but not under reflection, cf. [78]. The idea is to construct a Borel set $C \in \mathscr{B}(\mathbb{R})$ such that its image under the reflection $\varrho(x) := -x$ is not contained in

$$\mathscr{H} := \left\{ B \in \mathscr{B}(\mathbb{R});\ \exists (x_n)_{n\in\mathbb{N}} \subseteq \mathbb{R} \ :\ B \subseteq \bigcup_{n\in\mathbb{N}} (x_n + C) \right\},$$

i.e. $\varrho(C) \nsubseteq \bigcup_{n\in\mathbb{N}}(x_n + C)$ for any sequence $(x_n)_{n\in\mathbb{N}} \subseteq \mathbb{R}$. The set function

$$\mu(B) := \begin{cases} 0, & \text{if } B \in \mathscr{H}, \\ \infty, & \text{if } B \notin \mathscr{H} \end{cases}$$

is a Borel measure which is, by definition, invariant under translations. But μ is not invariant under reflection since $\mu(\varrho(C)) = \infty \neq 0 = \mu(C)$.

5.28 There is no Lebesgue measure in infinite dimension

In the Euclidean space $\mathbb{R}^n$ we can characterize Lebesgue measure as the unique Borel measure which is invariant under translations and which assigns the unit ball $B_1(0)$ its geometric volume; see e.g. [MIMS, p. 35]. In an infinite-dimensional space any such measure is trivial.

Let H be a separable Hilbert space and denote by $(e_n)_{n\in\mathbb{N}}$ some orthonormal base. We equip H with the Borel σ-algebra generated by the norm topology. Assume that there is a translation-invariant measure $\mu \not\equiv 0$ which assigns finite value to any ball $B_r(0)$. Since

$$\bigcupdot_{n\in\mathbb{N}} B_{1/2}(e_n) \subseteq B_2(0),$$

we get by σ-additivity and translation invariance

$$\mu(B_2(0)) \geqslant \sum_{n\in\mathbb{N}} \mu(B_{1/2}(e_n)) = \sum_{n\in\mathbb{N}} \mu(B_{1/2}(0)) = \infty,$$

contradicting the finiteness of μ on balls. Thus, $\mu \equiv 0$.

Comment There is a much stronger assertion due to A. Weil: *Let X be a Hausdorff topological group such that there is an invariant non-trivial Radon measure μ on the Borel σ-algebra $\mathscr{B}(X)$, then X is locally compact.* Since a locally compact normed space is necessarily finite-dimensional, cf. [151, Theorems 1.21, 1.22], Weil's theorem includes our counterexample. Weil's theorem is fairly easy to prove. Recall that a Radon measure is finite on compact sets $K \subseteq X$. We claim that the function $f(x) := \mu((x + K) \vartriangle K)$ is continuous; the symbol $A \vartriangle B := (A \setminus B) \cup (B \setminus A)$ denotes the symmetric difference. Fix $\epsilon > 0$ and use the regularity of μ to construct some open set $U \supseteq K$ such that $\mu(U \setminus K) < \epsilon$. There is some symmetric neighbourhood V of $0 \in X$ such that $K + V \subseteq U$. For all $x \in V$ we get

$$\begin{aligned} f(x) = \mu((x+K) \vartriangle K) &= \mu((x+K) \setminus K) + \mu(K \setminus (x+K)) \\ &= \mu((x+K) \setminus K) + \mu((-x+K) \setminus K) \\ &\leqslant 2\mu(U \setminus K) < 2\epsilon. \end{aligned}$$

Since $f(0) = 0$, this shows that f is continuous at $x = 0$ (in fact, f is everywhere continuous). From $\mu \not\equiv 0$ and the regularity of μ, we infer that there is a compact set K with $\mu(K) > 0$. The set

$$V_K = \{x \in X \,;\, \mu((x+K) \vartriangle K) < \mu(K)\}$$

is an open neighbourhood of $0 \in X$. From

$$V_K \subseteq \{x \in X \,;\, (x+K) \cap K \neq \emptyset\} \subseteq K - K$$

we see that the closure $\overline{V_K}$ is compact, i.e. V_K is a relatively compact neighbourhood of 0. Since X is a topological group, the shifts $y \longmapsto y + x$ and $y \longmapsto y - x$ are continuous for each $x \in X$, and it follows that $x + V_K$ is a relatively compact neighbourhood of $x \in X$, i.e. X is locally compact.

6

Range and Support of a Measure

Let $(X, \mathscr{A}, \mu)$ be an arbitrary measure space. The **range** of μ is the set

$$\mu(\mathscr{A}) = \{\mu(A);\ A \in \mathscr{A}\} \subseteq [0, \infty].$$

If $\#\mu(\mathscr{A}) = 2$, i.e. $0 < \mu(X) \leqslant \infty$ and $\mu(A) \in \{0, \mu(X)\}$ for all $A \in \mathscr{A}$, then μ is called **two-valued**. There are two-valued measures which are not point masses [☞ 6.9]. A set $A \in \mathscr{A}$ with $\mu(A) > 0$ is an **atom** of μ – not to be confused with an atom of the σ-algebra $\mathscr{A}$ [☞ (4.1)] – if

$$B \in \mathscr{A},\ B \subseteq A \ \Rightarrow\ \mu(B) = 0 \text{ or } \mu(A \setminus B) = 0. \tag{6.1}$$

For σ-finite measures spaces, there is the following equivalent characterization [☞ 6.11].

Lemma 6A *Let $(X, \mathscr{A}, \mu)$ be a σ-finite measure space. A set $A \in \mathscr{A}$ with strictly positive measure $\mu(A) > 0$ is an atom of μ if, and only if,*

$$B \in \mathscr{A},\ B \subseteq A,\ \mu(B) < \mu(A) \ \Rightarrow\ \mu(B) = 0. \tag{6.2}$$

If the measure μ has no atoms, then it is called **non-atomic** or **diffuse**. A measure μ is **(purely) atomic** if every set of positive measure contains an atom. Atoms have a considerable impact on properties of the range, e.g. $\mu(\mathscr{A})$ is convex if μ is non-atomic and finite, whereas convexity of the range fails for atomic measures.

If X is a topological space with Borel σ-algebra $\mathscr{B}(X)$, then the **(topological) support** of a measure μ on $(X, \mathscr{B}(X))$ is the complement of the union of all open μ-null sets, i.e.

$$(\operatorname{supp}\mu)^c = \bigcup_{U \text{ open},\ \mu(U)=0} U. \tag{6.3}$$

Thus, $\operatorname{supp}\mu$ is the intersection of all closed sets $F \subseteq X$ with $\mu(X \setminus F) = 0$. By definition,

$$x \in \operatorname{supp}\mu \iff \mu(U) > 0 \text{ for every open set } U \text{ containing } x.$$

If the union of all open μ-null sets has measure zero – this is, for instance, true if the topology has a countable basis – then $\mu(X \setminus \operatorname{supp}\mu) = 0$, i.e. $\operatorname{supp}\mu$ is the smallest closed set $F \subseteq X$ such that $\mu(X \setminus F) = 0$. In general, there need not exist such a smallest closed set [☞6.2] and $\operatorname{supp}\mu$ may not have full measure [☞6.3]. For the particular case that $X = \mathbb{R}^d$ and μ has a continuous density f with respect to Lebesgue measure, the support of μ equals the support of the function f, as defined in real analysis.

6.1 A measure where $\operatorname{supp}\mu \neq \bigcap\{B\,;\ \mu(B^c) = 0\}$

Let X be a topological space and $\mathscr{B}(X)$ the Borel σ-algebra. A measure μ on $(X, \mathscr{B}(X))$ is **supported in** or **carried by** a set $B \in \mathscr{B}(X)$ if $\mu(B^c) = 0$.

If X is separable with a countable dense subset $D = (x_n)_{n\in\mathbb{N}}$, then the finite measure $\mu := \sum_{n=1}^{\infty} 2^{-n}\delta_{x_n}$ has support $X = \overline{\{x_n\,;\ n \in \mathbb{N}\}} = \overline{D}$ but it is **carried by** D.

Comment In general, there is no smallest set which carries a measure μ (unless μ is atomic): consider Lebesgue measure on $\mathbb{R}$; then λ is carried by any set $\mathbb{R} \setminus \{x\}$ but $\bigcap_{x\in\mathbb{R}} \mathbb{R} \setminus \{x\} = \emptyset$.

6.2 A measure which has no minimal closed support

Consider $X = \mathbb{R}$ with the discrete topology [☞Example 2.1.(b)]. The corresponding Borel σ-algebra $\mathscr{B}(X)$ is the power set [☞4.14], and

$$\mu(A) := \begin{cases} 0, & \text{if } A \text{ is countable,} \\ \infty, & \text{otherwise,} \end{cases}$$

defines a measure on $(X, \mathscr{B}(X))$; the proof is similar to that in Example 5A. Since every set $A \subseteq X$ is open in the discrete topology, any subset of X is closed. Consequently, the set $F := X \setminus \{x\}$ is closed and $\mu(X \setminus F) = 0$ for each $x \in X$. Hence,

$$\operatorname{supp}\mu = \bigcap \{F \subseteq X\,;\ F \text{ closed and } \mu(X \setminus F) = 0\} = \emptyset.$$

This means that there is no smallest closed set $F \subseteq X$ such that $\mu(X \setminus F) = 0$; the support $\operatorname{supp}\mu$ is closed but does not satisfy $\mu(X \setminus \operatorname{supp}\mu) = 0$.

Comment A further example is given in Example 6.3.

6.3 Measures may have very small support

If X is a separable metric space, then the support of any Borel measure μ on $(X, \mathscr{B}(X))$ has full measure, i.e. $\mu(\operatorname{supp}\mu) = \mu(X)$, cf. [20, Proposition 7.2.9]. The following example shows that this property breaks down for measures on general topological spaces. Consider the ordinal space $\Omega = [0, \omega_1]$ [☞ Section 2.4] with the σ-algebra [☞ Example 4B]

$$\mathscr{A} := \{A \subseteq \Omega;\ A \cup \{\omega_1\} \text{ or } A^c \cup \{\omega_1\} \text{ contains an uncountable compact set}\}$$

and the Dieudonné measure [☞ Example 5B]

$$\mu(A) := \begin{cases} 1, & \text{if } A \cup \{\omega_1\} \text{ contains an uncountable compact set,} \\ 0, & \text{if } A^c \cup \{\omega_1\} \text{ contains an uncountable compact set.} \end{cases}$$

Since Ω is compact [☞ Section 2.4], we have $\mu(\Omega) = 1$, i.e. μ is a probability measure. We claim that $\operatorname{supp}\mu = \{\omega_1\}$. For any $\alpha < \omega_1$ the set $U = [0, \alpha + 1)$ is an open neighbourhood of α. Since α is a countable ordinal, U is countable, and so $\mu(U) = 0$. Consequently, $\alpha \notin \operatorname{supp}\mu$. Any open neighbourhood U of ω_1 contains an interval of the form $(\alpha, \omega_1]$ with $\alpha < \omega_1$. As $[\alpha + 1, \omega_1] \subseteq U$ is compact (because Ω is compact) and uncountable (since ω_1 is a limit ordinal), it follows that $\mu(U) = 1$, and this means that $\omega_1 \in \operatorname{supp}\mu$. Consequently,

$$\operatorname{supp}\mu = \{\omega_1\} \quad \text{and} \quad \mu(\operatorname{supp}\mu) = \mu(\{\omega_1\}) = 0.$$

If we denote by $\tilde{\mu}$ the restriction of μ to $[0, \omega_1)$ – which we equip with the trace σ-algebra [☞ Example 1.7.(e)] – then $\tilde{\mu}$ is still a probability measure and, by the above considerations, $\operatorname{supp}\tilde{\mu} = \emptyset$.

6.4 A measure μ such that the support of $\mu|_{\operatorname{supp}\mu}$ is strictly smaller than $\operatorname{supp}\mu$

Consider the probability measure μ on the ordinal space $[0, \omega_1]$ from [☞ 6.3]. Since $\operatorname{supp}\mu = \{\omega_1\}$ [☞ 6.3], it follows that $\mu(\operatorname{supp}\mu) = 0$. In particular, the support of $\mu|_{\operatorname{supp}\mu}$ is empty and hence a strict subset of $\operatorname{supp}\mu$.

Comment A slight modification of the above example yields a probability measure ν with the property that the support of $\nu|_{\operatorname{supp}\nu}$ is non-empty and a strict subset of $\operatorname{supp}\nu$. For fixed $\alpha \in \Omega$, $\alpha < \omega_1$, define

$$\nu(A) := \frac{1}{2}\delta_\alpha(A) + \frac{1}{2}\mu(A);$$

then $\operatorname{supp}\nu = \{\alpha, \omega_1\}$ and $\operatorname{supp}(\nu|_{\operatorname{supp}\nu}) = \{\alpha\} \subsetneq \operatorname{supp}\nu$.

6.5 A measure with $\operatorname{supp}\mu = \{c\}$ but $\mu \neq \delta_c$

Consider the measure μ constructed in Example 6.3: $\operatorname{supp}\mu = \{\omega_1\}$ and $\mu \neq \delta_{\omega_1}$.

6.6 Measures such that $\operatorname{supp}\mu + \operatorname{supp}\nu \subsetneqq \operatorname{supp}\mu * \nu$

The convolution of two σ-finite Borel measures μ and ν on the measurable space $(X, \mathscr{B}(X)) = (\mathbb{R}, \mathscr{B}(\mathbb{R}))$ – or on $(\mathbb{R}^d, \mathscr{B}(\mathbb{R}^d))$ or, indeed, on any topological group $(X, +)$ equipped with the Borel σ-algebra $\mathscr{B}(X)$ – is the image measure of the product measure $\mu \times \nu$ under the measurable map $(x, y) \longmapsto a(x, y) := x + y$ [☞ Definition 1.39], that is

$$\mu * \nu(B) = \iint_{X\times X} \mathbb{1}_B(x+y)\,\mu\times\nu(dx, dy) = \int_X \nu(B-x)\,\mu(dx)$$
$$= \int_X \mu(B-y)\,\nu(dy).$$

This definition is compatible with the 'usual' definition of the convolution of two functions [☞ p. 26].

The support of the measure $\mu * \nu$ is the closure of the set $\operatorname{supp}\mu + \operatorname{supp}\nu$.

Proposition 1 *If μ, ν are σ-finite Borel measures on $\mathbb{R}$ (or on a separable metric space), then*

$$\operatorname{supp}\mu * \nu = \overline{\operatorname{supp}\mu + \operatorname{supp}\nu} \tag{6.4}$$

where $\overline{S}$ denotes the closure of the set S.

Proof If $x \notin \overline{\operatorname{supp}\mu + \operatorname{supp}\nu}$, then there is an open neighbourhood U of x such that $(U - y) \cap \operatorname{supp}\nu = \emptyset$ for every $y \in \operatorname{supp}\mu$. Since $\mathbb{R}$ is a Polish space, the complement of the support has measure zero, and so $\nu(U - y) = 0$ for all $y \in \operatorname{supp}\mu$. Hence,

$$(\mu * \nu)(U) = \int_{\operatorname{supp}\mu} \nu(U-y)\,\mu(dy) = 0,$$

i.e. $x \notin \operatorname{supp}\mu * \nu$. This proves $\operatorname{supp}\mu * \nu \subset \overline{\operatorname{supp}\mu + \operatorname{supp}\nu}$. In order to see the converse inclusion, note that

$$\forall\epsilon > 0 \quad \forall x_1, x_2 \;:\; \mathbb{1}_{B_\epsilon(x_1+x_2)}(y+z) \geqslant \mathbb{1}_{B_{\epsilon/2}(x_1)}(y)\mathbb{1}_{B_{\epsilon/2}(x_2)}(z).$$

If $x_1 \in \operatorname{supp}\mu$ and $x_2 \in \operatorname{supp}\nu$, then

$$\mu * \nu(B_\epsilon(x_1+x_2)) = \iint \mathbb{1}_{B_\epsilon(x_1+x_2)}(y+z)\,\mu(dy)\,\nu(dz)$$

$$\begin{aligned} &\geqslant \int \mathbb{1}_{B_{\epsilon/2}(x_1)}(y)\,\mu(dy) \int \mathbb{1}_{B_{\epsilon/2}(x_2)}(z)\,\nu(dz) \\ &= \mu(B_{\epsilon/2}(x_1))\nu(B_{\epsilon/2}(x_2)) > 0, \end{aligned}$$

which shows that $x_1 + x_2 \in \operatorname{supp}\mu * \nu$. As $\operatorname{supp}\mu * \nu$ is closed, this gives

$$\overline{\operatorname{supp}\mu + \operatorname{supp}\nu} \subseteq \overline{\operatorname{supp}\mu * \nu} = \operatorname{supp}\mu * \nu. \qquad \square$$

If $\operatorname{supp}\mu$ and $\operatorname{supp}\nu$ are both compact, then the sum $\operatorname{supp}\mu + \operatorname{supp}\nu$ is closed, and thus $\operatorname{supp}\mu * \nu = \operatorname{supp}\mu + \operatorname{supp}\nu$. The following example shows that for general measures the support of $\mu * \nu$ can be larger than $\operatorname{supp}\mu + \operatorname{supp}\nu$, i.e. the closure on the right-hand side of (6.4) is really needed.

Consider on $(\mathbb{R}, \mathscr{B}(\mathbb{R}))$ the measures

$$\mu := \sum_{n\in\mathbb{N}} 2^{-n}\delta_n \quad \text{and} \quad \nu := \sum_{n\in\mathbb{N}} 2^{-n}\delta_{-n+1/n}.$$

From

$$\operatorname{supp}\mu = \mathbb{N} \quad \text{and} \quad \operatorname{supp}\nu = \{-n + 1/n\,;\ n \in \mathbb{N}\},$$

it follows that

$$\begin{aligned} \operatorname{supp}\mu + \operatorname{supp}\nu &= \{y + z\,;\ y \in \operatorname{supp}\mu, z \in \operatorname{supp}\nu\} \\ &= \{m - n + 1/n\,;\ m, n \in \mathbb{N}\}; \end{aligned}$$

in particular, $0 \notin \operatorname{supp}\mu + \operatorname{supp}\nu$. Moreover,

$$\begin{aligned} (\mu * \nu)((-r, r)) &= \int \nu((-r, r) - y)\,\mu(dy) \\ &= \sum_{n\in\mathbb{N}} 2^{-n}\nu((-r, r) - n) \geqslant \sum_{n\geqslant N} 2^{-2n} > 0, \end{aligned}$$

where $N \in \mathbb{N}$ is chosen such that $\frac{1}{N} < r$; thus, $0 \in \operatorname{supp}\mu * \nu$. In view of

$$\operatorname{supp}\mu + \operatorname{supp}\nu \subseteq \overline{\operatorname{supp}\mu + \operatorname{supp}\nu} \subseteq \operatorname{supp}\mu * \nu,$$

see below, we get $\operatorname{supp}\mu + \operatorname{supp}\nu \subsetneqq \operatorname{supp}\mu * \nu$.

6.7 Measures such that $\operatorname{supp}\mu * \nu \subsetneqq \overline{\operatorname{supp}\mu + \operatorname{supp}\nu}$

For Borel measures on $\mathbb{R}$ (or on a group X which is metrizable and separable), the support of the convolution $\mu * \nu$ is contained in $\overline{\operatorname{supp}\mu + \operatorname{supp}\nu}$ [☞ 6.6]. The following example shows that this inclusion can be strict in general topological spaces. Consider on the ordinal space $\Omega_0 = [0, \omega_1)$ [☞ 2.4] the Dieudonné measure μ [☞ 5B] on the σ-algebra $\mathscr{A}$ from Example 4B. We know that

$\operatorname{supp}\mu = \emptyset$ [☞ 6.3]. Pick any $\alpha \in [0, \omega_1)$ and consider the point mass δ_α at α. We have for every $A \in \mathscr{A}$,

$$\mu * \delta_\alpha(A) = \mu(A - \alpha) = \mu(A).$$

Thus,

$$\operatorname{supp}\mu * \delta_\alpha = \operatorname{supp}\mu = \emptyset \subsetneqq \{\alpha\} = \operatorname{supp}\delta_\alpha + \operatorname{supp}\mu = \overline{\operatorname{supp}\delta_\alpha + \operatorname{supp}\mu}.$$

Comment This phenomenon holds for any measure μ with $\operatorname{supp}\mu = \emptyset$. This, however, can never hold for non-trivial Borel measures on separable metric spaces [☞ 6.3].

6.8 A signed measure such that $\operatorname{supp}\mu^+ = \operatorname{supp}\mu^-$

Let $Q = (q_n)_{n\in\mathbb{N}}$ be an enumeration of $\mathbb{Q} \cap (0,1)$ and $R = (r_n)_{n\in\mathbb{N}}$ an enumeration of $(\mathbb{Q} + \sqrt{2}) \cap (0,1)$. We set

$$\mu(B) := \sum_{n=1}^{\infty} 2^{-n}\big(\delta_{q_n}(B) - \delta_{r_n}(B)\big), \quad B \in \mathscr{B}(0,1).$$

We have $\overline{Q} = \operatorname{supp}\mu^+$ and $\overline{R} = \operatorname{supp}\mu^-$, and since the rationals are dense in $(0,1)$, it follows that $\overline{R} = (0,1) = \overline{Q}$.

6.9 A two-valued measure which is not a point mass

Let $(X, \mathscr{A}, \mu)$ be a measure space such that $\{x\} \in \mathscr{A}$ for all $x \in X$. If $\mu(A) \in \{0,1\}$ for all $A \in \mathscr{A}$ this does not necessarily imply that μ is a point mass, i.e. $\mu = \delta_x$ for some $x \in X$.

Consider the measurable space $(\mathbb{R}, \mathscr{A})$ where $\mathscr{A}$ is the co-countable σ-algebra [☞ Example 4A],

$$\mathscr{A} := \{A \subseteq \mathbb{R};\ A \text{ or } A^c \text{ is at most countable}\}.$$

From Example 5A we know that the set function

$$\mu(A) := \begin{cases} 0, & \text{if } A \text{ is at most countable,} \\ 1, & \text{if } A^c \text{ is at most countable} \end{cases}$$

is a measure on $(\mathbb{R}, \mathscr{A})$ which obviously takes only two values – but it is not a point mass.

Another example of this type is the trivial measure $\tau(\emptyset) = 0$ and $\tau(A) = \infty$ if $A \neq \emptyset$, which can be defined on every measurable space $(X, \mathscr{A})$. Clearly, τ takes only the values 0 and ∞, but τ is not a point mass.

Comment If μ is a measure on $(\mathbb{R}, \mathscr{B}(\mathbb{R}))$, then $\mu(\mathscr{B}(\mathbb{R})) \subseteq \{0,1\}$ **does** imply $\mu = \delta_x$ for some $x \in \mathbb{R}$ [☞ 6.10].

6.10 A two-valued measure on a countably generated σ-algebra must be a point mass

Let $(X, \mathscr{A})$ be a measurable space which is countably generated, e.g. a Polish space X with its Borel σ-algebra. If μ is a measure on $(X, \mathscr{A})$ such that $\mu(A)$ is either 0 or $m := \mu(X) \in (0, \infty]$, then there exists some $x \in X$ such that $\mu = m\delta_x$, i.e.

$$\mu(A) = \begin{cases} m, & \text{if } x \in A, \\ 0, & \text{if } x \notin A, \end{cases} \tag{6.5}$$

for all $A \in \mathscr{A}$.

Proof Let $(G_n)_{n\in\mathbb{N}} \subseteq \mathscr{A}$ be such that $\mathscr{A} = \sigma(G_n, n \in \mathbb{N})$. Define

$$H_n := \begin{cases} G_n, & \text{if } \mu(G_n) = m, \\ G_n^c, & \text{if } \mu(G_n) = 0. \end{cases}$$

The set $H := \bigcap_{n\in\mathbb{N}} H_n \in \mathscr{A}$ satisfies $\mu(H) = m$; in particular, $H \neq \emptyset$. We claim that (6.5) holds for any $x \in H$.

Fix $x \in H$. First we verify (6.5) for $A = G_n$, $n \in \mathbb{N}$. There are two cases:

Case 1. $\mu(G_n) = 0$. Then $H_n = G_n^c$, and so $x \in H \subseteq H_n = G_n^c$. Hence, we see that $\mu(G_n) = 0 = m\delta_x(G_n)$.

Case 2. $\mu(G_n^c) = 0$. Then $H_n = G_n$, and so $x \in H \subseteq H_n = G_n$, implying that $\mu(G_n) = m = m\delta_x(G_n)$.

Next we show that

$$\Sigma := \{A \in \mathscr{A}\,;\ (6.5) \text{ holds}\}$$

is a σ-algebra. Clearly, $\emptyset \in \Sigma$, and so it suffices to check (i) its stability under complements and (ii) under countable unions.

(i) Let $A \in \Sigma$. If $\mu(A) = 0$ then $x \notin A$, i.e. $x \in A^c$ which implies $\mu(A^c) = m$. Thus, (6.5) holds for A^c. Similarly, if $\mu(A) = m$, then $x \in A$ which means that $\mu(A^c) = 0 = m\delta_x(A^c)$.

(ii) Let $(A_i)_{i\in\mathbb{N}} \subseteq \Sigma$ and set $A := \bigcup_{i\geqslant 1} A_i$. If $\mu(A_i) = 0$ for all $i \in \mathbb{N}$, i.e. $x \notin A_i$ for all $i \in \mathbb{N}$, then $x \notin A$, and so $\mu(A) = 0 = m\delta_x(A)$. If $\mu(A_i) = m$ for some $i \in \mathbb{N}$, i.e. $x \in A_i$ for some $i \in \mathbb{N}$, then $x \in A$ and

$$m \geqslant \mu(A) \geqslant \mu(A_i) = m.$$

Thus, $\mu(A) = m = m\delta_x(A)$.

Since $\mathscr{A}$ is generated by $(G_n)_{n\in\mathbb{N}} \subseteq \Sigma$, we conclude that

$$\mathscr{A} = \sigma(G_n, n \in \mathbb{N}) \subseteq \sigma(\Sigma) = \Sigma. \qquad \square$$

Comment The idea of the proof is taken from Rao [140, p. 14].

6.11 (Non-)equivalent characterizations of atoms of a measure

There are the following two equivalent possibilities to characterize the atoms [☞ p. 123] of a σ-finite measure μ:

Lemma *Let $(X, \mathscr{A}, \mu)$ be a σ-finite measure space and $A \in \mathscr{A}$ with $\mu(A) > 0$. The following properties are equivalent:*

(a) *A is an atom, i.e. if $B \in \mathscr{A}$ is such that $B \subseteq A$, then $\mu(B) = 0$ or $\mu(A \setminus B) = 0$.*
(b) *If $B \in \mathscr{A}$ is such that $B \subseteq A$ and $\mu(B) < \mu(A)$, then $\mu(B) = 0$.*

Proof Take $(X_n)_{n\in\mathbb{N}} \subseteq \mathscr{A}$ with $X_n \uparrow X$ and $\mu(X_n) < \infty$.

(b)⇒(a): If $B \in \mathscr{A}$ is such that $B \subseteq A$, then B and $A \setminus B$ are pairwise disjoint sets and their measures add up to $\mu(A)$. If $\mu(A) < \infty$ this implies that $\mu(B) < \mu(A)$ or $\mu(A \setminus B) < \mu(A)$. By (b), this gives that either $\mu(B) = 0$ or $\mu(A \setminus B) = 0$. If $\mu(A) = \infty$, then $B \cap X_n \subseteq A$ and $\mu(B \cap X_n) < \infty = \mu(A)$, and so, by (b), $\mu(B \cap X_n) = 0$, which implies that

$$\mu(B) \leqslant \sum_{n\in\mathbb{N}} \mu(B \cap X_n) = 0.$$

(a)⇒(b): Assume that $B \in \mathscr{A}$ with $B \subseteq A$ and $\mu(B) < \mu(A)$; then we see that $\mu(A \setminus B) = \mu(A) - \mu(B) > 0$, and it follows from (a) that $\mu(B) = 0$. $\square$

As can be seen from the proof of the lemma, the implication (a)⇒(b) holds for **any** measure. The following simple example shows that the converse is, in general, false for measures which are not σ-finite. Consider $X = \mathbb{N}$ with the power set $\mathscr{P}(X)$ and the trivial measure

$$\tau(A) = \begin{cases} 0, & \text{if } A = \emptyset, \\ \infty, & \text{otherwise.} \end{cases}$$

In particular, there is no set $A \neq \emptyset$ that satisfies property (a), whereas all sets $A \neq \emptyset$ satisfy (b).

6.12 A purely atomic measure such that $\mu \neq \sum_x \mu(\{x\})\delta_x$

Consider $X = \mathbb{R}$ with the co-countable σ-algebra $\mathscr{A}$ [☞ Example 4A] and the measure [☞ Example 5A]

$$\mu(A) := \begin{cases} 0, & \text{if } A \text{ is at most countable,} \\ 1, & \text{if } A^c \text{ is at most countable.} \end{cases}$$

Every set $A \in \mathscr{A}$ with $\mu(A) > 0$ is an atom. Indeed: take any $B \in \mathscr{A}$ with $B \subseteq A$, then either $\mu(B) = 0$ or $\mu(B) = 1$, i.e. $\mu(B) = 0$ or $\mu(A \setminus B) = \mu(A) - \mu(B) = 0$. Consequently, the measure μ is purely atomic. Since $\mu(\{x\}) = 0$ for all $x \in \mathbb{R}$, we see that

$$\mu(A) = 1 \neq 0 = \sum_{x \in \mathbb{R}} \mu(\{x\})\delta_x(A)$$

for any $A \in \mathscr{A}$ with countable complement.

6.13 A measure such that every set with positive measure is an atom

Let μ be a finite two-valued measure on some measurable space $(X, \mathscr{A})$. For any $A \in \mathscr{A}$ with $\mu(A) > 0$, we have $\mu(A) = \mu(X)$, and therefore the implication

$$B \subseteq A,\ \mu(B) < \mu(A) \implies \mu(B) = 0$$

is trivially satisfied, i.e. A is an atom of μ [☞ Lemma 6A]. An example of such a measure is e.g. the Dirac measure $\mu = \delta_x$, $x \in X$, on an arbitrary measurable space $(X, \mathscr{A})$; a more sophisticated example is the measure μ from Example 6.9.

6.14 An infinite sum of atomic measures which is non-atomic

Let I be an uncountable set, e.g. $I = [0, 1]$. Consider $X = \{0, 1\}^I$ with the product σ-algebra $\mathscr{A} = \mathscr{P}(\{0, 1\})^{\otimes I}$ [☞ Example 4C] and set

$$\mu(A) := \sum_{x \in X} \delta_x(A), \quad A \in \mathscr{A}.$$

Every non-empty set $A \in \mathscr{A}$ is uncountable [☞ 4.17], and so

$$\forall A \in \mathscr{A},\ A \neq \emptyset \ :\ \mu(A) = \infty.$$

This implies that μ is a non-atomic measure. Indeed, if $A \in \mathscr{A}$, $A \neq \emptyset$, then $\mu(B) > 0$ and $\mu(A \setminus B) > 0$ for any set $B \in \mathscr{A}$ with $B \subsetneq A$, $B \neq \emptyset$, i.e. A is not an atom. The Dirac measure δ_x is, however, purely atomic for each $x \in X$. Consequently, $\mu = \sum_{x \in X} \delta_x$ is a non-atomic measure which is a sum of atomic measures.

Comment Capek [32] showed that any **countable** sum of atomic measures is atomic. The idea for our counterexample is from [87], where further results in this direction can be found.

6.15 Any non-atomic finite σ-additive measure defined on $\mathscr{P}(\mathbb{R})$ is identically zero

Let μ be a finite measure on $(\mathbb{R}, \mathscr{P}(\mathbb{R}))$ such that $\mu(\{x\}) = 0$ for every $x \in \mathbb{R}$. We want to show that $\mu \equiv 0$.

Assume that the continuum hypothesis holds $\mathfrak{c} = 2^{\aleph_0} = \aleph_1$ [☞ p. 43]. By the axiom of choice there exists a well-ordering $\prec$ of $\mathbb{R}$ such that for every $y \in \mathbb{R}$ the set $[y] = \{x\,;\ x \prec y\}$ is at most countable. Thus, for every $y \in \mathbb{R}$ there exists an injective map $f(\cdot, y)\colon\ [y] \to \mathbb{N}$, that is

$$x \prec x' \prec y \ \Rightarrow\ f(x,y) \neq f(x',y).$$

Consider the sets $F_x^n := \{y \in \mathbb{R};\ x \prec y,\ f(x,y) = n\}$ where $x \in \mathbb{R}$ and $n \in \mathbb{N}$. We list these sets in an $\aleph_0 \times \aleph_1$-array

$$\begin{array}{cccccc}
F_{x_1}^1 & F_{x_2}^1 & F_{x_3}^1 & \dots & F_x^1 & \dots \\
F_{x_1}^2 & F_{x_2}^2 & F_{x_3}^2 & \dots & F_x^2 & \dots \\
F_{x_1}^3 & F_{x_2}^3 & F_{x_3}^3 & \dots & F_x^3 & \dots \\
\dots & \dots & \dots & \dots & \dots & \dots \\
F_{x_1}^n & F_{x_2}^n & F_{x_3}^n & \dots & F_x^n & \dots \\
\dots & \dots & \dots & \dots & \dots & \dots
\end{array}$$

which has the following properties:

(i) The sets in any row are disjoint, i.e. $F_x^n \cap F_{x'}^n = \emptyset$ for all $x \neq x'$ and $n \in \mathbb{N}$: since $F_x^n \cap F_{x'}^n = \{y \in \mathbb{R};\ x, x' \prec y,\ f(x,y) = f(x',y) = n\}$, we see from the injectivity of $f(\cdot, y)$ that this set is empty if $x \neq x'$.

(ii) The union of the sets in any column is co-countable, i.e. the set $\mathbb{R} \setminus \bigcup_{n\in\mathbb{N}} F_x^n$ is countable for all $x \in \mathbb{R}$: if $x \prec y$, we set $n := f(x,y)$ and get that $y \in F_x^n$. This shows that we have $\mathbb{R} \setminus \bigcup_{n\in\mathbb{N}} F_x^n \subseteq \{y;\ y \preceq x\}$, and the latter is a countable set.

Fix $n \in \mathbb{N}$. Since μ is σ-additive and $\mu(\mathbb{R}) < \infty$, there can be at most countably many x such that $\mu(F_x^n) > 0$. Thus, the array can contain at most countably many F_x^n such that $\mu(F_x^n) > 0$. Consequently, there is one column (indexed by x) such that $\mu(F_x^n) = 0$ for all $n \in \mathbb{N}$. By σ-additivity and property (ii) we get

$$\mu(\mathbb{R}) = \sum_{n=1}^{\infty} \mu(F_x^n) + \sum_{y \preceq x} \mu(\{y\}) = 0,$$

which means that $\mu \equiv 0$.

Comment This result is due to Ulam [190]; our presentation follows Oxtoby [133]. The original proof was stated for sets X of cardinality $\aleph_1$; this is the cardinality of all countable ordinal numbers ω_1. Using the 'usual' Zermelo–Fraenkel axioms, we know that $\aleph_0 < \aleph_1$ and that there is no other cardinal number between those two. In order to fit in $\mathfrak{c} = 2^{\aleph_0}$, the cardinality of the continuum, into the hierarchy of the $\aleph$s, we have to add the **continuum hypothesis** [☞ p. 43] to the Zermelo–Fraenkel axioms: $2^{\aleph_0} = \aleph_1$.

6.16 A measure on a discrete space which attains all values in $[0, \infty]$

Define on the measure space $(\mathbb{N}, \mathscr{P}(\mathbb{N}))$ the set function

$$\mu(A) = \sum_{n \text{ even}} 2^{-n/2}\delta_n(A) + \sum_{n \text{ odd}} \delta_n(A).$$

Since we can interchange the summation of positive sequences, it is clear that μ is σ-additive.

Let $x \in [0, 1)$ and write x as a dyadic fraction $x = 0.x_1x_2x_3 \dots$ where $(x_n)_{n\in\mathbb{N}}$ is a 0–1 sequence. Define the set $A = \{2n\,;\ x_n = 1\}$. Then we have $\mu(A) = x$. This shows that μ may attain any value in $[0, 1)$. Adding to A, say, k odd natural numbers, we see that μ takes all values in $[k, k + 1)$. Finally, $\mu(2\mathbb{N} + 1) = \infty$.

6.17 A measure whose range is not a closed set

Write $\mathbb{Q} \cap [1, \infty) = (q_n)_{n\in\mathbb{N}}$ and define a measure μ on $(\mathbb{N}, \mathscr{P}(\mathbb{N}))$ by

$$\mu(A) = \sum_{n\in\mathbb{N}} q_n\delta_n(A), \quad A \subseteq \mathbb{N}.$$

If $A \neq \emptyset$ is a finite set, then $\mu(A) \in \mathbb{Q} \cap [1, \infty)$; for infinite $A \subseteq \mathbb{N}$ we have $\mu(A) = \infty$. Hence, $\mu(\mathscr{P}(\mathbb{N})) \subseteq R := \{0, \infty\} \cup (\mathbb{Q} \cap [1, \infty))$. Moreover, there exists for any $r \in R$ a set $A \subseteq \mathbb{N}$ with $\mu(A) = r$. Thus, $\mu(\mathscr{P}(\mathbb{N})) = R$ which shows that the range of μ is not closed.

Comment 1 It is crucial that $(q_n)_{n\in\mathbb{N}}$ is an enumeration of $\mathbb{Q} \cap [1, \infty)$ rather than $\mathbb{Q}\cap[0, \infty)$. If, say, $(r_n)_{n\in\mathbb{N}}$ is an enumeration of $\mathbb{Q}\cap[0, \infty)$, then the range of the measure $\nu := \sum_{n\in\mathbb{N}} r_n\delta_n$ is $[0, \infty]$; this follows from the fact that any $x \geqslant 0$ can be written as $x = \lim_{n\to\infty} \sum_{k=1}^n x_k$ with $(x_k)_{k\in\mathbb{N}} \subseteq \mathbb{Q} \cap [0, \infty)$ satisfying $x_k \neq x_j$ for $k \neq j$. In particular, the range of ν is closed.

Comment 2 The range of every **finite** measure is closed, cf. Halmos [70]. This implies, in particular, that there is no measure μ with $\mu(\mathscr{A}) = [0, 1] \setminus \{\frac{1}{2}\}$.

6.18 A measure with countable range

[☞6.17]

6.19 A vector measure which is non-atomic but whose range is not convex

Let $(X, \mathscr{A})$ be a measurable space. Every signed measure $\mu : \mathscr{A} \to \mathbb{R}$ on $(X, \mathscr{A})$ has a Hahn decomposition of the form $\mu = \mu^+ - \mu^-$, where μ^+ and μ^- are positive measures on $(X, \mathscr{A})$ [☞Theorem 1.47]. The **total variation** of μ is defined by

$$|\mu|(A) := \mu^+(A) + \mu^-(A), \quad A \in \mathscr{A}.$$

We call $A \in \mathscr{A}$ an **atom** of the signed measure μ if A is an atom of the (positive) measure $|\mu|$, i.e. if $|\mu|(A) > 0$ and

$$B \in \mathscr{A}, B \subseteq A \Rightarrow |\mu|(B) \in \{0, |\mu|(A)\}.$$

Since the signed measure μ attains only finite values, $|\mu|(X) < \infty$, and this definition of atom is equivalent to [☞(6.1)].

Let $\nu = (\nu_1, \dots, \nu_d)$ be a σ-additive vector measure on $(X, \mathscr{A})$ with values in $\mathbb{R}^d$, i.e. each ν_i is a signed measure on $(X, \mathscr{A})$. Since ν is $\mathbb{R}^d$-valued, the total variation measure $|\nu| := \sum_{i=1}^d |\nu_i|$ is a finite measure. If the signed measures ν_i do not have atoms, then the range

$$\nu(\mathscr{A}) := \{\nu(A);\ A \in \mathscr{A}\} \subseteq \mathbb{R}^d$$

is a convex set, cf. [20, Theorem 9.12.38] or [151, Theorem 5.5]. This is a particular case of Lyapunov's convexity theorem [109], which holds for atomless vector measures taking values in general finite-dimensional Banach spaces. In particular, any (positive) atomless measure $\nu : (X, \mathscr{A}) \to [0, \infty)$ has a convex range. The following example from Uhl [188] shows that convexity of the range breaks down for vector measures with values in infinite-dimensional spaces.

Consider Lebesgue measure λ on the measurable space $([0,1], \mathscr{B}[0,1])$ and write $L^1 = L^1([0,1], \mathscr{B}[0,1], \lambda)$. The mapping

$$\nu : \mathscr{B}[0,1] \to L^1, \ B \longmapsto \mathbb{1}_B$$

defines a σ-additive vector measure on $([0,1], \mathscr{B}[0,1])$, that is, $\nu(\emptyset) = 0$ and

$$\nu\left(\dot{\bigcup_{n\geqslant 1}} B_n\right) = L^1\text{-}\lim_{n\to\infty} \sum_{i=1}^n \nu(B_i)$$

for any sequence $(B_n)_{n\in\mathbb{N}} \subseteq \mathscr{B}[0,1]$ of pairwise disjoint sets. The vector measure ν does not have atoms, i.e. there is no set $B \in \mathscr{B}[0,1]$ with $\nu(B) \neq 0$ such that the implication

$$A \in \mathscr{B}[0,1], A \subseteq B \Rightarrow \nu(A) = \nu(B) \quad \text{or} \quad \nu(A) = 0$$

holds. Indeed: fix $B \in \mathscr{B}[0,1]$ with $\mathbb{1}_B \neq 0$ in L^1; then $\lambda(B) > 0$. Since Lebesgue measure does not have atoms, there exists some $A \in \mathscr{B}[0,1]$ such that $A \subseteq B$ and $0 < \lambda(A) < \frac{1}{2}\lambda(B)$. Thus, $\nu(A) = \mathbb{1}_A \neq 0$ and $\nu(A) = \mathbb{1}_A \neq \mathbb{1}_B = \nu(B)$, i.e. B is not an atom. The range of ν is not convex: the mappings $\mathbb{1}_{[0,1/2]}$ and $\mathbb{1}_{(1/2,1]}$ are in the range of ν but their convex combination

$$\frac{1}{2}\mathbb{1}_{[0,1/2]} + \frac{1}{2}\mathbb{1}_{(1/2,1]} = \frac{1}{2}\mathbb{1}_{[0,1]}$$

is not in the range. Indeed, for any $B \in \mathscr{B}[0,1]$,

$$\left\|\frac{1}{2}\mathbb{1}_{[0,1]} - \nu(B)\right\|_{L^1} = \left\|\frac{1}{2}\mathbb{1}_{[0,1]} - \mathbb{1}_B\right\|_{L^1} = \frac{1}{2}\lambda([0,1] \setminus B) + \frac{1}{2}\lambda(B) = \frac{1}{2}.$$

Comment If ν is a finite (positive) measure on a measurable space $(X, \mathscr{A})$, then its range is bounded and closed – see the comment in Example 6.17 –, hence compact. Lyapunov [109] proved that the range of any atomless vector measure with values in a finite-dimensional Banach space is compact. The above counterexample shows that this result does not extend to infinite-dimensional spaces: the range of ν is closed but not compact; see Uhl [188]. We refer the reader to Diestel & Uhl [44, Chapter IX] for a thorough discussion of properties of the range of vector measures.

6.20 A non-trivial measure which assigns measure zero to all open balls

Consider $X = \mathbb{R}$ with the discrete metric d, i.e. $d(x,y) = 0$ if $x = y$ and $d(x,y) = \infty$ if $x \neq y$. The corresponding Borel σ-algebra $\mathscr{B}(X)$ is the power set [☞4.14]. Using a similar reasoning as in Example 5A, we see that the map

$$\mu(A) := \begin{cases} 0, & A \text{ is at most countable}, \\ \infty, & \text{otherwise}, \end{cases}$$

defines a measure on $(X, \mathscr{B}(X))$. The open balls $B_r(x)$ with respect to the metric d are singletons,

$$B_r(x) = \{x\}, \quad x \in X,\ r > 0,$$

and this implies $\mu(B_r(x)) = \mu(\{x\}) = 0$ for all $x \in X$ and $r > 0$.

Comment Since the compact sets are finite sets, μ assigns measure zero to all compact sets. In particular, μ is not inner compact regular [☞Chapter 9].

6.21 A signed measure $\mu : \mathscr{A} \to (-\infty, \infty]$ is uniformly bounded below

Let μ be a signed measure [☞Definition 1.45] on a measurable space $(X, \mathscr{A})$. Consider the set

$$\mathscr{G} := \left\{ A \in \mathscr{A} \,;\ \inf_{\mathscr{A} \ni B \subseteq A} \mu(B) = -\infty \right\}.$$

We have

$$\forall A \in \mathscr{G} \quad \forall a \in \mathbb{R} \quad \exists B \in \mathscr{G} \,:\, B \subseteq A \ \&\ \mu(B) \leqslant a. \tag{6.6}$$

Assume that (6.6) were not true, i.e. there is some $A \in \mathscr{G}$ and $a \in \mathbb{R}$ such that

$$\forall B \in \mathscr{G},\ B \subseteq A \,:\, \mu(B) > a. \tag{6.7}$$

Since $A \in \mathscr{G}$, there is some $A_0 \in \mathscr{A}$ such that $\mu(A_0) \leqslant a$. In view of (6.7) we see that $A_0 \notin \mathscr{G}$. This means that there is some $a_0 \in \mathbb{R}$ such that for all $B \subseteq A_0$, $B \in \mathscr{A}$ we have $\mu(B) \geqslant a_0$. Thus,

$$\forall C \in \mathscr{A},\ C \subseteq A \,:\, \mu(C) = \mu(C \setminus A_0) + \mu(C \cap A_0) \geqslant \mu(C \setminus A_0) + a_0.$$

As $A \in \mathscr{G}$, the estimate $\mu(C \setminus A_0) \leqslant \mu(C) - a_0$ for $C \subseteq A$, tells us that $A \setminus A_0 \in \mathscr{G}$. In particular, there is some $A_1 \subseteq A \setminus A_0$, $A_1 \in \mathscr{A}$, such that $\mu(A_1) \leqslant -1$. Since $A_0, A_1 \subseteq A$ are disjoint, we get

$$\mu(A_0 \uplus A_1) = \mu(A_0) + \mu(A_1) \leqslant a - 1 \leqslant a \overset{(6.7)}{\Longrightarrow} A_0 \uplus A_1 \notin \mathscr{G}.$$

Repeating these steps with A_0 replaced by $A_0 \uplus A_1$ furnishes a sequence of disjoint sets $(A_n)_{n\in\mathbb{N}} \subseteq \mathscr{A}$ such that $\mu(A_n) \leqslant -1$ for all $n \in \mathbb{N}$. Then, however,

$$\mu\left(\biguplus_{n\in\mathbb{N}} A_n\right) = \sum_{n\in\mathbb{N}} \mu(A_n) \leqslant \sum_{n\in\mathbb{N}} (-1) = -\infty$$

contradicts the fact that μ has values in $(-\infty, \infty]$. This proves (6.6).

Now assume that μ is not bounded below, i.e. $\inf_{A\in\mathscr{A}} \mu(A) = -\infty$. This means that $X \in \mathscr{G}$. Because of (6.6) we can find a decreasing sequence of sets $(A_n)_{n\in\mathbb{N}} \subseteq \mathscr{G} \subseteq \mathscr{A}$ such that $\mu(A_n) \leqslant -n$ for all $n \in \mathbb{N}$. Set $A := \bigcap_{n\in\mathbb{N}} A_n$. Since μ is σ-additive, we have

$$\mu(A_n) = \mu(A) + \sum_{i=n}^{\infty} \mu(A_i \setminus A_{i+1}).$$

Since the series converges, we get

$$\mu(A) = \lim_{n\to\infty} \mu(A_n) \leqslant \lim_{n\to\infty} (-n) = -\infty,$$

which is impossible since μ takes values in $(-\infty, \infty]$.

7

Measurable and Non-Measurable Sets

Let $(X, \mathscr{A})$ be a measurable space, i.e. a non-empty set X equipped with a σ-algebra $\mathscr{A}$. The sets in $\mathscr{A}$ are usually called **measurable sets**. Most examples in this chapter deal with measurable sets in $\mathbb{R}$ or $\mathbb{R}^d$, but we play with the notion of measurability by using different σ-algebras. Recall that the Borel sets $\mathscr{B}(\mathbb{R}^d)$ are generated by the open sets of $\mathbb{R}^d$, the Lebesgue sets $\mathscr{L}(\mathbb{R}^d)$ are obtained as completion of $\mathscr{B}(\mathbb{R}^d)$ with respect to Lebesgue measure λ, i.e.

$$\mathscr{L}(\mathbb{R}^d) = \sigma(\mathscr{B}(\mathbb{R}^d), \mathscr{N}^d),$$

where $\mathscr{N}^d = \{M \subseteq \mathbb{R}^d\,;\ \exists N \in \mathscr{B}(\mathbb{R}^d),\ \lambda(N) = 0,\ M \subseteq N\}$ is the family of all subsets of Borel measurable null sets. Finally, there is the family of **analytic** or **Souslin sets**. Originally, Souslin sets were defined as all sets that can be written in the following way:

$$A = \bigcup_{(i_k)_{k\in\mathbb{N}}\in\mathbb{N}^{\mathbb{N}}} \bigcap_{k\in\mathbb{N}} C_{i_1,\dots,i_k}$$

where $C_{i_1,\dots,i_k}$ is from some family of sets $\mathscr{C}$. This operation is called a Souslin scheme; notice that the union extends over all sequences $\mathbb{N}^{\mathbb{N}}$, i.e. it is an uncountable union which may destroy measurability even if the sets $C_{i_1,\dots,i_k}$ are from a σ-algebra. If $\mathscr{C}$ are the open balls in $\mathbb{R}^d$ with rational centres and rational radii, the family $\alpha(\mathscr{C})$ so defined are the analytic sets or Souslin sets in $\mathbb{R}^d$. Alternatively, one could replace $\mathscr{C}$ by the open sets, the closed sets or the compact sets of $\mathbb{R}^d$. Souslin's motivation was to fix the erroneous claim of Lebesgue [103] that projections of Borel sets in $\mathbb{R}^2$ are Borel sets in $\mathbb{R}$ [☞ 15.12]. Instead, he arrived at a new 'constructive' (i.e. without the axiom of choice) characterization of Borel sets: B is a Borel set if, and only if, both B and B^c are Souslin sets; see [170] and [20, Corollary 6.6.2].

The following inclusions hold, cf. [MIMS, Appendix G]:

$$\mathscr{B}(\mathbb{R}^d) = \sigma(\mathscr{C}) \subseteq \alpha(\mathscr{C}) \subseteq \sigma(\alpha(\mathscr{C})) \subseteq \mathscr{L}(\mathbb{R}^d) \subseteq \mathscr{P}(\mathbb{R}^d),$$

and all inclusions '$\subsetneq$' are strict [☞7.19, 4.20, 7.21, 7.22]. Some of these proofs are constructive (using, however, the axiom of choice) and some are based on cardinality considerations. One has

$$\#\mathscr{B}(\mathbb{R}^d) = \#\alpha(\mathscr{C}) = \#\sigma(\alpha(\mathscr{C})) = \mathfrak{c} < 2^{\mathfrak{c}} = \#\mathscr{P}(\mathbb{R}^d) = \#\mathscr{L}(\mathbb{R}^d)$$

[MIMS, Theorems 2.10, G.2, G.6, Lemma G.9] and Example 4.20; for further cardinality arguments involving σ-algebras [☞4.19, 4.20].

Modern texts use the following equivalent definitions of Souslin sets. A proof can be found in [49, Theorem 13.2.1].

Theorem 7A *A set* $A \subseteq \mathbb{R}^d$ *is Souslin if, and only if, one of the following equivalent conditions holds:*

(a) $A = f(\mathbb{R})$ *for some left-continuous function* $f : \mathbb{R} \to \mathbb{R}^d$.

(b) $A = g(\mathbb{N}^{\mathbb{N}})$ *for a Borel measurable (or continuous) function* $g : \mathbb{N}^{\mathbb{N}} \to \mathbb{R}^d$.

(c) $A = h(B)$ *for some Borel set* $B \in \mathscr{B}(X)$ *of a Polish space* X *and Borel measurable (or continuous) function* $h : B \to \mathbb{R}^d$.

Further characterizations of Souslin sets can be found in Example 7.19.

We will frequently use Baire's notion of the category of sets [☞Definition 2.3]. Let (X, d) be a metric space. Recall that a set $A \subseteq X$ is **nowhere dense**, if $\overline{A}$ has empty interior. A set is said to be of **first category** or **meagre** if it is a countable union of nowhere dense sets. A set is of **second category**, if it is not of first category. For more properties and examples we refer to p. 40 *et seq.* Baire's category divides sets into topologically small (first category) and big (second category) sets, but this notion of smallness may be completely different from measure-theoretic smallness, i.e. being a null set [☞7.8, 7.9, 7.10]. A very good introduction to this problem area is Oxtoby [133].

7.1 A dense open set in $(0,1)$ with arbitrarily small Lebesgue measure

Fix $\epsilon > 0$ and let $(q_n)_{n\in\mathbb{N}}$ be an enumeration of the rationals in $(0,1)$. For $n \in \mathbb{N}$ set $U_n := (q_n - \epsilon 2^{-n}, q_n + \epsilon 2^{-n}) \cap (0,1)$. Clearly, $U := \bigcup_{n\in\mathbb{N}} U_n \supseteq \mathbb{Q} \cap (0,1)$ is open and dense; moreover, by σ-subadditivity

$$\lambda(U) \leqslant \sum_{n=1}^{\infty} \lambda(U_n) = \sum_{n=1}^{\infty} \epsilon 2^{-n+1} = 2\epsilon.$$

7.2 A set of positive Lebesgue measure which does not contain any interval

The irrationals $\mathbb{R}\setminus\mathbb{Q}$ have strictly positive Lebesgue measure but do not contain any interval.

Comment [☞7.3] for a further example.

7.3 A Cantor-like set with arbitrary measure

Recall the construction of Cantor's middle-third set [☞Section 2.5]. We use the following variant of this construction: fix $\alpha \in [0,1)$ and take a sequence $(a_n)_{n\in\mathbb{N}_0}$ in $(0,1)$ such that $\sum_{n=0}^{\infty} a_n = 1-\alpha$. Set $C_0 := [0,1]$. Recursively, we obtain closed sets

C_1 by removing from the middle of C_0 an open interval I_2 of length a_0

C_2 by removing from the middles of the two parts of C_1 the open intervals I_{02} and I_{22}, each of length $\frac{1}{2}a_1$

C_3 by removing from the middles of the four parts of C_2 the open intervals $I_{002}, I_{022}, I_{202}$ and I_{222}, each of length $\frac{1}{4}a_2$

...

C_{n+1} by removing from the middles of the 2^n parts of C_n the open intervals $I_{t_n t_{n-1}\dots t_1 2}$, $t_i \in \{0,2\}$, each of length $2^{-n}a_n$.

and we define $C := C^{(\alpha)} := \bigcap_{n\in\mathbb{N}_0} C_n$. If $a_n = \frac{1}{3}\left(\frac{2}{3}\right)^n$ we get the classical Cantor set C. Because of the similarity of the construction, the sets $C^{(\alpha)}$ share the same topological properties as C: they are uncountable perfect subsets of $[0,1]$.

Let us look at their Lebesgue measure. By the continuity of measures, we have

$$\lambda(C^{(\alpha)}) = \lim_{n\to\infty} \lambda(C_{n+1}) = \lim_{n\to\infty}\left(1-\sum_{i=0}^{n} a_0\right) = \alpha.$$

Notice that, although $\lambda(C^{(0)}) = 0$, $C^{(0)}$ is not empty. We call any $C^{(\alpha)}$ with $\alpha > 0$ a **fat Cantor set** since it has strictly positive Lebesgue measure.

Often, one uses a **geometric fat Cantor set** where in the $(n+1)$st step, sets of length γ^{n+1} are removed from the centres of the 2^n remaining components. In this case, the sequence $a_n = \gamma(2\gamma)^n$ with $\gamma = \frac{1-\alpha}{3(1-\alpha)+\alpha}$ for $\alpha \in [0,1)$.

7.4 An uncountable set of zero measure

The classical middle-thirds Cantor set [☞ Section 2.5] is uncountable and has Lebesgue measure zero. The construction from Example 7.3 yields further sets with these two properties. The Cantor set $C = C^{(\alpha)}$ [☞ 7.3] is uncountable, in fact we may identify it with the set of triadic points

$$C' = \left\{ \sum_{n=1}^{\infty} \frac{t_n}{3^n} \,;\; t_n \in \{0, 2\} \right\}.$$

To do so, use the mapping $a_n \to 3^{-n-1}$ so that the right end point of the interval $I_{t_n t_{n-1} \dots t_2 2}$ $(t_1 = 2)$ appearing in the construction in Example 7.3 is identified with the triadic number $0.t_1 t_2 \dots t_n = \sum_{k=1}^{n} \frac{t_k}{3^k}$. Using Cantor's diagonal argument, it is easy to see that the triadic numbers are uncountable, hence C' and C are uncountable.

If $\alpha = 0$, e.g. if $a_n = 2^n 3^{-n-1}$, then $\lambda(C) = 0$, and the claim follows.

7.5 A Lebesgue null set $A \subseteq \mathbb{R}$ such that for every $\delta \in [0, 1]$ there exist $x, y \in A$ with $\delta = |x - y|$

A theorem by Steinhaus [☞ 7.28] states that the pointwise difference

$$A - A = \{x - y \,;\; x, y \in A\}$$

of every set $A \in \mathscr{B}(\mathbb{R})$ of strictly positive Lebesgue measure contains a neighbourhood of $x = 0$. In other words, $\lambda(A) > 0$ implies that there is a constant $r = r(A) > 0$ such that for any $\delta \in [0, r]$ there exist $x, y \in A$ with $\delta = |x - y|$. The condition $\lambda(A) > 0$ is not sharp, i.e. also the pointwise difference $A - A$ of a Lebesgue null set A may contain a neighbourhood of zero. An example is the classical middle-thirds Cantor set $C \subseteq [0, 1]$ [☞ Section 2.5]: it has Lebesgue measure zero and satisfies $C - C = [-1, 1]$ [☞ 7.30]. Another example is the set

$$A := \left\{ \sum_{k=2}^{\infty} \frac{a_k}{k!} \,;\; a_k \in \{0, \dots, k-2\} \right\} \subseteq [0, 1],$$

which was first studied by Erdös and Kakutani [55]. We will prove that A is a Lebesgue null set and that for every $\delta \in [0, \frac{1}{6})$ there exist $x, y \in A$ with $\delta = |x - y|$.

We start with some preparations.

Using a telescoping sum argument we see that

$$\forall n \in \mathbb{N} : \sum_{k=n+1}^{\infty} \frac{k-1}{k!} = \sum_{k=n+1}^{\infty} \left(\frac{1}{(k-1)!} - \frac{1}{k!} \right) = \frac{1}{n!}. \tag{7.1}$$

It follows from (7.1) that every point $x \in [0,1)$ can be written in the form

$$x = \sum_{k=2}^{\infty} \frac{a_k}{k!}, \qquad a_k \in \{0, \dots, k-1\}, \tag{7.2}$$

and this representation is unique except for countably many points.[1]

Now we return to the study of the set A. First, we will show that A is a Lebesgue null set. Consider

$$A_n := \{x \in [0,1) \,;\, a_k \neq k-1 \text{ in (7.2) for } k = 2, \dots, n\}, \quad n \geqslant 2.$$

There are $(n-1)!$ possibilities to choose the first $n-1$ coefficients $a_2, \dots, a_n$ of the points in A_n, i.e. there exists a set $F_n \subseteq [0,1)$ of cardinality $(n-1)!$ such that any $x \in A_n$ can be written as

$$x = y + \sum_{k=n+1}^{\infty} \frac{a_k}{k!}$$

for $y \in F_n$ and $a_k \in \{0, \dots, k-1\}$, $k \geqslant n+1$. Since, by (7.1),

$$\left| \sum_{k=n+1}^{\infty} \frac{a_k}{k!} \right| \leqslant \sum_{k=n+1}^{\infty} \frac{k-1}{k!} = \frac{1}{n!},$$

it follows that $x \in [y, y+\frac{1}{n!}]$. This means that A_n is contained in $(n-1)!$ intervals of length $\frac{1}{n!}$, and so $\lambda(A_n) \leqslant (n-1)! \cdot \frac{1}{n!} = \frac{1}{n}$. As $A_n \downarrow A$, we find that

$$\lambda(A) = \lim_{n \to \infty} \lambda(A_n) = 0.$$

Next we show that for every $\delta \in [0, \frac{1}{6})$ there exist $x, y \in A$ such that $\delta = |x-y|$. Fix $\delta \in [0, \frac{1}{6})$, then

$$\delta = \sum_{k=2}^{\infty} \frac{a_k}{k!}$$

with $a_2 = 0$, $a_3 = 0$ and $a_k \in \{0, \dots, k-1\}$, $k \geqslant 4$. We need to find some $x \in A$,

[1] If $x = \sum_{k=2}^{n-1} \frac{a_k}{k!} + \frac{a_n+1}{n!}$ for some $a_n \in \{0, \dots, n-2\}$, then (7.1) shows that $x = \sum_{k=2}^{n} \frac{a_k}{k!} + \sum_{k=n+1}^{\infty} \frac{k-1}{k!}$ is another admissible representation. To avoid this problem, we can assume that the coefficients a_k in (7.2) satisfy $a_k \neq k-1$ for infinitely many $k \geqslant 2$; this gives a unique representation for all $x \in [0,1)$.

which can be written in the form $x = \sum_{k=2}^{\infty} \frac{b_k}{k!}$, $0 \leqslant b_k \leqslant k-2$, such that $y := x + \delta = \sum_{k \geqslant 2} \frac{a_k + b_k}{k!} \in A$. Set

$$b_2 = b_3 = 0 \quad \text{and} \quad b_k := \begin{cases} 0, & \text{if } a_k \in \{0, \dots, k-3\}, \\ 1, & \text{if } a_k = k-1, \\ 2, & \text{if } a_k = k-2, \end{cases} \qquad k \geqslant 4,$$

and define $I := \{k \geqslant 4\,;\; a_k \leqslant k-3\}$, then

$$x + \delta = \sum_{k \geqslant 4, k \in I} \frac{a_k}{k!} + \sum_{k \geqslant 4, k \notin I} \frac{k}{k!}.$$

Since

$$\sum_{k \geqslant 4, k \notin I} \frac{k}{k!} = \sum_{k \geqslant 4, k \notin I} \frac{1}{(k-1)!} = \sum_{k \geqslant 3, k+1 \notin I} \frac{1}{k!},$$

it follows that $x + \delta = \sum_{k \geqslant 3} \frac{c_k}{k!}$ with

$$c_k := \begin{cases} a_k, & \text{if } k \in I, k+1 \in I, \\ a_k + 1, & \text{if } k \in I, k+1 \notin I, \\ 0, & \text{if } k \notin I, k+1 \in I, \\ 1, & \text{if } k \notin I, k+1 \notin I. \end{cases}$$

By definition, $a_k \leqslant k-3$ for $k \in I$, and so $c_k \in \{0, \dots, k-2\}$ for all $k \geqslant 3$. Consequently, $y = x + \delta \in A$.

Comment With a bit more effort, one can show that A contains a copy of any finite set with small diameter. More precisely, if $F \subseteq \mathbb{R}$ is a finite set of cardinality $n := \#F$ and its diameter $\delta := \sup_{x,y \in F} |x - y|$ is sufficiently small, in the sense that $\delta \leqslant r$ for a constant $r = r(n) > 0$, then there is some $t \in \mathbb{R}$ such that $t + F \subseteq A$. We have shown that A satisfies this property for $n = 2$ but, in fact, it holds for any $n \geqslant 2$, cf. [55]. By contrast, the Cantor set C satisfies this property only for $n = 2$. Let us sketch what goes wrong for $n = 3$. Take $0 < x < y < z < 1$ such that $\delta_1 := y - x \in \left(0, \frac{1}{3}\right)$. If $x, y \in C$, the construction of the Cantor set shows that x, y are either both in the 'left half' of the Cantor set, i.e. $x, y \in \left[0, \frac{1}{3}\right]$, or in the 'right half' of C, i.e. $x, y \in \left[\frac{2}{3}, 1\right]$. Choosing δ_1 close to $\frac{1}{3}$, it follows that the points x, y are close to the boundary of this interval. On the right of this interval, the Cantor set has a gap of length $\frac{1}{3}$. Consequently, we can choose $\delta_2 := z - y$ in such a way that $z = y + \delta_2 \notin C$, e.g. $\delta_2 = \frac{1}{2} \cdot \frac{1}{3}$.

This shows that C does not contain any copy of the set $F := \{0, \delta_1, \delta_1 + \delta_2\}$. Because of the self-similarity of C, this implies automatically that there are sets $F = \{x, y, z\}$ of arbitrary small diameter such that $t + F \not\subseteq C$ for all $t \in \mathbb{R}$.

7.6 A dense open set whose complement has positive measure

Let $C = C^{(\alpha)}$ be a fat Cantor set [☞7.3] in the measure space $([0,1], \mathscr{B}[0,1], \lambda)$ such that $\lambda(C) = \alpha$ for some $\alpha \in (0,1)$. Since C is a nowhere dense, closed set, its complement $B := [0,1] \setminus C$ is a dense open set in $[0,1]$. Thus, B is a dense open set whose complement has Lebesgue measure $\alpha > 0$.

7.7 A compact set whose boundary has positive Lebesgue measure

Let $C \subseteq [0,1]$ be a fat Cantor set [☞7.3] with Lebesgue measure $\lambda(C) > 0$. Since C is closed and bounded, it is compact. The set C has empty interior, and therefore the boundary ∂C of C equals C. Hence, C is a compact set whose boundary has measure $\lambda(\partial C) = \lambda(C) > 0$.

Comment The complement $U := [0,1] \setminus C$ is open and dense in $[0,1]$; therefore U is an open set whose boundary has strictly positive Lebesgue measure $\lambda(\partial U) = 1 - \lambda(U) = \lambda(C) > 0$.

7.8 A set of first category in $[0,1]$ with measure one

We are going to construct a Borel set B of first category in $[0,1]$ with $\lambda(B) = 1$. For this we use the fat Cantor sets $C^{(1-1/n)}$ constructed in Example 7.3 and set

$$B := \bigcup_{n \in \mathbb{N}} C^{(1-1/n)}.$$

This is a countable union of nowhere dense sets, hence it is of first category. Moreover,

$$\lambda(B) \geqslant \lambda(C^{(1-1/n)}) = 1 - \frac{1}{n} \xrightarrow[n\to\infty]{} 1.$$

Comment [☞7.9] for a further example: the set A^c in Example 7.9 is of first category and has measure one.

7.9 A set of second category with measure zero

We are going to construct a Borel set $A \subseteq [0,1]$ of second category such that $\lambda(A) = 0$. Let $(q_n)_{n\in\mathbb{N}}$ be an enumeration of $\mathbb{Q}$ and set

$$A = [0,1] \cap \bigcap_{n\in\mathbb{N}} U_n \quad \text{where} \quad U_n = \bigcup_{k\in\mathbb{N}} \left(q_k - 2^{-n-k}, q_k + 2^{-n-k}\right).$$

By σ-subadditivity and the continuity of measures,

$$\lambda(A) \leqslant 2\sum_{k=1}^{\infty} 2^{-n-k} = 2\cdot 2^{-n} \xrightarrow[n\to\infty]{} 0.$$

The complement $[0,1]\setminus A$ is of first category since

$$[0,1]\setminus A = \bigcup_{n\in\mathbb{N}} (U_n^c \cap [0,1])$$

and each U_n^c is closed and has empty interior: this follows from $\overline{U_n} \supseteq \overline{\mathbb{Q}} = \mathbb{R}$. Since we cannot write $[0,1] = A \cup ([0,1]\setminus A)$ as a union of two first-category sets, we see that A is of second category.

7.10 An uncountable, dense set of measure zero such that the complement is of first category

There is a dense set $A \subseteq [0,1]$ with the following properties: A has Lebesgue measure zero, it is of second category, i.e. A cannot be written as a countable union of nowhere dense sets, and the complement of A is of first category, [☞7.9] for the construction of such a set. It just remains to note that A is uncountable: if A were countable, then $A = \bigcup_{n\in\mathbb{N}}\{x_n\}$ for some sequence $(x_n)_{n\in\mathbb{N}} \subseteq [0,1]$. Since singletons are nowhere dense, this would mean that A can be written as a countable union of nowhere dense sets, in contradiction to the fact that A is of second category.

7.11 A null set which is not an F_σ-set

Denote by λ Lebesgue measure on $(\mathbb{R}, \mathscr{B}(\mathbb{R}))$. An F_σ-set is a set which can be written as a union of countably many closed sets. The set A constructed in Example 7.9 is a set of second category which is also a null set. We claim that it cannot be an F_σ-set. Assume, to the contrary, that $A = \bigcup_{n\in\mathbb{N}} F_n$ for closed sets F_n. Since A is a null set, the sets F_n are also null sets, but closed null sets are nowhere dense. Indeed, the interior has zero Lebesgue measure

$\lambda(F_n^\circ) = 0$, and since $\emptyset$ is the only open set with Lebesgue measure zero, this implies $(\overline{F_n})^\circ = F_n^\circ = \emptyset$. This means that we could write A as a union of nowhere dense sets, i.e. A would be a set of first category, which is not the case.

Comment 1 Since F_σ-sets are Borel measurable, any non-measurable subset of a Borel null set [☞7.22] is also an example of a (subset of a) null set which is not an F_σ-set.

Comment 2 We will see in Example 8.10 that the above construction is also an example of a (measurable) null set which cannot appear as the set of discontinuities of **any** function.

7.12 A Borel set which is neither F_σ nor G_δ

A subset of a metric space (X, d) is an **F_σ-set** if it can be written as a countable union of closed sets. It is called a **G_δ-set** if it is the countable intersection of open sets.

Set $F := \mathbb{Q} \cap [0, 1]^c$ and $G := [0, 1] \cap (\mathbb{R} \setminus \mathbb{Q})$. Then:

1° F is an F_σ-set as a union of countably many points.

2° F is not a G_δ-set. Indeed: suppose that $F = \bigcap_{n\geqslant 1} U_n$ for open sets U_n, then each U_n is dense in $\mathbb{R} \setminus (0, 1)$. If $(q_n)_{n\in\mathbb{N}}$ is an enumeration of F, then $V_n := (0, 1) \cup (U_n \setminus \{q_n\})$ is an open dense set in $\mathbb{R}$ and

$$\bigcap_{n\in\mathbb{N}} V_n = (0, 1) \cup \bigcap_{n\in\mathbb{N}} (U_n \setminus \{q_n\}) = (0, 1).$$

This contradicts Baire's category theorem [☞Theorem 2.4] which states that the intersection of countably many open dense sets in $\mathbb{R}$ is dense in $\mathbb{R}$.

3° $G = \bigcap_{q\in\mathbb{Q}} ((0, 1) \setminus \{q\})$ is a G_δ-set.

4° G is not an F_σ-set. Indeed, if G were an F_σ-set, then $\mathbb{R} \setminus G$ would be a G_δ-set, which gives a contradiction to Baire's category theorem using a very similar reasoning as in **2°**.

The union $B := F \cup G$ is a Borel set which is neither F_σ nor G_δ. Indeed, if $B = \bigcap_{n\in\mathbb{N}} U_n$ for open sets U_n, then

$$F = B \cap [0, 1]^c = \bigcap_{n\in\mathbb{N}} (U_n \cap [0, 1]^c),$$

i.e. F would be a G_δ-set in contradiction to the above considerations. Similarly, if $B = \bigcup_{n\in\mathbb{N}} F_n$ for closed sets F_n, then

$$G = B \cap [0, 1] = \bigcup_{n\in\mathbb{N}} (F_n \cap [0, 1])$$

would be F_σ, which is a contradiction.

Comment Example 4.21 contains a much stronger result: there are Borel sets that cannot be constructed from the open (or closed) sets by taking unions and complements countably many times.

7.13 Each Borel set is the union of a null set and a set of first category

Let $B \subseteq \mathbb{R}$ be a Borel set. Since there is nothing to show if $\lambda(B) = 0$, we may assume that $\lambda(B) > 0$. Fix $0 < b < \lambda(B)$ and observe that $\lambda(B \setminus \mathbb{Q}) = \lambda(B)$. Lebesgue measure λ is inner compact regular, and so there is a compact set $K = K(b) \subseteq B \setminus \mathbb{Q}$ such that $b < \lambda(K) \leqslant \lambda(B)$. By definition, K contains no rationals and hence no open interval, which means that K has empty interior; in other words: K is nowhere dense.

Recall that countable unions of nowhere dense sets are of first category (in the sense of Baire) [☞ Definition 2.3]. Thus, the set $K' := \bigcup_{b<\lambda(B),b\in\mathbb{Q}} K(b) \subseteq B$ is measurable, of first category and satisfies $\lambda(K') = \lambda(B)$; in particular, we have $\lambda(B \setminus K') = 0$.

Comment Our example shows that B can be decomposed into two 'small' sets: a topologically small set (first category) and a measure-theoretic small set (null set).

Because of the continuity of Lebesgue measure, the maps $r \longmapsto \lambda(K' \cap \overline{B_r(0)})$ and $r \longmapsto \lambda(K(b) \cap \overline{B_r(0)})$ are continuous. Therefore, our construction also shows that B contains subsets of first category, resp. compact nowhere dense sets, with any measure in $[0, \lambda(B)]$, resp. $[0, \lambda(B))$; note that the case $\lambda(B) = \infty$ is not excluded.

7.14 A set $B \subseteq \mathbb{R}$ such that $B \cap F \neq \emptyset$ and $B^c \cap F \neq \emptyset$ for any uncountable closed set F

There are continuum many ($\mathfrak{c}$) open subsets in $\mathbb{R}$, hence $\mathfrak{c}$ many uncountable closed subsets of $\mathbb{R}$. Let ω be the smallest ordinal number of cardinality $\mathfrak{c}$ – this number exists because of Zorn's lemma – and let $\{F_\alpha ; \alpha \prec \omega\}$ be a well-ordering of all uncountable closed subsets of $\mathbb{R}$. Using transfinite induction, we define the set B as follows:

- Let $x_0, y_0 \in F_0$ and $x_0 \neq y_0$.
- Assume that $x_\gamma, y_\gamma \in F_\gamma$, $x_\gamma \neq y_\gamma$ have been constructed for all $\gamma \prec \alpha$ and some $\alpha \prec \omega$.
- The set $A_\alpha := \{x_\gamma, y_\gamma ; \gamma \prec \alpha\}$ has cardinality strictly less than $\mathfrak{c}$ since $\alpha \prec \omega$ and ω is the smallest ordinal of cardinality $\mathfrak{c}$. Thus, $F_\alpha \setminus A_\alpha$ is uncountable. Pick $x_\alpha, y_\alpha \in F_\alpha \setminus A_\alpha$ with $x_\alpha \neq y_\alpha$.

Set $B := \{x_\alpha\,;\ \alpha \prec \omega\}$, then $B \cap F_\alpha \neq \emptyset$ as well as $F_\alpha \setminus B \neq \emptyset$ for all $\alpha \prec \omega$.

Comment This example is due to F. Bernstein.

7.15 A Borel set $B \subseteq \mathbb{R}$ such that $\lambda(B \cap I) > 0$ and $\lambda(B^c \cap I) > 0$ for all open intervals $I \neq \emptyset$

Enumerate all open intervals with rational endpoints $q < r$, $q, r \in [0,1] \cap \mathbb{Q}$ and call them $(I_n)_{n\in\mathbb{N}}$.

Since $\lambda(I_1) > 0$, there is a nowhere dense compact set $A_1 \subseteq I_1$ with $\lambda(A_1) > 0$, e.g. a 'fat' Cantor set [☞ 7.3]. The set $I_1 \setminus A_1$ contains intervals, and so there is another nowhere dense, compact set $B_1 \subseteq I_1 \setminus A_1$ with $\lambda(B_1) > 0$. Note that $A_1 \cap B_1 = \emptyset$.

Iterating this step we obtain nowhere dense sets $A_2 \subseteq I_2 \setminus (A_1 \cup B_1)$ and $B_2 \subseteq I_2 \setminus (A_1 \cup B_1 \cup A_2)$ such that $\lambda(A_2) > 0$ and $\lambda(B_2) > 0$.

Removing from any interval finitely many nowhere dense sets means that the leftover contains some interval. Thus, step $n+1$ furnishes nowhere dense sets $A_{n+1} \subseteq I_{n+1} \setminus \bigcup_{i=1}^n (A_i \cup B_i)$ and $B_{n+1} \subseteq I_{n+1} \setminus \bigcup_{i=1}^n \big((A_i \cup B_i) \cup A_{n+1}\big)$ such that $\lambda(A_{n+1}) > 0$ and $\lambda(B_{n+1}) > 0$.

Set $A' := \bigcup_{n=1}^\infty A_n$ and $B' := \bigcup_{n=1}^\infty B_n$. Any open interval $\emptyset \neq I \subseteq [0,1]$ contains some $I_n \subseteq I$ and, by construction, $A_n \uplus B_n \subseteq I_n$. This means, in particular,

$$\lambda(B' \cap I) \geqslant \lambda(B' \cap I_n) \geqslant \lambda(B_n) > 0,$$
$$\lambda((\mathbb{R} \setminus B') \cap I) \geqslant \lambda(A' \cap I) \geqslant \lambda(A_n) > 0.$$

Thus, the set $B := \bigcup_{k\in\mathbb{Z}} (B' + k)$ is the set we were looking for.

7.16 There is no Borel set B with $\lambda(B \cap I) = \frac{1}{2}\lambda(I)$ for all intervals I

Suppose there are $B \in \mathscr{B}(\mathbb{R})$ and $c \in (0,1)$ such that $\lambda(B \cap I) = c\lambda(I)$ for any open interval $I \subseteq \mathbb{R}$. Then

$$\lim_{r\to 0} \frac{1}{2r} \int_{(x-r,x+r)} \mathbb{1}_B(y)\,\lambda(dy) = \lim_{r\to 0} \frac{1}{2r} \lambda(B \cap (x-r, x+r)) = c$$

for all $x \in \mathbb{R}$. This contradicts Lebesgue's differentiation theorem [☞ 14C], which shows that

$$\lim_{r\to 0} \frac{1}{2r} \int_{(x-r,x+r)} \mathbb{1}_B(y)\,\lambda(dy) = \mathbb{1}_B(x) \in \{0,1\}$$

for Lebesgue almost all $x \in \mathbb{R}$.

Comment [☞7.15] for a set $B \in \mathscr{B}(\mathbb{R})$ satisfying

$$\exists c = c(I) \in (0,1) \ : \ \lambda(B \cap I) = c\lambda(I)$$

for all intervals I; in view of the above counterexample, it is impossible to construct B in such a way that c does not depend on I.

7.17 A non-Borel set B such that $K \cap B$ is Borel for every compact set K

Consider $S = \mathbb{R}$ with the Sorgenfrey topology [☞Example 2.1.(h)]. The corresponding Borel σ-algebra $\mathscr{B}(S)$ coincides with the usual (i.e. Euclidean) Borel σ-algebra $\mathscr{B}(\mathbb{R})$ [☞4.18]. Pick $B \in \mathscr{P}(S) \setminus \mathscr{B}(S) = \mathscr{P}(\mathbb{R}) \setminus \mathscr{B}(\mathbb{R})$ [☞7.20]. The compact sets in S are countable, and so $B \cap K$ is countable for any compact set $K \subseteq S$; in particular, $B \cap K \in \mathscr{B}(\mathbb{R}) = \mathscr{B}(S)$; but, by construction, we have $B \notin \mathscr{B}(S)$.

Comment The Sorgenfrey line is not locally compact, and one might suspect that this is the reason for this pathology. This is, however, not true: it is possible to construct a locally compact space X and a set $B \notin \mathscr{B}(X)$ such that $K \cap B$ is Borel for all compact sets K, cf. [93, p. 362]. In particular, [147, Lemma 9, p. 334] is false.

7.18 A convex set which is not Borel

Pick $A \subseteq (0, 2\pi)$ such that $A \notin \mathscr{B}(\mathbb{R})$, and define a convex set by

$$C := \{x \in \mathbb{R}^2 \,;\ |x| < 1\} \cup \{(\cos\phi, \sin\phi)^T \,;\ \phi \in A\} \subseteq \mathbb{R}^2,$$

i.e. C is the union of the open unit ball together with a non-measurable part of its boundary. Since the mapping

$$f : (\mathbb{R}, \mathscr{B}(\mathbb{R})) \to (\mathbb{R}^2, \mathscr{B}(\mathbb{R}^2)), \quad \phi \longmapsto f(\phi) := \begin{pmatrix} \cos\phi \\ \sin\phi \end{pmatrix}$$

is continuous, hence Borel measurable, it follows from $A = f^{-1}(C) \notin \mathscr{B}(\mathbb{R})$ that $C \notin \mathscr{B}(\mathbb{R}^2)$.

Comment In dimension $d = 1$ all convex sets are Borel sets; this follows from the fact that any convex set $C \subseteq \mathbb{R}$ is an interval. For every dimension $d \geqslant 1$ it holds that convex sets $C \subseteq \mathbb{R}^d$ are Lebesgue measurable. In fact, every convex set is μ^* measurable [☞(9.5)] if μ is any σ-finite product measure on $(\mathbb{R}^d, \mathscr{B}(\mathbb{R}^d))$. If $\mu = \lambda^d$, then the σ-algebra induced by $(\lambda^d)^*$ is just $\mathscr{L}(\mathbb{R}^d)$. A simple proof is given by Lang [95].

7.19 A Souslin set which is not Borel

Let (X, d) be a Polish space, i.e. a complete separable metric space. A set $A \subseteq X$ is called a **Souslin set** if it is the image of a Polish space under a continuous mapping. Souslin sets are also known as **analytic sets**. As we will see below, every Borel set $B \in \mathscr{B}(X)$ is a Souslin set. The converse, however, is not true: We are going to construct a Souslin set $A \subseteq \mathbb{R}$ which is not a Borel set. For the construction – and more generally, for the study of Souslin sets – the space $\mathbb{N}^{\mathbb{N}}$ of infinite sequences $t = (t(n))_{n\in\mathbb{N}} \subseteq \mathbb{N}$ plays an important role. Equipped with the metric

$$\rho(t, u) := \sum_{n\in\mathbb{N}} 2^{-n} \frac{|t(n) - u(n)|}{1 + |t(n) - u(n)|}, \qquad t, u \in \mathbb{N}^{\mathbb{N}},$$

the space $(\mathbb{N}^{\mathbb{N}}, \rho)$ becomes a separable complete metric space. We will first prove the existence of a non-Borel Souslin set $A \subseteq \mathbb{N}^{\mathbb{N}}$, and then we will 'transport' this set to the Euclidean space. We need the following characterizations of Souslin sets, cf. [171, Proposition 4.1.1] or [20, Theorem 6.7.2].

Theorem 1 *Let X be a Polish space and $A \subseteq X$. The following statements are equivalent:*

(a) *A is a Souslin set;*
(b) *A is the projection of a closed set in the space $X \times \mathbb{N}^{\mathbb{N}}$;*
(c) *A is the projection of a Borel set in $X \times \mathbb{R}$.*

Here, and below, the product $X \times Y$ of two metric spaces (X, d), (X', d') is considered as a metric space with the canonical metric

$$\tilde{d}((x, x'), (y, y')) := d(x, y) + d'(x', y'), \quad x, y \in X, \ x', y' \in X'.$$

As a direct consequence of Theorem 1, every Borel set $B \in \mathscr{B}(X)$ is a Souslin set since it can be written as a projection of the Borel set $B \times \{0\} \in \mathscr{B}(X \times \mathbb{R})$. The next result is a powerful tool to transport Souslin sets from one metric space to another, cf. [20, Theorem 6.7.3].

Theorem 2 *Let X, Y be Polish spaces and $f : X \to Y$ Borel measurable.*

(a) *If $A \subseteq X$ is a Souslin set, then $f(A) \subseteq Y$ is a Souslin set.*
(b) *If $B \subseteq Y$ is a Souslin set, then $f^{-1}(B) \subseteq X$ is a Souslin set.*

Compare this with the situation for Borel sets: by definition, the pre-image $f^{-1}(B)$ of a Borel set $B \in \mathscr{B}(Y)$ under a Borel measurable mapping $f : X \to Y$ is a Borel set in X; by contrast, the image $f(A)$ of $A \in \mathscr{B}(X)$ need not be Borel even if f is continuous [☞ 7.24].

We are now ready to prove the existence of Souslin sets which are not Borel measurable. We start with two auxiliary results which are of independent interest.[2]

Lemma 1 *Let (X, d) be a Polish space. There exists a closed set $F \subseteq X \times \mathbb{N}^{\mathbb{N}}$ such that every closed set $C \subseteq X$ coincides with a slice of F, i.e. there exists some $t \in \mathbb{N}^{\mathbb{N}}$ such that $C = \{x \in X\,;\ (x,t) \in F\} =: F_t$.*

Proof By assumption, there exists a countable dense set $D \subseteq X$. Let $(I_n)_{n\in\mathbb{N}}$ be an enumeration of the collection of balls $\{y\,;\ d(x,y) < r\}$ with $x \in D$ and $r \in \mathbb{Q} \cap (0,\infty)$. The set

$$F := \left\{(x,t) \in X \times \mathbb{N}^{\mathbb{N}}\,;\ x \notin \bigcup_{n=1}^{\infty} I_{t(n)}\right\}$$

has the desired properties. Indeed: if $C \subseteq X$ is closed, then $U = X \setminus C$ is open, and therefore

$$U = \bigcup_{k\,;\,I_k \subseteq U} I_k.$$

If we define recursively $t(n) := \inf\{k > t(n-1)\,;\ I_k \subseteq U\}$, $t_0 := 0$, then $(I_{t(n)})_{n\in\mathbb{N}}$ is an enumeration of all balls I_k with $I_k \subseteq U$, and so

$$X \setminus C = U = \bigcup_{n\in\mathbb{N}} I_{t(n)} = X \setminus F_t,$$

i.e. $C = F_t$. In order to prove that $F \subseteq X \times \mathbb{N}^{\mathbb{N}}$ is closed, we show that its complement is open.

If $(x,t) \notin F$, then $x \in \bigcup_{n=1}^{\infty} I_{t(n)}$, i.e. there is some $n \in \mathbb{N}$ such that $x \in I_{t(n)}$. Since $I_{t(n)}$ is open, $x' \in I_{t(n)}$ for small $d(x,x') < r$. For $t' \in \mathbb{N}^{\mathbb{N}}$ close to t (e.g. $\rho(t',t) < 2^{-2-t(n)}$) we have $t'(n) = t(n)$. Hence, $x' \in I_{t'(n)}$, which implies $(x',t') \notin F$ for (x',t') in the chosen neighbourhood of (x,t). □

Lemma 2 *Let (X, d) be a Polish space. There exists a Souslin set $A \subseteq X \times \mathbb{N}^{\mathbb{N}}$ such that any Souslin set $S \subseteq X$ is a slice of A.*

Proof Apply the previous lemma to $X \times \mathbb{N}^{\mathbb{N}}$ and denote by $F \subseteq X \times \mathbb{N}^{\mathbb{N}} \times \mathbb{N}^{\mathbb{N}}$ the corresponding closed set. Define

$$A := \{(x,t)\,;\ (x,u,t) \in F \text{ for some } u \in \mathbb{N}^{\mathbb{N}}\}.$$

[2] The assertions of Lemma 1 and 2 should be compared with the construction of universal sets [☞4.21] where whole families of sets appear as slices of a suitable 'product' set.

By Theorem 1, A is a Souslin set as a projection of a closed set in $X \times \mathbb{N}^{\mathbb{N}} \times \mathbb{N}^{\mathbb{N}}$. If $S \subseteq X$ is a Souslin set, then S is the projection of a closed set $C \subseteq X \times \mathbb{N}^{\mathbb{N}}$, i.e.

$$S = \{x \in X;\ (x, u) \in C \text{ for some } u \in \mathbb{N}^{\mathbb{N}}\}.$$

It follows from our choice of F that C is a slice of F, i.e. $C = F_t$ for some $t \in \mathbb{N}^{\mathbb{N}}$. Hence,

$$\begin{aligned} S &= \{x \in X;\ (x, u) \in C = F_t \text{ for some } u \in \mathbb{N}^{\mathbb{N}}\} \\ &= \{x \in X;\ (x, u, t) \in F \text{ for some } u \in \mathbb{N}^{\mathbb{N}}\} = A_t. \end{aligned}$$ □

Corollary 1 *There exists a Souslin set $S \subseteq \mathbb{N}^{\mathbb{N}}$ which is not a Borel set.*

Proof Denote by $A \subseteq \mathbb{N}^{\mathbb{N}} \times \mathbb{N}^{\mathbb{N}}$ the Souslin set from Lemma 2 for $X := \mathbb{N}^{\mathbb{N}}$, and consider

$$S := \{u \in \mathbb{N}^{\mathbb{N}};\ (u, u) \in A\} \subseteq \mathbb{N}^{\mathbb{N}}.$$

Since the diagonal map

$$\mathbb{N}^{\mathbb{N}} \to \mathbb{N}^{\mathbb{N}} \times \mathbb{N}^{\mathbb{N}},\ x \longmapsto F(x) := (x, x)$$

is continuous, hence Borel measurable, the pre-image $S = F^{-1}(A)$ is a Souslin set, cf. Theorem 2. Its complement $\mathbb{N}^{\mathbb{N}} \setminus S$ is not a Souslin set. Suppose, on the contrary, that it is a Souslin set. By the construction of A, there would exist some $t \in \mathbb{N}^{\mathbb{N}}$ such that $X \setminus S = A_t$, i.e.

$$\mathbb{N}^{\mathbb{N}} \setminus S = \{u \in \mathbb{N}^{\mathbb{N}};\ (u, t) \in A\}.$$

Combining this with the very definition of S, we find that

$$t \in S \iff (t, t) \in A \iff t \in \mathbb{N}^{\mathbb{N}} \setminus S,$$

which is impossible. Hence, $\mathbb{N}^{\mathbb{N}} \setminus S$ is not a Souslin set. In particular, S cannot be a Borel set: if S were Borel, then also its complement would be Borel and, thus, a Souslin set. □

Corollary 2 *There exists a Souslin set $A \subseteq \mathbb{R}$ which is not Borel.*

Proof The set of irrationals $\mathbb{R} \setminus \mathbb{Q}$ is homeomorphic to $\mathbb{N}^{\mathbb{N}}$, i.e. there exists a bijective mapping $f : \mathbb{N}^{\mathbb{N}} \to \mathbb{R} \setminus \mathbb{Q}$ such that f and f^{-1} are continuous, cf. [20, Theorem 6.1.6]. By Corollary 1, there is a Souslin set $S \subseteq \mathbb{N}^{\mathbb{N}}$ which is not a Borel set. Since f is continuous, hence Borel measurable, $A := f(S) \subseteq \mathbb{R} \setminus \mathbb{Q}$ cannot be a Borel set; otherwise $f^{-1}(A) = S$ would be Borel. Moreover, the Souslin set S is the image of a Polish space under a continuous mapping, and therefore $A = f(S)$ has the same property, i.e. it is a Souslin set. Hence, A is a non-Borel Souslin set in $\mathbb{R} \setminus \mathbb{Q}$. This implies that A is a Souslin set in $\mathbb{R}$ which is not Borel. □

Comment If $X \neq \emptyset$ is a complete metric space without isolated points, then X contains a subset homeomorphic to $\mathbb{N}^{\mathbb{N}}$, cf. [20, Lemma 6.1.16]. The proof of the above corollary shows that any such space contains a Souslin set which is not a Borel set.

7.20 A Lebesgue measurable set which is not Borel measurable

Let C be the middle-thirds Cantor set [☞Section 2.5] and let $f : [0,1] \to [0,1]$ be the Cantor function [☞Section 2.6]. Since f is continuous and increasing, there is a unique right-continuous generalized inverse $g : [0,1] \to [0,1]$ such that $f(g(y)) = y$; it is given by $g(y) := \inf\{x \in [0,1]\,;\ f(x) > y\}$. Since the flat parts of f become jumps of g, it is clear that $g([0,1]) \subseteq C$. Moreover, g is Borel measurable since it is right-continuous. Define

$$L := g(A) \quad \text{for some non-Lebesgue measurable } A \subseteq [0,1] \text{ [☞7.22]}.$$

Clearly, $g(A) \subseteq C$, so that L is a subset of the Borel null set C, and so $L = g(A)$ is Lebesgue measurable.

As g is injective, we have $g^{-1}(g(U)) = U$ for any set $U \subseteq [0,1]$. Moreover, since g is measurable, $g^{-1}(B)$ is Borel for each Borel set B. Assume that $L = g(A)$ were Borel, then we would have that $g^{-1}(L) = g^{-1}(g(A)) = A$ is a Borel set, which is a contradiction.

7.21 A Lebesgue measurable set which is not a Souslin set

Take a Souslin set $S \subseteq \mathbb{R}$ which is not a Borel set, [☞7.19] for the existence of the set as well as the definition of Souslin sets. Consider the complement $A := \mathbb{R}\setminus S$. Since any Souslin set in $\mathbb{R}$ is Lebesgue measurable, cf. [171, Theorem 4.3.1], it follows that A is Lebesgue measurable. But A is not a Souslin set. This is immediate from the fact that any Souslin set whose complement is a Souslin set is necessarily Borel measurable, cf. [20, Corollary 6.6.10].

7.22 A non-Lebesgue measurable set

Denote by $(\bar{\lambda}, \bar{\mathscr{B}}(\mathbb{R}))$ the completion of Lebesgue measure λ defined on $\mathscr{B}(\mathbb{R})$ [☞Example 1.7.(n)]. We will construct a non-Lebesgue measurable subset of $\mathbb{J} = [0,1)$. We call any two $x, y \in \mathbb{J}$ equivalent, $x \sim y$, if $x - y \in \mathbb{Q}$. The equivalence class containing x is denoted by

$$[x] = \{y \in \mathbb{J}\,;\ x - y \in \mathbb{Q}\} = (x + \mathbb{Q}) \cap \mathbb{J}.$$

By construction, $\mathbb{J}$ is partitioned by a family of mutually disjoint equivalence classes $[x_i]$, $i \in I$. Since I is uncountable, we have to use the axiom of choice in order to construct a set L which contains exactly one element, say m_i, from each of the classes $[x_i]$, $i \in I$. We will prove that L cannot be Lebesgue measurable.

Assume L were Lebesgue measurable. Since $[x] \cap L$, $x \in \mathbb{J}$, contains exactly one element, say m_{i_0} for some $i_0 = i_0(x) \in I$, we can find some $q \in \mathbb{Q}$ such that $x = m_{i_0} + q$. Obviously, $-1 < q < 1$. Thus

$$\mathbb{J} \subseteq L + (\mathbb{Q} \cap (-1,1)) \subseteq \mathbb{J} + (-1,1) \subseteq [-1,2),$$

which we can rewrite as

$$[0,1) \subseteq \bigcup_{q \in \mathbb{Q} \cap (-1,1)} (q + L) \subseteq [-1,2). \tag{7.3}$$

Moreover, $(r + L) \cap (q + L) = \emptyset$ for all $r \neq q$, $r, q \in \mathbb{Q}$. Otherwise, we would have $r + x = q + y$ for $x, y \in L$, so that $x \sim y$, which is impossible since L contains only one representative of each equivalence class. Therefore we can use the σ-additivity of the measure $\bar{\lambda}$ to find

$$1 = \bar{\lambda}[0,1) \leqslant \sum_{q \in \mathbb{Q} \cap (-1,1)} \bar{\lambda}(q + L) \leqslant \bar{\lambda}[-1,2) = 3. \tag{7.4}$$

Since $\bar{\lambda}$ is invariant under translations, we conclude that $\bar{\lambda}(q + L) = \bar{\lambda}(L)$ for all $q \in \mathbb{Q} \cap (-1,1)$. Thus, we see that

$$1 \leqslant \sum_{q \in \mathbb{Q} \cap (-1,1)} \bar{\lambda}(L) \leqslant 3, \tag{7.5}$$

which is not possible. This proves that L cannot be Lebesgue measurable.

A similar construction works for any set $\mathbb{I}$ of strictly positive Lebesgue measure: since Lebesgue measure is invariant under translations, we may assume that $\lambda(\mathbb{I} \cap [0,1)) > 0$. Now we can follow the above construction and find a non-measurable set inside $\mathbb{I} \cap [0,1)$. Using the bi-measurability of translations, we can transport this set, without losing non-measurability, to any place in the real line.

Comment This construction is due to Vitali; see also Lebesgue [105]. Lebesgue credits the first construction of non-measurable sets to G. Vitali '[...] l'existence d'ensembles non mesurables est certaine. C'est là un résultat qui n'est pas nouveau; M[onsieur] Vitali l'a obtenu en effet, comme conséquence de raisonnements idéalistes, dans une note *Sul problema della misure dei gruppi di punti di una retta* (Bologna 1905).'

It is known that all constructions of non-Lebesgue measurable sets must use the axiom of choice or some equivalent statement, cf. Solovay [169].

Vitali's non-measurable set is also an example of a set of second category (in the sense of Baire), i.e. a set which cannot be represented as a countable union of nowhere dense sets [☞ Definition 2.3]. Since $\bigcup_{q\in\mathbb{Q}\cap(-1,1)}(q+L) \supseteq [0,1)$ and since any interval is of second category [☞ Example 2.3], the sets $q+L$ cannot all be of first category. Since translation does not change the category of a set, L must be of second category.

7.23 Arbitrary unions of non-trivial closed balls need not be Borel measurable

Arbitrary unions of open sets are open, and so arbitrary unions of open balls are Borel measurable. This property is not true for closed balls. This is most easily seen in the Euclidean space $\mathbb{R}^d$ if we allow for closed balls with radius 0, i.e. singletons: any set $N \subseteq \mathbb{R}$ can be written as a union of singletons and since there are (necessarily uncountable) non-Borel measurable sets [☞ 7.20]; thus, the union of closed balls with radius $r \geqslant 0$ may not be Borel measurable. The following example shows that the union of closed balls with strictly positive radius may not be a Borel set.

Take a non-Borel measurable set $N \subseteq (1,\infty)$ [☞ 7.20]. For the closed balls

$$C_r := \left\{\binom{x}{y} \in \mathbb{R}^2;\ \sqrt{|x-r|^2+|y+r|^2} \leqslant r\right\} = \overline{B_r\left(\binom{r}{r}\right)} \subseteq \mathbb{R}^2, \quad r \in N,$$

consider the union

$$A = \bigcup_{r\in N} C_r,$$

see Fig. 7.1. Since $A \cap (\mathbb{R}\times\{0\}) = N\times\{0\} \notin \mathscr{B}(\mathbb{R}^2)$ and $\mathbb{R}\times\{0\}$ is a Borel set, it follows that $A \notin \mathscr{B}(\mathbb{R}^2)$.

Comment Arbitrary unions of non-trivial closed balls are **Lebesgue** measurable, cf. [7]. Note that it is crucial to assume that the balls are non-trivial, i.e. have strictly positive radius.

7.24 The image of a Borel set under a continuous mapping need not be Borel

Let (X,d) and (X',d') be metric spaces. The image of a Borel set $B \in \mathscr{B}(X)$ under a continuous mapping $f : X \to X'$ is not necessarily a Borel set in X'.

Take the Euclidean space $X' = \mathbb{R}$, and pick $X \subseteq \mathbb{R}$ with $X \notin \mathscr{B}(\mathbb{R})$ [☞ 7.20]. Consider X with the metric induced by the Euclidean metric. The mapping

$$\iota : X \to X',\ x \longmapsto x,$$

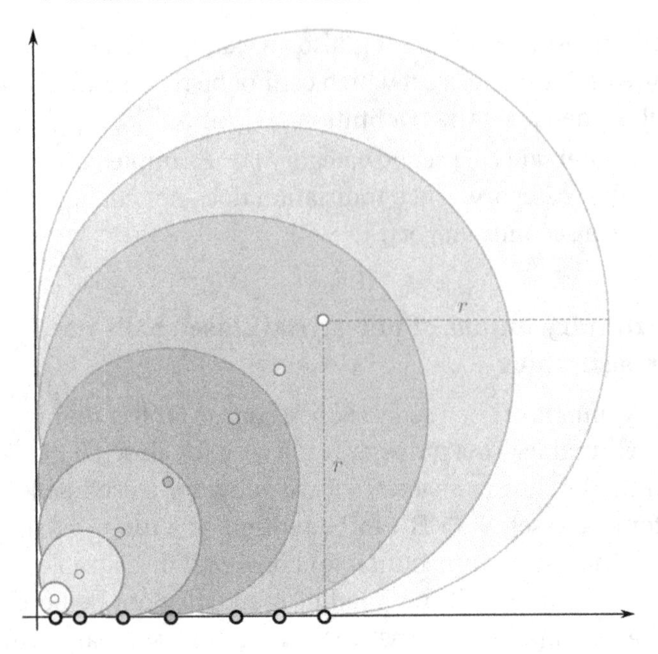

Figure 7.1 Each set $N \times \{0\} \subseteq \mathbb{R}^2$ can be written as the intersection of the horizontal axis with the union $\bigcup_{r \in N} C_r$ of all closed balls C_r centred at $\binom{r}{r} \in \mathbb{R}^2$ with radius $|r|$.

is injective and continuous. However, the image of $X \in \mathscr{B}(X)$ under ι is not a Borel set in $X' = \mathbb{R}$:

$$\iota(X) = X \notin \mathscr{B}(\mathbb{R}).$$

Even under the additional assumption that X, X' are Polish spaces, the image of a Borel set under a continuous mapping need not be Borel. For instance, if we consider $X = \mathbb{R}^2$ and $X' = \mathbb{R}$, then there is a Borel set $B \in \mathscr{B}(\mathbb{R}^2)$ whose image under the continuous projection $\mathbb{R}^2 \to \mathbb{R}, (x, y) \longmapsto x$ is not a Borel set [☞15.12]. In Example 7.26 we show that the continuous mapping $(x, y) \longmapsto f(x, y) = x + y$ does not preserve Borel sets: there are Borel sets A, B such that the sum $A + B = f(A \times B)$ is not Borel. Darst [38] even constructed a smooth function $f : \mathbb{R} \to \mathbb{R}$ which maps a Borel set onto a non-Borel set.

Comment 1 Let $(X, d), (X', d')$ be Polish spaces. If $f : X \to X'$ is injective and Borel measurable (e.g. continuous), then $f(B) \in \mathscr{B}(X')$ for any $B \in \mathscr{B}(X)$, cf. [20, Theorem 6.8.6]. Note that there is no contradiction between this result and our counterexamples: in the first example, the underlying space (X, d) is not Polish, but only a metric space, and in the second example the mapping under consideration is not injective. More generally, Purves [137] showed that a Borel measurable function $f : X \to X'$ satisfies $f(B) \in \mathscr{B}(X')$ for all $B \in \mathscr{B}(X)$ if, and only if, there are at most countably many values in the range of f which are attained uncountably often, i.e.

$$\#\{x' \in X' ;\ f^{-1}(\{x'\}) \text{ is uncountable}\} \leqslant \#\mathbb{N}.$$

Comment 2 If $f : \mathbb{R} \to \mathbb{R}$ is Borel measurable but not injective, then the image $f(B)$ is Lebesgue measurable for any $B \in \mathscr{B}(\mathbb{R})$. Indeed, any Borel set $B \in \mathscr{B}(\mathbb{R})$ is a Souslin set, and images of Souslin sets under Borel mappings are Souslin sets [☞7.19]. Hence, $f(B)$ is a Souslin set which, in turn, implies that $f(B)$ is Lebesgue measurable, cf. [171, Theorem 4.3.1].

7.25 The image of a Lebesgue set under a continuous mapping need not be Lebesgue measurable

Consider the projection $\pi_1 : \mathbb{R}^2 \to \mathbb{R}$, $(x, y) \longmapsto x$. Take a set $N \subseteq \mathbb{R}$ such that $N \notin \mathscr{L}(\mathbb{R})$ [☞7.22], and set $A := N \times \{0\}$. Since $(\mathbb{R}^2, \mathscr{L}(\mathbb{R}^2), \lambda^2)$ is a complete measure space, it follows from $A \subseteq \mathbb{R}\times\{0\}$ and $\lambda^2(\mathbb{R}\times\{0\}) = 0$ that A is Lebesgue measurable. But the image of A under the (Lipschitz) continuous mapping π_1 is not measurable:

$$\pi_1(A) = N \notin \mathscr{L}(\mathbb{R}).$$

Comment If $f : \mathbb{R}^d \to \mathbb{R}^n$ is continuous, then $f(B)$ is Lebesgue measurable for any Borel set $B \in \mathscr{B}(\mathbb{R}^d)$ [☞Comment 2 in 7.24].

In order to ensure that the image $f(A)$ of a Lebesgue set $A \in \mathscr{L}(\mathbb{R}^d)$ is again Lebesgue measurable, **Lusin's condition** (N) is necessary and sufficient; see [MIMS, p. 174]: f is said to have property (N), if

$$\text{for all null sets } N \in \mathscr{L}(\mathbb{R}^d) \text{ the image } f(N) \text{ is a null set in } \mathscr{L}(\mathbb{R}^n). \qquad (N)$$

One can show that α-Hölder continuous functions $f : \mathbb{R}^d \to \mathbb{R}^n$ enjoy the property (N) if $\alpha n \geqslant d$; see e.g. [MIMS, p. 175]. If $n = d = 1$, a function f is **absolutely continuous** if, and only if, (i) f is continuous, (ii) f is of bounded variation and (iii) f enjoys the property (N). This characterization is due to Banach and Zaretzki; see [125, Kapitel IX.3, Satz 4] and [153, Chapter IX.7]. Superpositions of functions enjoying the property (N) are discussed in [153, Chapter IX.8]. Further characterizations (due to Zaretzki) of absolute continuity of f and f^{-1} in terms of the mapping properties of the derivative f' are known, cf. [125, pp. 307–308]; see also [☞17.8].

7.26 The Minkowski sum $A + B$ of two Borel sets is not necessarily Borel

For two subsets A, B of a linear space X the **Minkowski sum** $A + B$ is defined by

$$A + B := \{a + b\,;\ a \in A,\ b \in B\}.$$

Let $A \in \mathscr{B}(\mathbb{R}^2)$ be such that its projection $\pi_1(A)$ onto the first coordinate is not Borel, [☞ 15.12] for the existence of such a set. For the set $B := \{0\}\times\mathbb{R} \in \mathscr{B}(\mathbb{R}^2)$ we have

$$\begin{aligned}(x,0) \in A + B &\iff \exists y \in \mathbb{R}, (a_1,a_2) \in A : (x,0) = (a_1,a_2) + (0,y)\\ &\iff x \in \pi_1(A),\end{aligned}$$

which implies that $(A+B)\cap(\mathbb{R}\times\{0\}) = \pi_1(A)\times\{0\}$ is not Borel. Hence, $A+B$ is not Borel.

Comment Erdös & Stone [56] proved the existence of Borel sets $A, B \in \mathscr{B}(\mathbb{R})$ such that $A+B \notin \mathscr{B}(\mathbb{R})$. Setting $C := \exp(A)$ and $D := \exp(B)$ gives an example of two Borel sets $C, D \in \mathscr{B}(\mathbb{R})$ whose product

$$C \cdot D := \{c \cdot d\,;\ c \in C,\ d \in D\}$$

is not Borel. For corresponding results on the Minkowski sum of Lebesgue measurable sets, see e.g. Sierpiński [163] and Rubel [148].

7.27 A Lebesgue null set B such that $B + B = \mathbb{R}$

Let $C \subseteq [0,1]$ be the classical Cantor middle-thirds set [☞ Section 2.5]. By construction, a point $x \in \mathbb{R}$ belongs to C if, and only if, its ternary expansion is of the form

$$x = \sum_{i=1}^{\infty} \frac{2a_i}{3^i}$$

with $a_i \in \{0,1\}$ for all $i \in \mathbb{N}$. We claim that

$$C + C := \{y + z\,;\ y,z \in C\} = [0,2].$$

Fix $x \in [0,1]$ and denote by $\sum_{i=1}^{\infty} x_i 3^{-i}$ its ternary expansion. Set $y = \sum_{i=1}^{\infty} y_i 3^{-i}$ and $z = \sum_{i=1}^{\infty} z_i 3^{-i}$ where

$$y_i := \begin{cases} 0, & \text{if } x_i = 0,\\ 1, & \text{if } x_i = 1 \text{ or } x_i = 2,\end{cases} \qquad \text{and} \qquad z_i := \begin{cases} 0, & \text{if } x_i = 0 \text{ or } x_i = 1,\\ 1, & \text{if } x_i = 2.\end{cases}$$

Since $y_i \in \{0,1\}$ and $z_i \in \{0,1\}$ for all $i \in \mathbb{N}$, it follows that $2y \in C$ and $2z \in C$. Moreover, $x = y + z$, i.e. $x \in \frac{1}{2}C + \frac{1}{2}C$. Thus, $[0,2] \subseteq C + C$. As $C \subseteq [0,1]$, the converse inclusion is trivial.

Set

$$B := \bigcup_{n\in\mathbb{Z}} (2n + C) = \bigcup_{n\in\mathbb{Z}} \{2n + x\,;\ x \in C\};$$

as a countable union of Lebesgue null sets, $B \in \mathscr{B}(\mathbb{R})$ is also a Lebesgue null set. Moreover, by the previous consideration,

$$B + B \supseteq \bigcup_{n\in\mathbb{Z}} ((2n + C) + C) = \bigcup_{n\in\mathbb{Z}} (2n + [0,2]) = \mathbb{R}.$$

Comment By a result due to Steinhaus, the sum $A + B$ of two Borel measurable sets with strictly positive Lebesgue measure contains a non-empty interval. It implies, in particular, that the difference of two fat Cantor sets contains an interval [☞7.28]. Our example illustrates that Steinhaus' theorem is not sharp: $A + B$ may contain a non-empty interval even if A, B are both Lebesgue null sets.

7.28 The difference of fat Cantor sets contains an interval

This follows at once from the following theorem due to Steinhaus [177]:

Theorem *Let $A \subseteq \mathbb{R}$ be a Borel set with strictly positive Lebesgue measure. Then the set of differences $A - A = \{a - b\,;\ a, b \in A\}$ contains some open neighbourhood of $x = 0$.*

Applying this theorem to a fat Cantor set with $\alpha < 1$ [☞7.3], we get the really surprising result that the difference of a nowhere dense set may contain a solid interval. In fact, the difference of two fat Cantor sets may contain the whole interval $[-1,1]$ [☞7.30].

Proof of the theorem Without loss we may assume that $\lambda(A) < \infty$. Since Lebesgue measure is open outer regular, there is some open set $U \supseteq A$ such that $\lambda(A) > \frac{3}{4}\lambda(U)$. We can represent U as a countable union of mutually disjoint open intervals, so there exists one of them, say I, such that $\lambda(A \cap I) > \frac{3}{4}\lambda(I)$. Set $\delta := \frac{1}{2}\lambda(I)$. For any $x \in (-\delta, \delta)$ the set $(x + I) \cup I$ is an interval of length less than $3\delta = \frac{3}{2}\lambda(I)$ such that

$$A \cap I \subseteq (x + I) \cup I \quad \text{and} \quad x + (A \cap I) \subseteq (x + I) \cup I.$$

From $\lambda(x+(A\cap I)) = \lambda(A\cap I) > \frac{3}{4}\lambda(I)$, we conclude that $(A\cap I)$ and $x+(A\cap I)$ have a non-empty intersection, i.e. there are $a, b \in A \cap I$ such that $a = x + b$, hence $x \in A - A$. Since $x \in (-\delta, \delta)$ is arbitrary, we get $(-\delta, \delta) \subseteq A$. □

There are various generalizations of Steinhaus' result.

(a) We may replace in the above theorem $\lambda(A) > 0$ by '*A is a measurable set of second category*' (see [133, Theorem 4.8]). Recall that a set is of second category (in the sense of Baire) if it cannot be represented as a countable union of nowhere dense sets. [☞7.29] for a related result.

(b) We may consider $A - B$ instead of $A - A$: if A, B are measurable sets with $\lambda(A) > 0$ and $\lambda(B) > 0$, then $A - B$ contains some non-empty open interval (see [16, Theorem 1.1.2]). Taking $B = -A$, we conclude that $A + A = \{a + b\,;\ a, b \in A\}$ contains a non-empty open interval.

(c) Here is a 'simple' proof of the previous statement. We show that for sets $A, B \in \mathscr{B}(\mathbb{R}^d)$, each having strictly positive Lebesgue measure, the sum $A + B = \{a + b\,;\ a \in A, b \in B\}$ contains an open ball. Without loss of generality, as can assume that A, B are bounded sets. Since the convolution $J(x) := \mathbb{1}_A * \mathbb{1}_B(x)$ is continuous, cf. [MIMS, p. 159], the set $U := \{J > 0\}$ is open. From

$$\int \mathbb{1}_A * \mathbb{1}_B(x)\,dx = \iint \mathbb{1}_A(x-y)\mathbb{1}_B(y)\,dy\,dx = \lambda(A)\lambda(B) > 0,$$

we see that $U \neq \emptyset$. Moreover, for $x \in U$ there is some $y \in B$ such that $x - y \in A$ – otherwise $\mathbb{1}_A(x-y)\mathbb{1}_B(y) = 0$ for all y which entails $J(x) = 0$ – and so $U \subseteq A + B$.

Comment 1 The sum $C + C$ of two classical (middle-thirds) Cantor sets is exactly the interval $[0, 2]$ [☞7.27]. This shows, in particular, that the sum $A+B$ of two Lebesgue null sets A, B may contain a non-empty interval. Consequently, Steinhaus' condition $\min\{\lambda(A), \lambda(B)\} > 0$ is sufficient but not necessary for $A + B$ to contain a non-empty interval.

Comment 2 There is a close relationship between Steinhaus' theorem and the notion of absolute continuity [☞Definition 1.42]: a finite measure μ on $(\mathbb{R}, \mathscr{B}(\mathbb{R}))$ is absolutely continuous with respect to Lebesgue measure if, and only if, for every compact set $K \subseteq \mathbb{R}$ with $\mu(K) > 0$ the difference $K - K$ contains an open neighbourhood of $x = 0$, cf. [166].

7.29 The sum of scaled Cantor sets is sometimes an interval

Let C' denote the Cantor-like set with $a_n = \left(\frac{1}{2}\right)^{n+1}$, $n \in \mathbb{N}_0$, [☞7.3], i.e. we remove at each step the **middle half**, i.e. the **two** middle quarters, of the respective intervals; the remaining mass has measure $\lambda(C') = 1 - \sum_{n=0}^{\infty} a_n = 0$. As in the classical middle-thirds case, we see that every $t \in C'$ can be written as a 4-adic fraction

$$t = 0.t_1 \ldots t_{n-1}t_n \cdots := \sum_{n=1}^{\infty} \frac{t_n}{4^n}, \quad t_i \in \{0, 3\}.$$

Observe that

$$\frac{2}{3}\left(C' + \frac{1}{2}C'\right) = \frac{2}{3}C' + \frac{1}{3}C' = \left\{x;\ x = \sum_{n=1}^{\infty} \left(\frac{2}{3}t_n + \frac{1}{3}s_n\right)4^{-n},\ s_i, t_i \in \{0, 3\}\right\}$$

contains all possible 4-adic fractions since $\left(\frac{2}{3}t_n + \frac{1}{3}s_n\right)$ can take any value from $\{0, 1, 2, 3\}$. Thus, $\frac{2}{3}\left(C' + \frac{1}{2}C'\right) = [0, 1]$ or $C' + \frac{1}{2}C' = \left[0, \frac{3}{2}\right]$.

One can show, however, see [174, pp. 376–377], that the set

$$\{\theta \in [0, 1];\ \lambda\left(C' + \theta C'\right) = 0\}$$

has Lebesgue measure one, i.e. the set $C' + \theta C'$ has Lebesgue measure zero for Lebesgue almost all θ.

Comment The sets $C' + \theta C'$ can be used to construct a so-called **Besicovitch set** (also: **Kakeya set**). These are compact subsets of $\mathbb{R}^2$ which have zero Lebesgue measure and contain translates of every unit line segment. A construction can be found in Stein & Shakarchi [174, Chapter 7.4].

7.30 The difference of fat Cantor sets is exactly $[-1, 1]$

Denote by C a fat Cantor set for some $\alpha \in [0, 1)$ [☞7.3]. We assume, moreover, that the sequence $(a_n)_{n\in\mathbb{N}}$ satisfies $a_n \leqslant \frac{1}{3}(1 - a_0 - \cdots - a_{n-1})$. The last condition means that we take out not more than $\frac{1}{3}$ of the remaining mass. For example, the sequence $a_n := \alpha 2^{n-1}3^{-n}$ will do nicely.

We will show that $C - C = [-1, 1]$. To do so, we use the geometric argument by Steinhaus [176]; see also [206, pp. 133–134] or [207, vol. 1, p. 235].

By construction [☞7.3], $C = \bigcap_{n=0}^{\infty} C_n$ where each C_n is made up of 2^n disjoint closed intervals. This means that we have $C \times C = \bigcap_{n=0}^{\infty}(C_n \times C_n)$ and each set $C_n \times C_n$ produces a chequerboard-like structure comprising $2^n \times 2^n$ squares.

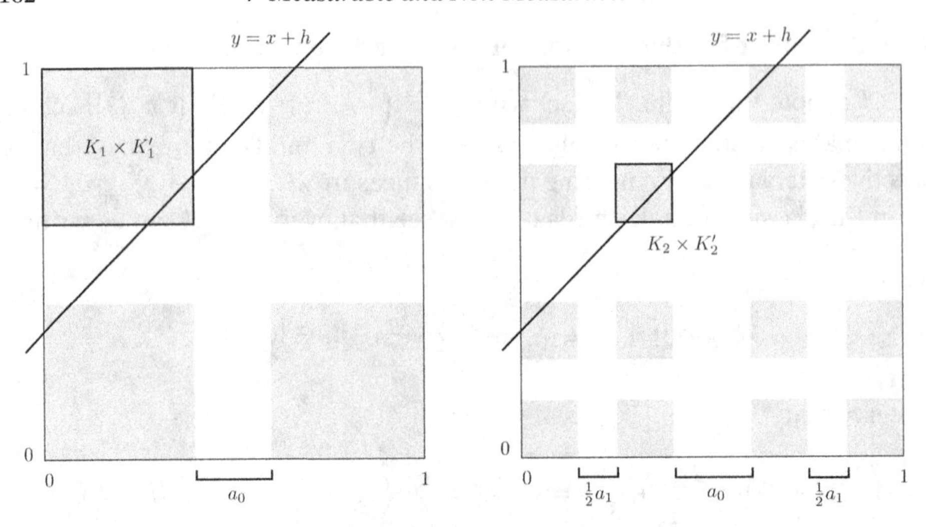

Figure 7.2 The figure shows the choice of the first two sets $K_1 \times K_1'$ and $K_2 \times K_2'$.

Fix $h \in (-1, 1)$. The line $y = x + h$ intersects $C_1 \times C_1$ in at least one square, say $K_1 \times K_1'$; this is ensured by the condition $a_n \geqslant \frac{1}{3} a_{n-1}$; see Fig. 7.2.

By self-similarity, $(K_1 \times K_1') \cap (C \times C)$ is essentially a smaller copy of $C \times C$, and we see that $y = x + h$ intersects $(K_1 \times K_1') \cap (C_2 \times C_2)$ in at least one square which we call $K_2 \times K_2'$, etc. This gives a nested sequence of compact sets

$$K_1 \times K_1' \supseteq K_2 \times K_2' \supseteq \cdots \supseteq K_n \times K_n' \supseteq \cdots$$

and, since the side length of K_n, K_n' is $2^{-n}(1 - a_0 - \cdots - a_{n-1}) \leqslant 2^{-n} \to 0$, the interval principle tells us that $\bigcap_{n=1}^{\infty} K_n \times K_n'$ contains exactly one point, say (x, y). Since this point also has to satisfy the linear relation $y = x + h$, we get $h = x - y \in C - C$. Hence, as $h \in (-1, 1)$ is arbitrary, $C - C = [-1, 1]$.

7.31 The Banach–Tarski paradox

The Banach–Tarski paradox [12] states that we can, in dimension $d = 3$ and higher, decompose the unit ball $B_1(0) \subseteq \mathbb{R}^d$ into finitely many disjoint sets $B_1(0) = \dot{\bigcup}_{i=1}^{n} A_i$, perform rigid motions ρ_i with each of them to get geometrically congruent sets $E_i = \rho_i(A_i)$, and reassemble these sets into a ball of double (or any other) radius $B_2(0) = \dot{\bigcup}_{i=1}^{n} E_i$.

Dekker & deGroot [42] extended the paper of Robinson [145] and showed that as few as $n = 5$ pieces suffice, with A_1 being a singleton and $A_2, \ldots, A_5$ connected and locally connected. Since invariance under rigid motions is one of the characteristic properties of Lebesgue measure or geometric volume, this means that **it is not possible to define a finitely additive version of Lebesgue measure on all subsets of $\mathbb{R}^d$ if $d \geqslant 3$.**

The origin of the Banach–Tarski paradox is Lebesgue's **problème de la mesure** [102, Chapitre VII.ii], which asks whether it is possible to find an extension of Lebesgue's measure $m: \mathcal{P}(\mathbb{R}^d) \to [0,\infty]$ such that (we follow Hausdorff [76, p. 401]):

(α) $m(A) = m(\rho(A))$ for all $A \subseteq \mathbb{R}^d$ and all rigid motions ρ;
(β) $m([0,1]^d) = 1$;
(γ) finite additivity: $m(A \uplus B) = m(A) + m(B)$ for all disjoint $A, B \subseteq \mathbb{R}^d$;
(δ) σ-additivity: $m(\biguplus_n A_n) = \sum_n m(A_n)$ for all disjoint $(A_n)_{n\in\mathbb{N}} \subseteq \mathcal{P}(\mathbb{R}^d)$.

Vitali's construction of a non-Lebesgue measurable set [☞7.22] from 1905 shows that one cannot have an extension satisfying all four properties (α)–(δ); the problem is, as we know, σ-additivity.

In a next step, Hausdorff constructed in the appendix to his *Grundzüge der Mengenlehre* [76, pp. 469–471] a partitioning of the unit sphere in $\mathbb{R}^3$, i.e. $\mathbb{S}^3 = \{x \in \mathbb{R}^3;\ x_1^2 + x_2^2 + x_3^2 = 1\}$, into four pieces $P \uplus S_1 \uplus S_2 \uplus S_3$ and two rotations ϕ, ψ by the angle π, resp. $\frac{2}{3}\pi$ (but different axes), such that P is countable and

$$\phi(S_1) = S_2 \cup S_3, \quad \psi(S_1) = \psi(S_2) \quad \text{and} \quad \psi\circ\psi(S_1) = S_3.$$

In particular, the sets S_1, S_2, S_3 and $S_2 \cup S_3$ are congruent (in the sense that they can be transformed into each other by a rigid motion); consequently **there cannot be a finitely additive extension of Lebesgue measure** in dimension $d = 3$ (or higher). Von Neumann showed in 1929 that it is the nature of the congruence group that is responsible for the different behaviour in $d = 1, 2$ vs. $d \geqslant 3$; see [127, pp. 39–41].

Banach showed in [10] that the measure problem for finitely additive measures m satisfying (α)–(γ) has (non-unique) solutions in dimensions $d = 1$ and $d = 2$, thereby establishing an auxiliary result which we know nowadays as the Hahn–Banach theorem. The final step, removing the countable exceptional set P in Hausdorff's construction, led to the Banach–Tarski paradox [12].

In [11] the measure problem was extended from Lebesgue measure: the only σ-additive measure on $\mathcal{P}(\mathbb{R}^d)$ such that $\mu(\{x\}) = 0$ for all x is the trivial measure $\mu \equiv 0$ [☞6.15].

A modern elementary exposition of the Banach–Tarski paradox is Stromberg [181], the monograph by Wagon [196] is one of the standard references. Interesting historical remarks can be found in Chatterji's commentary on Hausdorff's work in measure theory [75, pp. 788–800].

8

Measurable Maps and Functions

Let $(X, \mathscr{A})$ and $(Y, \mathscr{B})$ be measurable spaces. Recall that a map $f : X \to Y$ is **measurable** if $f^{-1}(\mathscr{B}) \subseteq \mathscr{A}$, i.e.

$$\forall B \in \mathscr{B} \ : \ f^{-1}(B) \in \mathscr{A}.$$

If we need to emphasize the σ-algebras, we write $f : (X, \mathscr{A}) \to (Y, \mathscr{B})$ or $\mathscr{A}/\mathscr{B}$ **measurable**. The pre-image $f^{-1}(B)$ is also denoted by $\{f \in B\}$. If $\mathscr{G}$ is a generator [☞ Example 1.7.(h) or p. 73] of $\mathscr{B}$, then f is measurable if, and only if, $f^{-1}(\mathscr{G}) \subseteq \mathscr{A}$ [☞ Rem. 1.9.(a)].

Unless mentioned otherwise, we use on $\mathbb{R}^d$ the Euclidean metric and denote by $\mathscr{B}(\mathbb{R}^d)$ the corresponding Borel σ-algebra. An $\mathscr{A}/\mathscr{B}(\mathbb{R}^d)$ measurable function $f : X \to \mathbb{R}^d$ is usually called **Borel measurable** or a **Borel function**. Since the intervals $(-\infty, a]$, $a \in \mathbb{Q}$, are a generator of $\mathscr{B}(\mathbb{R})$, a function $f : X \to \mathbb{R}$ is Borel measurable if, and only if,

$$\forall a \in \mathbb{Q} \ : \ \{f \leqslant a\} = f^{-1}(-\infty, a] \in \mathscr{A}.$$

The family of Borel measurable functions $f : X \to \mathbb{R}$ is a vector space which is closed under countable suprema and infima; in particular, $\liminf_{n\to\infty} f_n$, $\limsup_{n\to\infty} f_n$ and $\lim_{n\to\infty} f_n$ (if it exists) are Borel measurable for any sequence $(f_n)_{n\in\mathbb{N}}$ of Borel measurable functions. These properties break down in general measurable spaces: sums, products, suprema and limits of measurable functions need not be measurable [☞ 8.15, 8.16, 8.18].

A function $f : \mathbb{R}^k \to \mathbb{R}^d$ is **Lebesgue measurable** if the pre-image $f^{-1}(L)$ of every Lebesgue set $L \in \mathscr{L}(\mathbb{R}^d)$ is a Lebesgue set in $\mathbb{R}^k$ [☞ Example 1.7.(n)]. Continuous functions are always Borel measurable but may fail to be Lebesgue measurable [☞ 8.12]. Since the Lebesgue σ-algebra is strictly larger than the Borel σ-algebra [☞ 7.20,15.1], there exist functions that are Lebesgue measurable but not Borel measurable. Thus, Borel measurability does not imply Lebesgue measurability – and vice versa.

8.1 A measurable space where every map is measurable

Let X be any non-empty set and consider the σ-algebra $\mathscr{P}(X)$. Every function $f: (X, \mathscr{P}(X)) \to (\mathbb{R}, \mathscr{B}(\mathbb{R}))$ is measurable.

Similarly, if $(X, \mathscr{A})$ is an arbitrary measurable space and $Y \neq \emptyset$ is a further set, then every map $f: (X, \mathscr{A}) \to (Y, \{\emptyset, Y\})$ is measurable.

8.2 A measurable space where only constant functions are measurable

Let X be an arbitrary non-empty set and consider the σ-algebra $\{\emptyset, X\}$. A function $f: (X, \{\emptyset, X\}) \to (\mathbb{R}, \mathscr{B}(\mathbb{R}))$ is measurable if, and only if, f is constant. This follows from the fact that every level set $\{f = c\}$ must be either empty or X.

Comment As a consequence, the identity map $f: \mathbb{R} \to \mathbb{R}, x \longmapsto x$ is not $\mathscr{A}/\mathscr{B}(\mathbb{R})$ measurable for the trivial σ-algebra $\mathscr{A} = \{\emptyset, \mathbb{R}\}$ – or, more generally, whenever $\mathscr{A} \subsetneqq \mathscr{B}(\mathbb{R})$.

8.3 A non-measurable function whose modulus $|f|$ is measurable

Let $(X, \mathscr{A})$ be a measurable space such that $\mathscr{A} \neq \mathscr{P}(X)$. For $N \in \mathscr{P}(X) \setminus \mathscr{A}$ the function

$$f(x) := \mathbb{1}_N(x) - \mathbb{1}_{X \setminus N}(x)$$

is not Borel measurable since $\{f = 1\} = N \notin \mathscr{A}$. However, its modulus $|f| = 1$ is Borel measurable.

8.4 A non-measurable function whose level sets $\{x ; f(x) = \alpha\}$ are measurable

Consider the measurable space $(\mathbb{R}, \mathscr{B}(\mathbb{R}))$ and choose a non-measurable set $N \notin \mathscr{B}(\mathbb{R})$ [☞ 7.22]. The function

$$f(x) := e^x \mathbb{1}_N(x) - e^x \mathbb{1}_{\mathbb{R} \setminus N}(x)$$

is not Borel measurable since $f^{-1}(0, \infty) = N$. The level sets $\{x ; f(x) = \alpha\}$ are, however, either empty or singletons, hence Borel sets.

8.5 A measurable function which is not μ-a.e. constant on any atom

Let $(X, \mathscr{A}, \mu)$ be a measure space and let $A \in \mathscr{A}$ be an atom of μ [☞ (6.1)]. If $f: (X, \mathscr{A}) \to (\mathbb{R}, \mathscr{B})$ is a measurable function, then $A \cap \{f \in B\}$ is for every

$B \in \mathscr{B}$ a measurable subset of the atom A, and so either $\mu(A \cap \{f \in B\}) = 0$ or $\mu(A \cap \{f \notin B\}) = 0$. If the σ-algebra $\mathscr{B}$ is sufficiently rich – that is, if $\mathscr{B} \supseteq \mathscr{B}(\mathbb{R})$ – then this implies that f is μ-a.e. constant on A. By contrast, if $\mathscr{B}$ is too small, then $f^{-1}(\mathscr{B})$ does not carry much information on f, and f need not be μ-a.e. constant on A.

We start with the positive result for $\mathscr{B} = \mathscr{B}(\mathbb{R})$.

Proposition *Let $(X, \mathscr{A}, \mu)$ be a measure space. If $f : (X, \mathscr{A}) \to (\mathbb{R}, \mathscr{B}(\mathbb{R}))$ is a measurable function and $A \in \mathscr{A}$ is an atom of μ, then f is μ-a.e. constant on A, i.e. there exists a constant $c = c_A \in \mathbb{R}$ such that $\mu(\{x \in A;\ f(x) \neq c\}) = 0$.*

Proof For a fixed atom $A \in \mathscr{A}$ define $F(t) := \mu(\{x \in A;\ f(x) < t\})$. Since f is $\mathscr{A}/\mathscr{B}(\mathbb{R})$ measurable, we have $\{f < t\} \cap A \in \mathscr{A}$, and so F is well-defined. From $\{f < t\} \cap A \subseteq A$ and the fact that A is an atom, it follows that either $\mu(A \cap \{f < t\}) = 0$, i.e. $F(t) = 0$, or $\mu(A \cap \{f \geqslant t\}) = 0$, i.e. $F(t) = \mu(A)$. This means that F attains only the values 0 and $\mu(A)$. By definition, F is increasing and left-continuous, and so

$$F(t) = \mu(A)\mathbb{1}_{(c,\infty)}(t), \quad t \in \mathbb{R},$$

for some $c \in \mathbb{R}$. By construction, $\mu(A \cap \{f < c\}) = 0$ and

$$\mu(A \cap \{f > c\}) \leqslant \sum_{n \in \mathbb{N}} \mu(A \cap \{f \geqslant c + 1/n\}) = 0,$$

which proves that $\mu(A \cap \{f \neq c\}) = 0$. □

The following example shows that the proposition breaks down if we replace the Borel σ-algebra $\mathscr{B}(\mathbb{R})$ by a smaller σ-algebra on $\mathbb{R}$.

Consider on $X = \mathbb{R}$ the co-countable σ-algebra $\mathscr{A}$ [☞ Example 4A] and the measure [☞ Example 5A]

$$\mu(A) := \begin{cases} 0, & \text{if } A \text{ is at most countable,} \\ 1, & \text{if } A^c \text{ is at most countable.} \end{cases}$$

Note that the atoms of μ are the sets $A \in \mathscr{A}$ with $\mu(A) = 1$ [☞ 6.13], i.e. the sets with countable complements. The identity map $f : \mathbb{R} \to \mathbb{R}$, $x \longmapsto x$ is $\mathscr{A}/\mathscr{A}$ measurable, but strictly increasing and nowhere constant; in particular, f is not μ-a.e. constant on any atom.

8.6 A function $f(x, y)$ which is Borel measurable in each variable, but fails to be jointly measurable

Let λ^2 be Lebesgue measure on $(\mathbb{R}^2, \mathscr{B}(\mathbb{R}^2))$. Following Sierpiński [164] we construct a non-measurable set $N \subseteq [0, 1]^2$ such that for each line $\ell = \{x\} \times [0, 1]$

and $\ell = [0,1] \times \{y\}$ parallel to the x-, resp. y, axis the intersection with N contains at most one point. In particular, $N \cap \ell$ is a Borel set. This means that

$$x \longmapsto \mathbb{1}_N(x,y) \quad \text{and} \quad y \longmapsto \mathbb{1}_N(x,y)$$

are measurable, but $(x,y) \longmapsto \mathbb{1}_N(x,y)$ is not. In order to simplify things, we assume that the continuum hypothesis holds, i.e. $\mathfrak{c} = \aleph_1$ [☞ p. 43] which means that the ordinal number of $\mathbb{R}$ is the smallest uncountable ordinal.

In order to construct N we enumerate the compact sets $K \subseteq [0,1]^2$ with strictly positive Lebesgue measure; it is known that there are continuum ($\mathfrak{c}$) many such compact sets and we enumerate them as follows: K_α, $\alpha < \omega_1$. Now we use transfinite induction: let $\beta < \alpha < \omega_1$ and assume that for all $\beta < \alpha$ we have already found points $(x_\beta, y_\beta) \in K_\beta$ such that no two of the points $P_\alpha := \{(x_\beta, y_\beta)\}_{\beta<\alpha}$ belong to a line parallel to the coordinate axes. We claim

$$\exists (x_\alpha, y_\alpha) \in K_\alpha \setminus P_\alpha \;:\; (\{x_\alpha\} \times [0,1] \cup [0,1] \times \{y_\alpha\}) \cap P_\alpha = \emptyset.$$

Otherwise,

$$\forall (x,y) \in K_\alpha \setminus P_\alpha \;:\; (\{x\} \times [0,1] \cup [0,1] \times \{y\}) \cap P_\alpha \neq \emptyset.$$

This means that any section of K_α intersects at most countably many lines, i.e. less than $\mathfrak{c}$ many lines. By σ-subadditivity, K_α would be a null set, contradicting our assumption.

The set we are looking for is, therefore, $N = P_{\omega_1}$. We still have to show that N is not measurable. Assume that N were measurable. Then Tonelli's theorem shows that $\lambda^2(N) = 0$. Thus, $\lambda^2(N^c) > 0$, and by the inner compact regularity of Lebesgue measure, there must be a compact set $K \subseteq N^c$ such that $\lambda^2(K) > 0$. But this is, by construction of N, not possible: $\lambda^2(K) > 0$ entails $N \cap K \neq \emptyset$, i.e. $K \subseteq N^c$ fails to hold. In conclusion, N cannot be measurable.

8.7 Another function $f(x,y)$ which is Borel measurable in each variable, but fails to be jointly measurable

Let $A \subseteq [0,1]$ be such that $A \notin \mathscr{L}(\mathbb{R})$ [☞ 7.22], and set for $x, y \in \mathbb{R}$

$$f(x,y) := \begin{cases} 1, & \text{if } x = y \in A, \\ 0, & \text{otherwise.} \end{cases}$$

Since $f(x,\cdot) = 1_{\{x\}}$, resp. $f(x,\cdot) = 0$, depending on whether $x \in A$ or $x \notin A$, it follows that $y \longmapsto f(x,y)$ is Borel measurable for each $x \in \mathbb{R}$. Analogously, $x \longmapsto f(x,y)$ is also Borel measurable for each fixed $y \in \mathbb{R}$.

Because of the measurability of the mapping

$$(\mathbb{R}, \mathscr{B}(\mathbb{R})) \ni x \longmapsto h(x) := (x, x) \in (\mathbb{R}^2, \mathscr{B}(\mathbb{R}^2)),$$

f cannot be jointly measurable: if f were jointly measurable, then also the composition $f \circ h = \mathbb{1}_A$ would be measurable. This, however, is impossible because of our choice of A.

8.8 A function $f = (f_1, f_2)$ which is not measurable but whose components are measurable

Denote by S the Sorgenfrey line [☞ Example 2.1.(h)] and write $\mathscr{B}(S)$ for its Borel σ-algebra. Consider in the product $S \times S$ – equipped with the usual product topology – the antidiagonal $\Delta' = \{(x, -x);\ x \in S\}$. Since

$$\{(a, -a)\} = \Delta' \cap \big([a, a+1) \times [-a, -a+1)\big),$$

every point in Δ' is (relatively) open, hence every subset of Δ' is (relatively) open; this means that the induced topology on Δ' is the discrete topology. In particular, $\mathscr{B}(\Delta') = \mathscr{P}(\Delta')$.

Consider the maps $f_1, f_2 : (\mathbb{R}, \mathscr{B}(\mathbb{R})) \to (S, \mathscr{B}(S))$ given by $f_1(x) = x$ and $f_2(x) = -x$, respectively. Since both f_1^{-1} and f_2^{-1} map the sets $[x, z)$ into half-open intervals of the real line, f_1, f_2 are measurable. Consider

$$f := (f_1, f_2) : (\mathbb{R}, \mathscr{B}(\mathbb{R})) \to (\Delta', \mathscr{B}(\Delta')).$$

If we identify $(\Delta', \mathscr{B}(\Delta'))$ with $(\mathbb{R}, \mathscr{P}(\mathbb{R}))$ we can produce, e.g. using Example 7.22, some $N \subseteq \Delta'$ – this is by definition a Borel set of Δ' – such that $f^{-1}(N)$ is not in $\mathscr{B}(\mathbb{R})$: mind that every set $A \subseteq \Delta'$ can be uniquely described by its projection onto one of the coordinate axes. Thus, f cannot be measurable.

8.9 The set of continuity points of any function f is Borel measurable

Let (X, d) be a metric space, denote by $B_r(x) := \{y \in X;\ d(x, y) < r\}$ the open ball with radius $r > 0$ and centre $x \in X$ and let $\mathscr{B}(X)$ be the Borel sets. We are going to show that the continuity points C^f of any – not necessarily measurable – function $f : X \to \mathbb{R}$ is a G_δ-set, hence Borel measurable.

The modulus of continuity of a function $f : X \to \mathbb{R}$ is defined in the following way:

$$\omega^f(x) := \inf_{r>0} (\operatorname{diam} f(B_r(x))), \quad \operatorname{diam} B = \sup_{x,y \in B} |x - y|.$$

Since $r \longmapsto \operatorname{diam} f(B_r(x))$ is an increasing function, we may replace in the above definition $\inf_{r>0}$ with $\inf_{0<r<\delta}$.

Lemma 1 *The function f is continuous at x if, and only if, $\omega^f(x) = 0$.*

Proof '$\Rightarrow$': if f is continuous at the point x, then there is for every $\epsilon > 0$ some $r(\epsilon) > 0$ such that for all $r < r(\epsilon)$,

$$\left(\sup_{z\in B_r(x)} f(z) - f(x)\right) + \left(f(x) - \inf_{z\in B_r(x)} f(z)\right) < 2\epsilon.$$

Therefore,

$$\omega^f(x) \leqslant \sup_{z\in B_r(x)} f(z) - \inf_{z\in B_r(x)} f(z) < 2\epsilon \xrightarrow[\epsilon\to 0]{} 0.$$

'$\Leftarrow$': For all $r > 0$ and x, x' such that $d(x, x') < r$, we find

$$f(x) - f(x') \leqslant \sup_{z\in B_r(x)} f(z) - \inf_{z\in B_r(x)} f(z).$$

Since x and x' play symmetric roles, we conclude that

$$|f(x) - f(x')| \leqslant \sup_{z\in B_r(x)} f(z) - \inf_{z\in B_r(x)} f(z).$$

If $\omega^f(x) = 0$, then we can find for every $\epsilon > 0$ some $r(\epsilon)$ such that

$$|f(x) - f(x')| \leqslant \sup_{z\in B_r(x)} f(z) - \inf_{z\in B_r(x)} f(z) \leqslant \epsilon + \omega^f(x) = \epsilon$$

holds for all $r < r_\epsilon$ and $x' \in B_r(x)$. Thus, f is continuous at x. □

As a consequence of Lemma 1, we can characterize the set of continuity points C^f of the function f,

$$C^f = \bigcap_{\delta>0} \{\omega^f < \delta\} = \bigcap_{n\in\mathbb{N}} \left\{\omega^f < \frac{1}{n}\right\}.$$

The following lemma shows that the sets $\left\{\omega^f < \frac{1}{n}\right\}$ are open, hence C^f is a G_δ-set.

Lemma 2 *ω^f is upper semi-continuous, i.e. the set $\{\omega^f < \alpha\}$ is open for every $\alpha > 0$.*

Proof For every $x_0 \in \{\omega^f < \alpha\}$ there is some $r = r(\alpha) > 0$ such that

$$\sup_{z\in B_r(x_0)} f(z) - \inf_{z\in B_r(x_0)} f(z) < \alpha.$$

Pick $y \in B_{r/3}(x_0)$. Since $B_{r/3}(y) \subseteq B_r(x_0)$, we get

$$\omega^f(y) \leqslant \sup_{z\in B_{r/3}(y)} f(z) - \inf_{z\in B_{r/3}(y)} f(z) \leqslant \sup_{z\in B_r(x_0)} f(z) - \inf_{z\in B_r(x_0)} f(z) < \alpha.$$

This shows that $y \in \{\omega^f < \alpha\}$, and so $B_{r/3}(x_0) \subseteq \{\omega^f < \alpha\}$. □

Comment The fact that the continuity points C^f of $f : X \to \mathbb{R}$ are a G_δ-set does not mean that we can find an everywhere continuous function $u : X \to \mathbb{R}$ such that $u|_{C^f} = f|_{C^f}$. In **normal** topological spaces – these are spaces where two disjoint closed sets can be separated by disjoint open sets – a sufficient condition is

Tietze's extension theorem (Willard [200, Theorem 15.8]). *Let F be a closed subset of a normal space. If $f : F \to \mathbb{R}$ is continuous, then there exists a continuous function $u : X \to \mathbb{R}$ such that $f = u|_F$.*

There is a related result for **measurable** functions [☞Theorem 13A].

8.10 A set D for which there exists no function having D as its discontinuity set

Example 8.9 shows that the set of discontinuity points $D^f := X \setminus C^f$ of a function f is an F_σ-set, i.e. a countable union of closed sets. Therefore, the irrational numbers $D := \mathbb{R} \setminus \mathbb{Q}$ cannot be the discontinuity set of any function f.

Comment In a completely metrizable metric space (X, d) which contains a countable dense set D, the converse is also true: any F_σ-set is the discontinuity set of some function f on X. In order to see this, let $F = \bigcup_{n\in\mathbb{N}} F_n$ be some F_σ-set. Without loss of generality, we may assume that the sets F_n are increasing. Define

$$f(x) = \begin{cases} n(x)^{-1}, & \text{if } x \in F \cap D, \\ -n(x)^{-1}, & \text{if } x \in F \cap D^c, \\ 0, & \text{if } x \notin F, \end{cases}$$

where $n(x) = \min\{n \in \mathbb{N} ; x \in F_n\}$. We claim that the function f has F as its set of discontinuities. Since $f|_{F^c} \equiv 0$, f is continuous in the open interior of F^c. If $x \in F^c \cap \partial F = F^c \cap \overline{F}$ and if $(x_i)_{i\in\mathbb{N}} \subseteq F$ is a sequence with $x_i \to x$, then every subsequence of $\big(n(x_i)\big)_{i\in\mathbb{N}}$ is unbounded. Otherwise, there would be some $N \in \mathbb{N}$ and a sequence $i(k) \to \infty$ such that $n(x_{i(k)}) \leqslant N$. Then $x_{i(k)} \in F_N$ for all $k \in \mathbb{N}$ and

$$x = \lim_{k\to\infty} x_{i(k)} \in F_N \subseteq F,$$

which cannot be. Thus, $\lim_{i\to\infty} f(x_i) = 0 = f(x)$, and we conclude that f is continuous at each point $x \in F^c$.

We will now show that f is not continuous at $x \in F$. If $x \in F \cap D$, then $x \in F_{n(x)}$ and $x \notin F_{n(x)-1}$.

Case 1. x is in the open interior of $F_{n(x)}$. The dense set D is a countable union of singletons, hence a set of first category, and so it follows from the Baire category theorem [☞Theorem 2.4] that $D^c = X \setminus D$ is dense in X. Consequently, there is a sequence $x_i \in F \cap D^c$ such that $x_i \to x$ and $f(x_i) < 0$. Thus, $\limsup_{i\to\infty} f(x_i) \leqslant 0$ while $f(x) = n(x)^{-1} > 0$, i.e. f is not continuous at x.

Case 2. x is in the boundary of $F_{n(x)}$. Then every neighbourhood $U = U(x)$ intersects $F_{n(x)} \cap D^c$ and we can find again a sequence in D^c such that $x_i \to x$ and $f(x_i) \leqslant 0$, i.e. f is not continuous at x.

The argument for $x \in F \cap D^c$ is similar.

8.11 A bijective measurable function f such that f^{-1} is not measurable

Equip the space of integers $\mathbb{Z}$ with the σ-algebra $\mathscr{A}$ which consists of all sets $A \subseteq \mathbb{Z}$ such that

$$\forall n \in \mathbb{N} \ : \ 2n \in A \iff 2n+1 \in A.$$

From now on n denotes a positive integer $n \in \mathbb{N}$. Let us briefly verify the conditions (Σ_1)–(Σ_3) defining a σ-algebra [☞p. 7].

(Σ_1) Trivially, $\emptyset \in \mathscr{A}$.

(Σ_2) If $A \in \mathscr{A}$, then $2n \in A^c \iff 2n \notin A \iff 2n+1 \notin A \iff 2n+1 \in A^c$, so $A^c \in \mathscr{A}$.

(Σ_3) Let $(A_i)_{i\in\mathbb{N}} \subseteq \mathscr{A}$ and set $A = \bigcup_{i\in\mathbb{N}} A_i$. If $2n \in A$, then $2n \in A_{i_0}$ for some i_0, hence $2n+1 \in A_{i_0}$ and $2n+1 \in A$. Thus, $A \in \mathscr{A}$.

Consider the function $f : \mathbb{Z} \to \mathbb{Z}$, $f(n) := n+2$. This is a bijection and

$$\begin{aligned} 2n \in f^{-1}(A) &\iff 2n+2 \in A \iff 2(n+1)+1 \in A \\ &\iff 2n+1 \in f^{-1}(A) \end{aligned}$$

for every $A \in \mathscr{A}$. Thus, $f^{-1}(A) \in \mathscr{A}$, i.e. f is measurable. The set $A = \{0\}$ is in $\mathscr{A}$, but $\{2\} = f(A) \notin \mathscr{A}$, so f^{-1} is not measurable.

8.12 A continuous bijective function $f : [0,1] \to [0,1]$ which is not Lebesgue measurable

Equip $[0,1]$ with the Lebesgue sets $\mathscr{L}[0,1]$ and consider on $([0,1], \mathscr{L}[0,1])$ Lebesgue measure λ. Let C be the classical middle-thirds Cantor set and D a fat Cantor set with measure $\lambda(D) > 0$ [☞7.3]. By $\mathscr{I}$ and $\mathscr{J}$ we denote the family of open subsets that make up $[0,1] \setminus C$ and $[0,1] \setminus D$, respectively. We use the tagging with 0-2-sequences from Example 7.3.

We define an order isomorphism ϕ from $\mathcal{J}$ onto $\mathcal{I}$ by mapping $J \in \mathcal{J}$ to $I \in \mathcal{I}$ if I and J have the same tags. Note that $[0,1] \setminus D = \bigcup_{J \in \mathcal{J}} J$. We define a function $f : [0,1] \to [0,1]$ as follows:

- ▶ $f(0) = 0$;
- ▶ $\forall J \in \mathcal{J} \quad f : J \to I = \phi(J)$ is linear (map the left/right endpoint of J to the left/right endpoint of I and join them with a line segment);
- ▶ $\forall x \in D \setminus \{0\} \ : \ f(x) = \sup\{f(t);\ t < x,\ t \in [0,1] \setminus D\}$.

The function $f : [0,1] \to [0,1]$ is continuous, strictly increasing, hence bijective and Borel measurable; moreover, $f(D) = C$. Let $N \subseteq D$ be a non-Lebesgue measurable set [☞7.22], and let $M := f(N) \subseteq C$. Since M is a subset of a Borel null set, it is Lebesgue measurable and we see that $f^{-1}(M)$ is not a Lebesgue set.

Comment Since the function f is continuous, hence Borel measurable, our example also shows that Borel measurability does not imply Lebesgue measurability.

8.13 A Lebesgue measurable bijective map $f : \mathbb{R} \to \mathbb{R}$ whose inverse is not Lebesgue measurable

Denote by $C \subseteq [0,1]$ the classical (middle-thirds) Cantor set [☞Section 2.5] and let $\Phi : C \to [0,1]$ be a bijective mapping. By the construction of the Cantor set, the complement $(0,\infty) \setminus (C \cup \mathbb{N})$ can be written as a countable union of disjoint open non-empty intervals $(I_n)_{n \in \mathbb{N}}$. We define a bijective mapping $f : \mathbb{R} \to \mathbb{R}$ as follows:

- ▶ $f(x) = \Phi(x)$ for $x \in C$;
- ▶ $f(x) = x$ for $x \in I_0 := (-\infty, 0) \cup \{n \in \mathbb{N}\,;\ n \geqslant 2\}$;
- ▶ $f|_{I_n}$ is an affine function, i.e. $f(x) = ax+b$ with suitable coefficients $a = a_n$, $b = b_n$, such that $f(I_n) = (n, n+1)$.

First we check that f is Lebesgue measurable. For any $A \in \mathscr{L}(\mathbb{R})$, the pre-image $f^{-1}(A)$ is the union

$$\{x \in C\,;\ \Phi(x) \in A\} \cup \{x \in I_0\,;\ x \in A\} \cup \bigcup_{n \in \mathbb{N}} \{x \in I_n\,;\ a_n x + b_n \in A\}.$$

Since $\{x \in C\,;\ \Phi(x) \in A\}$ is contained in the Lebesgue null set C, the completeness of Lebesgue measure yields that this set is Lebesgue measurable. For the other sets the Lebesgue measurability follows from the fact that $I_0 \cap A \in \mathscr{L}(\mathbb{R})$ and that affine functions $x \longmapsto ax + b$ are Lebesgue measurable, for instance

$$\{x \in I_n;\ a_n x + b_n \in A\} = \underbrace{I_n}_{\in \mathscr{L}(\mathbb{R})} \cap \underbrace{\{x \in \mathbb{R};\ a_n x + b_n \in A\}}_{\in \mathscr{L}(\mathbb{R})} \in \mathscr{L}(\mathbb{R}).$$

Consequently, $f^{-1}(A) \in \mathscr{L}(\mathbb{R})$ for any $A \in \mathscr{L}(\mathbb{R})$, i.e. f is Lebesgue measurable. Pick some set $N \subseteq [0,1]$ which is not Lebesgue measurable [☞ 7.22]. If we set $A := f^{-1}(N)$, then $A \subseteq C$ and therefore $A \in \mathscr{L}(\mathbb{R})$. However, $(f^{-1})^{-1}(A) = f(A) = N \notin \mathscr{L}(\mathbb{R})$, which means that f^{-1} is not Lebesgue measurable.

8.14 Borel measurable bijective maps have Borel measurable inverses

Let $f : \mathbb{R} \to \mathbb{R}$ be bijective. From $f(B) = (f^{-1})^{-1}(B)$ it follows that the inverse f^{-1} is Borel measurable if, and only if,

$$B \in \mathscr{B}(\mathbb{R}) \ \Rightarrow\ f(B) \in \mathscr{B}(\mathbb{R}), \tag{8.1}$$

i.e. if the image of any Borel set under f is again Borel. This condition is satisfied for every injective Borel measurable map $f : \mathbb{R} \to \mathbb{R}$, cf. Example 7.24 and [20, Theorem 6.8.6]. Thus, bijective Borel measurable maps $f : \mathbb{R} \to \mathbb{R}$ have a Borel measurable inverse.

Comment 1 The situation is different if we consider Lebesgue measurability instead of Borel measurability [☞ 8.13].

Comment 2 A Borel measurable function $f : \mathbb{R} \to \mathbb{R}$ satisfying (8.1) is called **bi-measurable**. Darst [38] constructed a smooth function which fails to be bi-measurable. This illustrates that the bijectivity of f plays a crucial role in the above reasoning.

8.15 Sums and products of measurable functions need not be measurable

Let $(X, \mathscr{A})$ and $(\mathbb{R}, \mathscr{B})$ be measurable spaces and $f, g : X \to \mathbb{R}$ measurable functions. If $\mathscr{B} = \mathscr{B}(\mathbb{R})$, then the functions obtained as pointwise sum $f + g$, product $f \cdot g$ and multiplication with a scalar $\lambda \cdot f$, $\lambda \in \mathbb{R}$, are measurable. In general, i.e. if $\mathscr{B} \neq \mathscr{B}(\mathbb{R})$, then $f + g$, $f \cdot g$ and $\lambda \cdot f$ need not be measurable. This means that the space of $\mathscr{A}/\mathscr{B}$ measurable functions is, in general, not a linear space.

Let $(X, \mathscr{A})$ be a measurable space with $\mathscr{A} \subsetneq \mathscr{P}(X)$, and set

$$\mathscr{B} := \sigma((-4,4)) = \{\emptyset, (-4,4), \mathbb{R} \setminus (-4,4), \mathbb{R}\}.$$

For $A \in \mathcal{P}(X) \setminus \mathcal{A}$ define

$$f(x) := 3 \cdot \mathbb{1}_A(x), \quad x \in X.$$

As $f^{-1}(-4,4) = X$ and $f^{-1}(\mathbb{R} \setminus (-4,4)) = \emptyset$, it follows that f is $\mathcal{A}/\mathcal{B}$ measurable. However, neither $2f = f + f = 6 \cdot \mathbb{1}_A$ nor $f \cdot f = 9 \cdot \mathbb{1}_A$ are measurable since

$$(f+f)^{-1}(\mathbb{R} \setminus (-4,4)) = A \notin \mathcal{A} \quad \text{and} \quad (f \cdot f)^{-1}(\mathbb{R} \setminus (-4,4)) = A \notin \mathcal{A}.$$

Comment There is an interesting consequence of the fact that sums and differences of measurable functions need not be measurable: for measurable functions f, g the pre-images $\{f = g\}$, $\{f < g\}$ and $\{f > g\}$ may fail to be measurable. For another example of measurable functions whose sum is not measurable [☞ 15.16].

8.16 The limit of a sequence of measurable functions need not be measurable

The following counterexample is from Dudley [49, Proposition 4.2.3]. Consider the space $Y = [0,1]^{[0,1]}$ of all functions $\eta : [0,1] \to [0,1]$, $t \longmapsto \eta(t)$. We equip this space with the usual product topology, i.e. pointwise convergence [☞ Example 2.1.(i)], and the corresponding topological (Borel) σ-algebra $\mathcal{B}(Y)$. By Tychonoff's theorem, Y is a compact space. We take $(X, \mathcal{X}) = ([0,1], \mathcal{B}[0,1])$ and define $f_n : X \to Y$ by

$$f_n(x)(t) := (1 - n|x - t|)^+, \quad x \in X = [0,1],\ t \in [0,1].$$

Since $x \longmapsto f_n(x)(\cdot)$ is continuous, it is Borel measurable. Moreover, we have $\lim_{n\to\infty} f_n(x)(t) = \mathbb{1}_{\{x\}}(t)$, i.e. $f(x) = \lim_{n\to\infty} f_n(x) = \mathbb{1}_{\{x\}} \in Y$.

The limiting function $f = \mathbb{1}_{\{x\}}$ is not Borel measurable. Indeed, the union $U_A := \bigcup_{t\in A}\{\eta \in Y\,;\ \eta(t) > 0\}$ is an open set for any $A \subseteq [0,1]$, hence in $\mathcal{B}(Y)$, but we have

$$f^{-1}(U_A) = \{x \in X\,;\ f(x) \in U_A\} = \{x \in X\,;\ \mathbb{1}_{\{x\}} \in U_A\} = A.$$

Since A need not be a Borel set, f is not measurable.

Comment If $f_n : (X, \mathcal{X}) \to (Y, \mathcal{B}(Y))$ are measurable functions with values in a metric space (Y, d) equipped with the Borel σ-algebra induced by the metric d, the pointwise limit $f(x) = \lim_{n\to\infty} f_n(x)$ is measurable whenever it exists. For the particular case $Y = \mathbb{R}$ this follows from showing that inf and sup, hence lim inf and lim sup of a sequence of measurable functions are measurable, cf. [MIMS, p. 66]. For arbitrary metric spaces (Y, d) the measurability can be deduced from the identity

$$\left\{x;\ \lim_{n\to\infty} f_n(x) \in U\right\} = \bigcup_{m=1}^{\infty}\bigcup_{k=1}^{\infty}\bigcap_{n=k}^{\infty}\left\{x;\ d(f_n(x), Y\setminus U) > \tfrac{1}{m}\right\}$$

for all open sets $U \subseteq Y$. If $(Y, \mathscr{Y})$ is more general, the pointwise limit need not be measurable.

For a further example of a sequence of measurable functions whose limit is not measurable [☞ 8.18].

8.17 A sequence of measurable functions such that the set $\{x;\ \lim_{n\to\infty} f_n(x)$ exists$\}$ is not measurable

Let $(X, \mathscr{A})$ and $(Y, \mathscr{B})$ be measurable spaces and $f_n : X \to Y$, $n \in \mathbb{N}$, be measurable functions. If (Y, d) is a Polish space and $\mathscr{B} = \mathscr{B}(Y)$ the associated Borel σ-algebra, then the set

$$L := \left\{x \in X;\ \lim_{n\to\infty} f_n(x) \text{ exists}\right\}$$

is measurable. Indeed, by the completeness of Y, $(f_n(x))_{n\in\mathbb{N}}$ converges if, and only if, $(f_n(x))_{n\in\mathbb{N}}$ is a Cauchy sequence, and so

$$L = \bigcap_{m\geqslant 1}\bigcup_{k\geqslant 1}\bigcap_{n,N\geqslant k}\left\{x \in X;\ d(f_n(x), f_N(x)) \leqslant \frac{1}{m}\right\} \in \mathscr{A}.$$

If Y is not a Polish space, then L does not need to be measurable.

Consider $(X, \mathscr{A}) = (\mathbb{R}, \mathscr{B}(\mathbb{R}))$ and

$$Y := (\mathbb{R} \times \{1/n;\ n \in \mathbb{N}\}) \cup (C \times \{0\})$$

with the induced Borel σ-algebra (trace σ-algebra) $\mathscr{B} := \mathscr{B}(\mathbb{R}^2) \cap Y$ for some set $C \notin \mathscr{B}(\mathbb{R}) = \mathscr{A}$. The functions

$$f_n : (X, \mathscr{A}) \to (Y, \mathscr{B}), \quad x \longmapsto f_n(x) := \left(x, \tfrac{1}{n}\right)$$

are measurable. However,

$$\left\{x \in X;\ \lim_{n\to\infty} f_n(x) \text{ exists}\right\} = C \notin \mathscr{A}.$$

8.18 The supremum of measurable functions need to be measurable

Let $(X, \mathscr{A})$ and $(\mathbb{R}, \mathscr{B})$ be measurable spaces, and let $f_i : X \to \mathbb{R}$, $i \in I$, be a family of measurable functions. If I is at most countable and $\mathscr{B} = \mathscr{B}(\mathbb{R})$, then

$\sup_{i\in I} f_i$ is measurable. This follows from the fact that $\{(a,\infty);\ a \in \mathbb{R}\}$ is a generator of $\mathscr{B}(\mathbb{R})$ and

$$\left\{\sup_{i\in I} f_i > a\right\} = \bigcup_{i\in I}\{f_i > a\} \in \mathscr{A}, \quad a \in \mathbb{R};$$

here we use that $\#I \leqslant \#\mathbb{N}$. The following examples show that, in general, $\sup_{i\in I} f_i$ fails to be measurable if I is uncountable or $\mathscr{B} \neq \mathscr{B}(\mathbb{R})$.

The case where I is uncountable. Take $(X,\mathscr{A}) = (\mathbb{R},\mathscr{B}(\mathbb{R}))$ and $\mathscr{B} = \mathscr{B}(\mathbb{R})$. Pick $I \subseteq \mathbb{R}$ such that $I \notin \mathscr{B}(\mathbb{R})$ [☞7.22]. Note that I is uncountable, since every countable set is a Borel set. The mappings

$$f_i(x) := \mathbb{1}_{\{i\}}(x), \quad x \in \mathbb{R},$$

are measurable for each $i \in I$, but $\sup_{i\in I} f_i = \mathbb{1}_I$ fails to be measurable.

The case where $\mathscr{B} \neq \mathscr{B}(\mathbb{R})$. Consider $(X,\mathscr{A}) = (\mathbb{R},\mathscr{B}(\mathbb{R}))$ and $\mathscr{B} := \sigma((-1,1))$. Choose $A \subseteq \mathbb{R}$ such that $A \notin \mathscr{B}(\mathbb{R})$ and set

$$f_n(x) := \left(1 - \frac{1}{n}\right)\mathbb{1}_A(x), \quad x \in \mathbb{R},\ n \in \mathbb{N}.$$

Since $f_n^{-1}(-1,1) = \mathbb{R}$ and $f_n^{-1}(\mathbb{R} \setminus (-1,1)) = \emptyset$, it follows that f_n is measurable for each $n \in \mathbb{N}$. However,

$$f := \sup_{n\in\mathbb{N}} f_n = \mathbb{1}_A$$

is not measurable as $f^{-1}(\mathbb{R} \setminus (-1,1)) = A \notin \mathscr{B}(\mathbb{R})$. As $f = \lim_{n\to\infty} f_n$ this also shows that pointwise limits of measurable functions need not be measurable; see also Example 8.16.

8.19 Measurability is not preserved under convolutions

Let $f, g \in L^1(\mathbb{R},\mathscr{B}(\mathbb{R}),\lambda)$ be two integrable functions and denote by $f * g$ the convolution of f and g, i.e.

$$(f * g)(x) = \int_{\mathbb{R}} f(x-y)g(y)\,\lambda(dy), \quad x \in \mathbb{R}.$$

If $\mathscr{A} = \sigma(f(\cdot - y);\ y \in \mathbb{R}) \subseteq \mathscr{B}(\mathbb{R})$ is the smallest σ-algebra on $\mathbb{R}$ such that the translates $x \longmapsto f(x-y)$ are measurable for each fixed $y \in \mathbb{R}$, then the integrand $x \longmapsto f(x-y)g(y)$ is $\mathscr{A}$ measurable for all $y \in \mathbb{R}$ but the (parameter-dependent) integral $f * g$ may fail to be $\mathscr{A}$ measurable [☞16.9].

8.20 The factorization lemma fails for general measurable spaces

Let X be a set and $(Y, \mathscr{A})$ a measurable space. For a function $f : X \to Y$ the σ-algebra

$$\sigma(f) := \sigma\left(f^{-1}(A),\ A \in \mathscr{A}\right)$$

is the smallest σ-algebra $\mathscr{F}$ on X such that $f : (X, \mathscr{F}) \to (Y, \mathscr{A})$ is measurable. The **factorization lemma** (also known as the **Doob–Dynkin lemma**) states that a mapping $g : X \to \mathbb{R}$ is $\sigma(f)/\mathscr{B}(\mathbb{R})$ measurable if, and only if, there exists a measurable mapping $h : (Y, \mathscr{A}) \to (\mathbb{R}, \mathscr{B}(\mathbb{R}))$ such that $g(x) = h(f(x))$ for all $x \in X$, cf. [MIMS, p. 68].

This statement remains true for mappings g taking values in some standard topological space Z (e.g. a Polish space) equipped with the Borel σ-algebra: a map $g : X \to Z$ is $\sigma(f)/\mathscr{B}(Z)$ measurable if, and only if, $g = h \circ f$ for some measurable mapping $h : (Y, \mathscr{A}) \to (Z, \mathscr{B}(Z))$, making the diagram in Fig. 8.1 commutative, cf. [82, Theorem III.11.2]. This equivalence breaks down if we equip the codomain Z with a σ-algebra $\mathscr{B}$ which is smaller than the Borel σ-algebra. For example, take $\mathscr{B} := \{\emptyset, Z\}$ and let f be constant, i.e. $f(x) = c$, $x \in X$, for some fixed $c \in Y$. By construction, every $g : (X, \sigma(f)) \to (Z, \mathscr{B})$ is measurable, but the only maps $g : X \to Z$ admitting a representation of the form $g = h \circ f$ are the constant maps.

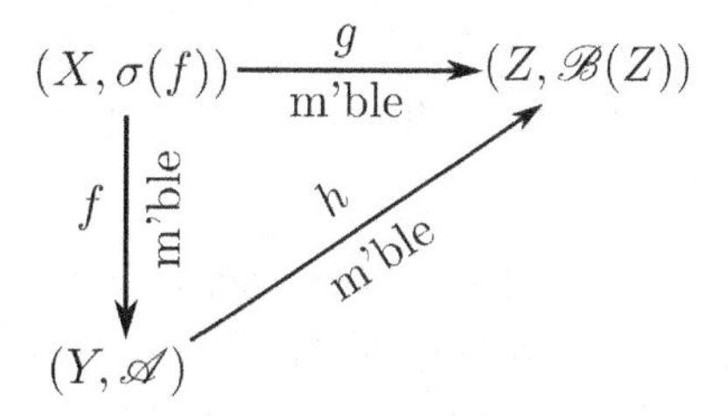

Figure 8.1 Factorization.

Comment This pathology happens because $g^{-1}(\mathscr{B})$ is too small to contain any essential information about g. If a function $g : X \to Z$ has a representation $g = h \circ f$ then the implication

$$f(x) = f(x') \Rightarrow g(x) = g(x')$$

holds for any $x, x' \in X$. This indicates that at least all singletons $\{z\}$, $z \in Z$, should belong to $\mathscr{B}$. In fact, this condition is almost sufficient for the factorization lemma to hold. If $\mathscr{B}$ is a σ-algebra on Z such that $\{z\} \in \mathscr{B}$, $z \in Z$, and $g : X \to Z$ is $\sigma(f)/\mathscr{B}$ measurable, then $g = h \circ f$ for a function $h : Y \to Z$ which is $\mathscr{A}'/\mathscr{B}$ measurable,

$$\mathscr{A}' := \sigma(\mathscr{A}, f(X)),$$

cf. [82, Theorem III.11.1]. Note that $\mathscr{A}' = \mathscr{A}$ if $f(X) \in \mathscr{A}$.

8.21 A Lebesgue measurable function $f: \mathbb{R} \to \mathbb{R}$ for which there is no Borel measurable function $g: \mathbb{R} \to \mathbb{R}$ such that $f \leqslant g$

For the construction we use the following lemma.

Lemma *Let (X, d) be a separable metric space with cardinality $\#X = \mathfrak{c}$, and denote by $\mathscr{B}(X)$ the Borel σ-algebra. The cardinality of the family of Borel measurable functions $u: X \to \mathbb{R}$ is that of the continuum.*

Proof Since all constant functions $u(x) := c$ are Borel measurable, the cardinality of the family of Borel measurable functions $u: X \to \mathbb{R}$ is at least as large as that of the continuum. If $u: X \to \mathbb{R}$ is Borel measurable, then u is uniquely determined by the countably many Borel sets $u^{-1}(q, \infty) \in \mathscr{B}(X)$, $q \in \mathbb{Q}$. Since X is separable, the associated Borel σ-algebra is countably generated, and it follows that $\#\mathscr{B}(X) = \mathfrak{c}$ [☞4.20]. Consequently, there are at most $\mathfrak{c}^{\#\mathbb{N}} = \mathfrak{c}$ many possibilities to choose the countably many pre-images $u^{-1}(q, \infty)$ from the Borel sets, which means that the cardinality of the family of Borel measurable functions is at most $\mathfrak{c}$. □

Let $C \subseteq [0, 1]$ be the classical Cantor set [☞Section 2.5]. Since C is uncountable, the lemma shows that there exists a bijective mapping $x \longmapsto h_x$ from C to the space of Borel measurable functions $u: C \to \mathbb{R}$. The mapping $f: \mathbb{R} \to \mathbb{R}$ defined by

$$f(x) := \begin{cases} h_x(x) + 1, & \text{if } x \in C, \\ 0, & \text{if } x \notin C, \end{cases}$$

is Lebesgue measurable. Indeed, if $L \in \mathscr{L}(\mathbb{R})$ is such that $0 \notin L$, then the pre-image $f^{-1}(L)$ is contained in C. Since the Cantor set has Lebesgue measure zero, it follows from the completeness of Lebesgue measure that any subset of C is Lebesgue measurable, and so $f^{-1}(L) \in \mathscr{L}(\mathbb{R})$. If $L \in \mathscr{L}(\mathbb{R})$ is such that $0 \in L$, then $f^{-1}(L) = (\mathbb{R} \setminus C) \cup C'$ for some $C' \subseteq C$, and hence $f^{-1}(L) \in \mathscr{L}(\mathbb{R})$.

Now suppose that there is a Borel measurable function $g: \mathbb{R} \to \mathbb{R}$ such that $f(x) \leqslant g(x)$ for all $x \in \mathbb{R}$. As $C \in \mathscr{B}(\mathbb{R})$, the restriction $g: C \to \mathbb{R}$ is Borel measurable, and therefore there exists some $x \in C$ with $h_x = g$. By the definition of f,

$$g(x) = h_x(x) < h_x(x) + 1 = f(x),$$

in contradiction to $f \leqslant g$.

Comment It is possible to construct a Lebesgue measurable function f such that there exists neither a Borel measurable function g nor a Borel measurable function h with $g(x) \geqslant f(x)$, resp. $h(x) \leqslant f(x)$, for all $x \in \mathbb{R}$. For further results in this direction see Kharazishvili [92, Chapter 7].

8.22 A positive Borel measurable function which cannot be approximated a.e. from below by step functions

Pick a Borel set $B \in \mathscr{B}[0,1]$ such that $\lambda(B \cap I) > 0$ and $\lambda(B^c \cap I) > 0$ for any interval $I \subseteq [0,1]$ with $\lambda(I) > 0$, i.e. for any interval which is not a singleton or empty [☞7.15]. Define

$$f(x) := \mathbb{1}_B(x), \quad x \in [0,1].$$

Let g be a positive step function, i.e. a function of the form

$$g(x) = \sum_{k=1}^{n} c_k \mathbb{1}_{I_k}(x), \quad x \in [0,1],$$

for $c_k \geqslant 0$ and pairwise disjoint intervals $I_k \neq \emptyset$ with cover the interval $[0,1]$. If $g \leqslant f$ (Lebesgue) almost everywhere, then $g(x) = 0$ for a.e. $x \notin B$, and so

$$c_k \mathbb{1}_{B^c \cap I_k} = 0 \quad \text{almost everywhere}$$

for $k = 1, \dots, n$. By virtue of our choice of B, we know that $\lambda(B^c \cap I_k) > 0$ for $k \in \{1, \dots, n\}$ with $\lambda(I_k) > 0$, and, hence, $c_k = 0$ for any such k. Thus, $g = 0$ a.e. In particular, there does not exist any sequence of positive step functions $(g_n)_{n \in \mathbb{N}}$ such that $g_n \uparrow f$ almost everywhere.

Comment Although there is no sequence of positive step functions $(g_n)_{n \in \mathbb{N}}$ which **increase** to f, it is possible to construct positive step functions $(g_n)_{n \in \mathbb{N}}$ such that $g_n \to f$ Lebesgue almost everywhere. Take any Borel measurable function $f : [0,1] \to [0,\infty)$. By slicing the range of f horizontally, we find a sequence $(f_n)_{n \in \mathbb{N}}$ of simple functions,

$$f_n(x) = \sum_{k=0}^{n2^n} k 2^{-n} \mathbb{1}_{A_k^n}(x), \quad x \in \mathbb{R},$$

such that $f_n \uparrow f$; see p. 6. Each of the Borel sets A_k^n can be approximated by intervals, i.e. for every $\epsilon > 0$ there exists a set I_k^n which is the union of finitely many intervals and which satisfies $\lambda(I_k^n \vartriangle A_k^n) \leqslant \epsilon$, cf. [MIMS, p. 38, Problem 5.12]. Approximating each of the sets $A_0^n, \dots, A_{n2^n}^n$ by finitely many intervals, e.g. for $\epsilon = \frac{1}{n+1} 2^{-2n}$, we get a positive step function g_n such that $\lambda(\{f_n \neq g_n\}) \leqslant 2^{-n}$. Since $\sum_{n \in \mathbb{N}} \lambda(\{f_n \neq g_n\}) < \infty$, the Borel–Cantelli lemma [☞Theorem 11A] shows that

$$N := \{x \in [0,1]\,;\ f_n(x) \neq g_n(x) \text{ for infinitely many } n \in \mathbb{N}\}$$

has Lebesgue measure zero. As

$$\forall x \in [0,1] \setminus N \ :\ \lim_{n\to\infty} g_n(x) = \lim_{n\to\infty} f_n(x) = f(x),$$

it follows that the sequence of step functions $(g_n)_{n\in\mathbb{N}}$ converges to f Lebesgue almost everywhere. Our construction contains the following more general result for arbitrary finite measure spaces $(X, \mathscr{A}, \mu)$.

Theorem *Let $(X, \mathscr{A}, \mu)$ be a finite measure space and $\mathscr{G}$ an algebra of sets, i.e. $X \in \mathscr{G}$ and $\mathscr{G}$ is stable under the formation of finite unions, intersections and complements. If $\mathscr{G}$ is a generator of $\mathscr{A}$, then there exists for every $\mathscr{A}$ measurable function $f : X \to [0,\infty)$ a sequence of simple functions $(f_n)_{n\in\mathbb{N}} \subseteq \mathcal{S}^+(\mathscr{G})$ with steps in $\mathscr{G}$,*

$$\mathcal{S}^+(\mathscr{G}) := \left\{ \sum_{i=1}^{n} c_i \mathbb{1}_{G_i}\,;\ n \in \mathbb{N},\ c_i \geqslant 0,\ G_i \in \mathscr{G} \right\},$$

such that $f_n \to f$ μ-almost everywhere.

8.23 $\mathbb{1}_{\mathbb{R}\setminus\mathbb{Q}}$ cannot be the pointwise limit of continuous functions

Assume there were continuous functions f_n with $\lim_{n\to\infty} f_n(x) = \mathbb{1}_{\mathbb{R}\setminus\mathbb{Q}}(x)$ for all $x \in \mathbb{R}$. Consider the closed sets $E_n := \left\{f_n \geqslant \frac{1}{2}\right\}$ and $F_N := \bigcap_{n=N}^{\infty} E_n$. Since $\lim_{n\to\infty} f_n(x) = \mathbb{1}_{\mathbb{R}\setminus\mathbb{Q}}(x)$, we have

$$x \in \mathbb{R} \setminus \mathbb{Q} \iff \exists N \in \mathbb{N} \quad \forall n \geqslant N \ :\ f_n(x) \geqslant \frac{1}{2} \iff x \in \bigcup_{N=1}^{\infty} \bigcap_{n=N}^{\infty} E_n.$$

Since the sets $F_N \subseteq \mathbb{R} \setminus \mathbb{Q}$ are closed and do not contain any open interval $I \neq \emptyset$, they are nowhere dense [☞ Definition 2.3]. Thus, $\mathbb{R} \setminus \mathbb{Q}$ would be of first category and so is $\mathbb{Q}$, hence $\mathbb{R}$ would be of first category, which is absurd, cf. Theorem 2.5 and Example 2.6.

Comment Functions which are pointwise limits of continuous functions are said to be in the **first Baire class**. Although $\mathbb{1}_{\mathbb{Q}}$ is not, we still have

$$\mathbb{1}_{\mathbb{Q}}(x) = \lim_{n\to\infty} \lim_{m\to\infty} \cos^m(2\pi n!x),$$

which follows from

$$\lim_{m\to\infty} \cos^m(2\pi n!x) = \begin{cases} 1, \text{ if } x = \frac{k}{n!}, & k \in \mathbb{Z}, \\ 0, \text{ if } x \in \mathbb{R} \setminus \mathbb{Q}. \end{cases}$$

This shows that $\mathbb{1}_{\mathbb{Q}}$ is in the **second Baire class**: functions which are limits of limits of continuous functions.

9

Inner and Outer Measure

Let $(X, \mathscr{A})$ be a measurable space. An (abstract) **outer measure** is a set function $\mu^* : \mathscr{P}(X) \to [0, \infty]$ that satisfies the properties (OM$_1$)–(OM$_3$) [☞ p. 28], i.e. $\mu^*(\emptyset) = 0$, μ^* is monotone increasing and σ-subadditive.[1] Outer measures are essential in the construction of measures [☞ Section 1.9]; the following covering outer measure was introduced by Carathéodory [34]; see also [35]. If $\mu : \mathscr{P}(X) \to [0, \infty]$ is any set function such that $\mu(\emptyset) = 0$ and $\mathscr{C} \subseteq \mathscr{P}(X)$ a family of sets containing $\emptyset \in \mathscr{C}$ (we will tacitly assume this from now on), then

$$(\mu|_{\mathscr{C}})^*(Q) := \inf\left\{\sum_{n=1}^{\infty} \mu(C_n);\ C_n \in \mathscr{C},\ Q \subseteq \bigcup_{n=1}^{\infty} C_n\right\}, \quad Q \subseteq X, \tag{9.1}$$

defines an outer measure.

For this definition, it is enough to assume that μ is defined only on $\mathscr{C}$; setting $\mu(D) := \infty$ for all $D \notin \mathscr{C}$ we can assume without loss of generality that μ is defined on $\mathscr{P}(X)$. Carathéodory's construction of μ^* is sometimes called 'Method I'; see e.g. [123, p. 47] or [MIMS, Section 18].[2] One has the following uniqueness theorem [123, Theorem 11.4].

Theorem 9A *Let X be some set and $\mathscr{C} \subseteq \mathscr{P}(X)$ a family with the property that X can be covered by a sequence $(C_n)_{n\in\mathbb{N}} \subseteq \mathscr{C}$. If $\mu, \nu : \mathscr{P}(X) \to [0, \infty]$ are any set functions with $\mu(\emptyset) = \nu(\emptyset) = 0$, then their Carathéodory extensions $\mu^* := (\mu|_{\mathscr{C}})^*$ and $\nu^* := (\nu|_{\mathscr{C}})^*$ satisfy*

$$\mu^*|_{\mathscr{C}} = \nu^*|_{\mathscr{C}} \implies \mu^* = \nu^*.$$

[1] Some authors, e.g. [57], call such set functions measures.

[2] In metric spaces, there is also a Method II [☞ Comment in 9.15]. This is similar to Method I, but it uses in a first step only coverings with sets C_n of diameter $\leqslant \epsilon$; in a second step, the limit $\epsilon \to 0$ – it is, in fact, a supremum – leads to an outer measure. The prime example for this procedure is Hausdorff measure, cf. [MIMS, Section 18].

A set $A \subseteq X$ is said to be μ^* **measurable**, if

$$\forall Q \subseteq X \;:\; \mu^*(Q) = \mu^*(A \cap Q) + \mu^*(A^c \cap Q). \tag{9.2}$$

We denote by $\mathscr{A}^*$ the family of all μ^* measurable sets. If the family $\mathscr{C}$ and the set function μ have some minimal structural properties – e.g. if $\mathscr{C}$ is a semi-ring [☞ p. 27] (or a ring or an algebra [☞ p. 73]) and μ is a measure relative to $\mathscr{C}$ – then $\mathscr{A}^*$ is sufficiently rich [☞ 9.13, 9.15], and we can combine μ^* measurability and σ-subadditivity of μ^* to show; see [MIMS, Proof of Theorem 6.1, pp. 41–45; Theorem 18.2]:

- $\mathscr{A}^*$ is a σ-algebra that contains $\mathscr{C}$ and $\sigma(\mathscr{C})$ [☞ 9.12]; the inclusion $\mathscr{C} \subseteq \mathscr{A}^*$ may fail if $\mathscr{C}$ is not a semi-ring [☞ 9.13].
- $\mu^*|_{\mathscr{A}^*}$ is a measure that extends the set function $\mu|_{\mathscr{C}}$.
- If μ is a σ-finite measure, then $\mu^*|_{\mathscr{A}^*}$ is the completion [☞ Example 1.7.(j)] of μ [☞ 9.9].

Without any structural assumptions on μ and $\mathscr{C}$, Carathéodory's construction always yields the largest outer measure μ^* below μ:

Theorem 9B (maximality of μ^*, [MIMS, p. 198]) *Let $(X, \mathscr{A})$ be a measurable space, $\mathscr{C} \subseteq \mathscr{P}(X)$, $\mu : \mathscr{C} \to [0, \infty]$ a set function with $\mu(\emptyset) = 0$ and $(\mu|_{\mathscr{C}})^*$ the outer measure associated with μ. If $\nu^* : \mathscr{P}(X) \to [0, \infty]$ is any outer measure, then*

$$\forall C \in \mathscr{C} \;:\; \nu^*(C) \leqslant \mu(C) \Rightarrow \forall Q \subseteq X \;:\; \nu^*(Q) \leqslant (\mu|_{\mathscr{C}})^*(Q),$$

i.e. μ^* is the largest outer measure such that $\mu^*|_{\mathscr{C}} \leqslant \mu|_{\mathscr{C}}$.

Let μ be a measure on a measurable space $(X, \mathscr{A})$ and $\mathscr{C}$ a family of sets which generates $\mathscr{A}$, i.e. $\sigma(\mathscr{C}) = \mathscr{A}$. Because of the monotonicity of the infimum, we have

$$(\mu|_{\mathscr{A}})^* \leqslant (\mu|_{\mathscr{C}})^*;$$

if the inequality $(\mu|_{\mathscr{C}})^*(A) \leqslant \mu(A)$ holds for all $A \in \mathscr{A}$, then Theorem 9B tells us that $(\mu|_{\mathscr{C}})^* = (\mu|_{\mathscr{A}})^*$.

Outer envelopes. There is a different way to extend a measure μ on $\mathscr{A}$ (relative to a family $\mathscr{C} \subseteq \mathscr{A}$) to all subsets of X:

$$(\mu|_{\mathscr{C}})^\circ(Q) := \inf\{\mu(C);\; C \supseteq Q,\; C \in \mathscr{C}\} \tag{9.3}$$

and $\mu^\circ := (\mu|_{\mathscr{A}})^\circ$. If the family $\mathscr{C}$ is stable under the formation of countable unions, then it is not hard to see that $(\mu|_{\mathscr{C}})^\circ$ is an outer measure such that

$(\mu|_{\mathscr{C}})^{\circ}(C) \leqslant \mu(C)$. Thus, by the maximality of the Carathéodory outer measure $(\mu|_{\mathscr{C}})^*$,

$$\forall Q \subseteq X \ : \ (\mu|_{\mathscr{C}})^{\circ}(Q) \leqslant (\mu|_{\mathscr{C}})^*(Q).$$

If $C \in \mathscr{C}$ is such that $C \supseteq Q$, we can use the trivial covering sequence $C, \emptyset, \emptyset, \ldots$ to see the opposite inequality $(\mu|_{\mathscr{C}})^*(Q) \leqslant (\mu|_{\mathscr{C}})^{\circ}(Q)$. This proves the following result.

Theorem 9C *If $(X, \mathscr{A}, \mu)$ is a measure space and $\mathscr{C} \subseteq \mathscr{A}$ a family of sets which is stable under the formation of countable unions, then*

$$(\mu|_{\mathscr{A}})^* = (\mu|_{\mathscr{A}})^{\circ} \leqslant (\mu|_{\mathscr{C}})^{\circ} = (\mu|_{\mathscr{C}})^*.$$

Remark 9D If the condition $(\mu|_{\mathscr{C}})^*(A) \leqslant \mu(A)$ mentioned in the previous paragraph holds for all $A \in \mathscr{A}$, then we conclude that $(\mu|_{\mathscr{C}})^* = (\mu|_{\mathscr{A}})^*$.

Inner measures. Instead of considering outer envelopes, we can approximate a measure μ (relative to a family $\mathscr{C} \subseteq \mathscr{A}$) from inside:

$$(\mu|_{\mathscr{C}})_*(Q) := \sup\{\mu(C);\ C \subseteq Q,\ C \in \mathscr{C}\}. \tag{9.4}$$

If $\mathscr{C}$ is stable under the formation of finite unions, then the set function $(\mu|_{\mathscr{C}})_*$, and in particular $\mu_* := (\mu|_{\mathscr{A}})_*$, is an (abstract) **inner measure**, i.e. the empty set has inner measure zero and the set functions are monotone increasing and σ-superadditive (for sequences $(Q_n)_{n\in\mathbb{N}}$ of pairwise disjoint sets).[3] Clearly,

$$(\mu|_{\mathscr{C}})_* \leqslant \mu_* \leqslant \mu^{\circ} = \mu^* \quad \text{and} \quad \forall A \in \mathscr{A} \ : \ \mu^*(A) = \mu_*(A) = \mu(A).$$

We have the following relation between inner measures and outer envelopes.

Theorem 9E ([35, p. 262 Satz 4, p. 266 Satz 7]) *Let $(X, \mathscr{A}, \mu)$ be a measure space, μ_* the inner measure and μ° the outer envelope (which coincides with the outer measure μ^*). Then*

$$\forall A \in \mathscr{A},\ \mu(A) < \infty \quad \forall Q \subseteq A \ : \ \mu_*(Q) = \mu(A) - \mu^{\circ}(A \setminus Q),$$
$$\forall Q, R \subseteq X,\ Q \cap R = \emptyset \ : \ \mu_*(R \uplus Q) \leqslant \mu_*(Q) + \mu^{\circ}(R) \leqslant \mu^{\circ}(Q \uplus R).$$

[3] There are other ways to define inner measures; see e.g. Royden [147, Section 12.6] who uses $\mu_*(Q) := \sup\{\mu(C) - \mu^*(C \setminus Q);\ C \in \mathscr{C},\ \mu^*(C \setminus Q) < \infty\}$ where $Q \subseteq X$ and $\mathscr{C}$ is an algebra. This set function coincides with our definition if $Q = A \in \mathscr{A}$ and $\mu(A) < \infty$. A good discussion is given in [19, Section 1.12(viii)] and [69, Chapter II.6]. Originally, inner measures were used to describe measurability, but this approach was quickly superseded by Carathéodory's outer measure method which is much less cumbersome; see the comments [33, Entry LXXXVII, pp. 276–277]. A good axiomatic description of inner measures on general structures is Ridder [141].

Regularity of measures. Let $(X, \mathscr{O})$ be a topological space, write $\mathscr{F}$ for the closed sets and $\mathscr{K}$ for the compact sets. A measure μ on a σ-algebra $\mathscr{A} \subseteq \mathscr{P}(X)$ is called **outer regular** if $\mathscr{B}(X) \subseteq \mathscr{A}$ and

$$\forall A \in \mathscr{A} \ : \ \mu(A) = \inf\{\mu(U);\ U \supseteq A,\ U \in \mathscr{O}\} \iff \mu = (\mu|_{\mathscr{O}})^{\circ}|_{\mathscr{A}}, \tag{9.5}$$

inner regular if $\mathscr{B}(X) \subseteq \mathscr{A}$ and

$$\forall U \in \mathscr{O} \ : \ \mu(U) = \sup\{\mu(F);\ F \subseteq U,\ F \in \mathscr{F}\} \iff \mu|_{\mathscr{O}} = (\mu|_{\mathscr{F}})_*|_{\mathscr{O}}, \tag{9.6}$$

and **inner compact regular** if $\mathscr{K} \subseteq \mathscr{A}$, $\mu(K) < \infty$ for $K \in \mathscr{K}$, and

$$\forall U \in \mathscr{O} \ : \ \mu(U) = \sup\{\mu(K);\ K \subseteq U,\ K \in \mathscr{K}\} \iff \mu|_{\mathscr{O}} = (\mu|_{\mathscr{K}})_*|_{\mathscr{O}}. \tag{9.7}$$

If the measure space is σ-compact and all compact sets are closed, e.g. if X is a metric space, then the notions of inner and inner compact regularity coincide.

Usually, regularity is discussed for locally finite measures on locally compact Polish spaces. Since Polish spaces are metrizable, compact sets are always closed; moreover locally finite measures are automatically σ-finite and finite on compact sets. Local compactness together with separability ensures σ-compactness. Therefore, many authors do not distinguish between inner and inner compact regularity. A measure is said to be **regular**, if it is both outer and inner compact regular.

On a σ-compact metric space (X, d) all measures which are finite on compact sets are inner (compact) regular. If there is an increasing sequence $U_n \uparrow X$, $U_n \in \mathscr{O}$ and $\mu(U_n) < \infty$, then μ is also outer regular, [MIMS, p. 439]. In particular, all finite measures on a metric space are outer regular.

Sometimes, inner (compact) regular means that (9.7) should hold for all Borel sets with finite measure. If (X, d) is a metric space and μ regular (i.e. (9.5) and (9.7) hold), then (9.7) extends to all $B \in \mathscr{B}(X)$ with $\mu(B) < \infty$. If (X, d) is σ-finite, we can even admit $\mu(B) = \infty$; see [MINT, Lemma A.9] or [60, Corollary 7.6].

9.1 An explicit construction of a non-measurable set

Let $X = [0,1] \times [0,1]$ and denote by λ Lebesgue measure on $([0,1], \mathscr{L}[0,1])$. Define on X the σ-algebra

$$\mathscr{A} := \{B \times [0,1];\ B \in \mathscr{L}[0,1]\}$$

and the measure $\mu(A) := \lambda(B)$ if $A = B \times [0,1] \in \mathscr{A}$. The corresponding outer measure is given by

$$\mu^*(Q) := \inf\{\mu(A);\ A \supseteq Q,\ A \in \mathscr{A}\}, \quad Q \subseteq [0,1] \times [0,1].$$

Clearly, $\mu^*([0,1] \times I) = 1$ for any interval $I \subseteq [0,1]$. Let $Q = [0,1] \times \left[0, \frac{1}{4}\right]$. Moreover, we have

$$\begin{aligned}\mu^*(Q) = 1 \neq 2 &= \mu^*\left([0,1] \times \left[0, \tfrac{1}{4}\right]\right) + \mu^*\left([0,1] \times \left(\tfrac{1}{4}, \tfrac{1}{2}\right]\right) \\ &= \mu^*\left([0,1] \times \left[0, \tfrac{1}{2}\right] \cap Q\right) + \mu^*\left([0,1] \times \left[0, \tfrac{1}{2}\right] \setminus Q\right),\end{aligned}$$

and this shows that $[0,1] \times \left[0, \frac{1}{2}\right]$ is not μ^* measurable.

9.2 A set which is not Lebesgue measurable with strictly positive outer and zero inner measure

Let L be the set constructed in Example 7.22. We can repeat the calculations in (7.3)–(7.5) made in Example 7.22, where we replace the completion $\bar{\lambda}$ by the outer measure λ^*. Note that $\lambda^*(x + A) = \lambda^*(A)$ is again invariant under translations and σ-subadditive:

$$1 = \lambda^*[0,1) \leqslant \sum_{q \in \mathbb{Q} \cap (-1,1)} \lambda^*(q + L) = \sum_{q \in \mathbb{Q} \cap (-1,1)} \lambda^*(L).$$

This is only possible if $\lambda^*(L) > 0$.

On the other hand, L cannot contain a compact set of strictly positive Lebesgue measure. Indeed, if there were some compact set $K \subseteq L$ with $\bar{\lambda}(K) > 0$, then $(r + K) \cap (q + K) = \emptyset$ for all $r \neq q$, $r, q \in \mathbb{Q}$, and the second inclusion in (7.3) would give the contradiction

$$\infty = \sum_{q \in \mathbb{Q} \cap (-1,1)} \bar{\lambda}(q + K) \leqslant 3.$$

Consequently, $\lambda(K) = 0$ for any compact set $K \subseteq L$. Since Lebesgue measure is inner compact regular, it follows that $\lambda(B) = 0$ for any Borel set $B \subseteq L$, and so

$$\lambda_*(L) = \sup\{\lambda(B)\,;\; B \subseteq L, B \in \mathscr{B}(\mathbb{R})\} = 0.$$

9.3 A decreasing sequence $A_n \downarrow \emptyset$ such that $\lambda^*(A_n) = 1$

Let λ be Lebesgue measure on $[0,1)$ and construct, as in Example 7.22, a non-Lebesgue measurable set L. For an enumeration $(q_n)_{n \in \mathbb{N}}$ of $\mathbb{Q} \cap [0,1)$ we define

$$L_n := \left(L \cup (L + q_1) \cup \cdots \cup (L + q_n)\right) \cap [0,1) \quad \text{and} \quad A_n := [0,1) \setminus L_n.$$

Since $L_n \uparrow [0,1)$, see the construction in Example 7.22, the sequence A_n decreases to $\emptyset$.

The construction of L in Example 7.22 also shows that L does not contain a compact set with strictly positive Lebesgue measure, [☞9.2] for details, and this property is inherited by L_n. By the inner compact regularity of Lebesgue measure, this implies $\lambda(B) = 0$ for any Borel set $B \subseteq L_n$. Hence, $\lambda_*(L_n) = 0$ and [☞Theorem 9E]

$$\lambda^*(A_n) = \lambda([0,1)) - \lambda_*(L_n) = 1.$$

9.4 A set such that $\lambda_*(E) = 0$ and $\lambda^*(E \cap B) = \lambda(B) = \lambda^*(B \setminus E)$ for all $B \in \mathscr{B}(\mathbb{R})$

Let λ_* and λ^* be the inner and outer measure [☞p. 183] associated with Lebesgue measure λ, i.e.

$$\lambda_*(Q) := \sup\{\lambda(B);\ B \subseteq Q,\ B \in \mathscr{B}(\mathbb{R})\},$$
$$\lambda^*(Q) := \inf\{\lambda(B);\ B \supseteq Q,\ B \in \mathscr{B}(\mathbb{R})\},$$

for $Q \subseteq \mathbb{R}$. Note that λ^* coincides with the Carathéodory extension of Lebesgue measure λ [☞Theorem 9C]. Later on, we will need the following observation: if $Q \subseteq \mathbb{R}$ and $B \in \mathscr{B}(\mathbb{R})$ are such that $Q \subseteq B$, then [☞Theorem 9C]

$$\lambda^*(Q) = \lambda(B) - \lambda_*(B \setminus Q). \tag{9.8}$$

Similar to the construction of a non-Lebesgue measurable set in Example 7.22, we define an equivalence relation for $x, y \in \mathbb{R}$

$$x \sim y \iff \exists n, m \in \mathbb{Z}\ :\ x - y = n + m\sqrt{2}$$

and, by the axiom of choice, we select a set E_0 containing exactly one representative of each equivalence class. Define

$$E := \left\{e_0 + 2n + m\sqrt{2};\ e_0 \in E_0,\ m, n \in \mathbb{Z}\right\}.$$

For any $B \in \mathscr{B}(\mathbb{R})$ with $B \subseteq E$ the difference $B - B = \{x - y;\ x, y \in B\}$ does not contain points of the form $2n + 1 + m\sqrt{2}$ with $n, m \in \mathbb{Z}$, hence there is no open interval inside $B - B$; because of Steinhaus' theorem [☞7.28] this means that $\lambda(B) = 0$, thus, $\lambda_*(E) = 0$.

We claim that $\mathbb{R} \setminus E = E + 1$. In order to see this, we observe that for any x and its representative $x_0 \in E_0$ we know that $x - x_0 = n + m\sqrt{2}$ for suitable $n, m \in \mathbb{Z}$. Thus,

$$x = x_0 + n + m\sqrt{2} \in \begin{cases} E, & \text{if } n \text{ even}, \\ E + 1, & \text{if } n \text{ odd}. \end{cases}$$

Since E_0 contains exactly one representative of each equivalence class, the representative x_0 is unique, and so $E \cap (E + 1) = \emptyset$, i.e. $E^c = E + 1$. We conclude that $\lambda_*(E^c) = \lambda_*(E+1) = 0$. By (9.8) and the monotonicity of λ_*, we find for all $B \in \mathscr{B}(\mathbb{R})$

$$\lambda^*(B \cap E) = \lambda(B) - \lambda_*(B \setminus E) = \lambda(B)$$

and

$$\lambda^*(B \setminus E) = \lambda^*(B \setminus (B \cap E)) = \lambda(B) - \lambda_*(B \cap E) = \lambda(B).$$

9.5 Lebesgue measure beyond the Lebesgue sets

Let λ be Lebesgue measure on $(\mathbb{R}, \mathcal{L}(\mathbb{R}))$ and pick some $N \notin \mathcal{L}(\mathbb{R})$ [☞7.22]. Assume that $\lambda^*(N) = \beta < \infty$. The family

$$\mathscr{A} := \{C = (A \cap N) \cup (B \cap N^c);\ A, B \in \mathscr{L}(\mathbb{R})\}$$

is a σ-algebra. In fact:

(Σ_1) if we take $A = B = \emptyset$, we get $\emptyset \in \mathscr{A}$;
(Σ_2) if $C = (A \cap N) \cup (B \cap N^c)$, then

$$C^c = (A^c \cup N^c) \cap (B^c \cup N) = (A^c \cap N) \cup (B^c \cap N^c) \in \mathscr{A};$$

(Σ_3) if $C_n = (A_n \cap N) \cup (B_n \cap N^c)$, $n \in \mathbb{N}$, then

$$\bigcup_{n \in \mathbb{N}} C_n = (A \cap N) \cup (B \cap N^c) \in \mathscr{A} \quad \text{with} \quad A = \bigcup_{n \in \mathbb{N}} A_n,\ B = \bigcup_{n \in \mathbb{N}} B_n.$$

Note that $\mathscr{L}(\mathbb{R}) \subseteq \mathscr{A}$. From the definition of the outer measure $\lambda^*(N)$ we know that there is some $N^* \in \mathscr{L}(\mathbb{R})$ such that $N \subseteq N^*$ and $\lambda(N^*) = \lambda^*(N)$. Let $\gamma \in [0, \beta]$. The set function

$$\mu(C) := \frac{\gamma}{\beta}\lambda(A \cap N^*) + \left(1 - \frac{\gamma}{\beta}\right)\lambda(B \cap N^*) + \lambda(B \setminus N^*)$$

defined for the set $C = (A \cap N) \cup (B \cap N^c) \in \mathscr{A}$ is obviously a measure – once we know that μ is well-defined, i.e. independent of the representation of C in terms of $A, B \in \mathscr{L}(\mathbb{R})$; in this case, we have $\mu(N) = \gamma$ and $\mu(A) = \lambda(A)$ for all $A \in \mathscr{L}(\mathbb{R})$.

We will now show that μ is well-defined. Recall that $\lambda^*|_{\mathscr{L}(\mathbb{R})} = \lambda$. Using the additivity of λ and the monotonicity of the outer measure λ^*, we get

$$\lambda(A \cap N^*) = \lambda(N^* \setminus (N^* \setminus A)) = \lambda(N^*) - \lambda(N^* \setminus A) \leqslant \lambda^*(N) - \lambda^*(N \setminus A).$$

Since λ^* is subadditive and monotone, we use $N = (N \cap A) \uplus (N \setminus A)$ to obtain

$$\lambda(A \cap N^*) \leqslant \lambda^*(A \cap N) + \lambda^*(N \setminus A) - \lambda^*(N \setminus A) = \lambda^*(A \cap N) \leqslant \lambda(A \cap N^*).$$

This shows that $\lambda^*(A \cap N) = \lambda(A \cap N^*)$ holds for all $A \in \mathscr{L}(\mathbb{R})$. Thus, if

$$C = (A \cap N) \cup (B \setminus N) = (A' \cap N) \cup (B' \setminus N),$$

we get $C \cap N = A \cap N = A' \cap N$, hence

$$\lambda(A \cap N^*) = \lambda^*(A \cap N) = \lambda^*(A' \cap N) = \lambda(A' \cap N^*).$$

A similar argument yields

$$\lambda(B \setminus N^*) = \lambda^*(B \setminus N) = \lambda^*(B' \setminus N) = \lambda(B' \setminus N^*),$$

and we conclude that the measure μ is well-defined

Comment The above extension procedure can be repeated finitely often, but it is not possible to construct, using transfinite induction, a σ-additive extension of Lebesgue measure for all sets from $\mathscr{P}(\mathbb{R})$. This is the famous *problème de la mesure* of Lebesgue [102]. Vitali [☞ 7.22] was the first to construct a non-Lebesgue measurable set, proving that Lebesgue's problem has no solution. A nice account is given in the short note by Banach and Kuratowski [11]; see also Example 7.31. If one settles for finite additivity instead of σ-additivity, Lebesgue's measure problem can be solved in dimension 1 and 2, but from dimension 3 onwards, there is the Banach–Tarski paradox [☞ 7.31].

9.6 The Carathéodory extension λ^* of $\lambda|_{[0,1)}$ is not continuous from above

The finite measure $\lambda_{[0,1)}$ is continuous from above, but this property is not inherited by its Carathéodory extension λ^* [☞ 9.3].

9.7 An outer measure which is not continuous from below

The mapping

$$\mu^* : \mathscr{P}(\mathbb{N}) \to [0, \infty],\ A \longmapsto \mu^*(A) := \begin{cases} 0, & \text{if } A = \emptyset, \\ 1, & \text{if } 1 \leqslant \#A < \infty, \\ \infty, & \text{if } \#A = \infty, \end{cases}$$

defines an outer measure on $X = \mathbb{N}$ [☞ (OM$_1$)–(OM$_3$) on p. 28]. However, μ^* is not continuous from below: take $A_n := \{1, \dots, n\}$, then $A_n \uparrow \mathbb{N}$ but $\mu^*(A_n) = 1$ does not converge to $\mu^*(\mathbb{N}) = \infty$.

Comment Let (X, d) be a metric space. If μ^* is a metric outer measure on X, i.e. an outer measure with the additional property that

$$A, B \subseteq X,\ d(A, B) = \inf_{x \in A, y \in B} d(x, y) > 0 \Rightarrow \mu^*(A \cup B) = \mu^*(A) + \mu^*(B),$$

then μ^* is continuous from below, cf. [MIMS, p. 201].

9.8 A measure μ such that its outer measure μ^* is not additive

Define on $X = \{0, 1\}$ the σ-algebra $\mathscr{A} = \{\emptyset, X\}$ and the measure $\mu(\emptyset) = 0$ and $\mu(X) = 1$. If μ^* is **any** outer extension, then $\mu^*(\emptyset) = 0$ while $\mu^*(A) = 1$ for all sets $A \in \{\{0\}, \{1\}, X\}$; in particular,

$$1 = \mu^*(X) = \mu^*(\{0\} \cup \{1\}) \neq \mu^*(\{0\}) + \mu^*(\{1\}) = 2.$$

9.9 A measure space such that $(X, \mathscr{A}^*, \mu^*|_{\mathscr{A}^*})$ is not the completion of $(X, \mathscr{A}, \mu)$

Consider $X = \mathbb{R}$ with the Borel σ-algebra $\mathscr{B}(\mathbb{R})$ and the counting measure $\mu = \zeta_{\mathbb{R}}$. Since $\emptyset$ is the only set $B \in \mathscr{B}(\mathbb{R})$ with $\mu(B) = 0$, the measure space $(\mathbb{R}, \mathscr{B}(\mathbb{R}), \zeta_{\mathbb{R}})$ is complete. Moreover, by the definition of the Carathéodory extension μ^*,

$$\mu^*(A) = \zeta_{\mathbb{R}}(A), \quad A \subseteq X.$$

If $A \subseteq X$ and $E \subseteq X$ are subsets of X, then $A \cap E$ and $A^c \cap E$ are disjoint. Hence,

$$\mu^*(E) = \mu^*(A \cap E) + \mu^*(A^c \cap E),$$

which gives $\mathscr{A}^* = \mathscr{P}(X)$. Consequently, $(X, \mathscr{A}^*, \mu^*|_{\mathscr{A}^*}) = (\mathbb{R}, \mathscr{P}(\mathbb{R}), \zeta_{\mathbb{R}})$ is not the completion of $(X, \mathscr{A}, \mu) = (\mathbb{R}, \mathscr{B}(\mathbb{R}), \zeta_{\mathbb{R}})$.

Comment This pathology happens because the counting measure is not σ-finite. If μ is a σ-finite measure on a measurable space $(X, \mathscr{A})$, then the measure space $(X, \mathscr{A}^*, \mu^*|_{\mathscr{A}^*})$ is the completion of $(X, \mathscr{A}, \mu)$, cf. [MIMS, pp. 50–51, Problem 6.4].

9.10 A measure space where $\mu_*(E) = \mu^*(E)$ does not imply measurability of E

Let $X = \{a, b, c\}$ and define a set function on $\mathscr{P}(X)$ by

$$\mu^*(X) = 2, \quad \mu^*(\emptyset) = 0, \quad \mu^*(A) = 1 \text{ for all other sets.}$$

One can check directly that μ^* is an outer measure. The σ-algebra of μ^* measurable sets $\mathscr{A}^*$ consists of all sets $A \subseteq X$ such that

$$\forall E \subseteq X \,:\, \mu^*(E) = \mu^*(E \cap A) + \mu^*(E \cap A^c). \tag{9.9}$$

If $A \notin \{\emptyset, X\}$, then (9.9) fails – take $E = \{x, y\}$ for $x \in A$ and $y \in X \setminus A$. Consequently, $\mathscr{A}^* = \{\emptyset, X\}$. The outer measure μ^* induces an inner measure μ_*,

$$\mu_*(E) = \sup\{\mu^*(A) - \mu^*(A \setminus E);\ A \in \mathscr{A}^*,\ \mu^*(A \setminus E) < \infty\},$$

and since $\mathscr{A}^* = \{\emptyset, X\}$, it follows that

$$\mu_*(E) = \mu^*(X) - \mu^*(X \setminus E) = 2 - 1 = 1 \ \text{ if } E \notin \{\emptyset, X\}$$

while $\mu_*(X) = 2$ and $\mu_*(\emptyset) = 0$.

This shows that $\mu_* = \mu^*$. Since $\mathscr{A}^*$ is trivial, all non-trivial sets have the same inner and outer measures, but they are not measurable.

Comment This pathology can occur only if the outer measure μ^* is not induced by a measure m relative to a semi-ring $\mathscr{C}$.

9.11 A non-Lebesgue measurable set with identical inner and outer measure

Let λ be Lebesgue measure on $(\mathbb{R}, \mathscr{L}(\mathbb{R}))$. Take $N \subseteq [0,1]$ such that $N \notin \mathscr{L}(\mathbb{R})$ [☞ 7.22], and define $A := N \cup (-\infty, 0)$. Since the inner and outer measures are monotone, we have

$$\lambda^*(A) \geqslant \lambda^*(-\infty, 0) = \infty \quad \text{and} \quad \lambda_*(A) \geqslant \lambda_*(-\infty, 0) = \infty,$$

and so $\lambda^*(A) = \infty = \lambda_*(A)$. But the set A is not Lebesgue measurable because otherwise $N = A \cap [0,1]$ would be Lebesgue measurable.

9.12 A measure such that every set is μ^* measurable

Let $(X, \mathscr{A}, \mu)$ be a measure space. Let $\mu^*(Q) := \inf\{\mu(A);\ Q \subseteq A,\ A \in \mathscr{A}\}$ be the outer measure induced by μ. Recall that a set A is μ^* measurable, if

$$\forall Q \subseteq X \,:\, \mu^*(Q) = \mu^*(Q \cap A) + \mu^*(Q \setminus A). \tag{9.10}$$

We denote by $\mathscr{A}^*$ the σ-algebra of all μ^* measurable sets.

Consider any uncountable set X and let $\mathscr{A}$ be the co-countable σ-algebra [☞ Example 4A], i.e. $A \in \mathscr{A}$ if either A or A^c is at most countable. Let μ be the counting measure on $(X, \mathscr{A})$.

Let us first determine μ^*: if $Q \in \mathscr{A}$, then $\mu^*(Q) = \mu(Q)$. If $Q \notin \mathscr{A}$, then both Q and Q^c are uncountable. In this case $\mu^*(Q) = \infty$ since every measurable $A \supseteq Q$ must be uncountable, so $\mu^*(Q) = \inf\{\mu(A);\ A \in \mathscr{A},\ A \text{ uncountable}\} = \infty$. Let $A, Q \subseteq X$ be any two sets. We check (9.10) by discussing all possible cases.

Q countable. Since both $Q \cap A$ and $Q \setminus A = Q \setminus (Q \cap A)$ are countable, the values of μ and μ^* coincide, and (9.10) is obvious.

Q uncountable. Since $Q = (Q \cap A) \cup (Q \setminus A)$ at least one of the sets $Q \cap A$ and $Q \setminus A$ is uncountable, and (9.10) reads $\infty = n + \infty$ with $n \in \mathbb{N}_0 \cup \{\infty\}$.

This means that (9.10) holds for any $A \subseteq X$, i.e. $\mathscr{A}^* = \mathscr{P}(X)$.

9.13 A measure μ relative to $\mathcal{S}$ such that every non-empty set in $\mathcal{S}$ fails to be μ^* measurable

If $\mu : \mathcal{S} \to [0, \infty]$ is a measure relative to a semi-ring $\mathcal{S} \subseteq \mathscr{P}(X)$, then (the proof of) Carathéodory's theorem [☞Theorem 1.41] shows that every set $S \in \mathcal{S}$ is μ^* measurable, i.e.

$$\forall S \in \mathcal{S}, \quad \forall E \subseteq X \ : \ \mu^*(E) = \mu^*(E \cap S) + \mu^*(E \setminus S).$$

The following example from [139, Problem 2.2.6] illustrates that the statement breaks down if $\mathcal{S}$ is not a semi-ring.

Consider $X = \mathbb{N}$ and the family

$$\mathcal{S} := \{A \subseteq \mathbb{N};\ A = \emptyset \text{ or } \#A = 2\}.$$

The set function $\mu(A) := \#A$ defines a measure relative to $\mathcal{S}$, and the corresponding outer measure μ^* is given by

$$\mu^*(A) := \begin{cases} \#A + 1, & \text{if } \#A \text{ is odd,} \\ \#A, & \text{otherwise.} \end{cases}$$

If $S \in \mathcal{S}$, $S \neq \emptyset$, then $S = \{m, n\}$ for some $m \neq n \in \mathbb{N}$. Pick $k \in \mathbb{N} \setminus S$ and set $E = \{k, m\}$, then

$$\mu^*(E) = \mu^*(\{k, m\}) = 2,$$

while

$$\mu^*(E \cap S) + \mu^*(E \setminus S) = \mu^*(\{m\}) + \mu^*(\{k\}) = 2 + 2,$$

which shows that S is not μ^* measurable.

9.14 An additive set function μ on a semi-ring such that μ^* is not an extension of μ

If μ is a measure relative to a semi-ring $\mathcal{S} \subseteq \mathcal{P}(X)$, then the Carathéodory extension μ^* defined by

$$\mu^*(A) = \inf\left\{\sum_{n=1}^{\infty} \mu(S_n);\ S_n \in \mathcal{S},\ \bigcup_{n\in\mathbb{N}} S_n \supseteq A\right\}$$

is an extension of μ, i.e. $\mu^*|_{\mathcal{S}} = \mu$ [☞Theorem 1.41]. The following example shows that this property may break down if μ is not σ-additive on $\mathcal{S}$ but only finitely additive.

Consider $X = \mathbb{N}$ and the semi-ring

$$\mathcal{S} := \{S \subseteq \mathbb{N}\,;\ \#S < \infty \text{ or } \#S^c < \infty\}.$$

The set function

$$\mu(S) := \begin{cases} 0, & \text{if } \#S < \infty, \\ 1, & \text{if } \#S^c < \infty, \end{cases}$$

is finitely additive and $\mu(\{n\}) = 0$ for any $n \in \mathbb{N}$. Since every set $A \subseteq \mathbb{N}$ can be covered by countably many singletons, the definition of the Carathéodory extension gives $\mu^*(A) = 0$ for all $A \subseteq \mathbb{N}$. In particular, $\mu^*(\mathbb{N}) = 0 < 1 = \mu(\mathbb{N})$, and so $\mu^*|_{\mathcal{S}} \neq \mu$.

9.15 An outer measure constructed on the intervals $[a, b)$ such that not all Borel sets are measurable

Consider the semi-ring of open intervals $\mathcal{C} = \{[a, b)\,;\ a < b\}$ on $\mathbb{R}$ and the set function $m(C) := (\operatorname{diam} C)^\alpha$, $\alpha \in (0, 1)$; here, $\operatorname{diam} C = \sup_{x,y\in C} |x - y|$ is the diameter of the set C. The set function m fails to be a measure relative to $\mathcal{C}$: for any $\mathcal{C}$-cover $([a_i, b_i))_{i\in\mathbb{N}}$ of $[0, 1)$ Lebesgue measure satisfies

$$1 = \lambda[0, 1) \leqslant \lambda\Big(\bigcup_{i=1}^{\infty}[a_i, b_i)\Big) \leqslant \sum_{i=1}^{\infty}(b_i - a_i),$$

and by the subadditivity of $x \longmapsto x^\alpha$, $\alpha \in (0, 1)$, this gives

$$\sum_{i=1}^{\infty} m[a_i, b_i) = \sum_{i=1}^{\infty}(b_i - a_i)^\alpha \geqslant \left(\sum_{i=1}^{\infty}(b_i - a_i)\right)^\alpha \geqslant 1.$$

Thus, $\mu^*[0, 1) \geqslant 1$. Since $\{[0, 1)\}$ is a trivial $\mathcal{C}$-cover of $[0, 1)$, i.e. $\mu^*[0, 1) \leqslant 1$, we get $\mu^*[0, 1) = 1$. If $\dot{\bigcup}_{i=1}^{\infty}[a_i, b_i) = [0, 1)$ is a non-trivial partition, then the above inequality is strict, and therefore m is not σ-additive on $\mathcal{C}$.

The same reasoning applies to $[-1,0)$ and shows $\mu^*[-1,0) = 1$. Finally, observe that $\{[-1,1)\}$ is a $\mathscr{C}$-cover of $[-1,1)$; thus $\mu^*[-1,1) \leqslant m[-1,1) = 2^\alpha$.

Let $A = [0,1)$ and $E = [-1,1)$. The previous calculations show

$$\mu^*(E \cap A) + \mu^*(E \setminus A) = \mu^*[0,1) + \mu^*[-1,0) = 2 > 2^\alpha \geqslant \mu^*[-1,1) = \mu^*(E),$$

which means that $[0,1)$ is not μ^* measurable; in particular, $\mathscr{B}(\mathbb{R})$ is not contained in $\mathscr{A}^*$.

Comment In this example we encounter again the pathology from Example 9.10 where the defining set function m is not a measure relative to $\mathscr{C}$ (a pre-measure). In metric spaces, a small change in the definition of the outer measure avoids this type of pathology: instead of considering all covers in the definition of μ^*, we use only covers comprising sets of diameter up to $\epsilon > 0$, and let $\epsilon \to 0$

$$\mu^*(E) := \sup_{\epsilon>0} \inf \left\{ \sum_{n=1}^\infty m(C_n);\ C_n \in \mathscr{C},\ E \subseteq \bigcup_{n\in\mathbb{N}} C_n,\ \operatorname{diam} C_n \leqslant \epsilon \right\}.$$

A full discussion can be found in [MIMS, pp. 199–202].

9.16 There exist non-μ^* measurable sets if, and only if, μ^* is not additive on $\mathscr{P}(X)$

Let μ^* be an outer measure on X. Recall that the μ^* measurable sets are the family

$$\mathscr{A}^* = \{A \subseteq X;\ \forall Q \subseteq X\ :\ \mu^*(Q) = \mu^*(Q \cap A) + \mu^*(Q \setminus A)\}.$$

From the definition it is clear that $\mathscr{A}^* = \mathscr{P}(X)$ holds if, and only if, μ^* is additive on $\mathscr{P}(X)$.

Comment Since μ^* is σ-subadditive, the additivity and σ-additivity of μ^* are, in fact, equivalent. If μ^* is σ-additive, then there is nothing to show. Assume that μ^* is additive. Let $(Q_n)_{n\in\mathbb{N}}$ be any sequence of pairwise disjoint sets. Then we get from the monotonicity and the additivity of μ^* that

$$\mu^*\left(\bigcupdot_{n=1}^\infty Q_n\right) \geqslant \mu^*\left(\bigcupdot_{n=1}^N Q_n\right) = \sum_{n=1}^N \mu^*(Q_n)$$

for any $N \in \mathbb{N}$. Letting $N \to \infty$ shows that μ^* is σ-superadditive, hence σ-additive.

9.17 An outer regular measure which is not inner compact regular

Let λ be Lebesgue measure on $[0, 1]$ and construct, as in Example 7.14, a set $N \subseteq [0, 1]$ such that $N \cap F \neq \emptyset$ and $F \setminus N \neq \emptyset$ for every uncountable closed set F. In particular, every compact set $K \subseteq N$ is countable, so $\lambda(K) = 0$, and we get $\lambda_*(N) = 0$. The same argument applies to $[0, 1] \setminus N$, and so $\lambda_*([0, 1] \setminus N) = 0$, hence $\lambda^*(N) > 0$. This shows that N is not a Lebesgue measurable set.

Consider on N the trace σ-algebra $\mathcal{L}(N) := N \cap \mathcal{L}[0, 1]$ [☞ Example 1.7.(e)] and the measure μ on $\mathcal{L}(N)$ given by

$$\forall M_N \in \mathcal{L}(N) \ : \ \mu(M_N) := \lambda(M)$$

where $M \in \mathcal{L}[0, 1]$ with $M_N = N \cap M$ [☞ 5.9]. Since all compact sets K contained in N, hence in $M_N \in \mathcal{L}(N)$, are countable, we have $\mu(K) = \lambda(K) = 0$, and

$$\mu_*(M_N) = \sup\{\mu(K);\ K \subseteq M_N \ \text{ compact}\} = 0.$$

This means that μ cannot be inner compact regular. Recall that a set U_N is relatively open for N if there is an open set U in $[0, 1]$ such that $U_N = N \cap U$. So,

$$\begin{aligned} \mu(M_N) &= \lambda(M) \\ &= \inf\{\lambda(U);\ U \supseteq M, \quad U \text{ is open in } [0, 1]\} \\ &= \inf\{\mu(U_N);\ U_N \supseteq M_N, \quad U_N \text{ is relatively open in } N\}, \end{aligned}$$

which shows that μ is outer regular.

Comment For a further example of a measure which is outer regular but not inner compact regular [☞ 9.21].

9.18 An inner compact regular measure which is neither inner nor outer regular

Consider $X = \mathbb{R}$ with the co-finite topology [☞ Example 2.1.(d)], i.e. $U \subseteq \mathbb{R}$ is open if, and only if, $U = \emptyset$ or $\mathbb{R} \setminus U$ is finite. In other words: a set is closed if, and only if, it is finite or $\mathbb{R}$. The associated Borel σ-algebra $\mathscr{B}(X)$ coincides with the co-countable σ-algebra [☞ Example 4A]

$$A \in \mathscr{B}(X) \iff \#A \leqslant \#\mathbb{N} \text{ or } \#A^c \leqslant \#\mathbb{N}.$$

Consequently,

$$\mu(A) := \begin{cases} 0, & \text{if } \#A \leqslant \#\mathbb{N}, \\ 1, & \text{if } \#A^c \leqslant \#\mathbb{N}, \end{cases}$$

is a measure on $(X, \mathscr{B}(X))$ [☞Example 5A]. Since any set $A \subseteq X$ is compact in the co-finite topology, see e.g. [172, Example 18], it is trivial that μ is inner compact regular. Any open set $U \neq \emptyset$ has finite complement, and so $\mu(U) = 1$. This means that

$$\inf\{\mu(U);\ U \supseteq A,\ U \text{ open}\} = 1 > \mu(A)$$

for any set $A \in \mathscr{B}(X)$, $A \neq \emptyset$, with $\mu(A) = 0$, i.e. μ is not outer regular. Similarly, if $U \neq \emptyset$ is open, then $\#F < \infty$ for every closed set $F \subseteq U$, and

$$\sup\{\mu(F);\ F \subseteq U,\ F \text{ closed}\} = 0 < 1 = \mu(U),$$

i.e. μ is not inner regular.

Comment 1 If (X, d) is a metric space, then every finite Borel measure on $(X, \mathscr{B}(X))$ is outer regular [MIMS, p. 439]. Our example shows that the statement does not extend to general topological spaces.

Comment 2 If X is a topological space such that every compact set is closed, e.g. a Hausdorff or metric space, then

$$\sup\{\mu(F);\ F \subseteq U,\ F \text{ compact}\} \leqslant \sup\{\mu(F);\ F \subseteq U,\ F \text{ closed}\},$$

and therefore inner compact regularity implies inner regularity for such topological spaces. As our example shows, this implication breaks down if there are compact sets which are not closed.

9.19 A measure which is neither inner nor outer regular

Consider the Sorgenfrey line S [☞Example 2.1.(h)], i.e. the real line $\mathbb{R}$ equipped with the topology generated by the half-open intervals $\{[x, y);\ x < y\}$, and denote by $\mathscr{B}(S)$ the associated Borel σ-algebra. Define a measure on $(S, \mathscr{B}(S))$ by

$$\mu(A) := \begin{cases} 0, & \text{if } \#A \leqslant \#\mathbb{N}, \\ \infty, & \text{otherwise.} \end{cases}$$

Since any compact set $K \subseteq S$ is at most countable, cf. Example 2.1.(h), we have $\mu(K) = 0$, and therefore

$$0 = \sup\{\mu(K);\ K \subseteq A,\ K \text{ compact}\} < \mu(A)$$

for any uncountable set $A \in \mathscr{B}(S)$. This means that μ is not inner compact regular. To see that μ is not outer regular, consider e.g. the set $A := \mathbb{N} \in \mathscr{B}(S)$. Any open set U with $U \supseteq A$ has uncountably many elements, and so

$$\mu(A) = 0 < \infty = \inf\{\mu(U);\ U \supseteq A,\ U \text{ open}\}.$$

9.20 A measure which is inner regular but not inner compact regular

Let S be the Sorgenfrey line [☞Example 2.1.(h)]. The corresponding Borel σ-algebra $\mathscr{B}(S)$ is the 'usual' Borel σ-algebra $\mathscr{B}(\mathbb{R})$, and so we may consider Lebesgue measure λ as measure on $(S, \mathscr{B}(S))$. We claim that λ – as a measure on $(S, \mathscr{B}(S))$ – is inner regular but not inner compact regular. Every compact set $K \subseteq S$ has at most countably many elements. In particular, compact sets in S have Lebesgue measure zero, which implies that

$$\sup\{\lambda(K)\,;\; K \subseteq A, K \text{ compact in } S\} = 0 < \lambda(A)$$

for any $A \in \mathscr{B}(S)$ with $\lambda(A) > 0$. Consequently, $(S, \mathscr{B}(S), \lambda)$ is not inner compact regular. It remains to show that $(S, \mathscr{B}(S), \lambda)$ is inner regular. Since the real line $\mathbb{R}$ with the Euclidean metric is σ-compact, the inner regularity of λ on $(\mathbb{R}, \mathscr{B}(\mathbb{R}))$ yields

$$\lambda(B) = \sup\{\lambda(F)\,;\; F \subseteq B \text{ closed in } \mathbb{R}\}, \quad B \in \mathscr{B}(\mathbb{R}),$$

cf. [MIMS, p. 439]. The Sorgenfrey topology is finer than the Euclidean topology, i.e. every open set in $\mathbb{R}$ is open in S. In other words, every closed set in $\mathbb{R}$ is closed in S, too. As $\mathscr{B}(S) = \mathscr{B}(\mathbb{R})$, it follows that

$$\lambda(B) = \sup\{\lambda(F)\,;\; F \subseteq B \text{ closed in } S\}, \quad B \in \mathscr{B}(S).$$

In particular, λ is inner regular on $(S, \mathscr{B}(S))$.

Comment If X is a σ-compact Hausdorff space, then the notions of inner regularity and inner compact regularity coincide for measures which are finite on compact sets. Our example shows that the σ-compactness is crucial. Indeed, the Sorgenfrey line S is a Hausdorff space but it is not σ-compact because every compact set in S is at most countable.

9.21 The regularity of a measure depends on the topology

Denote by S the Sorgenfrey line [☞Example 2.1.(h)], i.e. the real line $\mathbb{R}$ equipped with the topology generated by the half-open intervals $\{[x, y)\,;\; x < y\}$. Since the associated Borel σ-algebra $\mathscr{B}(S)$ coincides with the Borel σ-algebra $\mathscr{B}(\mathbb{R})$ [☞4.18], Lebesgue measure λ is a measure on $(S, \mathscr{B}(S))$. The measure space $(S, \mathscr{B}(S), \lambda)$ is not inner compact regular [☞9.20]. However, $(S, \mathscr{B}(S), \lambda)$ is outer regular. Indeed, the topology generated by the half-open intervals is finer than the topology generated by the open intervals, i.e. any set which is open in $\mathbb{R}$ – equipped with the Euclidean metric – is also open in S, and since $(\mathbb{R}, \mathscr{B}(\mathbb{R}), \lambda)$ is outer regular, this implies immediately that $(S, \mathscr{B}(S), \lambda)$ is outer regular.

If we equip the real line $\mathbb{R}$ with the Euclidean metric, then Lebesgue measure λ is inner compact regular and outer regular on $(\mathbb{R}, \mathscr{B}(\mathbb{R}))$.

9.22 A regular Borel measure whose restriction to a Borel set is not regular

Consider the product $[0,1] \times X$ of the Euclidean space $[0,1]$ equipped with the Euclidean metric and $X = [0,1]$ with the discrete metric, i.e. $d(x,y) = 0$ if $x = y$ and $d(x,y) = \infty$ if $x \neq y$. We use on $[0,1] \times X$ the canonical product metric which is given by

$$\rho((s,x),(t,y)) := |s-t| + d(x,y), \quad s,t, \in [0,1],\ x,y \in X.$$

For $A \subseteq [0,1] \times X$ and $x \in X$ denote by $A^x := \{t \in [0,1];\ (t,x) \in A\}$ the (horizontal) slice of A. Denote by λ Lebesgue measure on $([0,1], \mathscr{B}[0,1])$ and set

$$\mu(B) := \begin{cases} \sum_{x\in X} \lambda(B^x), & \text{if } \{x \in X;\ B^x \neq \emptyset\} \text{ is at most countable,} \\ \infty, & \text{otherwise.} \end{cases}$$

Note that μ is well-defined since $B^x \in \mathscr{B}[0,1]$ for any $B \in \mathscr{B}([0,1] \times X)$ – just observe that $t \longmapsto (t,x)$ is continuous, hence Borel measurable. We claim that μ is a regular Borel measure on $([0,1] \times X, \mathscr{B}([0,1] \times X))$.

1° μ is a measure. $\mu(\emptyset) = 0$ is obvious. Let $(B_i)_{i\in\mathbb{N}} \subseteq \mathscr{B}([0,1]\times X)$ be pairwise disjoint and $B := \bigcup_{i\in\mathbb{N}} B_i$. If $B^x \neq \emptyset$ for uncountably many $x \in X$, then there is a number $i \in \mathbb{N}$ such that $\{x \in X;\ B_i^x \neq \emptyset\}$ is uncountable, hence

$$\mu(B) = \infty = \sum_{k\in\mathbb{N}} \mu(B_k).$$

Now suppose that $I := \{x \in X;\ B^x \neq \emptyset\}$ is at most countable. Then the set $I_k := \{x \in X;\ B_k^x \neq \emptyset\}$ is at most countable, and $I = \bigcup_{k\in\mathbb{N}} I_k$. Since the slices B_k^x are pairwise disjoint and $B^x = \biguplus_{k\in\mathbb{N}} B_k^x$, we get

$$\mu(B) = \sum_{x\in I} \lambda(B^x) = \sum_{x\in I} \sum_{k\in\mathbb{N}} \lambda(B_k^x) = \sum_{k\in\mathbb{N}} \sum_{x\in I} \lambda(B_k^x) = \sum_{k\in\mathbb{N}} \sum_{x\in I_k} \lambda(B_k^x)$$
$$= \sum_{k\in\mathbb{N}} \mu(B_k).$$

2° μ is finite on compact sets. If $K \subseteq [0,1]\times X$ is compact, then $K^x \neq \emptyset$ only for finitely many $x \in X$. Indeed, the projection $\pi_2 : [0,1]\times X \to X, (t,x) \longmapsto x$ is continuous, and therefore

$$\pi_2(K) = \{x \in X;\ \exists t \in [0,1]\ :\ (t,x) \in K\} = \{x \in X;\ K^x \neq \emptyset\}$$

is compact in X. Since all singletons $\{x\}$ are open in the discrete topology, compact sets in X are finite, and so $K^x \neq \emptyset$ for at most finitely many $x \in X$. Moreover, each slice K^x is compact in $[0,1]$. Hence,

$$\mu(K) = \sum_{x \in F} \lambda(K^x) < \infty$$

for the finite set $F := \pi_2(K) \subseteq X$.

3° μ is inner compact regular. Let $U \subseteq [0,1] \times X$ be open. By the definition of the metric ρ, there is for every $(t,x) \in U$ some $\epsilon > 0$ such that $(t-\epsilon, t+\epsilon) \subseteq U^x$, and so all slices U^x are open. Set $I := \{x \in X\,;\; U^x \neq \emptyset\}$ and fix $\delta > 0$. If I is countable, take an enumeration $(x_n)_{n\in J}$ of I with $J := \{n \in \mathbb{N}\,;\; n \leqslant \#I\}$. Since U^{x_n} is open and $\lambda(U^{x_n}) < \infty$, the inner compact regularity of Lebesgue measure shows that there is a compact set $K_n \subseteq U^{x_n}$ such that $\lambda(U^{x_n} \setminus K_n) \leqslant \delta 2^{-n}$. Then

$$C_k := \bigcup_{n=1}^{k} K_n \times \{x_n\} \subseteq U, \quad k \in J,$$

is compact and

$$\sup_{k\in J} \mu(C_k) = \sup_{k\in J} \sum_{n=1}^{k} \lambda(K_n) \geqslant \sup_{k\in J} \sum_{n=1}^{k} (\lambda(U^{x_n}) - \delta 2^{-n}) \geqslant \mu(U) - \delta.$$

As $\delta > 0$ is arbitrary, this proves inner compact regularity. In the other case, i.e. if I is uncountable, we have $\mu(U) = \infty$. Since U^x is open, its Lebesgue measure $\lambda(U^x)$ is strictly positive for any $x \in X$ with $U^x \neq \emptyset$. Consequently, $I = \{x \in X\,;\; \lambda(U^x) > 0\}$ is uncountable, and so is

$$\left\{x \in X\,;\; \lambda(U^x) > \frac{1}{i}\right\}$$

for large i. If we choose a sequence of distinct points $(x_n)_{n\in\mathbb{N}}$ from this set, then there exist compact sets $K_n \subseteq U^{x_n}$ with $\lambda(K_n) \geqslant 1/i$. In consequence, $C_k := \bigcup_{n=1}^{k} K_n \times \{x_n\} \subseteq U$ is compact and

$$\sup_{k\in\mathbb{N}} \mu(C_k) = \sup_{k\in\mathbb{N}} \sum_{n=1}^{k} \lambda(K_n) \geqslant \sup_{k\in\mathbb{N}} \frac{k}{i} = \infty = \mu(U).$$

4° μ is outer regular. Fix $B \in \mathscr{B}([0,1] \times X)$ and set $I := \{x \in X\,;\; B^x \neq \emptyset\}$. If I is uncountable, then $\{x \in X\,;\; U^x \neq \emptyset\}$ is uncountable for any set $U \supseteq B$, i.e. $\mu(B) = \infty = \mu(U)$, and this proves outer regularity. It remains to consider the case that I is at most countable. Let $(x_n)_{n\in J}$ be an enumeration of I with $J := \{n \in \mathbb{N}\,;\; n \leqslant \#I\}$. By outer regularity of Lebesgue measure, we can choose an open set $U_n \supseteq B^{x_n}$ such that $\lambda(U_n \setminus B^{x_n}) \leqslant \delta 2^{-n}$ for fixed $\delta > 0$. The set $U := \bigcup_{n\in J} U_n \times \{x_n\} \supseteq B$ is open in $[0,1]\times X$ and

$$\mu(U) = \sum_{n\in J} \lambda(U_n) \leqslant \sum_{n\in J} \lambda(B^{x_n}) + \delta \sum_{n\in J} 2^{-n} \leqslant \mu(B) + \delta.$$

This finishes the proof that μ is a regular measure on $([0,1]\times X, \mathscr{B}([0,1]\times X))$. Finally, we show that the restriction of μ to the Borel set $B := \{0\}\times[0,1]$ is **not** regular. Since $B^x \neq \emptyset$ for uncountable many $x \in X$, we have $\mu(B) = \infty$. Any compact set $K \subseteq B$ is of the form $K = \{0\} \times F$ for a finite set $F \subseteq X$; see step **2°** above. In particular,

$$\mu(K) = \sum_{x\in F} \lambda(\{0\}) = 0,$$

which implies that

$$\sup\{\mu(K);\ K \subseteq B \text{ compact}\} = 0 < \infty = \mu(B),$$

i.e. μ is not inner compact regular.

Comment The measure μ reminds one of the measure $B \longmapsto \int_X \lambda(B^x)\,\zeta(dx)$ where ζ is the counting measure on X; this is a candidate for the (not unique) product measure of the non-σ-finite counting measure ζ and Lebesgue measure λ [☞ 16.1]. Note, however, that this product measure differs from μ on, e.g. the diagonal in $[0,1]\times X$ which consists of uncountably many slices which are Lebesgue null sets.

The idea for the definition of μ is taken from [60, Problem 7.13]. The above pathology occurs because μ is not σ-finite. There is the following positive result:

Theorem *Let X be a locally compact metric space. If μ is a σ-finite regular measure on $(X, \mathscr{B}(X))$, then the restriction $\mu_A(B) := \mu(A \cap B)$ is a regular measure on $(X, \mathscr{B}(X))$ for any $A \in \mathscr{B}(X)$.*

Let us outline the proof. It is straightforward to verify that μ_A is a measure which is finite on compact sets. In order to prove inner compact regularity, we use that the σ-finite regular measure μ is inner compact regular on all **Borel** sets, i.e.

$$\mu(B) = \sup\{\mu(K);\ K \subseteq B, K \text{ compact}\}, \qquad B \in \mathscr{B}(X), \tag{9.11}$$

cf. [60, Corollary 7.6] or [MINT, Lemma A.9]. Fix $B \in \mathscr{B}(X)$ and $\epsilon > 0$, and choose a sequence $(X_n)_{n\in\mathbb{N}} \subseteq \mathscr{B}(X)$ such that $\mu(X_n) < \infty$ and $X_n \uparrow X$. If $\mu(A \cap B) < \infty$, then there exists by (9.11) a compact set $K \subseteq A \cap B$ such that $\mu(K) \geqslant \mu(A\cap B) - \epsilon$, and so

$$\mu_A(K) = \mu(A \cap K) = \mu(K) \geqslant \mu(A \cap B) - \epsilon = \mu_A(B) - \epsilon.$$

If $\mu(A \cap B) = \infty$, then we apply the previous consideration to $B \cap X_n$ to get a compact set $K_n \subseteq A \cap B$ with

$$\mu_A(K_n) \geqslant \mu_A(B \cap X_n) - \epsilon.$$

Thus,

$$\sup_{n \in \mathbb{N}} \mu_A(K_n) \geqslant \sup_{n \in \mathbb{N}} \mu_A(B \cap X_n) - \epsilon = \mu_A(B) - \epsilon.$$

For the proof of the outer regularity, we observe that for any $B \in \mathcal{B}(X)$ and $\epsilon > 0$ there exists an open set $U \supseteq B$ with $\mu(U \setminus B) \leqslant \epsilon$, cf. [60, Proposition 7.7]. Consequently, if $B \in \mathcal{B}(X)$, then there is an open set $U \supseteq B \cap A$ with $\mu(U \setminus (B \cap A)) \leqslant \epsilon$, and so

$$\mu_A(U) = \mu_A(B) + \mu_A(U \setminus (B \cap A)) \leqslant \mu_A(B) + \epsilon.$$

10

Integrable Functions

The two guiding principles for the construction of integrals are: (i) $f \longmapsto \int f$ should be a positive linear form and (ii) $A \longmapsto \int \mathbb{1}_A$ should be a measure. If we start from a measure μ on a measurable space $(X, \mathscr{A})$, then this means that the integral is necessarily of the form

$$\int \mathbb{1}_A \, d\mu := \mu(A) \quad \text{and} \quad \int \sum_{i=1}^{n} y_i \mathbb{1}_{A_i} \, d\mu := \sum_{i=1}^{n} y_i \mu(A_i)$$

whenever $y_i \geqslant 0$ and $A_i \in \mathscr{A}$. Since the linear form is positive, it is natural to look at upper envelopes of measurable step functions (elementary functions); the Sombrero lemma [☞Theorem 1.10] and Beppo Levi's theorem [☞1.13] tell us how to define the integral of positive measurable functions:

$$\int f \, d\mu = \sup_{\phi \leqslant f} \int \phi \, d\mu = \lim_{n \to \infty} \int \phi_n \, d\mu \in [0, \infty],$$

where ϕ is a generic elementary function such that $0 \leqslant \phi \leqslant f$, resp. $(\phi_n)_{n \in \mathbb{N}}$, is any sequence of positive elementary functions approximating f from below, i.e. $0 \leqslant \phi_n \leqslant f$ and $\sup_{n \in \mathbb{N}} \phi_n = \lim_{n \to \infty} \phi_n = f$. Note that we admit the value $+\infty$ in the definition of the integral of positive measurable functions. Of course, if $\int f \, d\mu < \infty$, then $\mu(\{f = \infty\}) = 0$ [☞(1.11)].

Again by linearity, we are led to define

$$\forall f = f^+ - f^-, \ f \text{ measurable} \ : \ \int f \, d\mu := \int f^+ \, d\mu - \int f^- \, d\mu;$$

in order to ensure the finiteness of the difference of integrals, we require that both $\int f^+ \, d\mu$ and $\int f^- \, d\mu$ are finite, hence $\int f \, d\mu \in \mathbb{R}$. Some authors consider so-called extended integrals where either $\int f^+ \, d\mu$ or $\int f^- \, d\mu$ has to be finite, resulting in $\int f \, d\mu \in (-\infty, \infty]$ or $\int f \, d\mu \in [-\infty, \infty)$. The space of ($\overline{\mathbb{R}}$-valued) integrable functions is denoted by $\mathscr{L}_{\mathbb{R}}(\mu)$ (resp. $\mathscr{L}_{\overline{\mathbb{R}}}(\mu)$). By $L^1(\mu)$ we denote the space of all equivalence classes of integrable functions, i.e. we identify those $f, g \in \mathscr{L}_{\overline{\mathbb{R}}}$ which satisfy $\mu(\{f \neq g\}) = 0$; since $f \in \mathscr{L}_{\overline{\mathbb{R}}}(\mu)$ is a.e. $\mathbb{R}$-valued, the elements of $L^1(\mu)$ are real-valued 'functions' – as usual we identify the equivalence classes with their (real-valued) representatives.

For complex-valued functions f, we can use the decomposition

$$f = (\operatorname{Re} f)^+ - (\operatorname{Re} f)^- + i\left((\operatorname{Im} f)^+ - (\operatorname{Im} f)^-\right)$$

and linearity to define the integral

$$\int f\,d\mu := \int (\operatorname{Re} f)^+\,d\mu - \int (\operatorname{Re} f)^-\,d\mu + i\left(\int (\operatorname{Im} f)^+\,d\mu - \int (\operatorname{Im} f)^-\,d\mu\right);$$

we require that all integrals are finite. Since there exists a bi-measurable map $(\mathbb{R}^2, \mathscr{B}(\mathbb{R}^2)) \to (\mathbb{C}, \mathscr{B}(\mathbb{C}))$, we know that $f : X \to \mathbb{C}$ is measurable if, and only if, all of the functions $(\operatorname{Re} f)^{\pm}, (\operatorname{Im} f)^{\pm} : X \to [0, \infty]$ are measurable. It is not hard to see that $f \longmapsto \int f\,d\mu$ is a $\mathbb{C}$-linear form and

$$\operatorname{Re}\int f\,d\mu = \int \operatorname{Re} f\,d\mu, \quad \operatorname{Im}\int f\,d\mu = \int \operatorname{Im} f\,d\mu, \quad \overline{\int f\,d\mu} = \int \overline{f}\,d\mu$$

as well as $\left|\int f\,d\mu\right| \leqslant \int |f|\,d\mu$; see e.g. [MIMS, Appendix D].

Let $(X, \mathscr{O})$ be a topological space and μ a measure on $(X.\mathscr{B}(X))$. We call a function $f : X \to \mathbb{R}$ **locally integrable** if every $x \in X$ has an open neighbourhood $U = U(x)$ such that $x \longmapsto f(x)\mathbb{1}_U(x)$ is integrable; in particular, $f\mathbb{1}_K$ is integrable for every compact set $K \in \mathscr{B}(X)$ (recall that compact sets need not be closed!). If (X, d) is a locally compact metric space, then f is locally integrable if, and only if, $f\mathbb{1}_K$ is integrable for every compact set K.

10.1 An integrable function which is unbounded in every interval

Consider on $([0, 1], \mathscr{B}[0, 1])$ Lebesgue measure and define the following function

$$u(x) := \begin{cases} 0, & \text{if } x \in [0, 1] \setminus \mathbb{Q}, \\ q, & \text{if } x = \frac{p}{q} \in [0, 1] \cap \mathbb{Q},\ \gcd(p, q) = 1, \end{cases}$$

which resembles the reciprocal of Thomae's function [☞ 3.8], cf. Fig. 10.1. As $\{u \neq 0\}$ is countable, u is Borel measurable and $\int_{[0,1]} u(x)\,dx = 0$. In every subinterval $(a, b) \subseteq [0, 1]$ we can find rationals $\frac{p}{q} \in (a, b)$ with arbitrarily large q and $\gcd(p, q) = 1$, which means that the function u is locally unbounded.

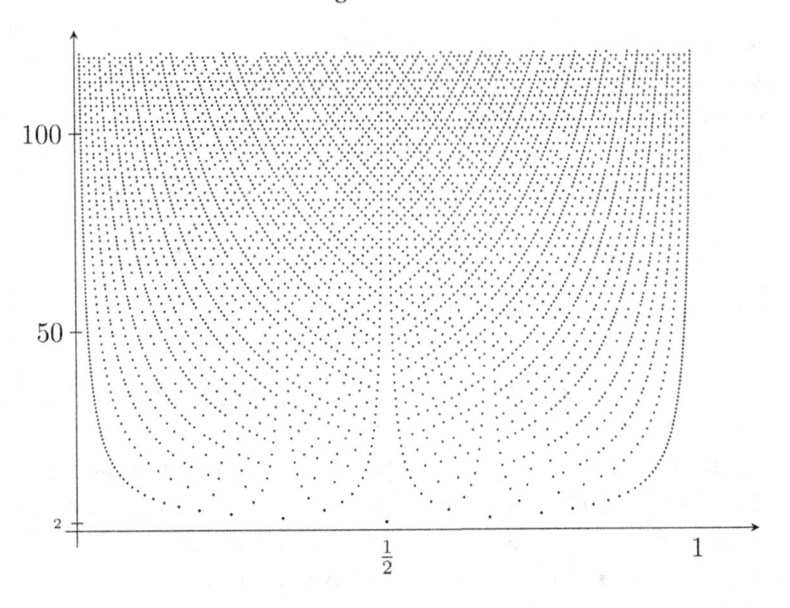

Figure 10.1 Plot of image points $u(x)$ for the 'reciprocal Thomae function' u from Example 10.1 and $x = \frac{p}{q} \in \mathbb{Q} \cap (0,1)$ with $\gcd(p,q) = 1$, $q \leqslant 120$.

10.2 A continuous integrable function such that $\lim_{|x|\to\infty} f(x) \neq 0$

Take $(\mathbb{R}, \mathscr{B}(\mathbb{R}), \lambda)$ as measure space and define the tent functions

$$t_n(x) = n^2\left(x + \frac{1}{n}\right)\mathbb{1}_{(-1/n,0]}(x) - n^2\left(x - \frac{1}{n}\right)\mathbb{1}_{(0,1/n)}(x), \quad x \in \mathbb{R},\ n \in \mathbb{N}.$$

Clearly, t_n is continuous and $\int_{\mathbb{R}} t_n(x)\,dx = 1$. Define

$$f(x) := \sum_{n=1}^{\infty} \frac{1}{n^2} t_{2^n}(x - 2^n), \quad x \in \mathbb{R}.$$

Since the terms of the sum have disjoint supports, it is clear that f is continuous. Moreover,

$$\int_{\mathbb{R}} f(x)\,dx = \sum_{n=1}^{\infty} \frac{1}{n^2} \int_{\mathbb{R}} t_{2^n}(x)\,dx < \infty,$$

while $f(2^n) = 2^n n^{-2}$ tends to infinity as $n \to \infty$.

10.3 A continuous function vanishing at infinity which is not in L^p for any $p > 0$

Consider $((0,\infty), \mathscr{B}(0,\infty))$ with Lebesgue measure λ. The function

$$f(x) := \frac{1}{1 + |\log x|}, \quad x > 0,$$

is continuous and vanishes at infinity, i.e. $\lim_{|x|\to\infty} f(x) = 0$. Observing that $1 + |\log x| \leqslant |x|^{1/p}$ for all $|x| > 1$, we get

$$\int_{(0,\infty)} |f(x)|^p \, \lambda(dx) \geqslant \int_{(1,\infty)} |x|^{-1} \, \lambda(dx) = \infty$$

for any $p > 0$.

10.4 A non-integrable function such that $\lim_{r\to\infty} r\mu(\{|f| > r\}) = 0$

Let $(X, \mathscr{A}, \mu)$ be a measure space and $f : X \to \mathbb{R}$ an integrable function. By Markov's inequality,

$$r\mu(\{|f| \geqslant r\}) \leqslant r \int_{\{|f|\geqslant r\}} \frac{|f(x)|}{r} \, \mu(dx) = \int_{\{|f|\geqslant r\}} |f(x)| \, \mu(dx), \quad r > 0,$$

and therefore it follows from the dominated convergence theorem that

$$\lim_{r\to\infty} r\mu(\{|f| \geqslant r\}) = 0 \tag{10.1}$$

for every integrable function f. The converse is not true, i.e. (10.1) does not imply the integrability of f.

This is most easily seen in infinite measure spaces: every bounded function satisfies (10.1), but boundedness does not imply integrability; consider e.g. the function $f(x) = 1/x$ on $((1,\infty), \mathscr{B}(1,\infty))$ with Lebesgue measure.

Here is a counterexample in a finite measure space. Equip $((0,1), \mathscr{B}((0,1))$ with Lebesgue measure λ, and set

$$f(x) := \frac{1}{|x|} \frac{1}{|\log x|}, \quad x \in (0,1).$$

Then $\int_{(0,1)} f \, d\lambda = \infty$ but

$$\begin{aligned}
\limsup_{r\to\infty} r\lambda(\{|f| \geqslant r\}) &= \limsup_{r\to\infty} r\lambda\left(\left\{x \in [0,1/r]\,;\, \frac{1}{|x|}\frac{1}{|\log x|} \geqslant r\right\}\right) \\
&\leqslant \limsup_{r\to\infty} r\lambda\left(\left\{x \in [0,1/r]\,;\, \frac{1}{|x|}\frac{1}{|\log r|} \geqslant r\right\}\right) \\
&= \limsup_{r\to\infty} \frac{1}{r|\log r|} = 0.
\end{aligned}$$

Comment If the measure space is σ-finite, then we may evaluate the integral of a measurable function $f : X \to \mathbb{R}$ using the distribution function,

$$\int_X |f|\, d\mu = \int_0^\infty \mu(\{|f| > r\})\, dr;$$

it is clear that the condition $\mu(\{|f| > r\}) = o(r^{-1-\epsilon})$, $\epsilon > 0$, is sufficient to guarantee that the integral $\int |f|\, d\mu$ is finite. On the other hand, (10.1) shows that $\mu(\{|f| > r\}) = o(r^{-1})$ is necessary for the finiteness of the integral. This argument remains true for non-sigma finite measure spaces; see the comment in Example 10.5.

10.5 Characterizing integrability in terms of series

Let $(X, \mathscr{A}, \mu)$ be a measure space. If $f : X \to [0, \infty)$ is a positive measurable function, then it follows from the pointwise inequality

$$\sum_{n=1}^\infty \mathbb{1}_{\{f>n\}}(x) \leqslant f(x) \leqslant \sum_{n=0}^\infty \mathbb{1}_{\{f>n\}}(x), \quad x \in X,$$

and the monotone convergence theorem that

$$\sum_{n=1}^\infty \mu(\{f > n\}) \leqslant \int_X f\, d\mu \leqslant \sum_{n=0}^\infty \mu(\{f > n\}).$$

For finite measure spaces this implies that the equivalence

$$\int_X |f|\, d\mu < \infty \iff \sum_{n \in I} \mu(\{|f| > n\}) < \infty \tag{10.2}$$

holds for any measurable function f and $I = \mathbb{N}$ or $I = \mathbb{N}_0$. If μ is an infinite measure, then

$$\sum_{n=0}^\infty \mu(\{|f| > n\}) < \infty \Rightarrow \int |f|\, d\mu < \infty \Rightarrow \sum_{n=1}^\infty \mu(\{|f| > n\}) < \infty$$

but the converse implications do not hold. Consider on $(\mathbb{R}, \mathscr{B}(\mathbb{R}))$ Lebesgue measure λ.

$\sum_{n=1}^\infty \mu(\{|f| > n\}) \not\Rightarrow \int |f|\, d\mu < \infty$: the constant function $f(x) := \frac{1}{2}$, $x > 0$, satisfies $\sum_{n=1}^\infty \lambda(\{|f| > n\}) = 0 < \infty$ but $\int |f|\, d\lambda = \infty$.

$\int |f|\, d\mu < \infty \not\Rightarrow \sum_{n=0}^\infty \mu(\{|f| > n\}) < \infty$: the function $f(x) := 1/(x^2 + 1)$ is integrable on $\mathbb{R}$ but the series $\sum_{n=0}^\infty \lambda(\{|f| > n\})$ is divergent because $\lambda(\{|f| > 0\}) = \lambda(\mathbb{R}) = \infty$.

Comment For infinite measure spaces, the equivalence (10.2) fails because $\mu(\{|f|>n\})$, $n\in\mathbb{N}_0$, does not capture enough information on the behaviour of f for the small values of the range. It is, however, possible to characterize integrability in terms of the measure of the superlevel sets $\{|f|>r\}$, $r>0$. Indeed, if $(X,\mathscr{A},\mu)$ is a σ-finite measure space then

$$\int_X |f|\,d\mu = \int_{(0,\infty)} \mu(\{|f|>r\})\,\lambda(dr) \in [0,\infty] \tag{10.3}$$

[☞ Example 1.36.(b)]; in particular,

$$\int_X |f|\,d\mu < \infty \iff \int_{(0,\infty)} \mu(\{|f|>r\})\,\lambda(dr) < \infty. \tag{10.4}$$

The characterization can be extended to measure spaces which are not necessarily σ-finite.

To this end, consider the measure $\tilde{\mu}(A) := \mu(A\cap\{|f|>0\})$ on the measurable space $(X',\mathscr{A}') = (\{|f|>0\},\{|f|>0\}\cap\mathscr{A})$. If one of the expressions in (10.4) is finite, then $\tilde{\mu}$ is σ-finite [☞ 10.20], and it follows that

$$\begin{aligned}\int_X |f|\,d\mu = \int_{X'} |f|\,d\tilde{\mu} &= \int_{(0,\infty)} \tilde{\mu}(\{|f|>r\})\,\lambda(dr)\\ &= \int_{(0,\infty)} \mu(\{|f|>r\})\,\lambda(dr);\end{aligned}$$

if both expressions in (10.4) are infinite, then there is nothing to show. Consequently, (10.3) and (10.4) remain valid for measures which are not necessarily σ-finite.

10.6 A non-integrable function such that $f(x-1/n)$ is integrable for all $n\in\mathbb{N}$

Consider Lebesgue measure λ on $(0,1)$. The function $f(x)=1/(1-x)$ fails to be integrable

$$\int_{(0,1)} f(x)\,\lambda(dx) = \int_{(0,1)} \frac{1}{x}\,\lambda(dx) = \infty,$$

but

$$\int_{(0,1)} f(x-1/n)\,\lambda(dx) = \int_{(0,1)} \frac{1}{1-x+1/n}\,\lambda(dx) \leqslant n\int_{(0,1)} \lambda(dx) < \infty$$

for all $n\in\mathbb{N}$.

10.7 An integrable function such that $f(x - 1/n)$ fails to be integrable for all $n \in \mathbb{N}$

[☞18.33]

Comment There is no such function f for Lebesgue measure λ on $(\mathbb{R}, \mathscr{B}(\mathbb{R}))$. Because of the invariance of Lebesgue measure under translations, we have

$$\int_{\mathbb{R}} |f(x)| \, \lambda(dx) = \int_{\mathbb{R}} |f(x-h)| \, \lambda(dx), \quad h \in \mathbb{R},$$

for any Borel measurable function $f : \mathbb{R} \to \mathbb{R}$. In particular, $f \in L^1(\lambda)$ if, and only if, $f(\cdot - h) \in L^1(\lambda)$.

10.8 An improperly Riemann integrable function which is not Lebesgue integrable

Consider the integral $\int_0^\infty \frac{\sin x}{x} \, dx$. This improper Riemann integral exists and has the value $\frac{\pi}{2}$. Indeed: since $\frac{1}{x} = \int_0^\infty e^{-tx} \, dt$ and $\operatorname{Im} e^{ix} = \sin x$, Fubini's theorem shows

$$\begin{aligned}\int_0^a \frac{\sin x}{x} \, dx = \int_0^a \int_0^\infty e^{-tx} \sin x \, dt \, dx &= \int_0^\infty \int_0^a e^{-tx} \operatorname{Im} e^{ix} \, dx \, dt \\ &= \int_0^\infty \operatorname{Im} \left[\int_0^a e^{-(t-i)x} \, dx \right] dt.\end{aligned}$$

The inner integral yields

$$\begin{aligned}\operatorname{Im} \left[\int_0^a e^{-(t-i)x} \, dx \right] = \operatorname{Im} \left[\frac{e^{-(t-i)x}}{i-t} \right]_0^a &= \operatorname{Im} \left[\frac{e^{(i-t)a} - 1}{i-t} \right] \\ &= \operatorname{Im} \left[\frac{(e^{(i-t)a} - 1)(-i-t)}{1+t^2} \right].\end{aligned}$$

Using dominated convergence we get

$$\begin{aligned}\int_0^a \frac{\sin x}{x} \, dx &= \int_0^\infty \frac{-te^{-ta} \sin a - e^{-ta} \cos a + 1}{1+t^2} \, dt \\ &\overset{s=ta}{=} \int_0^\infty \frac{-se^{-s} \sin a}{a^2 + s^2} \, ds + \int_0^\infty \frac{-e^{-ta} \cos a + 1}{1+t^2} \, dt \\ &\xrightarrow[a \to \infty]{\text{dom. conv.}} \int_0^\infty \frac{1}{1+t^2} \, dt = \arctan t \Big|_0^\infty = \frac{\pi}{2}.\end{aligned}$$

On the other hand, $\frac{1}{x}\sin x$ is not Lebesgue integrable on $(0,\infty)$.

Here are two quick ways to see this. Using the elementary estimate

$$\frac{|\sin x|}{x} \geqslant \frac{1}{(n+1)\pi}|\sin x|, \quad x \in [n\pi,(n+1)\pi),$$

we deduce

$$\int_0^\infty \frac{|\sin x|}{x}\,dx \geqslant \sum_{n=1}^N \frac{1}{(n+1)\pi}\int_{n\pi}^{(n+1)\pi} |\sin x|\,dx = \frac{1}{\pi}\int_0^\pi \sin x\,dx \cdot \sum_{n=1}^N \frac{1}{n+1}$$

for every $N \in \mathbb{N}$. Since the harmonic series diverges, this shows that $\frac{1}{x}\sin x$ is not Lebesgue integrable on $(0,\infty)$.

Alternatively, we can use the fact that $(\sin t)^2 + (\cos t)^2 = 1$ to see for every $N \geqslant 2$:

$$\begin{aligned}\int_{\pi/2}^N \frac{dt}{t} &= \int_{\pi/2}^N \frac{(\sin t)^2}{t}\,dt + \int_{\pi/2}^N \frac{(\cos t)^2}{t}\,dt \\ &= \int_{\pi/2}^N \frac{(\sin t)^2}{t}\,dt + \int_0^{N-\pi/2} \frac{(\cos(x+\pi/2))^2}{x+\pi/2}\,dx.\end{aligned}$$

Since $\cos(x+\pi/2) = -\sin x$ and $(\sin x)^2 \leqslant |\sin x|$, we conclude that

$$\int_{\pi/2}^N \frac{dt}{t} \leqslant 2\int_0^\infty \frac{|\sin x|}{x}\,dx,$$

proving that the latter integral diverges. Thus, $\frac{1}{x}\sin x$ is not Lebesgue integrable on $(0,\infty)$.

10.9 A function such that $\lim_{n\to\infty}\int_0^n f\,d\lambda$ exists and is finite but $\int_0^\infty f\,d\lambda$ does not exist

Consider the measurable space $((0,\infty),\mathscr{B}(0,\infty))$ and define $f(x) := (-1)^i i^{-1}$ for $x \in (i-1,i]$, $i \in \mathbb{N}$. Clearly, f is measurable, and we have for each $n \in \mathbb{N}$,

$$\int_0^n |f|\,d\lambda = \sum_{i=1}^n \frac{1}{i} < \infty$$

as well as

$$\lim_{n\to\infty}\int_0^n f\,d\lambda = \sum_{i=1}^\infty \frac{(-1)^i}{i} = \log 2,$$

but the integral $\int_0^\infty |f|\,d\lambda = \sum_{i=1}^\infty \frac{1}{i} = \infty$ diverges.

Comment A similar example can be constructed using $f(x) = \sin(2\pi x)$. Clearly, $\int_0^n |\sin(2\pi x)|\,\lambda(dx) \leqslant n$, and by periodicity, $\int_0^n \sin(2\pi x)\,\lambda(dx) = 0$. The Lebesgue integral over $[0,\infty)$ does not exist as $\int_0^\infty |\sin(2\pi x)|\,\lambda(dx) = \infty$.

A more elaborate example is the integral sine from Example 10.8. The above example has the advantage that it carries over to any σ-finite measure space $(X, \mathscr{A}, \mu)$ with total mass $\mu(X) = \infty$: if $A_0 := \emptyset \subsetneqq A_1 \subsetneqq A_2 \subsetneqq A_3 \subsetneqq \ldots$ is a strictly increasing sequence of sets such that $A_i \uparrow X$ and $\mu(A_i \setminus A_{i-1}) > 0$, $i \in \mathbb{N}$, then the function $g(x) := (-1)^i i^{-1} \mu(A_i \setminus A_{i-1})^{-1}$, $x \in A_i \setminus A_{i-1}$, behaves like the function f on $(0,\infty)$.

10.10 A function which is nowhere locally integrable

[☞ 18.12].

10.11 Integrable functions f, g such that $f \cdot g$ is not integrable

Consider $X = (0,1)$ with the Borel σ-algebra $\mathscr{B}(0,1)$ and Lebesgue measure λ. Then $f(x) := x^{-1/2}$, $x \in (0,1)$, is integrable but $f \cdot f = 1/x$ is not in L^1.

Comment 1 The above example is in stark contrast to the situation for the Riemann integral. Since, by definition, Riemann integrable functions are bounded, the family of Riemann integrable functions is an algebra.

Comment 2 Let $(X, \mathscr{A}, \mu)$ be any measure space. If $p, q \in [1,\infty]$ are conjugate indices, i.e. $p^{-1} + q^{-1} = 1$, then Hölder's inequality [☞ Theorem 1.18]

$$\int_X |fg|\,d\mu \leqslant \|f\|_{L^p} \|g\|_{L^q}$$

holds for all $f \in L^p$, $g \in L^q$. In particular, $f \in L^p$ and $g \in L^q$ implies $fg \in L^1$. The above counterexample illustrates that we cannot expect the inequality to hold for $p^{-1} + q^{-1} \neq 1$.

10.12 A function such that $f \notin L^p$ for all $p \in [1,\infty)$ but $fg \in L^1$ for all $g \in L^q, q \geqslant 1$

Let $X \neq \emptyset$ be some set and $\mathscr{A}$ a σ-algebra on X. Then

$$\mu(A) := \begin{cases} 0, & \text{if } A = \emptyset, \\ \infty, & \text{if } A \neq \emptyset,\ A \in \mathscr{A}, \end{cases}$$

defines a measure on $(X, \mathscr{A})$ and $L^p(X, \mathscr{A}, \mu) = \{0\}$ for any $p \in [1, \infty)$ [☞18.3]. In particular, the constant mapping $f := 1$ does not belong to any L^p with $p \in (0, \infty)$. Yet $fg = 0 \in L^1$ for any $g \in L^q$, $q \in [1, \infty)$.

Comment Let $(X, \mathscr{A}, \mu)$ be any measure space. If $f \in L^p(X, \mathscr{A}, \mu)$ for some $p \in [1, \infty]$, then it follows from Hölder's inequality that

$$\int_X |fg| \, d\mu \leqslant \|g\|_{L^q} \|f\|_{L^p} < \infty$$

where $q \in [1, \infty]$ is the conjugate index. In particular,

$$f \in L^p \Rightarrow \sup_{\|g\|_{L^q} \leqslant 1} \int_X |fg| \, d\mu < \infty.$$

Our example shows that the converse is, in general, false; [☞18.19] for more details.

10.13 $f \in L^p$ for all $p < q$ does not imply $f \in L^q$

Consider on $(\mathbb{R}, \mathscr{B}(\mathbb{R}))$ Lebesgue measure λ and fix $q \in (0, \infty)$. The function

$$f(x) := \frac{1}{x^{1/q}} \mathbb{1}_{(0,1)}(x), \quad x \in \mathbb{R},$$

satisfies

$$\int_{\mathbb{R}} |f(x)|^p \, \lambda(dx) = \int_{(0,1)} x^{-p/q} \, \lambda(dx) < \infty$$

for all $0 \leqslant p < q$, but

$$\int_{\mathbb{R}} |f(x)|^q \, \lambda(dx) = \int_{(0,1)} x^{-1} \, \lambda(dx) = \infty.$$

Comment In a similar fashion we can construct a function g such that $g \in L^p$ for all $p > q$ but $g \notin L^q$; consider e.g.

$$g(x) = \frac{1}{x^{1/q}} \mathbb{1}_{(1,\infty)}(x), \quad x \in \mathbb{R}.$$

10.14 A function such that $f \in L^p$ for all $p < \infty$ but $f \notin L^\infty$

Consider on $((0,1), \mathscr{B}(0,1))$ Lebesgue measure λ and define

$$f(x) := \log x, \quad x \in (0,1).$$

Fix $p > 0$. Since the logarithm grows slower than any (fractional) power, we have

$$\forall \epsilon > 0 \quad \exists r_\epsilon > 0 \quad \forall x \in (0, r_\epsilon) \; : \; |\log x| \leqslant x^{-\epsilon}.$$

From this estimate we find for $\epsilon = 1/(2p)$ some $r = r_{1/(2p)}$ such that

$$\begin{aligned}\int_{(0,1)} |f(x)|^p \, \lambda(dx) &= \int_{(0,r)} |\log x|^p \, \lambda(dx) + \int_{(r,1)} |\log x|^p \, \lambda(dx) \\ &\leqslant \int_{(0,r)} |x|^{-p/(2p)} \, \lambda(dx) + |\log r|^p \int_{(r,1)} \lambda(dx) < \infty.\end{aligned}$$

Hence, $f \in L^p$ for all $p > 0$. Since f is continuous on $(0, 1)$ and $f(x) \to -\infty$ as $x \to 0$, we have $\|f\|_{L^\infty} = \infty$.

Comment See the comment to the following example.

10.15 A function such that $f \in L^\infty$ but $f \notin L^p$ for all $p < \infty$

Consider on $([0, \infty), \mathscr{B}[0, \infty))$ Lebesgue measure and $f \equiv 1$. Clearly, f is in L^∞ but f fails to be in any of the spaces L^p if $p < \infty$.

Comment Let $(X, \mathscr{A}, \mu)$ be any measure space. It is possible to show that $\lim_{p\to\infty} \|f\|_{L^p} = \|f\|_{L^\infty} \in [0, \infty]$ is always true for $f \in \bigcap_{p\geqslant 1} L^p(\mu)$; see [MIMS, p. 134, Exercise 13.21]. Our example shows that it is essential that $f \in L^p(\mu)$ for sufficiently large values of p.

10.16 A function which is in exactly one space L^p

Consider $(\mathbb{R}, \mathscr{B}(\mathbb{R}))$ endowed with Lebesgue measure λ, and fix $p \in (0, \infty)$. The function

$$f(x) := \frac{1}{x^{1/p}} \frac{1}{(\log x)^{2/p} + 1} \mathbb{1}_{(0,\infty)}(x)$$

is in $L^q(\mathbb{R})$ if, and only if, $p = q$.

Since f is bounded on compact subsets of $(0, \infty)$, we have

$$f \in L^q(\mathbb{R}) \iff \int_{(0,1/2)\cup(2,\infty)} \left| \frac{1}{x^{1/p}} \frac{1}{(\log x)^{2/p}} \right|^q \lambda(dx) < \infty. \tag{10.5}$$

Recall that

$$\forall \epsilon > 0 \quad \exists C_\epsilon > 0 \quad \forall x \geqslant 1 \; : \; |\log x| \leqslant C_\epsilon |x|^\epsilon.$$

For $q < p$ we find that

$$\int_{(2,\infty)} \frac{1}{x^{q/p}} \frac{1}{|\log x|^{2q/p}} \lambda(dx) \geqslant \frac{1}{C_\epsilon^{2q}} \int_{(2,\infty)} \frac{1}{|x|^{q/p+2q\epsilon/p}} \lambda(dx) = \infty$$

if we choose $\epsilon > 0$ so small that $q/p + 2q\epsilon/p < 1$. Hence, by (10.5), $f \notin L^q(\mathbb{R})$ for $q < p$. Similarly, for $q > p$,

$$\int_{(0,1/2)} \frac{1}{x^{q/p}} \frac{1}{|\log x|^{2q/p}} \lambda(dx) \geqslant \frac{1}{C_\epsilon^{2q}} \int_{(0,1/2)} \frac{1}{x^{q/p-2q\epsilon/p}} \lambda(dx) = \infty$$

for small $\epsilon > 0$, and so $f \notin L^q(\mathbb{R})$ for $q > p$. For $q = p$ we note that

$$\begin{aligned}
\int_{(2,\infty)} \frac{1}{x} \frac{1}{(\log x)^2} \lambda(dx) &= \sup_{n\in\mathbb{N}} \int_{(2,n)} \frac{1}{x} \frac{1}{(\log x)^2} \lambda(dx) \\
&= \sup_{n\in\mathbb{N}} \left[-\frac{1}{\log x} \right]_2^n \\
&= \sup_{n\in\mathbb{N}} \left(\frac{1}{\log 2} - \frac{1}{\log n} \right) \\
&= \frac{1}{\log 2} < \infty;
\end{aligned}$$

a similar computation shows $\int_{(0,1/2)} (x(\log x)^2)^{-1} \lambda(dx) < \infty$, and therefore we conclude from (10.5) that $f \in L^p(\mathbb{R})$.

10.17 Convolution is not associative

Let λ be Lebesgue measure on $\mathbb{R}$. If f is a bounded function and $g \in L^1$, then the **convolution** of f and g is defined by

$$(f * g)(x) := \int_{\mathbb{R}} f(x-y)g(y)\,dy, \quad x \in \mathbb{R},$$

[☞ p. 26]. The following example shows that the convolution is not associative. Set

$$f(x) := \mathbb{1}_{(0,\infty)}(x), \quad g(x) := \mathbb{1}_{(-1,0)}(x) - \mathbb{1}_{(0,1)}(x), \quad \text{and} \quad h(x) := \mathbb{1}_{\mathbb{R}}(x).$$

From

$$(f * g)(x) = \int_{-\infty}^{\infty} \mathbb{1}_{(0,\infty)}(x-y)g(y)\,dy = \int_{-\infty}^{x} g(y)\,dy$$

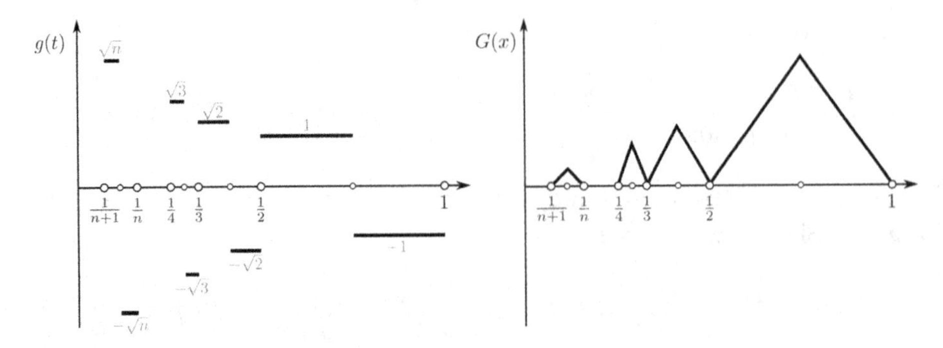

Figure 10.2 The function g and its primitive G in Example 10.18.

we find that

$$(f * g)(x) = \begin{cases} 0, & \text{if } x \notin [-1,1], \\ x+1, & \text{if } x \in (-1,0], \\ 1-x, & \text{if } x \in (0,1). \end{cases}$$

Moreover,

$$(g * h)(x) = \int_{-\infty}^{\infty} g(y)\,dy = 0, \quad x \in \mathbb{R}.$$

For the three-fold convolutions we have

$$((f * g) * h)(x) = \int_{-\infty}^{\infty} (f * g)(y)\,dy = \int_{-1}^{0} (y+1)\,dy + \int_{0}^{1} (1-y)\,dy = 1$$

and

$$(f * (g * h))(x) = (f * 0)(x) = 0$$

for all $x \in \mathbb{R}$, which shows that $(f * g) * h \neq f * (g * h)$.

Comment By Tonelli's theorem, this pathology cannot happen if f, g and h are **positive** functions. In other words: if μ, ν and ρ are (positive) measures with density f, g and h, respectively, then $\mu * (\nu * \rho) = (\mu * \nu) * \rho$. Our example shows that this need not hold any longer for signed measures, i.e. if the densities are not necessarily positive.

10.18 An example where integration by substitution goes wrong

Consider Lebesgue measure on $\mathbb{R}$ and define a function $g : \mathbb{R} \to \mathbb{R}$ by

$$g(t) = \begin{cases} -\sqrt{n}, & \text{if } \frac{1}{2}\left(\frac{1}{n} + \frac{1}{n+1}\right) < t \leqslant \frac{1}{n}, \quad n \in \mathbb{N}, \\ +\sqrt{n}, & \text{if } \frac{1}{n+1} < t \leqslant \frac{1}{2}\left(\frac{1}{n} + \frac{1}{n+1}\right), \quad n \in \mathbb{N}, \\ 0, & \text{if } x \notin (0,1], \end{cases}$$

cf. Fig. 10.2. As $\sum_{n=1}^{\infty} \sqrt{n}\left(\frac{1}{n} - \frac{1}{n+1}\right) = \sum_{n=1}^{\infty} \frac{1}{(n+1)\sqrt{n}} < \infty$, the function g is Lebesgue integrable, and therefore the primitive $G(x) := \int_0^x g(t)\,dt$ exists.

Since g is with respect to the mid-point of each interval $\left[\frac{1}{n+1}, \frac{1}{n}\right]$ an odd function, we have

$$\int_{\frac{1}{n+1}}^{\frac{1}{n}} g(t)\,dt = 0, \quad n \in \mathbb{N};$$

hence

$$G\left(\tfrac{1}{n}\right) = G\left(\tfrac{1}{n+1}\right) = 0 \text{ and } G\left(\tfrac{1}{2}\left(\tfrac{1}{n} + \tfrac{1}{n+1}\right)\right) = \frac{1}{2}\sqrt{n}\left(\tfrac{1}{n} - \tfrac{1}{n+1}\right) = \frac{1}{2}\frac{\sqrt{n}}{n(n+1)};$$

in between these points, $x \longmapsto G(x)$ is a piecewise linear function.

Let $f(x) = |x|^{-1/2}$ and $F(x) = 2\sqrt{|x|}$. On each interval $\left[\frac{1}{n+1}, \frac{1}{n}\right]$ the function $f(G(t))$ is bounded below by $\sqrt{2}n^{1/4}\sqrt{n+1}$, thus

$$\begin{aligned}\int_0^1 f(G(t))|g(t)|\,dt &\geqslant \sum_{n=1}^{\infty} \sqrt{2}n^{1/4}\sqrt{n+1}\sqrt{n}\left(\frac{1}{n} - \frac{1}{n+1}\right) \\ &= \sqrt{2}\sum_{n=1}^{\infty} \frac{n^{1/4}}{\sqrt{n(n+1)}} = \infty,\end{aligned}$$

which shows that $f(G(t))g(t)$ is not integrable over $[0,1]$. On the other hand, $\frac{d}{dt}F(G(t)) = f(G(t))g(t)$ for Lebesgue almost all t, i.e. the function $f(G(t))g(t)$ has a primitive.

10.19 There is no non-constant function such that $\int_{\mathbb{R}^d\setminus\{0\}} \int_{\mathbb{R}^d} |f(x+y) - f(x)||y|^{-d-1}\,dx\,dy < \infty$

Let $f : \mathbb{R}^d \to \mathbb{R}$ be a Borel measurable function such that

$$I := \int_{\mathbb{R}^d\setminus\{0\}} \int_{\mathbb{R}^d} \frac{|f(x+y) - f(x)|}{|y|^{d+1}}\,dx\,dy < \infty.$$

If we set

$$I(R) := \int_{0<|y|\leqslant R} \int_{\mathbb{R}^d} \frac{|f(x+y)-f(x)|}{|y|^{d+1}}\, dx\, dy,$$

then, by the translation invariance of Lebesgue measure,

$$\begin{aligned} I(R) &\leqslant \int_{0<|y|\leqslant R} \int_{\mathbb{R}^d} \frac{|f(x+y)-f(x+y/2)|}{|y|^{d+1}}\, dx\, dy \\ &\quad + \int_{0<|y|\leqslant R} \int_{\mathbb{R}^d} \frac{|f(x+y/2)-f(x)|}{|y|^{d+1}}\, dx\, dy \\ &= 2\int_{0<|y|\leqslant R} \int_{\mathbb{R}^d} \frac{|f(x+y/2)-f(x)|}{|y|^{d+1}}\, dx\, dy. \end{aligned}$$

Changing variables according to $z = \frac{y}{2}$ gives

$$I(R) \leqslant \int_{0<|z|\leqslant R/2} \int_{\mathbb{R}^d} \frac{|f(x+z)-f(x)|}{|z|^{d+1}}\, dx\, dz = I(R/2).$$

Since we can iterate this step, it follows from the monotone convergence theorem that

$$I(R) \leqslant \lim_{k\to\infty} I(R2^{-k}) = 0.$$

As $R > 0$ is arbitrary, an application of Beppo Levi's theorem yields $I = 0$. Hence,

$$|f(x+y)-f(x)| = 0$$

for Lebesgue almost every $(x, y) \in \mathbb{R}^d \times \mathbb{R}^d$. By Tonelli's theorem, there exists an $x \in \mathbb{R}^d$ such that $|f(x+y)-f(x)| = 0$ for Lebesgue almost every $y \in \mathbb{R}^d$, i.e. f is Lebesgue almost everywhere constant.

Comment 1 The assertion of our example remains valid, if we consider functions f defined only on an connected, open set $\Omega \subseteq \mathbb{R}^d$. A discussion of this case and of its consequences can be found in Brézis [25].

Comment 2 A measurable function $f : \mathbb{R}^d \to \mathbb{R}$ is in the so-called **Besov space** $B^s_{pp}(\mathbb{R}^d)$, $s \in (0,1)$ and $p \geqslant 1$, if

$$\|f\|_{B^s_{pp}} = \|f\|_{L^p} + \left(\int_{\mathbb{R}^d} \int_{\mathbb{R}^d} \frac{|f(x)-f(y)|^p}{|x-y|^{sp+d}}\, dx\, dy \right)^{1/p} < \infty.$$

Our example shows that, if we tried to define B^1_{11} through the above expression, then $B^1_{11} = \{0\}$. In fact, if $s = n + \sigma$, $n \in \mathbb{N}_0$ and $\sigma \in [0,1)$, then for all multi-indices $\alpha \in \mathbb{N}_0^d$

$$f \in B^s_{pp} \iff \partial^\alpha f \in L^p \quad \forall |\alpha| < n \quad \text{and} \quad \partial^\alpha f \in B^\sigma_{pp} \quad \forall |\alpha| = n.$$

A standard reference on Besov spaces and more general function spaces are the books by Triebel [187].

10.20 A measure space which has no strictly positive function $f \in L^1$

Take any measure space $(X, \mathscr{A}, \mu)$ which is not σ-finite. Recall that σ-finiteness means that there is a sequence $(X_n)_{n\in\mathbb{N}} \subseteq \mathscr{A}$ such that $X_n \uparrow X$ and $\mu(X_n) < \infty$ for all $n \in \mathbb{N}$.

In fact, the following characterization of σ-finite measure spaces holds.

Lemma *Let $(X, \mathscr{A}, \mu)$ be a measure space. There exists a strictly positive function $f \in L^1(\mu)$ if, and only if, $(X, \mathscr{A}, \mu)$ is σ-finite.*

Proof If there is a strictly positive function $f \in L^1(\mu)$, then $X_n := \{f > 1/n\}$ defines a sequence of increasing sets $(X_n)_{n\in\mathbb{N}} \subseteq A$ such that $X_n \uparrow \{f > 0\} = X$ and, by the Markov inequality,

$$\mu(X_n) \leqslant n \int_X |f|\, d\mu < \infty,$$

hence $(X, \mathscr{A}, \mu)$ is σ-finite. This proves '$\Rightarrow$'. Now assume that $(X, \mathscr{A}, \mu)$ is σ-finite. Without loss of generality, we may assume that $\mu(X) > 0$, and therefore there exists a sequence $(X_n)_{n\in\mathbb{N}} \subseteq \mathscr{A}$ such that $X_n \uparrow X$ and $0 < \mu(X_n) < \infty$ for all $n \in \mathbb{N}$. If wet set $a_n := 2^{-n}/\mu(X_n)$, $n \in \mathbb{N}$, then

$$f(x) := \sum_{n\in\mathbb{N}} a_n \mathbb{1}_{X_n}(x), \quad x \in X,$$

defines a strictly positive measurable function and

$$\int_X |f(x)|\, \mu(dx) = \sum_{n\in\mathbb{N}} a_n \mu(X_n) = \sum_{n\in\mathbb{N}} 2^{-n} < \infty. \qquad \square$$

10.21 In infinite measure spaces there is no function $f > 0$ with $f \in L^1$ and $1/f \in L^1$

Let $(X, \mathscr{A}, \mu)$ be a measure space with $\mu(X) = \infty$. If $f : X \to [0, \infty)$ is a measurable function, then, by the Cauchy–Schwarz inequality,

$$\mu(X) = \int_X \sqrt{f(x)} \frac{1}{\sqrt{f(x)}}\, \mu(dx) \leqslant \sqrt{\int_X f(x)\, \mu(dx)} \sqrt{\int_X \frac{1}{f(x)}\, \mu(dx)}.$$

Consequently, if there exists some $f > 0$ with $f \in L^1(\mu)$ and $1/f \in L^1(\mu)$, then $\mu(X) < \infty$.

Comment For finite measure spaces it is trivial that there exists a function $f > 0$ with $f \in L^1(\mu)$ and $1/f \in L^1(\mu)$ – just take $f = 1$.

10.22 There is no continuous function $f \geqslant 0$ with $\int f^n \, d\lambda = 1$ for all $n \in \mathbb{N}$

We begin with the following auxiliary result which is of independent interest.

Proposition *Let $(X, \mathscr{A}, \mu)$ be a measure space and $f \geqslant 0$ a measurable function. Then*

$$\forall n \in \mathbb{N} \ : \ \int_X f(x)^n \, \mu(dx) = 1 \tag{10.6}$$

if, and only if, $f = \mathbb{1}_A$ a.e. for a set $A \in \mathscr{A}$ with $\mu(A) = 1$.

Proof[1] If $f = \mathbb{1}_A$ for $A \in \mathscr{A}$ with $\mu(A) = 1$, then (10.6) trivially holds. Now assume that f satisfies (10.6) and define $A := \{f = 1\}$. We can write (10.6) in the following form:

$$1 = \int_{\{f<1\}} f(x)^n \, \mu(dx) + \mu(A) + \int_{\{f>1\}} f(x)^n \, \mu(dx). \tag{10.7}$$

Letting $n \to \infty$, by Beppo Levi's theorem the second integral would go to infinity – unless $\mu(\{f > 1\}) = 0$. By the dominated convergence theorem, the first integral tends to zero, and we conclude that $\mu(A) = 1$. If we insert this again into (10.7), we see that $\mu(\{f < 1\}) = 0$, hence $f = \mathbb{1}_A$ a.e. □

Now consider Lebesgue measure λ on $(\mathbb{R}, \mathscr{B}(\mathbb{R}))$, and suppose that there is a continuous function $f \geqslant 0$ with $\int_{\mathbb{R}} f^n \, d\lambda = 1$ for all $n \in \mathbb{N}$. By the above proposition, $f(x) \in \{0, 1\}$ for almost all $x \in \mathbb{R}$ and $\{f = 0\} \neq \emptyset$, $\{f = 1\} \neq \emptyset$. Pick $x, y \in \mathbb{R}$ such that $f(x) = 0$ and $f(y) = 1$. By the intermediate value theorem, there is some $\eta \in \mathbb{R}$ with $f(\eta) = \frac{1}{2}$, and the continuity of f gives $0 < f < 1$ in a neighbourhood of η. Since the neighbourhood has positive Lebesgue measure, we have reached a contradiction.

Comment Let $(X, \mathscr{A}, \mu)$ be a probability space. The above proposition shows that the constant map $f \equiv 1$ is the only function (up to modification on null sets) with $\int_X f^n \, d\mu = 1$ for all $n \in \mathbb{N}$. There is a stronger version of this result: if $f : X \to \mathbb{R}$ is a measurable function, then $f = 1$ almost everywhere if, and only if,

$$\int f\,d\mu = 1 \quad \text{and} \quad \int f^2\,d\mu = 1. \tag{10.8}$$

Indeed, if f satisfies (10.8), then we find for $m := \int f\,d\mu$ that

$$\int_X (f(x) - m)^2\,\mu(dx) = \int_X f(x)^2\,\mu(dx) - m^2 = 1 - 1 = 0,$$

and so $f = m$ almost everywhere, i.e. $f = 1$. The converse is trivial.

10.23 A measure space where $\int_A f\,d\mu = \int_A g\,d\mu$ (for all A) does not entail $f = g$ a.e.

Consider on $(\mathbb{R}, \mathscr{B}(\mathbb{R}))$ Lebesgue measure λ and the measure

$$\mu(B) = \begin{cases} \lambda(B), & \text{if } B \in \mathscr{B}(\mathbb{R}),\ B \subseteq [0,\infty), \\ \infty, & \text{for all other } B \in \mathscr{B}(\mathbb{R}), \end{cases}$$

and the sequence of functions $f_i(x) := \mathbb{1}_{[0,\infty)}(x) + i\mathbb{1}_{(-\infty,0)}(x)$, $i \in \mathbb{N}$. Clearly,

$$\int_B f_i(x)\,\mu(dx) = \int_B f_k(x)\,\mu(dx), \quad B \in \mathscr{B}(\mathbb{R}),$$

for all $i, k \in \mathbb{N}$, but $\{f_i \neq f_k\} = (-\infty, 0)$ for $i \neq k$.

Comment The problem is that the measure μ has a non-trivial atom $(-\infty, 0)$ which is where things go wrong. If $(X, \mathscr{A}, \mu)$ has the **finite subset property**, i.e. for every $A \in \mathscr{A}$ with $\mu(A) > 0$ there is some $A_0 \subseteq A$, $A_0 \in \mathscr{A}$, such that $0 < \mu(A_0) < \infty$[2] – this is e.g. the case if the measure space is σ-finite – then one has for any two measurable functions $f, g : X \to [0, \infty]$

$$\forall A \in \mathscr{A} : \int_A f\,d\mu = \int_A g\,d\mu \implies \mu(\{f \neq g\}) = 0.$$

This becomes clear from the following argument: let $A = \{f > g\}$ and assume that $\mu(A) > 0$. If there is some measurable set $A_0 \subseteq A$ with $0 < \mu(A_0) < \infty$, then we have for any $n \in \mathbb{N}$

$$\int_{\{f>g\}\cap\{n>g\}\cap A_0} (f - g)\,d\mu = \int_{\{f>g\}\cap\{n>g\}\cap A_0} f\,d\mu - \underbrace{\int_{\{f>g\}\cap\{n>g\}\cap A_0} g\,d\mu}_{<\infty} = 0,$$

which means that $\mu(\{f > g\} \cap \{n > g\} \cap A_0) = 0$. As $A_0 \subseteq \{g < f\} \subseteq \{g < \infty\}$, we find by letting $n \to \infty$ that $\mu(A_0) = 0$, which is a contradiction.

[2] If the measure space is σ-finite, then this property excludes atoms with infinite measure [☞ Lemma 6A].

10.24 A vector function which is weakly but not strongly integrable

Let $\mathcal{H}$ be a separable Hilbert space with scalar product $(g, h) \longmapsto \langle g, h \rangle$ and norm $\| \cdot \|$. For an orthonormal basis $(h_n)_{n\in\mathbb{N}_0}$, we define $f : [0, 1] \to \mathcal{H}$ by

$$f(t) = \sum_{n=0}^{\infty} \frac{2^n}{n+1} \mathbb{1}_{(2^{-n-1}, 2^{-n}]}(t) h_n.$$

Let $\Phi := \sum_{n=0}^{\infty} \phi_n h_n'$ be an element of $\mathcal{H}'$ with the dual basis $(h_n')_{n\in\mathbb{N}_0}$. By Bessel's identity, $\sum_{n=0}^{\infty} \phi_n^2 < \infty$, and it follows from $\Phi h_n = \phi_n$ that

$$\Phi f(t) = \sum_{n=0}^{\infty} \frac{2^n}{n+1} \mathbb{1}_{(2^{-n-1}, 2^{-n}]}(t) \Phi h_n = \sum_{n=0}^{\infty} \frac{2^n}{n+1} \mathbb{1}_{(2^{-n-1}, 2^{-n}]}(t) \phi_n.$$

The strong integral does not exist, since

$$\int_0^1 \|f(t)\| \, dt = \sum_{n=0}^{\infty} \frac{2^n}{n+1} \int_0^1 \mathbb{1}_{(2^{-n-1}, 2^{-n}]}(t) \, dt = \frac{1}{2} \sum_{n=0}^{\infty} \frac{1}{n+1} = \infty,$$

whereas the weak integral exists

$$\begin{aligned} \int_0^1 |\Phi f(t)| \, dt &= \sum_{n=0}^{\infty} \frac{2^n}{n+1} |\phi_n| \int_0^1 \mathbb{1}_{(2^{-n-1}, 2^{n}]}(t) \, dt \\ &= \frac{1}{2} \sum_{n=0}^{\infty} \frac{1}{n+1} |\phi_n| \\ &\leqslant \frac{1}{2} \left[\sum_{n=0}^{\infty} \frac{1}{(n+1)^2} \right]^{\frac{1}{2}} \left[\sum_{n=0}^{\infty} \phi_n^2 \right]^{\frac{1}{2}} < \infty. \end{aligned}$$

11

Modes of Convergence

Let us briefly recall the principal types of convergence on an arbitrary measure space $(X, \mathscr{A}, \mu)$. A sequence of measurable functions $f_n : X \to \mathbb{R}$ is said to converge to a measurable function $f : X \to \mathbb{R}$

$$
\begin{aligned}
&\text{almost everywhere (a.e.), if} && \mu\left(\left\{x;\ \lim_{n\to\infty} |f_n(x) - f(x)| > 0\right\}\right) = 0\\
&\text{almost uniformly (a.u.), if} && \forall\epsilon > 0 \quad \exists A \in \mathscr{A},\ \mu(X \setminus A) < \epsilon\\
& && \lim_{n\to\infty} \sup_{x\in A} |f_n(x) - f(x)| = 0\\
&\text{in measure, if} && \forall\epsilon > 0 :\ \lim_{n\to\infty} \mu\left(\{x : |f_n(x) - f(x)| > \epsilon\}\right) = 0\\
&\text{in probability, if} && \forall F \in \mathscr{A},\ \mu(F) < \infty \quad \forall\epsilon > 0 :\\
& && \lim_{n\to\infty} \mu\left(F \cap \{x : |f_n(x) - f(x)| > \epsilon\}\right) = 0\\
&\text{in } L^p(\mu) \text{ for } 1 \leqslant p < \infty, \text{ if} && \lim_{n\to\infty} \int |f_n - f|^p \, d\mu = 0\\
&\text{in } L^\infty(\mu) \text{ if} && \lim_{n\to\infty} \underbrace{\inf\{c;\ \mu\{|f_n - f| > c\} = 0\}}_{=\operatorname{esssup}|f_n - f|} = 0\\
&\text{weakly in } L^p,\ 1 \leqslant p \leqslant \infty, \text{ if} && \forall g \in (L^p)^*(\mu) :\ \lim_{n\to\infty} \int f_n g \, d\mu = \int f g \, d\mu.
\end{aligned}
$$

There is no general agreement in the literature on ‘convergence in measure’ and ‘convergence in probability’; unfortunately, these two concepts are pretty much incompatible unless one considers finite measure spaces. An overview of the connections between these modes of convergence is given in Fig. 1.4, in Chapter 1 and in Tables 11.1–11.3 below.

An important tool in the study of convergence is the (easy part of the) **Borel–Cantelli lemma**.

Theorem 11A (Borel) *Let $(X, \mathscr{A}, \mu)$ be a measure space. For every sequence of measurable sets $(A_n)_{n\in\mathbb{N}} \subseteq \mathscr{A}$ the following implication holds:*

$$\sum_{i=1}^{\infty} \mu(A_i) < \infty \Rightarrow \mu\left(\bigcap_{n=1}^{\infty}\bigcup_{i=n}^{\infty} A_i\right) = 0.$$

Usually, the Borel–Cantelli lemma is stated for probability measures (the 'difficult' part is based on the notion of pairwise independence; see [MIMS, p. 296]); the above implication requires, however, only monotonicity of the measure and σ-subadditivity. The condition $x \notin \bigcap_{n=1}^{\infty}\bigcup_{i=n}^{\infty} A_i$ means that $\sum_{i=1}^{\infty} \mathbb{1}_{A_i}(x) < \infty$, i.e. x is contained in at most finitely many of the sets $A_1, A_2, \dots$.

Convergence in measure implies convergence in probability, and we can describe convergence in probability by a **subsequence principle**.

Lemma 11B (subsequence principle) *Let $(X, \mathscr{A}, \mu)$ be a measure space and $(f_n)_{n\in\mathbb{N}}$ be a sequence of measurable functions.*

(a) $f_n \xrightarrow{\text{probab.}} f$ *if, and only if, for every set $F \in \mathscr{A}$ with $\mu(F) < \infty$, every subsequence $(f'_n)_{n\in\mathbb{N}} \subseteq (f_n)_{n\in\mathbb{N}}$ has a further subsequence $(f''_n)_{n\in\mathbb{N}}$ such that $f''_n \mathbb{1}_F \xrightarrow{\text{a.e.}} f\mathbb{1}_F$.*

(b) *If $(X, \mathscr{A}, \mu)$ is σ-finite, then $f_n \xrightarrow{\text{probab.}} f$ if, and only if, every subsequence $(f'_n)_{n\in\mathbb{N}} \subseteq (f_n)_{n\in\mathbb{N}}$ has a further subsequence $(f''_n)_{n\in\mathbb{N}}$ such that $f''_n \xrightarrow{\text{a.e.}} f$.*

Part (a) is a standard Borel–Cantelli argument, e.g. [49, Theorem 9.2.1], the second part relies on a diagonalization argument which allows us to remove the dependence on the set F in the choice of the sub-subsequence in Part (a); see [MIMS, p. 273, Problem 22.9]. This breaks down, if we give up σ-finiteness [☞ 11.7].

Convergence in measure cannot be characterized using a.e. convergent (sub-) subsequences: on $((0,\infty), \mathscr{B}(0,\infty), \lambda)$ the sequence $f_n(x) = \mathbb{1}_{(n,n+1)}(x)$ converges a.e. to $f = 0$, hence every sub- and sub-sub-sequence converges a.e. to 0, but f_n does not converge in measure, [☞ 11.5, Example (e) and Table 11.1].

With a similar argument Fréchet [62] showed that a.e. convergence is, in general, not metrizable [☞ 11.17].

11.1 Classical counterexamples to a.e. convergence vs. convergence in probability

Consider Lebesgue measure λ on the measurable space $((0,1], \mathscr{B}(0,1])$. We split $(0,1]$ into 2^n disjoint intervals $I_{n,k} := ((k-1)2^{-n}, k2^{-n}]$, $n \in \mathbb{N}$, $k = 1, \dots, 2^n$, and define $f_{n,k}(x) := \mathbb{1}_{I_{n,k}}(x)$. We arrange the functions lexicographically in one sequence

$$f_{1,1}, f_{1,2}, f_{2,1}, f_{2,2}, f_{2,3}, f_{2,4}, f_{3,1}, \dots$$

which we enumerate as $(F_i)_{i\in\mathbb{N}}$. By construction,

$$\forall x \in (0,1] \ : \ \liminf_{i\to\infty} F_i(x) = 0 < 1 = \limsup_{i\to\infty} F_i(x),$$

while

$$\lim_{i\to\infty} \lambda(\{x \in (0,1]\,;\ F_i(x) > \epsilon\}) = \lim_{i\to\infty} \lambda(\{F_i = 1\}) = \lim_{n\to\infty} 2^{-n} = 0$$

for all $\epsilon \in (0,1)$. Almost the same calculation shows

$$\lim_{i\to\infty} \int_{(0,1]} |F_i|^p \, d\lambda = \lim_{i\to\infty} \lambda(\{F_i = 1\}) = 0,$$

i.e. the sequence $(F_i)_{i\in\mathbb{N}}$ is nowhere pointwise convergent, but it converges in measure and in L^p-sense to 0.

Comment It is easy to see that $(F_i)_{i\in\mathbb{N}}$ contains a subsequence which converges a.e. to 0, for example $(f_{n,3})_{n\geqslant 4}$.

11.2 Pointwise convergence does not imply convergence in measure

Consider Lebesgue measure λ on $(\mathbb{R}, \mathscr{B}(\mathbb{R}))$, and set $f_n := \mathbb{1}_{[n,n+1]}$. The sequence f_n converges pointwise to $f = 0$, but

$$\forall \epsilon \in (0,1) \ : \ \lambda\left(\{|f_n| > \epsilon\}\right) = \lambda\left([n, n+1]\right) = 1.$$

Comment This pathology cannot happen for convergence in probability, as pointwise convergence implies convergence in probability. Indeed, for an arbitrary measurable set F with $\mu(F) < \infty$ we have

$$\begin{aligned}\mu(\{|f_n - f| > \epsilon\} \cap F) &= \mu(\{\min(1, |f_n - f|) > \epsilon\} \cap F)\\ &\leqslant \frac{1}{\epsilon} \int_F \min(1, |f_n - f|) \, d\mu \xrightarrow[n\to\infty]{} 0\end{aligned}$$

by dominated convergence. Notice that $\mathbb{1}_F$ can be used as integrable majorant as $\mu(F) < \infty$.

11.3 L^p-convergence does not imply L^r-convergence for $r \neq p$

For a general measure space $(X, \mathscr{A}, \mu)$ the inclusion $L^p(\mu) \subseteq L^r(\mu)$ may fail for any $r \neq p$ [☞ 18.1]. This implies, in particular, that L^p-convergence of a sequence $(f_n)_{n\in\mathbb{N}}$ does not imply L^r-convergence of $(f_n)_{n\in\mathbb{N}}$. Indeed, for any $f \in L^p(\mu)\backslash L^r(\mu)$ the sequence $f_n := \frac{1}{n} f$ converges in $L^p(\mu)$ to zero, but $(f_n)_{n\in\mathbb{N}}$ does not converge in $L^r(\mu)$ since $\|f_n\|_{L^r(\mu)} = \infty$ for all $n \in \mathbb{N}$.

Comment If μ is a **finite** measure, then L^p-convergence implies L^r-convergence for all $r \in [1, p]$. This is an immediate consequence of the inequality

$$\|f\|_{L^r(\mu)} \leqslant \mu(X)^{\frac{1}{r}-\frac{1}{p}} \|f\|_{L^p(\mu)}, \quad r \in [1, p],\ p \in [1, \infty],$$

which follows from Hölder's inequality.

11.4 Classical counterexamples related to weak convergence in L^p

Let λ be Lebesgue measure on $[0, \infty)$ and let $L^p = L^p(\lambda)$ for some $1 \leqslant p < \infty$. Denote by $q \in (1, \infty]$ the conjugate index, i.e. $p^{-1} + q^{-1} = 1$. None of the following sequences $(f_n)_{n\in\mathbb{N}} \subseteq L^p$ converges weakly in L^p to zero, i.e.

$$\exists g \in L^q : \int_{[0,\infty)} f_n(x)g(x)\,dx \text{ does not converge to zero.} \tag{11.1}$$

(a) $f_n(x) := n\mathbb{1}_{[0,1/n]}(x)$ converges to zero a.e. and in measure, but it does not converge weakly to zero. To see this, take $g(x) = \mathbb{1}_{[0,1]}(x)$ in (11.1).

(b) $f_n(x) := n^{-1}\mathbb{1}_{[0,e^n]}(x)$ converges to zero uniformly, but it does not converge weakly to zero. To see this, take $g(x) = x^{-1}\mathbb{1}_{[1,\infty)}$ and observe that $g \in L^q$ for any $p \geqslant 1$ and that

$$\int_0^\infty f_n(x)g(x)\,dx = \frac{1}{n}\int_1^{e^n} \frac{dx}{x} = 1$$

does not converge to zero.

On the other hand, weak convergence in L^p does not imply convergence in probability; this means, in particular, that it does not entail uniform convergence, almost everywhere convergence or convergence in L^p. The sequence $f_n(x) = \cos(nx)\mathbb{1}_{[0,2\pi]}(x)$ is in L^p for every $p \geqslant 1$ and, by the Riemann–Lebesgue lemma [MIMS, p. 222], we have

$$\forall g \in L^q : \lim_{n\to\infty} \int_{[0,\infty)} f_n(x)g(x)\,dx = 0,$$

i.e. $f_n \to 0$ weakly. Since $\int_0^\infty f_n^2(x)\,dx = \pi$ for all $n \in \mathbb{N}$, the sequence f_n

cannot converge to zero in probability. Suppose, on the contrary, that $f_n \to 0$ in probability, then there exists a subsequence f_{n_k} which converges to zero a.e. By dominated convergence, we get $\int_0^\infty f_{n_k}^2(x)\,dx \to 0$, which is not possible.

11.5 The convergence tables

In order to show the interrelations between various modes of convergence, a few counterexamples suffice. Throughout λ denotes one-dimensional Lebesgue measure and $n \in \mathbb{N}$ is a natural number. The following examples are adapted from the previous Examples 11.1–11.4 and Example 12.1.

(a) $f_n(x) = \frac{1}{n}\mathbb{1}_{[0,e^n]}(x)$ on $([0,\infty), \mathscr{B}[0,\infty), \lambda)$.
(b) $f_n(x) = n\mathbb{1}_{\left(0,\frac{1}{n}\right)}(x)$ on $([0,1], \mathscr{B}[0,1], \lambda)$.
(c) $f_n(x) = x^n$ on $([0,1], \mathscr{B}[0,1], \lambda)$.
(d) $f_{i,n}(x) = \mathbb{1}_{\left(\frac{i-1}{n},\frac{i}{n}\right)}(x)$, $i = 1, \dots, n$, lexicographically ordered, on $([0,1], \mathscr{B}[0,1], \lambda)$.
(e) $f_n(x) = \mathbb{1}_{(n,n+1)}(x)$ on $((0,\infty), \mathscr{B}(0,\infty), \lambda)$.
(f) $f_n(x) = \cos(2\pi n x)$ on $([0,1], \mathscr{B}[0,1], \lambda)$.

Tables 11.1–11.3 contain an overview on the relations between the various modes of convergence.

11.6 The limit in probability is not necessarily unique

Consider Lebesgue measure λ on $(\mathbb{R}, \mathscr{B}(\mathbb{R}))$. Set

$$f(x) := \begin{cases} 1, & \text{if } x \in [0,1], \\ \infty, & \text{if } x \notin [0,1], \end{cases}$$

and denote by $\mu = f\lambda$ the measure with density f w.r.t. Lebesgue measure. For any set $F \in \mathscr{B}(\mathbb{R})$ with $\mu(F) < \infty$, we have $\lambda(F \cap [0,1]^c) = 0$, i.e. $F \subseteq [0,1]$ 'up to a (Lebesgue) null set'. Thus,

$$\mu(\{x \in \mathbb{R};\ |g(x)| > \epsilon\} \cap F) = \lambda(\{x \in [0,1];\ |g(x)| > \epsilon\} \cap F)$$

for all measurable functions $g : \mathbb{R} \to \mathbb{R}$ and all $\epsilon > 0$. If $(f_n)_{n\in\mathbb{N}}$ is some sequence with $f_n \to f$ in probability w.r.t. Lebesgue measure, then this gives $f_n \to g$ in probability w.r.t. μ for **any** measurable function g satisfying $f|_{[0,1]} = g|_{[0,1]}$. Hence, the limit in probability is not unique (μ-almost everywhere).

Table 11.1 *Relation between modes of convergence: general measure spaces. The letters* (a)–(f) *refer to the sequences in Example 11.5, indicating that the respective implication fails.*

	Does this imply convergence in						
	unif.	a.unif.	μ a.e.	probab.	meas.	$L^p(\mu)$	weak
$f_n \xrightarrow{\text{unif}} f$	yes	yes	yes	yes	yes	(a)	(a)
$f_n \xrightarrow{\text{a.u.}} f$	(c)	yes	yes	yes	yes	(a)	(a)
$f_n \xrightarrow{\text{a.e.}} f$	(c)	(e)	yes	yes	(e)	(b)	(b)
$f_n \xrightarrow{\text{prob}} f$	(c)	(d)	(d)	yes	(e)	(b)	(b)
$f_n \xrightarrow{\text{meas}} f$	(c)	(d)	(d)	yes	yes	(b)	(b)
$f_n \xrightarrow{L^p} f$	(c)	(d)	(d)	yes	yes	yes	yes
$f_n \xrightarrow{\text{weak}} f$	(c)	(d)	(d)	(f)	(f)	(f)	yes

Table 11.2 *Relation between modes of convergence: finite measure spaces. The letters* (a)–(f) *refer to the sequences in Example 11.5, indicating that the respective implication fails.*

	Does this imply convergence in						
	unif.	a.unif.	μ a.e.	probab.	meas.	$L^p(\mu)$	weak
$f_n \xrightarrow{\text{unif}} f$	yes	yes	yes	yes	yes	yes	yes
$f_n \xrightarrow{\text{a.u.}} f$	(c)	yes	yes	yes	yes	(b)	(b)
$f_n \xrightarrow{\text{a.e.}} f$	(c)	yes	yes	yes	yes	(b)	(b)
$f_n \xrightarrow{\text{prob}} f$	(c)	(d)	(d)	yes	yes	(b)	(b)
$f_n \xrightarrow{\text{meas}} f$	(c)	(d)	(d)	yes	yes	(b)	(b)
$f_n \xrightarrow{L^p} f$	(c)	(d)	(d)	yes	yes	yes	yes
$f_n \xrightarrow{\text{weak}} f$	(c)	(d)	(d)	(f)	(f)	(f)	yes

Comment This pathology cannot happen if the measure space is σ-finite:

Table 11.3 *Relation between modes of convergence: integrable majorant: $|f_n| \leqslant g$, $g \in L^p$. The letters* (a)–(f) *refer to the sequences in Example 11.5, indicating that the respective implication fails.*

	Does this imply convergence in						
	unif.	a.unif.	μ a.e.	probab.	meas.	$L^p(\mu)$	weak
$f_n \xrightarrow{\text{unif}} f$	yes	yes	yes	yes	yes	yes	yes
$f_n \xrightarrow{\text{a.u.}} f$	(c)	yes	yes	yes	yes	yes	yes
$f_n \xrightarrow{\text{a.e.}} f$	(c)	yes	yes	yes	yes	yes	yes
$f_n \xrightarrow{\text{prob}} f$	(c)	(d)	(d)	yes	yes	yes	yes
$f_n \xrightarrow{\text{meas}} f$	(c)	(d)	(d)	yes	yes	yes	yes
$f_n \xrightarrow{L^p} f$	(c)	(d)	(d)	yes	yes	yes	yes
$f_n \xrightarrow{\text{weak}} f$	(c)	(d)	(d)	(f)	(f)	(f)	yes

if $(X, \mathscr{A}, \mu)$ is a σ-finite measure space and a sequence $(f_n)_{n\in\mathbb{N}}$ converges in probability to f, then f is unique up to a μ-null set, cf. [MIMS, p. 261].

11.7 A sequence converging in probability without having an a.e. converging subsequence

Let $(X, \mathscr{A}, \mu)$ be a measure space. If $(f_n)_{n\in\mathbb{N}}$ is a sequence which converges in measure to a function f, then an application of the Borel–Cantelli lemma [☞ Theorem 11A] shows that there is a subsequence $(f_{n(k)})_{k\in\mathbb{N}}$ converging almost everywhere to f.

If the measure space is not σ-finite, then this result does not extend to 'convergence in probability', i.e. $f_n \to f$ in probability need not imply $f_{n(k)} \to f$ almost everywhere for a suitable subsequence $(f_{n(k)})_{k\in\mathbb{N}}$. The reason is that limits in probability are not necessarily unique [☞ 11.6].

Take a measure space $(X, \mathscr{A}, \mu)$ which is not σ-finite, and a sequence $(f_n)_{n\in\mathbb{N}}$ such that $f_n \to f$ and $f_n \to g$ in probability for functions f, g satisfying $\mu(\{f \neq g\}) > 0$. If it were possible to select a.e. converging subsequences, there would exist a subsequence $(f_{n(k)})_{k\in\mathbb{N}}$ such that $f_{n(k)} \to f$ almost everywhere. As $f_{n(k)} \to g$ in probability, we could choose a further subsequence $(f'_{n(k)})_{k\in\mathbb{N}}$ with $f'_{n(k)} \to g$ almost everywhere. Hence, $f = g$ almost everywhere, which contradicts our choice of f and g.

Comment For σ-finite measure spaces the above pathology cannot occur. If $(X, \mathscr{A}, \mu)$ is σ-finite and $f_n \to f$ in probability, then $f_{n(k)} \to f$ almost everywhere for a subsequence $(f_{n(k)})_{k\in\mathbb{N}}$ of $(f_n)_{n\in\mathbb{N}}$. Let us sketch the idea of the proof. Since the measure space is σ-finite, there exists some $g \in L^1(\mu)$ with $g > 0$ almost everywhere [☞10.20]. It follows from $f_n - f \to 0$ in probability that $\min\{|f_n - f|, 1\} \to 0$ in probability, cf. [MIMS, p. 259], which in turn implies that $h_n := g\cdot\min\{|f_n - f|, 1\} \to 0$ in probability. As $|h_n| \leqslant |g| \in L^1(\mu)$ this gives $h_n \to 0$ in $L^1(\mu)$, cf. [MIMS, p. 260]. By the Riesz–Fischer theorem, there exists a subsequence $(h_{n(k)})_{k\in\mathbb{N}}$ converging to zero almost everywhere. As $g > 0$ this means that $\min\{|f_{n(k)} - f|, 1\} \to 0$ almost everywhere, i.e. $|f_{n(k)} - f| \to 0$ almost everywhere.

11.8 A sequence converging in probability without having any subsequence converging in measure

Consider on $(\mathbb{R}, \mathscr{B}(\mathbb{R}), \lambda)$ the 'sliding hump' $f_n(x) := \mathbb{1}_{[n,n+1)}(x)$. If $m \neq n$, then

$$\forall \epsilon \in (0,1) \ : \ \lambda(\{|f_n - f_m| > \epsilon\}) = 2,$$

which shows that $(f_n)_{n\in\mathbb{N}}$ does not have any subsequence which converges in measure. The sequence f_n converges to zero Lebesgue almost everywhere and so, in particular, $f_n \to 0$ in probability [☞Tab. 11.1].

Comment This pathology cannot occur if $(f_n)_{n\in\mathbb{N}}$ has an integrable envelope. Let $(X, \mathscr{A}, \mu)$ be a measure space and $(f_n)_{n\in\mathbb{N}} \subseteq L^1(\mu)$ such that $f_n \to f$ in probability. If there exists some $g \in L^1(\mu)$ such that $|f_n| \leqslant g \in L^1(\mu)$ for all $n \in \mathbb{N}$, then $f_n \to f$ in measure. Indeed, since $(f_n)_{n\in\mathbb{N}}$ is bounded by an integrable function, it follows from (a version of) the dominated convergence theorem, [MIMS, p. 260], that $f_n \to f$ in $L^1(\mu)$. Hence, by Markov's inequality,

$$\forall \epsilon > 0 \ : \ \mu(\{|f_n - f| > \epsilon\}) \leqslant \frac{1}{\epsilon}\|f_n - f\|_{L^1(\mu)} \xrightarrow[n\to\infty]{} 0,$$

i.e. $f_n \to f$ in measure. If the measure space $(X, \mathscr{A}, \mu)$ is σ-finite, then Vitali's convergence theorem [☞Theorem 1.31] shows that the assumption that $(f_n)_{n\in\mathbb{N}}$ has an integrable envelope can be relaxed to uniform integrability of $(f_n)_{n\in\mathbb{N}}$.

11.9 A sequence such that $\int f_n(x)\,dx \to 0$ but $(f_n)_{n\in\mathbb{N}}$ has no convergent subsequence

Let λ be Lebesgue measure on $((-\pi,\pi],\mathscr{B}(-\pi,\pi])$ and consider the sequence of functions $f_n(x) = \sin(nx)$. We have

$$0 \leqslant |f_n| \leqslant 1 \quad \text{and} \quad \lim_{n\to\infty} \int_{-\pi}^{\pi} f_n(x)\,dx = 0.$$

However, there is no subsequence n_i such that $\lim_{i\to\infty} \sin(n_i x)$ exists for irrational x.

11.10 A sequence converging a.e. and in measure but not almost uniformly

Egorov's theorem states that $f_n \to f$ a.e. implies $f_n \to f$ almost uniformly [☞p. 221] for any **finite** measure space, cf. [19, Theorem 2.2.1]. The following example shows, in particular, that the assumption $\mu(X) < \infty$ cannot be dropped.

Consider Lebesgue measure λ on $((0,\infty),\mathscr{B}(0,\infty))$, and set

$$s_0 := 0 \quad \text{and} \quad s_n := \sum_{i=1}^{n} \frac{1}{i}, \qquad n \in \mathbb{N}.$$

The sequence $f_n := \mathbb{1}_{[s_n, s_{n+1})}$ converges in measure to zero and $f_n(x) \to 0$ for all $x > 0$. However, f_n does not converge almost uniformly to 0. Fix some set $A \in \mathscr{B}(0,\infty)$ with $\lambda(A) < 1$. For any $N \in \mathbb{N}$ there exists some $x > s_N$ such that $x \notin A$. By the construction of $(f_n)_{n\in\mathbb{N}}$, there is a number $n > N$ such that $f_n(x) = 1$, and so

$$\sup_{n\geqslant N} \sup_{y\in X\setminus A} |f_n(y) - 0| \geqslant 1$$

for any $N \geqslant 1$ which shows that f_n does not converge uniformly to 0 on $X \setminus A$. This proves that $(f_n)_{n\in\mathbb{N}}$ does not converge almost uniformly to 0.

11.11 Egorov's theorem fails for infinite measures

[☞11.10]

11.12 Egorov's theorem does not hold for nets

Let $(X,\mathscr{A},\mu)$ be a finite measure space, and f_n, $n \in \mathbb{N}$, and f real-valued measurable functions. Egorov's theorem [19, Theorem 2.2.1] states that $f_n \to f$ almost everywhere implies $f_n \to f$ almost uniformly [☞p. 221]. Here we present an example due to Walter [197] which shows that Egorov's theorem fails to hold for nets. More precisely, we construct a family of measurable functions $f_t : (0,1) \to \mathbb{R}$, $t \in [2,\infty)$, such that the pointwise limit

$$f(x) = \lim_{t\to\infty} f_t(x), \quad x \in (0,1),$$

exists but f_t does not converge almost uniformly to f. A further example of this phenomenon is given in [22, Exercise IV.5.4].

Consider on $(X, \mathscr{A}) = ((0,1), \mathscr{L}(0,1))$ Lebesgue measure λ, and pick a set $L \subseteq (0, \frac{1}{2})$ such that $L \notin \mathscr{L}(0,1)$ and $(r+L) \cap (q+L) = \emptyset$ for all rational numbers $r \neq q$ [☞7.22]. Since L is not Lebesgue measurable, it has strictly positive outer Lebesgue measure $\alpha := \lambda^*(L) > 0$. The sets

$$L_n := \frac{1}{n} + L \subseteq (0,1), \quad n \geqslant 2,$$

are pairwise disjoint and $\lambda^*(L_n) = \alpha > 0$ by the translation invariance of Lebesgue measure. For $t \in I := [2,\infty)$ define

$$f_t(x) := \begin{cases} 1, & \text{if } x \in L_n \text{ and } t = x + n \text{ for some } n \geqslant 2, \\ 0, & \text{otherwise.} \end{cases}$$

In other words, given $t \in [2,\infty)$ we write $t = x + n$ for $n \in \mathbb{N}$ and $r \in (0,1)$, and set $f_t = \mathbb{1}_{\{r\}\cap L_n}$, i.e. $f_t \equiv 0$ if $r \notin L_n$ and $f_t = \mathbb{1}_{\{r\}}$ if $r \in L_n$.

In particular, there exists for each $t \in I$ at most one $x \in (0,1)$ such that $f_t(x) \neq 0$, and so $x \longmapsto f_t(x)$ is measurable. Moreover, since the sets L_n are pairwise disjoint, there is for each fixed $x \in (0,1)$ at most one $t \in I$ such that $f_t(x) \neq 0$, and this implies

$$\lim_{t\to\infty} f_t(x) = 0, \quad x \in (0,1).$$

Now assume that $f_t \to 0$ uniformly on $(0,1) \setminus N$ for some $N \in \mathscr{L}(0,1)$. Then there exists a number $n \in \mathbb{N}$ such that

$$\forall x \in (0,1) \setminus N \ : \ \sup_{t\geqslant n} |f_t(x)| \leqslant \frac{1}{2}.$$

From

$$\sup_{t\geqslant n} |f_t(x)| \geqslant |f_{n+x}(x)| = 1, \quad x \in L_n,$$

it follows that $N \supseteq L_n$. Hence, $\lambda(N) = \lambda^*(N) \geqslant \lambda^*(L_n) = \alpha > 0$. This shows that for $\epsilon \in (0,\alpha)$ there does not a exist a set $N \in \mathscr{L}(0,1)$ with $\lambda(N) < \epsilon$ such that $f_t \to 0$ uniformly on $(0,1) \setminus N$. Consequently, f_t does not converge almost uniformly to 0.

Comment Let $f_t : X \to \mathbb{R}$, $t \in I := [1, \infty)$, be a family of measurable functions on a finite measure space $(X, \mathscr{A}, \mu)$ such that $f(x) := \lim_{t\to\infty} f_t(x)$ exists for all $x \in X$, i.e.

$$g_n(x) := \sup_{t\geqslant n} |f_t(x) - f(x)| \xrightarrow[n\to\infty]{} 0, \quad x \in X.$$

If g_n is measurable for each $n \in \mathbb{N}$, then it follows from Egorov's theorem that $g_n \to 0$ almost uniformly, which implies that $f_t \to f$ almost uniformly. For the family of functions which we have constructed above, the supremum g_n fails to be measurable,

$$g_n(x) = \sup_{t\geqslant n} |f_t(x)| = \begin{cases} 1, & \text{if } x \in \bigcup_{k\geqslant n} L_k, \\ 0, & \text{otherwise,} \end{cases}$$

and therefore Egorov's theorem does not apply.

11.13 A uniformly convergent sequence of L^1-functions which is not convergent in L^1

Consider Lebesgue measure λ on $(\mathbb{R}, \mathscr{B}(\mathbb{R}))$ and define

$$g_n(x) := \frac{1}{2n}\mathbb{1}_{[-n^3,n^3]}(x) \quad \text{and} \quad f_n(x) := \sum_{i=1}^{n} \frac{g_i(x)}{i^2}, \quad x \in \mathbb{R}.$$

Since $0 \leqslant g_n \leqslant 1$, the sequence f_n converges uniformly and

$$\int g_n \, d\lambda = \frac{2n^3}{2n} = n^2 \quad \text{and} \quad \int f_n \, d\lambda = \sum_{i=1}^{n} \frac{1}{i^2} \int g_i \, d\lambda = n.$$

This example works on any σ-finite (but not finite) measure space. For finite measures uniform convergence of L^1-functions does imply L^1-convergence; see Table 11.2.

11.14 Convergence in measure is not stable under products

Consider $((0, \infty), \mathscr{B}(0, \infty))$ with Lebesgue measure λ, and set

$$f_n(x) := x + \frac{1}{n} \quad \text{and} \quad f(x) := x.$$

Since $|f_n(x) - f(x)| = \frac{1}{n}$ for all $x > 0$, it follows that

$$\lim_{n\to\infty} \lambda(\{|f_n - f| > \epsilon\}) = 0$$

for all $\epsilon > 0$, i.e. $f_n \to f$ in measure. However, f_n^2 does not converge in measure to f^2 as

$$\lambda(\{|f_n^2 - f^2| \geqslant 2\}) \geqslant \lambda(\{x \in (0,\infty)\,;\ 2nx \geqslant 2\}) = \infty.$$

Comment If the limit f is **almost bounded**, i.e. if for every $\epsilon > 0$ there is some $C = C_\epsilon < \infty$ such that $\mu(\{|f| > C_\epsilon\}) < \epsilon$, then $f_n \xrightarrow{\text{meas}} f$ does imply $f_n^2 \xrightarrow{\text{meas}} f^2$. Moreover the limit (in measure) of a sequence of almost bounded functions is almost bounded; see [61]. Our example shows that convergence in measure does not imply that the functions f_n and f need to be (almost) bounded.

11.15 A measure space where convergence in measure and uniform convergence coincide

Let $(X, \mathscr{A}, \mu)$ be a measure space. If $(f_n)_{n\in\mathbb{N}}$ is a sequence of measurable functions which converges uniformly to f, then $\{x\,;\ |f_n(x) - f(x)| > \epsilon\}$ is empty if $n \geqslant n(\epsilon)$ for some sufficiently large $n(\epsilon) \in \mathbb{N}$. Thus, uniform convergence always implies convergence in measure (and in probability).

Now let μ be counting measure on $(\mathbb{Z}, \mathscr{P}(\mathbb{Z}))$ and let $(f_n)_{n\in\mathbb{N}}$ be a sequence of functions $f_n : \mathbb{Z} \to \mathbb{R}$ which converges in measure to f. By assumption,

$$\forall \epsilon > 0 \ :\ \#\{x \in \mathbb{Z}\,;\ |f_n(x) - f(x)| > \epsilon\} \xrightarrow[n\to\infty]{} 0.$$

Since

$$(a_n)_{n\in\mathbb{N}} \subseteq \mathbb{N} \cup \{\infty\},\ a_n \xrightarrow[n\to\infty]{} 0 \Rightarrow \#\{n \in \mathbb{N}\,;\ a_n \neq 0\} < \infty,$$

this means that there is for any $\epsilon > 0$ a constant $N = N(\epsilon) \in \mathbb{N}$ such that

$$\forall n \geqslant N \ :\ \#\{x \in \mathbb{Z}\,;\ |f_n(x) - f(x)| > \epsilon\} = 0,$$

i.e.

$$\forall n \geqslant N, \quad \forall x \in \mathbb{Z} \ :\ |f_n(x) - f(x)| \leqslant \epsilon.$$

Comment A measure μ on a measurable space $(\Omega, \mathscr{A})$ is called **strictly positive** if $\mu(A) > 0$ for all $A \in \mathscr{A}$, $A \neq \emptyset$. Equivalently, $A = \emptyset$ is the only set of measure zero. For finite measures μ it can be shown that 'convergence in measure' is equivalent to 'pointwise convergence everywhere' if, and only if, μ is strictly positive, cf. Marczewski [113, Theorem III]. Moreover, any strictly positive measure is purely atomic, cf. [113].

11.16 A measure space where strong and weak convergence of sequences in L^1 coincide

Let $(X, \mathscr{A}, \mu)$ be a measure space and set $L^1 = L^1(X, \mathscr{A}, \mu)$. If $f_n \to f$ in L^1, then $f_n \rightharpoonup f$ weakly in L^1, i.e.

$$\forall g \in (L^1)^* \ : \ \lim_{n\to\infty} \int_X f_n g \, d\mu = \int_X fg \, d\mu.$$

The converse is, in general, false [☞11.4]. There are, however, measure spaces where the notion of weak and strong convergence of sequences in L^1 coincide.

Proposition *Let $(X, \mathscr{A}, \mu)$ be a measure space such that $\mu(\{x\}) > 0$ for every $x \in X$. Then $f_n \to f$ in L^1 if, and only if, $f_n \rightharpoonup f$ weakly in L^1.*

Proof It suffices to prove that weak convergence implies strong convergence. Assume that $f_n \rightharpoonup f$ weakly in L^1. Since f_n and f are integrable, we have $f_n(x) = 0 = f(x)$ for any $x \in X$ with $\mu(\{x\}) = \infty$. For all other $x \in X$, i.e. $x \in X$ with $0 < \mu(\{x\}) < \infty$, it follows from weak convergence that

$$\mu(\{x\}) f_n(x) = \int_{\{x\}} f_n(y) \, \mu(dy) \xrightarrow[n\to\infty]{} \int_{\{x\}} f(y) \, \mu(dy) = \mu(\{x\}) f(x),$$

and so $f_n(x) \to f(x)$ for all $x \in X$. In particular, $f_n \to f$ in probability. Together with the weak convergence $f_n \rightharpoonup f$ in L^1 this gives $\|f_n - f\|_{L^1} \to 0$, cf. [51, Theorem IV.8.12]. □

Comment The assumption $\mu(\{x\}) > 0$, $x \in X$, can be replaced by the weaker assumption that any set $A \in \mathscr{A}$ with $0 < \mu(A) < \infty$ can be written as a union of atoms, cf. [205, Theorem 52.4, p. 376].

11.17 Convergence a.e. is not metrizable

On every metric space (X, d) the so-called **subsequence principle** holds: a sequence $(x_n)_{n\in\mathbb{N}} \subseteq X$ converges to $x \in X$ (w.r.t. d) if, and only if, any subsequence of $(x_n)_{n\in\mathbb{N}}$ has a further subsequence converging to x (w.r.t. d).

Let $(X, \mathscr{A}, \mu)$ be a measure space. Assume that μ-almost everywhere convergence is metrizable, i.e. that there is a metric d on X such that $d(f_n, f) \to 0$ if, and only if, $f_n \to f$ μ-almost everywhere. Take a sequence $(f_n)_{n\in\mathbb{N}}$ such that f_n converges in measure (or in probability) to f. Clearly, any subsequence $(f'_n)_{n\in\mathbb{N}}$ of $(f_n)_{n\in\mathbb{N}}$ also converges in measure (in probability) to f, and therefore it follows from the Borel–Cantelli lemma [☞Theorem 11A] that we can extract a subsequence $(f''_n)_{n\in\mathbb{N}}$ of $(f'_n)_{n\in\mathbb{N}}$ which converges almost everywhere to f; thus, $d(f''_n, f) \to 0$. Applying the subsequence principle for metric spaces we conclude that $d(f_n, f) \to 0$, i.e. $f_n \to f$ almost everywhere.

Thus, if almost everywhere convergence is metrizable, then convergence in measure implies almost everywhere convergence. On many measure spaces this implication is false [☞ 11.1], and therefore almost everywhere convergence is, in general, not metrizable.

Comment This result is due to Fréchet [62]. In fact, a.e. convergence is metrizable if, and only if, it coincides with convergence in probability. A sufficient criterion is that μ is of the form $\mu = \sum_{i=1}^{\infty} m_i \delta_{a_i}$, a necessary and sufficient criterion is Marczewski's theorem [☞ comment to 11.15].

Both convergence in measure and convergence in probability are metrizable. A possible choice is the following metric (again due to Fréchet [62]) for functions $f, g : X \longmapsto \mathbb{R}$:

$$d(f,g) := \inf\{\epsilon + \eta\,;\ \mu(\{|f-g| \geqslant \eta\}) \leqslant \epsilon\} \in [0,\infty]$$

($\inf \emptyset = \infty$) and its simplification due to Ky Fan [58]:

$$\tilde{d}(f,g) := \inf\{\epsilon\,;\ \mu(\{|f-g| \geqslant \epsilon\}) \leqslant \epsilon\} \in [0,\infty].$$

These metrics are usually considered only for convergence in probability; we follow the exposition of Malliavin [110, Section 5.2] who works in general measure spaces. The values $d(f,g)$ and $\tilde{d}(f,g)$ remain unchanged if we modify f and g on sets of measure zero. Thus, d and $\tilde{d}$ are complete metrics on the space $L^0(\mu)$ (of equivalence classes modulo μ null sets) of all measurable functions; note, however, that $L^0(\mu)$ is not described by $f \in L^0(\mu) \iff d(f,0) < \infty$. Consider, for example $f(x) = x$ on $((0,\infty), \mathscr{B}(0,\infty), \lambda)$: we have $d(f,0) = \infty$; worse, it may happen that $d(f_n, f) \to 0$ while $d(f,0) = d(f_n, 0) = \infty$ [☞ 11.14]. In **finite** measure spaces the distance functions

$$(f,g) \longmapsto \int \min\{|f-g|, 1\}\, d\mu \quad \text{or} \quad (f,g) \longmapsto \int \frac{|f-g|}{1+|f-g|}\, d\mu$$

are further, equivalent metrics for convergence in probability.

Non-metrizability of a.e. convergence does not come as a surprise. Pointwise convergence of functions $f_n : X \to \mathbb{R}$ corresponds to the canonical product topology on the infinite product $\mathbb{R}^X$. From topology it is known that infinite products of the form Y^X, where Y has at least two points, are metrizable if, and only if, X is at most countable, cf. [53, Corollary 4.2.4] or [200, Theorem 22.3]. If μ is such that $X \setminus N$ is uncountable for every μ null set N, then convergence μ a.e. should not be metrizable.

12

Convergence Theorems

Let $(X, \mathscr{A}, \mu)$ be a measure space and $(f_n)_{n\in\mathbb{N}}$ a sequence of real-valued measurable functions. Interchanging limit and integral,

$$\lim_{n\to\infty} \int_X f_n \, d\mu = \int_X \left(\lim_{n\to\infty} f_n \right) d\mu, \tag{12.1}$$

is possible in each of the following cases:

- $(f_n)_{n\in\mathbb{N}}$ is an increasing sequence of positive functions (Beppo Levi [☞ Theorem 1.13]);
- $(f_n)_{n\in\mathbb{N}} \subseteq L^1(\mu)$ is an increasing sequence with $\sup_{n\in\mathbb{N}} \int f_n \, d\mu < \infty$ (monotone convergence [☞ Theorem 1.24]);
- $(f_n)_{n\in\mathbb{N}} \subseteq L^1(\mu)$ converges μ-almost everywhere and there is some function $g \in L^1(\mu)$ such that $|f_n| \leqslant g$ for all $n \in \mathbb{N}$ (dominated convergence [☞ Theorem 1.26]).

None of the above conditions is necessary for (12.1), e.g. it may be possible to interchange limit and integration even if $(f_n)_{n\in\mathbb{N}}$ does not have an integrable envelope [☞ 12.11]. **Fatou's lemma** [☞ Theorem 1.25] states that the inequality

$$\int_X \liminf_{n\to\infty} f_n \, d\mu \leqslant \liminf_{n\to\infty} \int_X f_n \, d\mu$$

holds for any sequence of measurable functions $f_n \geqslant 0$. In general, the inequality is strict [☞ 12.3] and fails to hold for non-positive integrands [☞ 12.2].

Another important result on the convergence of integrals is **Vitali's convergence theorem** [☞ 1.31]: if $(f_n)_{n\in\mathbb{N}}$ is a sequence of functions such that $f_n \to f$ in measure, then $(f_n)_{n\in\mathbb{N}}$ converges in $L^1(\mu)$ if, and only if, $(f_n)_{n\in\mathbb{N}}$ is **uniformly integrable** [☞ Definition 1.31]. This equivalence breaks down if we drop the assumption that $(f_n)_{n\in\mathbb{N}}$ converges in measure [☞ 12.15]. It follows from Vitali's convergence theorem that the identity

$$\lim_{n\to\infty}\int_X f_n\,d\mu = \int_X f\,d\mu = \int_X \left(\lim_{n\to\infty} f_n\right)d\mu$$

holds for any uniformly integrable sequence $(f_n)_{n\in\mathbb{N}}$ such that $f_n \to f$ in measure. If $(f_n)_{n\in\mathbb{N}} \subseteq L^1(\mu)$ is a sequence of functions bounded by an integrable function, $|f_n| \leqslant g \in L^1(\mu)$, then $(f_n)_{n\in\mathbb{N}}$ is uniformly integrable and $f_n \to f$ a.e. implies $f_n \to f$ in measure [☞ Tab. 11.3]. This shows that Vitali's convergence theorem is a generalization of the dominated convergence theorem; [☞ 12.11, 12.12] describe situations where Vitali's convergence theorem applies but the dominated convergence theorem is not applicable.

12.1 Classical counterexamples to dominated convergence

In none of the following cases does one have $\lim_{n\to\infty}\int f_n(x)\,dx = \int \lim_{n\to\infty} f_n(x)\,dx$:

(a) $f_n(x) = n(n+1)\mathbb{1}_{(1/(n+1),1/n)}(x)$ satisfies

$$f_n \geqslant 0,\quad \int_0^1 f_n(x)\,dx = 1,\quad \lim_{n\to\infty} f_n(x) = 0,\quad \lim_{n\to\infty}\|f_n\|_\infty = \infty;$$

(b) $f_n(x) = \frac{1}{2}n\left(\mathbb{1}_{(1/2-1/n,1/2)}(x) + \mathbb{1}_{(1/2,1/2+1/n)}(x)\right)$ satisfies

$$f_n \geqslant 0,\quad \int_0^1 f_n(x)\,dx = 1,\quad \lim_{n\to\infty} f_n(x) = 0,\quad \lim_{n\to\infty}\|f_n\|_\infty = \infty;$$

(c) $f_n(x) = n^2\mathbb{1}_{(0,1/n)}(x)$ satisfies

$$f_n \geqslant 0,\quad \int_0^1 f_n(x)\,dx = n,\quad \lim_{n\to\infty} f_n(x) = 0,\quad \lim_{n\to\infty}\|f_n\|_\infty = \infty;$$

(d) $f_n(x) = n^{-1}\mathbb{1}_{(0,n^2)}(x)$ satisfies

$$f_n \geqslant 0,\quad \int_{(0,\infty)} f_n(x)\,dx = n,\quad \lim_{n\to\infty} f_n(x) = 0,\quad \lim_{n\to\infty}\|f_n\|_\infty = 0.$$

12.2 Fatou's lemma may fail for non-positive integrands

Consider Lebesgue measure λ on $((0,\infty), \mathscr{B}(0,\infty))$ and define

$$f_n(x) = e^{-2x} - ne^{-nx},\quad x \in (0,\infty).$$

We have

$$\int_0^\infty \left(e^{-2x} - ne^{-nx}\right) dx = \left[-\tfrac{1}{2}e^{-2x}\right]_0^\infty - \left[-e^{-nx}\right]_0^\infty = \tfrac{1}{2} - 1 = -\tfrac{1}{2},$$

whereas

$$\liminf_{n\to\infty} f_n(x) = e^{-2x} \quad \text{and} \quad \int_0^\infty e^{-2x}\,dx = \tfrac{1}{2}.$$

This shows that $\liminf_{n\to\infty} \int_0^\infty f_n(x)\,dx = -\frac{1}{2} < \frac{1}{2} = \int_0^\infty \liminf_{n\to\infty} f_n(x)\,dx$.

Comment The problem is that the sequence $f_n(x)$ does not admit an integrable minorant which would be needed to make Fatou's lemma work. In fact, the largest pointwise minorant would be $\inf_{n\in\mathbb{N}} f_n(x) = e^{-2x} - \sup_{n\in\mathbb{N}} ne^{-nx}$, but $\sup_{n\in\mathbb{N}} ne^{-nx}$ is not integrable. Using the monotonicity in n, we get for all $x > 0$

$$\sup_{n\in\mathbb{N}} ne^{-(n-1)x} \geqslant \sup_{n\in\mathbb{N}} \sup_{\lambda\in(n-1,n]} \lambda e^{-\lambda x} = \left(\sup_{\lambda>0} \lambda x e^{-\lambda x}\right)\frac{1}{x} = \frac{1}{ex};$$

hence $\sup_{n\in\mathbb{N}} ne^{-nx} \geqslant (ex)^{-1}e^{-x}$, and this function is not integrable near the origin.

12.3 Fatou's lemma may lead to a strict inequality

Consider Lebesgue measure λ on $(\mathbb{R}, \mathscr{B}(\mathbb{R}))$ and define $f_n(x) := \mathbb{1}_{[n,n+1]}(x)$. Clearly, $\liminf_{n\to\infty} f_n(x) = 0$, $\int f_n(x)\,dx = \int_{[n,n+1]} dx = 1$, and

$$0 = \int \liminf_{n\to\infty} f_n(x)\,dx < \liminf_{n\to\infty} \int f_n(x)\,dx = 1.$$

The same phenomenon appears for finite measures, consider e.g. $X = [0,2]$ with Lebesgue measure and the sequence

$$f_{2n}(x) = \mathbb{1}_{[0,1]}(x) \quad \text{and} \quad f_{2n+1}(x) := \mathbb{1}_{[1,2]}(x).$$

12.4 The monotone convergence theorem needs a lower integrable bound

Consider Lebesgue measure λ on $((0,1), \mathscr{B}(0,1))$ and define

$$f(x) = \frac{1}{x} \quad \text{and} \quad f_n(x) := -\frac{1}{n}f(x).$$

Clearly, $\lim_{n\to\infty} f_n(x) = 0$ for every x, $f_1 \leqslant f_2 \leqslant f_3 \leqslant \ldots$ but $\int f_n(x)\,\lambda(dx) = -\infty$ for all $n \in \mathbb{N}$. Thus,

$$\lim_{n\to\infty} \int_{(0,1)} f_n(x)\,\lambda(dx) = -\infty \neq 0 = \int_{(0,1)} \lim_{n\to\infty} f_n(x)\,\lambda(dx).$$

12.5 A series of functions such that integration and summation do not interchange

Consider on the measure space $((0,\infty), \mathscr{B}(0,\infty), \lambda)$ the sequence of functions

$$f_n(x) := e^{-nx} - 2e^{-2nx}, \quad n \in \mathbb{N},\ x > 0.$$

We have

$$\sum_{n=1}^{\infty} f_n(x) = \frac{e^{-x}}{1-e^{-x}} - \frac{2e^{-2x}}{1-e^{-2x}} = \frac{1}{e^x - 1} - \frac{2}{e^{2x}-1} = \frac{1}{e^x+1}.$$

Therefore,

$$\int_0^\infty \left[\sum_{n=1}^{\infty} f_n(x)\right] dx = \int_0^\infty \frac{dx}{e^x+1} = \log \frac{e^x}{e^x+1}\bigg|_0^\infty = \log 2,$$

while

$$\int_0^\infty f_n(x)\,dx = \frac{1}{n} - \frac{2}{2n} = 0 \quad \text{and} \quad \sum_{n=1}^{\infty} \int_0^\infty f_n(x)\,dx = 0.$$

Comment If the functions $f_n(x)$ are positive, one can use either the Beppo Levi theorem or Tonelli's theorem to conclude that

$$\sum_{n=1}^{\infty} \int_0^\infty f_n(x)\,dx = \int_0^\infty \sum_{n=1}^{\infty} f_n(x)\,dx.$$

In the above example, f_n is not positive. As one would expect, we have that

$$\int_0^\infty |f_n(x)|\,dx = \frac{1}{n}\int_0^1 |1-2y|\,dy = \frac{1}{2n}$$

implies

$$\sum_{n=1}^{\infty} \int_0^\infty |f_n(x)|\,dx = \sum_{n=1}^{\infty} \frac{1}{2n} = \infty.$$

12.6 Riesz's convergence theorem fails for $p = \infty$

Riesz's convergence theorem asserts that on every measure space $(X, \mathscr{A}, \mu)$ one has

$$f_n \xrightarrow[n\to\infty]{\text{a.e.}} f \text{ and } \lim_{n\to\infty} \|f_n\|_{L^p} = \|f\|_{L^p} < \infty \Rightarrow f_n \xrightarrow[n\to\infty]{L^p} f$$

for all $p \in [1, \infty)$, cf. [MIMS, p. 123]. This is no longer true if $p = \infty$. Consider the measure space $([0,1], \mathscr{B}[0,1], \lambda + \delta_1)$ and the sequence $f_n(x) := x^n$. Obviously, $\lim_{n\to\infty} f_n(x) = \mathbb{1}_{\{1\}}(x)$ and $\lim_{n\to\infty} \|f_n\|_{L^\infty} = 1 = \|\mathbb{1}_{\{1\}}\|_{L^\infty}$ whereas the convergence $f_n \to f$ cannot be uniform.

12.7 A sequence such that $f_n \to 0$ pointwise but $\int_I f_n \, d\lambda \to \lambda(I)$ for all intervals

Consider Lebesgue measure λ on $([0,1], \mathscr{B}[0,1])$ and set

$$f_n(x) := n^2 \sum_{k=0}^{n-1} \mathbb{1}_{\left[\frac{k}{n}, \frac{k}{n}+\frac{1}{n^3}\right)}(x), \quad x \in [0,1], \; n \in \mathbb{N}.$$

Since

$$\sum_{n\in\mathbb{N}} \lambda(\{|f_n| > 0\}) \leqslant \sum_{n\in\mathbb{N}} n\frac{1}{n^3} < \infty,$$

it follows from the Borel–Cantelli lemma [☞ Theorem 11A] that $f_n(x) \to 0$ for Lebesgue almost all $x \in [0,1]$. In order to prove that $\int_I f_n \, d\lambda \to \lambda(I)$ for all intervals I, we first show that the sequence of measures $\mu_n := f_n \lambda$ converges weakly to λ, i.e.

$$\forall g \in C[0,1] \; : \; \lim_{n\to\infty} \int_{[0,1]} g(x) \, \mu_n(dx) = \int_{[0,1]} g(x) \, \lambda(dx). \tag{12.2}$$

Pick $g \in C[0,1]$ and $\epsilon > 0$. Since g is uniformly continuous on $[0,1]$ there is some $N \in \mathbb{N}$ such that

$$|x - y| \leqslant \frac{1}{N} \Rightarrow |g(x) - g(y)| \leqslant \epsilon.$$

As

$$\int_{[0,1]} g(x) \, \mu_n(dx) - \frac{1}{n} \sum_{k=0}^{n-1} g(k/n) = n^2 \sum_{k=0}^{n-1} \int_{k/n}^{k/n+1/n^3} (g(x) - g(k/n)) \, \lambda(dx)$$

this implies

$$\left|\int_{[0,1]} g(x)\,\mu_n(dx) - \frac{1}{n}\sum_{k=0}^{n-1} g(k/n)\right| \leqslant \epsilon n^2 \sum_{k=0}^{n-1} \frac{1}{n^3} = \epsilon$$

for all $n \geqslant N$. The function $g \in C[0,1]$ is Riemann integrable and its Riemann integral coincides with the Lebesgue integral [☞ Theorem 1.28]. Consequently, the Riemann sums $n^{-1}\sum_{k=0}^{n-1} g(k/n)$ converge to $\int_{[0,1]} g(x)\,\lambda(dx)$ as $n \to \infty$. Hence,

$$\lim_{n\to\infty}\int_{[0,1]} g(x)\,\mu_n(dx) = \lim_{n\to\infty}\frac{1}{n}\sum_{k=0}^{n-1} g(k/n) = \int_{[0,1]} g(x)\,\lambda(dx),$$

which proves (12.2). By the portmanteau theorem [☞ Theorems 19A, 19C] it now follows that $\mu_n(a,b) \to \lambda(a,b)$ for any $0 \leqslant a < b \leqslant 1$, i.e.

$$\int_{(a,b)} f_n(x)\,\lambda(dx) \xrightarrow[n\to\infty]{} \lambda(a,b) = b - a.$$

12.8 $\int_I f_n\,d\lambda \to \int_I f\,d\lambda$ for all intervals I does not imply $\int_B f_n\,d\lambda \to \int_B f\,d\lambda$ for all Borel sets B

Let C be the geometric fat Cantor set [☞ Example 7.3] with $a_n = \frac{1}{4}\frac{2^n}{4^n}$, that is $C = \bigcap_{n\in\mathbb{N}_0} C_n$ where C_n is obtained by removing from the middle of each of the 2^{n-1} parts of C_{n-1} an open interval of length $2^{-(n-1)}a_{n-1} = 4^{-n}$. By construction, $[0,1]\setminus C_n$ consists of $\sum_{i=1}^{n} 2^{i-1}$ intervals; if we subdivide these intervals further, we see that $[0,1]\setminus C_n$ contains $s_n := \sum_{i=0}^{n-1} 4^{n-1-i}2^i$ disjoint open intervals of length 4^{-n}, i.e. there exist $0 \leqslant x_1^{(n)} < \cdots < x_{s_n}^{(n)} \leqslant 1$ such that

$$[0,1]\setminus C_n \supseteq \biguplus_{k=1}^{s_n}\left(x_k^{(n)}, x_k^{(n)} + 4^{-n}\right).$$

As $[0,1]\setminus C$ is dense in $[0,1]$, it follows that $\Delta_n := \max_{k\leqslant s_n}|x_k^{(n)} - x_{k-1}^{(n)}| \to 0$ as $n \to \infty$. Define a sequence of functions

$$f_n(x) := 4^n \sum_{k=1}^{s_n-1} (x_{k+1}^{(n)} - x_k^{(n)})\mathbb{1}_{\left(x_k^{(n)}, x_k^{(n)}+4^{-n}\right)}(x), \quad x \in [0,1].$$

We claim that the sequence of measures $\mu_n := f_n\lambda$, $n \in \mathbb{N}$, on $([0,1], \mathscr{B}[0,1])$ converges weakly to Lebesgue measure λ, i.e.

$$\forall g \in C[0,1] : \lim_{n\to\infty}\int_{[0,1]} g\,d\mu_n = \int_{[0,1]} g\,d\lambda. \tag{12.3}$$

Fix $g \in C[0,1]$ and $\epsilon > 0$. Since g is uniformly continuous on $[0,1]$ there is some $N \in \mathbb{N}$ such that

$$|x-y| \leqslant 4^{-N} \implies |g(x)-g(y)| \leqslant \epsilon.$$

As

$$\int_{[0,1]} g \, d\mu_n - \sum_{k=1}^{s_n-1} g(x_k^{(n)}) \left(x_{k+1}^{(n)} - x_k^{(n)}\right)$$
$$= 4^n \sum_{k=1}^{s_n-1} \left(x_{k+1}^{(n)} - x_k^{(n)}\right) \int_{x_k^{(n)}}^{x_k^{(n)}+4^{-n}} \left(g(x) - g(x_k^{(n)})\right) \lambda(dx)$$

it follows that

$$\left| \int_{[0,1]} g \, d\mu_n - \sum_{k=1}^{s_n-1} g(x_k^{(n)}) \right| \leqslant 4^n \sum_{k=1}^{s_n-1} \epsilon 4^{-n} \left(x_{k+1}^{(n)} - x_k^{(n)}\right) \leqslant \epsilon$$

for all $n \geqslant N$. The continuous function $g \in C[0,1]$ is Riemann integrable and its Riemann integral coincides with the Lebesgue integral [☞Theorem 1.28]. If we consider the partition $\Pi_n = \{x_1^{(n)} < \cdots < x_{s_n}^{(n)}\}$ of $[0,1]$, then its mesh size $\Delta_n = \max_{k \leqslant s_n} |x_k^{(n)} - x_{k-1}^{(n)}|$ tends to zero and the left (resp. right) end point $x_1^{(n)}$ (resp. $x_{s_n}^{(n)}$) converges to 0 (resp. 1) as $n \to \infty$. Consequently, the Riemann sums $\sum_{k=1}^{s_n-1} g(x_k^{(n)})(x_{k+1}^{(n)} - x_k^{(n)})$ converge to $\int_{[0,1]} g(x) \, \lambda(dx)$ as $n \to \infty$. Hence,

$$\lim_{n\to\infty} \int_{[0,1]} g(x) \, \mu_n(dx) = \lim_{n\to\infty} \sum_{k=1}^{s_n-1} g(x_k^{(n)}) \left(x_{k+1}^{(n)} - x_k^{(n)}\right) = \int_{[0,1]} g(x) \, \lambda(dx).$$

This finishes the proof of (12.3). An application of the portmanteau theorem [☞Theorems 19A, 19C] now yields $\mu_n(a,b) \to \lambda(a,b)$ for all $a < b$, i.e.

$$\forall a < b \; : \; \lim_{n\to\infty} \int_{(a,b)} f_n \, d\lambda = \int_{(a,b)} 1 \, d\lambda.$$

On the other hand, we have by the very construction of f_n that $f_n(x) = 0$ for all $x \in C$. Since the fat Cantor set C has measure $\lambda(C) = 1 - \sum_{n \in \mathbb{N}_0} a_n = \frac{1}{2}$, it follows that

$$\lim_{n\to\infty} \int_C f_n \, d\lambda = 0 \neq \frac{1}{2} = \int_C 1 \, d\lambda.$$

Comment By the portmanteau theorem [☞Theorem 19C], weak convergence $\mu_n = f_n \lambda \to \lambda$ on $([0,1], \mathscr{B}[0,1])$ implies that

$$\lim_{n\to\infty} \int_B f_n \, d\lambda = \lambda(B)$$

for all $B \in \mathscr{B}[0,1]$ with $\lambda(\partial B) = 0$. The above example illustrates that the convergence, in general, fails to hold for Borel sets B whose boundary has strictly positive Lebesgue measure. Indeed, the fat Cantor set C is closed and has empty interior, and so $\partial C = C$, which implies $\lambda(\partial C) = \lambda(C) = \frac{1}{2} > 0$.

12.9 The classical convergence theorems fail for nets

Consider the ordinal space $\Omega = [0, \omega_1]$ [☞Section 2.4] with the co-countable σ-algebra $\mathscr{A}$ [☞Example 4A] and the measure

$$\mu(A) := \begin{cases} 0, & \text{if } A \text{ is at most countable,} \\ 1, & \text{if } A^c \text{ is at most countable} \end{cases}$$

[☞Example 5A], and define $f_\alpha := \mathbb{1}_{(0,\alpha)}$ for $\alpha < \omega_1$. The net $(f_\alpha)_{\alpha<\omega_1}$ is dominated by the integrable function $\mathbb{1}_{[0,\omega_1]}$ and converges pointwise to $f := \mathbb{1}_{(0,\omega_1)}$, i.e. $f_\alpha(\beta)$ converges to $f(\beta)$ for each $\beta \in \Omega$. Indeed, if $0 < \beta < \omega_1$, then $f_\alpha(\beta) = 1 = f(\beta)$ for each $\alpha > \beta$; for $\beta = \omega_1$ and $\beta = 0$ the convergence follows from $f_\alpha(\beta) = 0 = f(\beta)$. Since ω_1 is the first uncountable ordinal, we have $\int_\Omega f_\alpha \, d\mu = 0$ for all $\alpha < \omega_1$ and $\int_\Omega f \, d\mu = 1$. In particular, $\int_\Omega f_\alpha \, d\mu$ does not converge to $\int_\Omega f \, d\mu$, i.e. the dominated convergence theorem does not apply.

This example also shows that both the monotone convergence theorem and Fatou's lemma fail for nets.

Comment Let X be a topological space and $\mathscr{A}$ a σ-algebra on X. A measure μ on $(X, \mathscr{A})$ is called **τ-additive** (or **τ-continuous**), if for every increasing net of open sets $(U_i)_{i\in I} \subseteq \mathscr{A}$ with $\bigcup_{i\in I} U_i \in \mathscr{A}$ the equality

$$\mu\Big(\bigcup_{i\in I} U_i\Big) = \sup_{i\in I} \mu(U_i)$$

holds. Every inner compact regular Borel measure (a so-called **Radon measure**) and every Borel measure on a separable metric space is τ-additive [20, Proposition 7.2.2] or [43, III.49, p. 108]; in particular, d-dimensional Lebesgue measure is τ-additive. For τ-additive measures there is a variant of the monotone convergence theorem, which holds for nets of functions, cf. [20, Lemma 7.2.6]. Our example shows that the measure μ on the ordinal space is not τ-additive.

12.10 The continuity lemma 'only' proves sequential continuity

Let $(X, \mathscr{A}, \mu)$ be a measure space and $I = [a, b] \subseteq \mathbb{R}$ a parameter set which we consider as a metric space with the Euclidean metric. If $f : I \times X \to \mathbb{R}$ is a function such that $f(t, \cdot)$ is measurable for each $t \in I$, $f(\cdot, x)$ is continuous for each $x \in X$ and $|f(t, x)| \leqslant g(x) \in L^1(\mu)$, then

$$F : I \longmapsto \mathbb{R},\ t \longmapsto F(t) := \int_X f(t, x)\, \mu(dx)$$

is continuous [☞ Example 1.27]. This result is known as the **continuity lemma** for parameter-dependent integrals. The following example shows that the assertion breaks down if we replace I by some general (topological) space.

Let λ be Lebesgue measure on $([0, 1], \mathscr{B}[0, 1])$ and $\Omega = [0, 1]^{[0,1]}$ be the family of all functions $\omega : [0, 1] \to [0, 1]$ equipped with the canonical product topology [☞ Example 2.1.(i)] and the corresponding Borel σ-algebra $\mathscr{B}(\Omega)$. The map $f : [0, 1] \times \Omega \to [0, 1]$, $f(t, \omega) := \omega(t)$ is obviously continuous in each variable, but the parameter-dependent integral

$$F : \Omega \to [0, 1], \quad F(\omega) := \int_0^1 f(t, \omega)\, dt = \int_0^1 \omega(t)\, dt$$

is discontinuous. Indeed, for any choice of $t_1, \dots, t_n \in [0, 1]$ we can find some $\omega \in \Omega$ such that $\omega(t_i) = 0$ and $F(\omega) > \frac{1}{2}$, and so there is no open neighbourhood U of $\omega_0 = 0$ such that $|F(\omega) - F(\omega_0)| < \frac{1}{2}$ for all $\omega \in U$, i.e. F is not continuous at $\omega_0 = 0$. The dominated convergence theorem shows that $F(\cdot)$ is sequentially continuous, i.e. $\lim_{n\to\infty} \omega_n = \omega$ implies that $\lim_{n\to\infty} F(\omega_n) = F(\omega)$.

Comment 1 This example highlights the fact that Ω is not metrizable under pointwise convergence. This means, in particular, that the notions of continuity and sequential continuity do not coincide. A related example where one has continuity along sequences but not along nets arises in connection with the Fourier transform [☞ 19.12].

Comment 2 There are generalizations of the continuity lemma for general spaces satisfying certain topological assumptions; see e.g. Nussbaum [130]. In contrast to our example for $I = [a, b]$, these generalized versions frequently require separate continuity in both variables, i.e. both $f(t, \cdot)$ and $f(\cdot, x)$ need to be continuous. Under the even stronger assumption that $f : I \times X \to \mathbb{R}$ is jointly continuous and bounded, it can be shown that $t \longmapsto \int_X f(t, x)\, \mu(dx)$ is continuous whenever I is some Hausdorff space and μ is a Borel measure on a separable metric space X, cf. [20, Proposition 7.14.8].

12.11 A sequence f_n converging to 0 in L^1 without integrable envelope – the 'sliding hump'

For a positive continuous $f : [0,1] \to (0,\infty)$ define

$$f_n(x) := \frac{1}{n} f(x-n)\mathbb{1}_{[n,n+1]}(x), \quad x \in \mathbb{R}.$$

Obviously, $\lim_{n\to\infty} f_n(x) = 0$ uniformly for all $x \in \mathbb{R}$ and

$$\int_{\mathbb{R}} |f_n(x) - 0|\, dx = \frac{1}{n}\int_0^1 f(x)\, dx \xrightarrow[n\to\infty]{} 0.$$

The smallest majorant is $g(x) := \sup_{n\in\mathbb{N}} f_n(x)$ and its integral is

$$\int_{\mathbb{R}} g(x)\, dx = \sum_{n=1}^{\infty} \frac{1}{n}\int_0^1 f(x)\, dx = \infty.$$

12.12 A sequence $(f_n)_{n\in\mathbb{N}}$ which is uniformly integrable but $\sup_n |f_n|$ is not integrable

Consider on $(\mathbb{R}, \mathscr{B}(\mathbb{R}))$ Lebesgue measure λ and set

$$f_n(x) := \mathbb{1}_{(n,n+1/n)}(x), \quad n \in \mathbb{N},\ x \in \mathbb{R}.$$

Since the intervals $(n, n+1/n)$ are disjoint, we have $\sup_{n\in\mathbb{N}} f_n = \sum_{n\in\mathbb{N}} f_n$, and so

$$\int_{\mathbb{R}} \sup_{n\in\mathbb{N}} f_n(x)\,\lambda(dx) = \sum_{n\in\mathbb{N}} \int_{\mathbb{R}} f_n(x)\,\lambda(dx) = \sum_{n\in\mathbb{N}} \frac{1}{n} = \infty.$$

Yet $(f_n)_{n\in\mathbb{N}}$ is uniformly integrable: to see this, define for fixed $N \in \mathbb{N}$ the function $g(x) := \mathbb{1}_{[0,N]}(x)$. Then

$$\forall n < N : \int_{|f_n|>g} |f_n|\, d\lambda = 0$$

and

$$\forall n \geqslant N : \int_{|f_n|>g} |f_n|\, d\lambda \leqslant \int |f_n|\, d\lambda = \frac{1}{n} \leqslant \frac{1}{N},$$

i.e.

$$\sup_{n\in\mathbb{N}} \int_{|f_n|>g} |f_n|\, d\lambda \leqslant \frac{1}{N}.$$

As g is integrable, this proves that $(f_n)_{n\in\mathbb{N}}$ is uniformly integrable.

12.13 A sequence which is not uniformly integrable but $f_n \to 0$ and $\int f_n \, d\lambda \to 0$

Let λ be Lebesgue measure on $[0,1]$ and consider $f_n(x) := n\mathbb{1}_{(0,1/n)} - n\mathbb{1}_{(1/n,2/n)}$. Obviously, $\lim_{n\to\infty} f_n(x) = 0$ and $\lim_{n\to\infty} \int_{[0,1]} f_n \, d\lambda = 0$, but $(f_n)_{n\in\mathbb{N}}$ is not uniformly integrable:

$$\forall n > R : \int_{|f_n|>R} |f_n| \, d\lambda = \int_{[0,2/n]} n \, d\lambda = 2.$$

Comment 1 It is essential in this example that f_n takes both positive and negative values. If $f_n \geqslant 0$, $\int f_n \, d\lambda \to 0$ and $f_n \to 0$ a.e., then $f_n \to 0$ in L^1 and, by Vitali's convergence theorem [☞ Theorem 1.31], the sequence $(f_n)_{n\in\mathbb{N}}$ must be uniformly integrable.

Comment 2 The sequence $(f_n)_{n\in\mathbb{N}}$ is **tight**, i.e.

$$\lim_{R\to\infty} \sup_{n\in\mathbb{N}} \lambda(\{|f_n| \geqslant R\}) = 0.$$

Consequently, the above example shows that tightness plus convergence to zero in L^1 and a.e. does not imply uniform integrability.

12.14 An L^1-bounded sequence which is not uniformly integrable

[☞ 12.13].

12.15 A uniformly integrable sequence which does not converge in L^1

Let $(X, \mathscr{A}, \mu)$ be a σ-finite measure space, and let $(f_n)_{n\in\mathbb{N}}$ be a sequence converging in probability to a function f. Vitali's convergence theorem [☞ 1.31] states that $(f_n)_{n\in\mathbb{N}}$ is uniformly integrable if, and only if, $f_n \to f$ in L^1. The following example shows that this equivalence breaks down if we drop the assumption that $(f_n)_{n\in\mathbb{N}}$ converges in probability.

Pick an integrable function $g \geqslant 0$ such that $\int g \, d\mu > 0$, and set

$$f_n(x) := (-1)^n g(x), \quad n \in \mathbb{N}, \ x \in X.$$

Since g is integrable, the sequence $(f_n)_{n\in\mathbb{N}}$ is uniformly integrable. By construction,

$$\int |f_n - f_{n+1}| \, d\mu = 2 \int g \, d\mu > 0,$$

which shows that $(f_n)_{n\in\mathbb{N}}$ is not a Cauchy sequence in $L^1(\mu)$; hence, $(f_n)_{n\in\mathbb{N}}$

does not converge in $L^1(\mu)$. Note that $0 < \int g\,d\mu < \infty$ ensures that there is some set $A \in \mathscr{A}$ with $\mu(A) < \infty$ and

$$\mu(A \cap \{|f_n - f_{n+1}| > \epsilon\}) = \mu(A \cap \{2g > \epsilon\}) > 0,$$

i.e. f_n does not converge in probability.

12.16 An L^1-bounded sequence which fails to be uniformly integrable on any set of positive measure

Consider on $((0,1), \mathscr{B}(0,1))$ Lebesgue measure λ. Let $(q_n)_{n\in\mathbb{N}}$ be an enumeration of $\mathbb{Q} \cap (0,\infty)$ and define

$$f_{i,k}(x) := \frac{1}{q_i}\mathbb{1}_{(q_k-q_i,q_k+q_i)\cap(0,1)}(x), \quad x \in (0,1),\ i,k \in \mathbb{N}.$$

Since $\int_{(0,1)} |f_{i,k}|\,d\lambda \leqslant 2$ for all $i,k \in \mathbb{N}$, the sequence $(f_{i,k})_{i,k\in\mathbb{N}}$ is bounded in L^1. To prove that it fails to be uniformly integrable on any set of positive Lebesgue measure, fix $A \in \mathscr{B}(0,1)$ with $\lambda(A) > 0$. By Lebesgue's differentiation theorem [☞ 14C], we have

$$\mathbb{1}_A(x) = \lim_{r\downarrow 0} \frac{1}{2r} \int_{(x-r,x+r)} \mathbb{1}_A(y)\,\lambda(dy)$$

for Lebesgue almost every $x \in (0,1)$. As $\lambda(A) > 0$ this implies that there exist $x \in A$ and $N \in \mathbb{N}$ such that

$$\forall r \in (0, N^{-1}) : \frac{1}{2r}\lambda(A \cap (x-r, x+r)) \geqslant \frac{1}{2}.$$

Pick $n \geqslant N$ and choose sequences $(r_i)_{i\in\mathbb{N}} \subseteq \mathbb{Q} \cap (0,1)$ and $(x_k)_{k\in\mathbb{N}} \subseteq \mathbb{Q} \cap (0,1)$ with $r_i \uparrow r := 1/n$ and $x_k \to x$. Since $1/r_i \geqslant n$ we get

$$\begin{aligned}\sup_{i,k\in\mathbb{N}} \int_{\{|f_{i,k}|\geqslant n\}\cap A} |f_{i,k}|\,d\lambda &\geqslant \sup_{i,k\in\mathbb{N}} \int_A \frac{1}{r_i}\mathbb{1}_{(x_k-r_i,x_k+r_i)\cap(0,1)}\,d\lambda \\ &\geqslant \frac{1}{r}\lambda(A \cap (x-r,x+r)) \geqslant 1.\end{aligned}$$

Comment The example is due to Ball and Murat [8].

13

Continuity and a.e. Continuity

A function $f : \mathbb{R} \to \mathbb{R}$ is **(Lebesgue) almost everywhere** continuous if the set D_f of discontinuity points of f is a Lebesgue null set. Since D_f is a Borel set for any function $f : \mathbb{R} \to \mathbb{R}$ [☞ 8.9], its Lebesgue measure $\lambda(D_f)$ is well-defined – even if f is not measurable. In fact, there are a.e. continuous functions which are not Borel measurable [☞ 13.8]. This is a first indication that properties of a.e. continuous functions differ substantially from those of continuous functions. Further differences will become apparent during the course of this chapter, e.g. strict positivity on a dense subset does not imply positivity everywhere [☞ 13.12], the range of an a.e. continuous function may be countably infinite [☞ 13.7], and the composition of a.e. continuous functions may be nowhere continuous [☞ 13.10].

Beware that an a.e. continuous function need not be a.e. equal to a continuous function [☞ 13.1, 13.3]. Conversely, a function may be nowhere continuous although it is a.e. equal to a continuous function [☞ 13.2]. At the same time, **Lusin's theorem** shows that any Borel measurable function $f : \mathbb{R} \to \mathbb{R}$ coincides with some continuous function up to a set of arbitrarily small (but strictly positive) Lebesgue measure.

Theorem 13A (Lusin's theorem [MIMS, p. 195]) *Let X be a metric space. If μ is a regular measure on $(X, \mathscr{B}(X))$ and $f : X \to \mathbb{R}$ Borel measurable, then there exists for every $\epsilon > 0$ a continuous function $u : X \to \mathbb{R}$ such that $\mu(\{u \neq f\}) \leqslant \epsilon$.*

Let us emphasize that Lusin's theorem does **not** imply that f is continuous on $\{u = f\}$. Borel measurable functions may be nowhere continuous, e.g. Dirichlet's jump function $f(x) = \mathbb{1}_{\mathbb{Q}\cap[0,1]}(x)$, but this is not in contradiction to Theorem 13A.

13.1 An a.e. continuous function which does not coincide a.e. with any continuous function

Let λ be Lebesgue measure on $(\mathbb{R}, \mathscr{B}(\mathbb{R}))$ and define $f(x) = \mathbb{1}_{[0,\infty)}(x)$. Since f is discontinuous only at $x = 0$, it is obviously λ a.e. continuous. On the other hand, it cannot coincide a.e. with any continuous function.

Assume, to the contrary, that g is continuous and $f = g$ on a set $\mathbb{R} \setminus N$ such that $\lambda(N) = 0$. Since $\mathbb{R} \setminus N$ is dense in $\mathbb{R}$, we could find sequences $x_n < 0 < y_n$ converging to zero and such that

$$g(x_n) = f(x_n) = 0 < 1 = f(y_n) = g(y_n).$$

Letting $n \to \infty$ yields $g(0) = \lim_n g(x_n) = 0$ and $g(0) = \lim_n g(y_n) = 1$, which is impossible.

13.2 A nowhere continuous function which equals a.e. a continuous function

Dirichlet's jump function $f(x) := \mathbb{1}_{\mathbb{Q}\cap[0,1]}(x)$, $x \in [0, 1]$, is nowhere continuous but $f = 0$ a.e.

13.3 A function f such that every g with $f = g$ a.e. is nowhere continuous

Equip $(\mathbb{R}, \mathscr{B}(\mathbb{R}))$ with Lebesgue measure λ. Let $B \in \mathscr{B}(\mathbb{R})$ be a Borel set with $\lambda(B \cap I) > 0$ and $\lambda(B \cap I^c) > 0$ for all open intervals $I \neq \emptyset$ [☞ 7.15], and set

$$f(x) := \mathbb{1}_B(x), \quad x \in \mathbb{R}.$$

If $g : \mathbb{R} \to \mathbb{R}$ is a Borel measurable function such that $f = g$ Lebesgue almost everywhere, then

$$\lambda(\{g = 1\} \cap I) = \lambda(\{f = 1\} \cap I) = \lambda(B \cap I) > 0 \implies \{g = 1\} \cap I \neq \emptyset$$

and

$$\lambda(\{g = 0\} \cap I) = \lambda(\{f = 0\} \cap I) = \lambda(B^c \cap I) > 0 \implies \{g = 0\} \cap I \neq \emptyset$$

for all intervals I. Consequently, g takes in every interval I the values 0 and 1, and this means that g cannot have any continuity points.

Comment By Lusin's theorem [☞ 13A], there exist for every $\epsilon > 0$ a closed set F and a continuous function u such that $\lambda(F \cap \{f \neq u\}) \leqslant \epsilon$. Our example illustrates that this does not imply that f (or any of its modifications) has any continuity points.

13.4 A function which is everywhere sequentially continuous but nowhere continuous

[☞ 12.10]

13.5 An a.e. continuous function whose discontinuity points are dense

Thomae's function [☞ 3.8] has this property: it is continuous at all irrationals and discontinuous at all rationals.

Comment It is possible to construct a **bijective** function $f: \mathbb{R} \to \mathbb{R}$ which is continuous at all irrationals and discontinuous at all rationals, cf. [88].

13.6 An a.e. discontinuous function whose continuity points are dense

For $n \in \mathbb{N}$ let $C_n \subseteq [0,1]$ be a fat Cantor set [☞ 7.3] with Lebesgue measure $\lambda(C_n) = 1 - 2^{-n}$, and set

$$A := \bigcup_{n \in \mathbb{N}} C_n.$$

From $A \supseteq C_n$ it is immediate that $\lambda(A) = 1$. Since $[0,1] \setminus C_n$ is dense and open for each $n \in \mathbb{N}$, it follows from Baire's theorem [☞ Theorem 2.4] that

$$[0,1] \setminus A = \bigcap_{n \in \mathbb{N}} [0,1] \setminus C_n$$

is dense in $[0,1]$. We will show that

$$f(x) := \sum_{n \in \mathbb{N}} 2^{-n} \mathbb{1}_{C_n}(x), \quad x \in [0,1],$$

is continuous at every $x \in [0,1] \setminus A$ and discontinuous at every $x \in A$. Since $\lambda(A) = 1$ and $[0,1] \setminus A$ is dense, this means that f is an a.e. discontinuous function whose continuity points are dense.

Take $x \in A$, then $x \in C_n$ for some $n \in \mathbb{N}$, and so $f(x) \geqslant 2^{-n}$. As $[0,1] \setminus A$ is dense, there exists in any neighbourhood of x some $y \in [0,1] \setminus A$, i.e. $f(y) = 0$. Thus, $|f(x) - f(y)| \geqslant 2^{-n}$, which shows that f is discontinuous at x.

It remains to prove continuity at $x \in [0,1] \setminus A$. For fixed $\epsilon > 0$ choose $N \in \mathbb{N}$ such that $\sum_{n>N} 2^{-n} \leqslant \epsilon$. Since

$$x \in [0,1] \setminus A = \bigcap_{n \in \mathbb{N}} [0,1] \setminus C_n \subseteq \bigcap_{n=1}^{N} [0,1] \setminus C_n$$

and $\bigcap_{n=1}^{N}[0,1]\setminus C_n$ is open, being a finite intersection of open sets, there exists some $\delta > 0$ such that

$$(x-\delta, x+\delta) \subseteq \bigcap_{n=1}^{N}[0,1]\setminus C_n.$$

Consequently,

$$|f(y)-f(x)| = |f(y)| \leqslant \sum_{n>N} 2^{-n} \leqslant \epsilon$$

for all $y \in (x-\delta, x+\delta)$, i.e. f is continuous at x.

Comment This example is from [182]. Since the set of discontinuity points of any function $f\colon [0,1] \to \mathbb{R}$ is an F_σ-set, it is **not** possible to construct a function f which is discontinuous at all irrational x and continuous at all rational numbers [☞ 8.9, 8.10].

13.7 The composition of two a.e. continuous functions which is nowhere continuous

Define two functions $f, g\colon \mathbb{R} \to [0,1]$ as

$$g(x) = \begin{cases} 0, & \text{if } x \notin \mathbb{Q}, \\ 1, & \text{if } x = 0, \\ \dfrac{1}{|n|}, & \text{if } x = \dfrac{m}{n} \text{ and } \gcd(m,n) = 1 \end{cases}$$

and

$$f(x) = \begin{cases} 1, & \text{if } x = \dfrac{1}{n},\ n \in \mathbb{N}, \\ 0, & \text{for all other } x. \end{cases}$$

These functions are Lebesgue a.e. continuous, [☞ 3.8] for the proof of a.e. continuity of g. However, $f \circ g = \mathbb{1}_{\mathbb{Q}}$ is nowhere continuous.

Comment This pathology cannot happen if f is continuous and g is a.e. continuous: in this case $f \circ g$ is a.e. continuous. In order to see this, we write C^h, resp. D^h, for the continuity, resp. discontinuity, points of the function h. Since $C^f = \mathbb{R}$, it is obvious that $C^g \subseteq C^{f\circ g}$; thus, $D^g \supseteq D^{f\circ g}$ and the claim follows from the fact that D^g is a null set.

13.8 An a.e. continuous function which is not Borel measurable

[☞ 3.18]

13.9 A bounded Borel measurable function such that $f(x + 1/n) \to f(x)$ fails to hold on a set of positive measure

For each $n \in \mathbb{N}$ we split the interval $[0, 1)$ into 4^n half-open intervals,

$$\left[\frac{k}{4^n}, \frac{k+1}{4^n}\right), \quad k = 0, \dots, 4^n - 1,$$

and remove from each an interval of length $1/4^{2n}$,

$$I_{k,n} := \left[\frac{k}{4^n}, \frac{k+1}{4^n} - \frac{1}{4^{2n}}\right), \quad k = 0, \dots, 4^n - 1.$$

Set $K_n := \bigcup_{k=0}^{4^n-1} I_{k,n}$ and $K := \bigcap_{n=1}^{\infty} K_n$, and consider $f(x) := \mathbb{1}_K(x)$. We claim that $f(x + 1/n)$ fails to converge to $f(x)$ on a set of positive Lebesgue measure. Since

$$\lambda([0, 1) \setminus K_n) \leqslant 4^n \frac{1}{4^{2n}} = \frac{1}{4^n}$$

implies that

$$\lambda([0, 1) \setminus K) \leqslant \sum_{n \in \mathbb{N}} \frac{1}{4^n} < 1,$$

we have $\lambda(K) > 0$. Consequently, it suffices to show that there exist for every $x \in K$ infinitely many $n \in \mathbb{N}$ such that $x + n^{-1} \notin K$. Fix $x \in K$. By definition, $x \in K_n$ for all $n \in \mathbb{N}$, and so the intersection of $I := [x, x + 4^{-n})$ with K^c contains an interval of length $1/4^{2n}$. Since $x + m^{-1} \in I$ for all $m > 4^n$ and

$$\left|\frac{1}{m} - \frac{1}{m+1}\right| = \frac{1}{m(m+1)} < \frac{1}{4^{2n}}, \quad m > 4^n,$$

it follows that there is at least one number $m > 4^n$ such that $x + m^{-1} \in I \cap K^c$.

Comment An obvious modification of this construction shows that there exists for every $\epsilon \in (0, 1)$ a bounded Borel measurable function $f : [0, 1) \to \mathbb{R}$ such that

$$\lambda\left(\left\{x \in [0, 1);\ \lim_{n\to\infty} f(x + 1/n) \neq f(x)\right\}\right) \geqslant 1 - \epsilon.$$

13.10 A nowhere constant function which is a.e. continuous and has countable range

Let $f : I \to \mathbb{R}$ be a continuous function on an interval $I \neq \emptyset$. If f is not constant, then its range $f(I)$ has the cardinality of $\mathbb{R}$.

Indeed, if f takes at least two values, say y_0 and y_1, then the intermediate value theorem implies $[y_0, y_1] \subseteq f(I)$, and so $\#f(I) = \mathfrak{c}$.

The situation is substantially different for functions which are only almost everywhere continuous. Consider, for instance, $((0,1), \mathscr{B}(0,1))$ with Lebesgue measure λ, then $f = \mathbb{1}_{(0,1/2)}$ is almost everywhere continuous and $\#f((0,1)) = 2$. This example is, however, a bit artificial: though f is non-constant, it is constant almost everywhere. For a more interesting example, consider

$$f(x) := \begin{cases} 0, & \text{if } x \in (0,1) \setminus \mathbb{Q}, \\ q^{-1}, & \text{if } x = \frac{p}{q} \in \mathbb{Q} \cap (0,1), \gcd(p,q) = 1. \end{cases}$$

f is Lebesgue almost everywhere continuous [☞3.8], it is nowhere constant, and the range $f((0,1))$ is countable: $\#f((0,1)) = \#\mathbb{Q} = \#\mathbb{N}$.

13.11 A continuous function such that $f(x) \in \mathbb{Q}$ a.e. and f is not constant on any interval

Constant functions $f \equiv c$ are the only continuous functions $f : [0,1] \to \mathbb{R}$ such that $f(x) \in \mathbb{Q}$ for all $x \in [0,1]$. This is an immediate consequence of the intermediate value theorem. The following example illustrates that the situation is different if we require only $f(x) \in \mathbb{Q}$ for (Lebesgue) almost all $x \in \mathbb{R}$.

Let $F : [0,1] \to [0,1]$ be the Cantor function [☞Section 2.6] and construct a tent function from F by setting

$$T(x) := F(1+x)\mathbb{1}_{(-1,0)}(x) + F(1-x)\mathbb{1}_{[0,1)}(x), \quad x \in \mathbb{R}.$$

T has the following properties: it is continuous, its support equals $[-1,1]$ and $T(x) \in \mathbb{Q}$ for Lebesgue almost all $x \in \mathbb{R}$. Denote by $(I_n)_{n \in \mathbb{N}_0}$ an enumeration of all open intervals (p,q), $0 < p < q < 1$, with rational end points. We set

$$g_k(x) := \frac{1}{3^k} T\left(\frac{x - x_k}{|y_k - x_k|}\right) \quad \text{and} \quad f_n(x) := \sum_{k=0}^{n} g_k(x), \quad x \in [0,1],$$

where $(x_k)_{k \in \mathbb{N}_0}$ and $(y_k)_{k \in \mathbb{N}_0}$ are recursively chosen in the following way: for $k = 0$ pick any $x_0 \neq y_0 \in I_0$. In step k choose $x_k, y_k \in I_k$, $x_k \neq y_k$, such that

$|x_k - y_k| \leqslant 3^{-k}$ and $f_{k-1}(x_k) \geqslant f_{k-1}(y_k)$. We claim that

$$f(x) := \lim_{n\to\infty} f_n(x) = \sum_{k=0}^{\infty} g_k(x), \quad x \in [0,1],$$

has all desired properties. As $\|g_k\|_\infty \leqslant 3^{-k}$, the series on the right-hand side converges uniformly for $x \in [0,1]$, and therefore f is well-defined and, being the uniform limit of continuous functions, continuous. Since $\operatorname{supp} T = [-1,1]$, the support of each g_k is an interval of length $2|y_k - x_k| \leqslant 2 \cdot 3^{-k}$. Hence, $\lambda(\operatorname{supp} g_k) \leqslant 2 \cdot 3^{-k}$. Combining

$$\{f \notin \mathbb{Q}\} \subseteq \bigcup_{k=1}^{n} \{g_k \notin \mathbb{Q}\} \cup \bigcup_{k=n+1}^{\infty} \{g_k \neq 0\}, \quad n \in \mathbb{N},$$

with the fact that each g_k takes rational values almost everywhere, we get

$$\lambda(\{f \notin \mathbb{Q}\}) \leqslant \sum_{k=1}^{n} \lambda(\{g_k \notin \mathbb{Q}\}) + \sum_{k=n+1}^{\infty} \lambda(\{g_k \neq 0\}) \leqslant 2 \sum_{k=n+1}^{\infty} 3^{-k} \xrightarrow[n\to\infty]{} 0.$$

It remains to prove that f is not constant on any non-empty interval. For a given interval $(a,b) \subseteq (0,1)$ choose $n \in \mathbb{N}$ such that $I_n \subseteq (a,b)$. Denote by $x_n, y_n \in I_n$ the points selected in the construction of f. By construction, $g_n(x_n) = 3^{-n}$ and $f_{n-1}(x_n) \geqslant f_{n-1}(y_n)$. As $g_k \geqslant 0$ for all $k \geqslant 0$, this gives

$$f(x_n) \geqslant f_n(x_n) = f_{n-1}(x_n) + g_n(x_n) \geqslant f_{n-1}(y_n) + 3^{-n}.$$

From $g_n(y_n) = 0$ and $\|g_k\|_\infty \leqslant 3^{-k}$, it follows that

$$f(y_n) = f_{n-1}(y_n) + \sum_{k=n+1}^{\infty} g_k(y_n) \leqslant f_{n-1}(y_n) + \sum_{k=n+1}^{\infty} 3^{-k} \leqslant f_{n-1}(y_n) + \frac{3^{-n}}{2}.$$

Thus, $f(x_n) > f(y_n)$, and this proves that f is not constant on $I_n \subseteq (a,b)$.

13.12 A continuous function which is strictly positive on $\mathbb{Q}$ but fails to be strictly positive almost everywhere

Consider Lebesgue measure λ on $(\mathbb{R}, \mathscr{B}(\mathbb{R}))$, and let $(q_n)_{n\in\mathbb{N}}$ be an enumeration of $\mathbb{Q}$. If we define

$$U := \bigcup_{n\in\mathbb{N}} (q_n - 2^{-n}, q_n + 2^{-n}),$$

then U contains all rationals and, by the subadditivity of Lebesgue measure,

$$\lambda(U) \leqslant 2 \sum_{n=1}^{\infty} 2^{-n} = 2.$$

Set $F := \mathbb{R} \setminus U$ and

$$f(x) := d(x, F) := \inf_{z \in F} d(x, z), \quad x \in \mathbb{R}.$$

The function f is (Lipschitz) continuous: for $x, y \in \mathbb{R}$ and $z \in F$ we have $d(x, z) \leqslant d(x, y) + d(y, z)$, and so

$$\inf_{z \in F} d(x, z) \leqslant d(x, y) + \inf_{z \in F} d(y, z) \quad \text{and} \quad d(x, F) - d(y, F) \leqslant d(x, y).$$

Since x and y play symmetric roles, we may interchange them and get

$$|d(x, F) - d(y, F)| \leqslant d(x, y).$$

Moreover, $f(q_n) \geqslant 2^{-n}$ implies that f is strictly positive on the rationals. From $\lambda(\{f = 0\}) \geqslant \lambda(U^c) = \infty$ we see that f is not strictly positive almost everywhere.

13.13 A measurable function which is zero almost everywhere but whose graph is dense

Let $(q_n)_{n \in \mathbb{N}}$ be an enumeration of $\mathbb{Q}$. Define a function $f : \mathbb{R} \to \mathbb{R}$ recursively as follows. Choose a countable set $D_1 \subseteq \mathbb{R}$ such that D_1 is dense in $\mathbb{R}$, and set $f(x) :=_1$ for all $x \in D_1$. Iterating this step, we can construct a countable set $D_n \subseteq \mathbb{R} \setminus (D_1 \cup \cdots \cup D_{n-1})$ which is dense in $\mathbb{R}$ and set

$$f(x) := q_n, \quad x \in D_n, \ n \in \mathbb{N}.$$

This defines the function f on $D := \bigcup_{n \in \mathbb{N}} D_n$; for $x \in D^c$ set $f(x) = 0$. Since D is countable, f is Borel measurable and zero almost everywhere. Moreover, by the construction of f, the pre-image $f^{-1}(\{r\})$ is dense in $\mathbb{R}$ for any $r \in \mathbb{Q}$. This implies that the graph of f is dense in $\mathbb{R}^2$. Indeed: fix $(x, y) \in \mathbb{R}^2$ and choose $(r_n)_{n \in \mathbb{N}} \subseteq \mathbb{Q}$ such that $r_n \to y$. Since $f^{-1}(\{r_n\})$ is dense in $\mathbb{R}$, there is some $x_n \in f^{-1}(\{r_n\})$ with $|x - x_n| \leqslant n^{-1}$. Thus,

$$(x, y) = \lim_{n \to \infty} (x_n, r_n) = \lim_{n \to \infty} (x_n, f(x_n)).$$

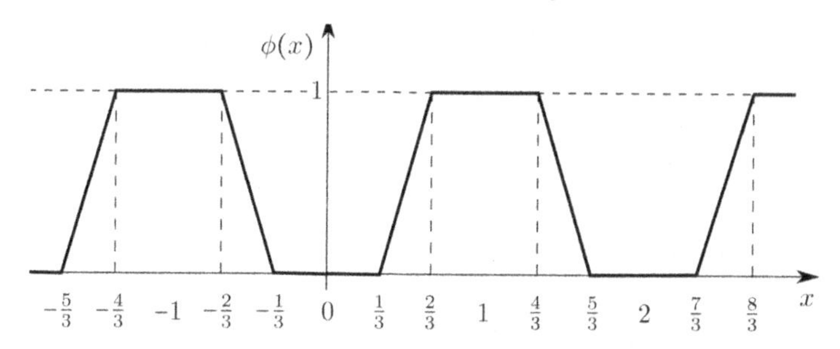

Figure 13.1 Graph of the 2-periodic function $\phi : \mathbb{R} \to [0, 1]$.

13.14 A continuous function $f : [0, 1] \to \mathbb{R}^2$ whose image has positive Lebesgue measure

Denote by $\phi : \mathbb{R} \to [0, 1]$ the 2-periodic function satisfying

$$\phi(x) = \begin{cases} 0, & 0 \leqslant \text{if } |x| \leqslant \frac{1}{3}, \\ 3|x| - 1, & \text{if } \frac{1}{3} < |x| < \frac{2}{3}, \\ 1, & \text{if } \frac{2}{3} \leqslant |x| \leqslant 1, \end{cases}$$

see Fig. 13.1.

Define a function $f : [0, 1] \to \mathbb{R}^2$, $t \longmapsto f(t) = (f_1(t), f_2(t))$, by

$$f_1(t) := \frac{1}{2} \sum_{k=0}^{\infty} \frac{\phi(3^{2k} t)}{2^k} \quad \text{and} \quad f_2(t) := \frac{1}{2} \sum_{k=0}^{\infty} \frac{\phi(3^{2k+1} t)}{2^k}.$$

Since ϕ is continuous and the series converge uniformly in $t \in [0, 1]$, it follows that f is continuous. In order to prove that the image $f([0, 1])$ has positive Lebesgue measure, we show that f maps the classical (middle-thirds) Cantor set C onto $[0, 1]^2$, i.e. $f(C) = [0, 1]^2$. By the construction of the Cantor set [☞ Section 2.5], C is the set of points $t \in [0, 1]$ with ternary expansion

$$t = \sum_{i=1}^{\infty} \frac{2a_i}{3^i},$$

where $a_i \in \{0, 1\}$ for all $i \in \mathbb{N}$. Consequently,

$$3^m t = 2 \sum_{i=1}^{m} a_i 3^{m-i} + \frac{2}{3} a_{m+1} + \sum_{i=m+2}^{\infty} 2a_i 3^{m-i}$$

for all $m \in \mathbb{N}_0$. Since the sum $2\sum_{i=1}^{m} a_i 3^{m-1}$ is an even integer and

$$0 \leqslant \sum_{i=m+2}^{\infty} 2a_i 3^{m-i} \leqslant 2\sum_{n=2}^{\infty} \frac{1}{3^n} = \frac{1}{3},$$

it follows from the definition of ϕ that

$$\phi(3^m t) = \begin{cases} 0, & \text{if } a_{m+1} = 0, \\ 1, & \text{if } a_{m+1} = 1, \end{cases}$$

i.e. $\phi(3^m t) = a_{m+1}$. Hence,

$$f_1(t) = \frac{1}{2}\sum_{k=0}^{\infty} \frac{a_{2k+1}}{2^k} = \frac{a_1}{2} + \frac{a_3}{2^2} + \cdots\,,$$

$$f_2(t) = \frac{1}{2}\sum_{k=0}^{\infty} \frac{a_{2k+2}}{2^k} = \frac{a_2}{2} + \frac{a_4}{2^2} + \cdots\,.$$

Since any point $x \in [0,1]$ has a binary expansion of the form $x = \frac{1}{2}\sum_{k=0}^{\infty} c_k 2^{-k}$, $c_k \in \{0,1\}$, we conclude that

$$f(C) = [0,1] \times [0,1].$$

Comment 1 The above construction is due to Schoenberg [158]. Continuous mappings $f : [0,1] \to \mathbb{R}^n$ with the property that the image $f([0,1])$ has positive Lebesgue measure are called **space-filling curves**.

Comment 2 With some more effort, we can construct a continuous function f which maps $[0,1]$ onto the infinite-dimensional **Hilbert cube** $[0,1]^{\mathbb{N}}$. Partition $\mathbb{N}$ into countably many sequences $(n(k,j))_{k\in\mathbb{N}_0} \subseteq \mathbb{N}$, and set

$$f_j(t) := \frac{1}{2}\sum_{k=0}^{\infty} \frac{\phi\left(3^{n(k,j)-1}t\right)}{2^k}, \qquad t \in [0,1],\ j \in \mathbb{N}.$$

If $t \in C$ has the ternary expansion $t = \sum_{i=1}^{\infty} 2a_i 3^{-i}$ with $a_i \in \{0,1\}$, then by our previous considerations

$$f_j(t) = \frac{1}{2}\sum_{k=0}^{\infty} \frac{a_{n(k,j)}}{2^k} = \frac{a_{n(1,j)}}{2} + \frac{a_{n(2,j)}}{2^2} + \cdots. \tag{13.1}$$

Now let $x \in [0,1]^{\mathbb{N}}$, i.e. $x = (x_j)_{j\in\mathbb{N}} \subseteq [0,1]$, and write $x_j = \frac{1}{2}\sum_{k=0}^{\infty} c(k,j)2^{-k}$ with $c(k,j) \in \{0,1\}$. For every $i \in \mathbb{N}$ there exist unique numbers $k \in \mathbb{N}_0$, $j \in \mathbb{N}$

such that $i = n(k, j)$; set $a_i := c(k, j)$. Then (13.1) shows that

$$t := \sum_{i=1}^{\infty} \frac{2a_i}{3^i} \in C$$

satisfies $f_j(t) = x_j$ for all $j \in \mathbb{N}$. Consequently, $f := (f_1, f_2, ...)$ maps the set $C \subseteq [0,1]$ onto $[0,1]^{\mathbb{N}}$. Note that, by construction, $f_j(0) = 0$ and $f_j(1) = 1$ for all $j \in \mathbb{N}$.

13.15 The image of a Lebesgue null set under a continuous bijective mapping need not have Lebesgue measure zero

Denote by $F : [0,1] \to [0,1]$ the Cantor function [☞Section 2.6] and define

$$f(x) := x + F(x), \qquad x \in [0,1].$$

Since F is monotone increasing and $x \longmapsto x$ is strictly increasing, it follows that f is strictly increasing. Consequently, f is a continuous bijective mapping from $[0,1]$ to $[f(0), f(1)] = [0,2]$. We claim that f maps the Cantor set C onto a set of positive Lebesgue measure. To see this, let $(I_n)_{n\in\mathbb{N}}$ be an enumeration of the intervals that are removed in the construction of the Cantor middle-thirds set [☞Section 2.5]. Since F is constant on I_n, the image $f(I_n)$ is an interval with the same length as I_n. Moreover, the sets $f(I_n)$, $n \in \mathbb{N}$, are pairwise disjoint and so

$$\lambda(f([0,1] \setminus C)) = \lambda\left(\bigcup_{n\in\mathbb{N}} f(I_n)\right) = \sum_{n\in\mathbb{N}} \lambda(f(I_n)) = \sum_{n\in\mathbb{N}} \lambda(I_n) = 1.$$

Thus,

$$\lambda(f(C)) = \lambda([0,2]) - \lambda(f([0,1] \setminus C)) = 1.$$

Comment For a further example [☞13.14].

13.16 Lusin's theorem fails for non-regular measures

Let $\mathscr{B}(X)$ be the Borel σ-algebra of a metric space X. If μ is a regular measure [☞p. 185] on $(X, \mathscr{B}(X))$, then there exists for every Borel measurable function $f : X \to \mathbb{R}$ and every $\epsilon > 0$ a continuous mapping $u : X \to \mathbb{R}$ such that $\mu(\{u \neq f\}) \leqslant \epsilon$ (Lusin's theorem [☞13A]). The following example illustrates that the regularity of μ plays a crucial role.

Consider the counting measure $\zeta_{\mathbb{R}}$ on $(\mathbb{R}, \mathscr{B}(\mathbb{R}))$ – this measure is not regular – and the Borel measurable function $f(x) := \mathbb{1}_{\mathbb{Q}}(x)$. For every continuous mapping $u : \mathbb{R} \to \mathbb{R}$ the set $\{u \neq f\}$ is infinite, and therefore $\zeta_{\mathbb{R}}(\{u \neq f\}) = \infty$.

Comment Similar phenomena can happen for **finite** measures. Denote by $\Omega = [0, \omega_1]$ the ordinal space [☞ Section 2.4] with the σ-algebra

$$\mathscr{A} := \{A \subseteq \Omega;\ A \cup \{\omega_1\} \text{ or } A^c \cup \{\omega_1\} \text{ contains an uncountable compact set}\}$$

[☞ Example 4B] and the Dieudonné measure

$$\mu(A) := \begin{cases} 0, & \text{if } A \cup \{\omega_1\} \text{ contains an uncountable compact set,} \\ 1, & \text{if } A^c \cup \{\omega_1\} \text{ contains an uncountable compact set} \end{cases}$$

[☞ Example 5B]. Thus,

$$\nu := \mu + \delta_{\omega_1}$$

is a finite measure on $(\Omega, \mathscr{A})$. Consider the measurable function $f := \mathbb{1}_{\{\omega_1\}}$, and let $u : \Omega \to \mathbb{R}$ be some continuous mapping. By the definition of the topology on Ω, u is eventually constant, i.e. $u|_{[\alpha_u, \omega_1]} = u(\omega_1)$ for some $\alpha_u < \omega_1$ [☞ Section 2.4]. If $u(\omega_1) = 1$, then

$$\nu(\{u \neq f\}) \geqslant \nu[\alpha_u, \omega_1) = \mu[\alpha_u, \omega_1) = 1.$$

If $u(\omega_1) = 0$, then

$$\nu(\{u \neq f\}) \geqslant \nu(\{\omega_1\}) = \delta_{\omega_1}(\{\omega_1\}) = 1.$$

Hence, $\nu(\{u \neq f\}) \geqslant 1$ for any continuous mapping u.

13.17 The convolution of two integrable functions may be discontinuous

Consider on $(\mathbb{R}, \mathscr{B}(\mathbb{R}), \lambda)$ the integrable function

$$f(x) = \frac{1}{\sqrt{x}} \mathbb{1}_{(0,1)}(x).$$

We will construct a function $g \in L^1$ such that the convolution

$$(f * g)(x) = \int_{\mathbb{R}} f(x - y) g(y)\, dy$$

is bounded but discontinuous on a dense subset of $\mathbb{R}$.

We will need a few properties of the convolution $f * f$. Since the support of f is contained in $[0, 1]$, the support of $f * f$ is contained in $[0, 1] + [0, 1] = [0, 2]$, i.e. $(f * f)(x) = 0$ for all $x \notin [0, 2]$. It is easy to see from the definition that $(f * f)(0) = 0$. If $x \in (0, 1]$, then

$$(f * f)(x) = \int_0^x \frac{1}{\sqrt{x - y}} \frac{1}{\sqrt{y}}\, dy = \int_0^1 \frac{1}{\sqrt{1 - u}} \frac{1}{\sqrt{u}}\, du;$$

use $y = xu$ and $dy = xdu$ for the second equality. Since $2\arctan\sqrt{\frac{x}{1-x}}$ is a primitive of $\frac{1}{\sqrt{x(1-x)}}$ and since the latter function is Riemann integrable on $\left[\frac{1}{n}, 1-\frac{1}{n}\right]$, it follows from the monotone convergence theorem that

$$\begin{aligned}(f * f)(x) &= \lim_{n\to\infty} \int_{1/n}^{1-1/n} \frac{1}{\sqrt{1-u}}\frac{1}{\sqrt{u}}\,du \\ &= \lim_{n\to\infty} 2\arctan\sqrt{\frac{x}{1-x}}\Bigg|_{x=1/n}^{1-1/n} = \pi, \quad x \in (0,1].\end{aligned}$$

For $x \in (1,2)$ a similar calculation shows that

$$(f * f)(x) = \int_{x-1}^{1} \frac{1}{\sqrt{x-y}}\frac{1}{\sqrt{y}}\,dy = 2\arctan\sqrt{\frac{1}{x-1}} - 2\arctan\sqrt{x-1},$$

and so $(f * f)$ decays on $(1,2)$ continuously from π to 0. Consequently, $(f * f)$ is continuous on $\mathbb{R} \setminus \{0\}$ and at $x = 0$ there is a jump of height π, cf. Fig. 13.2. Let $(q_n)_{n\in\mathbb{N}}$ be a dense subset of $\mathbb{R}$, e.g. the rationals, and define

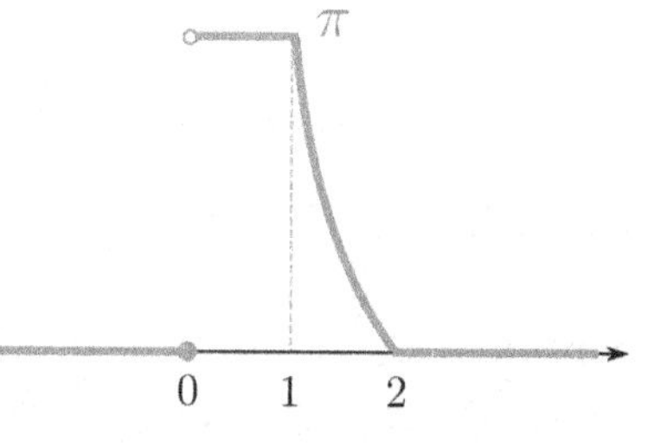

Figure 13.2 Plot of $f * f$.

$$g(x) := \sum_{n\in\mathbb{N}} \frac{1}{n^2} f(x - q_n), \quad x \in \mathbb{R}.$$

Because of the translation invariance of Lebesgue measure, we have

$$\int_{\mathbb{R}} |g(x)|\,dx \leqslant \sum_{n\in\mathbb{N}} \frac{1}{n^2} \int_{\mathbb{R}} |f(x)|\,dx < \infty,$$

i.e. $g \in L^1$. Since $f \geqslant 0$ and $g \geqslant 0$, the Beppo Levi theorem [☞(1.8)] gives

$$\begin{aligned}(f * g)(x) &= \sum_{n\in\mathbb{N}} \frac{1}{n^2} \int_{\mathbb{R}} f(x - y - q_n) f(y)\,dy \\ &= \sum_{n\in\mathbb{N}} \frac{1}{n^2} (f * f)(x - q_n).\end{aligned} \tag{13.2}$$

As we have seen, the convolution $f * f$ is uniformly bounded by π, and so this identity shows in particular that

$$\|f * g\|_\infty = \sup_{x\in\mathbb{R}} |(f * g)(x)| \leqslant \pi \sum_{n\in\mathbb{N}} \frac{1}{n^2} < \infty.$$

Finally we show that $f * g$ is discontinuous at every $x = q_k$, $k \in \mathbb{N}$. Since the functions $(f * f)(\cdot - x_n)$ are positive and continuous at $x = q_k$ for $k \neq n$, cf. Fig. 13.2, it follows from Fatou's lemma, applied for the counting measure, that

$$\sum_{n \neq k} (f * f)(x - q_n) \leqslant \liminf_{i \to \infty} \sum_{n \neq k} (f * f)(x - q_n + 1/i).$$

From

$$(f * f)(y - q_k + 1/i) = \pi \quad \text{and} \quad (f * f)(x - q_k) = 0,$$

and (13.2), we conclude that

$$\begin{aligned}(f * g)(x) = \sum_{n \neq k} (f * f)(x - q_n) &\leqslant \liminf_{i \to \infty} \sum_{n \neq k} (f * f)(x - q_n + 1/i) \\ &= \liminf_{i \to \infty} \left((f * g)(x + 1/i) - \pi\right),\end{aligned}$$

i.e.

$$\liminf_{i \to \infty} (f * g)(x + 1/i) \geqslant (f * g)(x) + \pi.$$

This shows that $(f * g)$ is not (right-)continuous at $x = q_k$.

Comment 1 The discontinuities of $f * g$ are caused by 'explosions' of f and g. More precisely, if the convolution $f * g$ of two integrable functions is discontinuous at some $x \in \mathbb{R}$, then there exist points $x_1 \leqslant x$ and $x_2 \leqslant x$ with $x_1 + x_2 = x$ such that f is unbounded in a neighbourhood of x_1 and g is unbounded in a neighbourhood of x_2, cf. [121].

Comment 2 If $f \in L^p$ and $g \in L^q$ for conjugate exponents $p \in [1, \infty]$ and $q \in [1, \infty]$, then the convolution $f * g$ is continuous, cf. [MIMS, p. 159].

14

Integration and Differentiation

The development of many modern theories of integration may be seen as a 'quest for a unified treatment of derivative and integral'. For a long time, integration and differentiation were thought of as being inverse operations, but this is only true for certain (rather restricted) classes of functions or for ever more general definitions of integral and derivative. The counterexamples of this section are, therefore, a historical tour from the beginnings (Newton and Leibniz) to the modern time of Lebesgue – and beyond.

Let $f : [a, b] \to \mathbb{R}$ be a differentiable function. If the derivative f' is Riemann integrable – e.g. if $f \in C_b^1(\mathbb{R})$ – then the fundamental theorem of calculus shows that $f(x) - f(a) = \int_a^x f'(t)\,dt$ for all $x \in [a, b]$, i.e. the Riemann integral acts as an antiderivative in this class of functions. There are various generalizations of this classical result. For increasing functions one has the following version of the fundamental theorem of calculus.

Theorem 14A *Let $f : [a, b] \to \mathbb{R}$ be an increasing function on a compact interval. There is a Lebesgue null set N such that f' exists and is integrable outside N, and one has for all $[c, d] \subseteq [a, b]$,*

$$\int_{[c,d]} f'(t)\,\lambda(dt) = \int_{[c,d]\setminus N} f'(t)\,\lambda(dt) \leqslant f(d) - f(c).$$

A proof can be found in Kestelman [90, Theorem 258] or Bogachev [19, Corollary 5.2.7]. The inequality in Theorem 14A can be strict, i.e. $\int_{[c,d]} f'\,d\lambda$ may be strictly smaller than $f(d) - f(c)$ [☞ 14.4, 14.5]. The reason for this pathology is that the a.e. derivative of a function f does not encode any information on the monotonicity of f, e.g. f may be strictly increasing but $f' = 0$ a.e. [☞ 14.5] or f may be nowhere increasing but $f' > 1$ a.e. [☞ 14.6]. The following variant of the fundamental theorem of calculus is from Kestelman [90, Theorem 264].

Theorem 14B *Let $f : [a,b] \to \mathbb{R}$ be a continuous function such that f' exists and is finite except at the points from an at most countable set N. If f' is Lebesgue integrable, then $f(x) - f(a) = \int_{[a,x]} f' \, d\lambda$ for all $x \in [a,b]$.*

We will see that this result is sharp in the sense that the integrability of f' is indeed needed [☞14.2] and that the countability of the exceptional set N is crucial [☞14.4]. In conclusion, the notions of antiderivative and Lebesgue integral do not coincide, even if all appearing objects are well-defined.

There is another important theorem which links integration and differentiation.

Theorem 14C (Lebesgue's differentiation theorem [MIMS, p. 315]) *Let λ^d be Lebesgue measure on $(\mathbb{R}^d, \mathscr{B}(\mathbb{R}^d))$. If $f \in L^1(\lambda^d)$, then Lebesgue almost every $x \in \mathbb{R}^d$ is a **Lebesgue point** of f, i.e.*

$$\lim_{r \to 0} \frac{1}{\lambda^d(B_r(x))} \int_{B_r(x)} |f(y) - f(x)| \, \lambda^d(dy) = 0. \tag{14.1}$$

In particular,

$$\lim_{r \to 0} \frac{1}{\lambda^d(B_r(x))} \int_{B_r(x)} f(y) \, \lambda^d(dy) = f(x) \quad \text{a.e.}$$

In dimension $d = 1$, Lebesgue's differentiation theorem shows that

$$\forall f \in L^1(\lambda) : \lim_{h \to 0} \frac{1}{2h} \int_{x-h}^{x+h} f(y) \, dy = f(x) \quad \text{a.e.},$$

which generalizes the well-known identity $\frac{d}{dx} \int^x f(y) \, dy = f(x)$ for continuous functions f. Theorem 14C remains valid if the open balls $B_r(x)$ are replaced by Borel sets $A_r = A_r(x)$ which shrink in a controlled way to $\{x\}$ (so-called **nicely shrinking** sets), but for general $A_r(x)$ with $A_r(x) \downarrow \{x\}$ convergence in (14.1) may fail to hold [☞14.19]. Lebesgue's differentiation theorem can be extended to a large class of Borel measures on $(\mathbb{R}^d, \mathscr{B}(\mathbb{R}^d))$ [☞14.20]. A thorough treatment of differentiation vs. integration can be found, for example, in the monograph by Pfeffer [136].

14.1 A non-Riemann integrable function f which has a primitive

A Riemann integrable function f is necessarily defined on a bounded interval and its range is bounded [☞3.2, 3.3]. On the other hand, a primitive is 'just' a function F such that F' exists and equals f. The following example shows that the notions of 'antiderivative' and Riemann integral need not coincide.

Set $F(x) := x^{\frac{3}{2}} \sin \frac{1}{x}$ for $0 < x \leqslant 1$ and $F(0) = 0$. It is not difficult to see that F' exists on $[0, 1]$ and that

$$f(x) := F'(x) = \begin{cases} -x^{-\frac{1}{2}} \cos \frac{1}{x} + \frac{3}{2} x^{\frac{1}{2}} \sin \frac{1}{x}, & \text{if } 0 < x \leqslant 1, \\ 0, & \text{if } x = 0. \end{cases}$$

Although the function f is defined on the compact interval $[0, 1]$, it is easy to see that $f((2\pi n)^{-1}) = -\sqrt{2\pi n}$ for all $n \in \mathbb{N}$, i.e. f is unbounded, hence not integrable in the sense of Riemann.

14.2 A function f which is differentiable, but f' is not integrable

The function $f : \mathbb{R} \to \mathbb{R}$ defined by

$$f(x) := \begin{cases} x^2 \sin \frac{1}{x^2}, & \text{if } x \neq 0, \\ 0, & \text{if } x = 0, \end{cases}$$

is differentiable at each point $x \in \mathbb{R}$. However, its derivative

$$f'(x) := \begin{cases} 2x \sin \frac{1}{x^2} - 2\frac{1}{x} \cos \frac{1}{x^2}, & \text{if } x \neq 0, \\ 0, & \text{if } x = 0, \end{cases}$$

is neither Lebesgue nor Riemann integrable as $\int_0^1 |f'(x)|\, dx = \infty$ and f' is unbounded near $x = 0$.

Comment One of Lebesgue's motivations when constructing his integral was to overcome certain shortcomings of the Riemann integral, e.g. the requirements that Riemann integrable functions need to be bounded, (Lebesgue) a.e. continuous, and that one usually needs uniform convergence in most convergence theorems for Riemann integrals. Moreover, Lebesgue's theory has a fairly general form of the fundamental theorem of integral calculus: if F has a bounded derivative F' on a compact interval $[a, b]$, then F' is Lebesgue integrable and $F(x) - F(a) = \int_{[a,x]} F'\, d\lambda$ for $x \in [a, b]$. The general situation is described in Theorem 14B. Example 14.2 shows that the **assumption** that F' be integrable is indeed necessary.

There are further generalizations of Lebesgue's integral, e.g. the Denjoy integral, which overcome this problem; see e.g. Gordon [66] or Kestelman [90].

14.3 Volterra's version of Example 14.2

V. Volterra [195] constructed the following variant of Example 14.2: let C be any geometric fat Cantor set [☞7.3] with $\alpha = \lambda(C) > 0$ and set $f(x) := x^2 \sin x^{-1}$, $x > 0$. If $I = (a, b)$ is a component (that is, a maximal connected interval) within $[0, 1] \setminus C$, we define $c := c(a, b) = \sup\left\{0 < x \leqslant \frac{1}{2}(b - a);\ f'(x) = 0\right\}$.

Consider the following function $F : [0, 1] \to \mathbb{R}$:

$$F(x) := \begin{cases} 0, & \text{if } x \in C, \\ f(x - a), & \text{if } a < x < a + c, \\ f(c), & \text{if } a + c \leqslant x \leqslant b - c, \\ f(b - x), & \text{if } b - c < x < b. \end{cases}$$

Clearly, F' exists and is continuous on $[0, 1] \setminus C$. Moreover, F is continuous on $[0, 1]$ and $F'(x) = 0$ for all $x \in C$. Indeed, let $x \in C$ and $h > 0$. If $x + h \in C$, then $F(x + h) - F(x) = 0$. Otherwise, $x + h$ is in some component (a, b) such that $x \leqslant a < x + h < b$. In this case we have

$$|F(x + h) - F(x)| = |F(x + h)| \leqslant (x + h - a)^2 \leqslant h^2,$$
$$\text{implying} \quad \lim_{h \to 0} \frac{F(x + h) - F(x)}{h} = 0.$$

This shows that $F'(x)$ exists for all $x \in [0, 1]$ and satisfies $|F'(x)| \leqslant 3$.

The function F' is discontinuous at the end points of each component (a, b), hence also at their limiting points. This means that $F'(x)$ is not continuous for all $x \in C$. Since we know that $\lambda(C) > 0$ and, by self-similarity of the set C, also $\lambda(C \cap [0, x]) > 0$, the function F' cannot be Riemann integrable over any interval $[0, x]$, $0 < x < 1$.

Since F' is Lebesgue integrable and $F'|_C = 0$, the fundamental theorem of integral calculus [☞14B] gives

$$\begin{aligned} \int_{[0,x]\setminus C} F'(t)\,\lambda(dt) &= \int_{[0,x]\setminus C} F'(t)\,\lambda(dt) + \int_{[0,x]\cap C} F'(t)\,\lambda(dt) \\ &= \int_{[0,x]} F'(t)\,\lambda(dt) = F(x). \end{aligned}$$

Comment Lebesgue discusses this example in his doctoral dissertation [101] from 1902. It seems that this was one of the motivations for him to come up with this new theory of integration.

14.4 A continuous function such that f' exists almost everywhere and is integrable but the fundamental theorem of calculus fails

We are going to construct a continuous, increasing function $f : [0,1] \to [0,1]$ such that $f(0) < f(1)$ and $f'(x) = 0$ for Lebesgue almost all $x \in [0,1]$. Since f' is defined outside a Lebesgue null set N', we may set $f'(x) := 0$ for $x \in N'$; thus, $f' \equiv 0$, and

$$\int_{[0,1]\setminus N'} f'\,d\lambda = \int_{[0,1]} f'\,d\lambda = 0 < f(1) - f(0),$$

i.e. the fundamental theorem of calculus fails.

This shows that the best one can expect is Theorem 14A and that the countability of the exceptional set N appearing in Theorem 14B is crucial; it is not enough to assume that N is a Lebesgue null set.

The Cantor function [☞ Section 2.6] has the desired properties, but we prefer the following simpler construction.

Let $(q_i)_{i\in\mathbb{N}}$ be an enumeration of $\mathbb{Q} \cap [0,1)$ and set

$$g(0) = 0 \quad \text{and} \quad g(y) = \sum_{i:q_i<y} 2^{-i}, \quad y \in (0,1].$$

Since the rational numbers are dense in $[0,1]$, the function g is strictly increasing from $g(0) = 0$ to $g(1) = 1$. Moreover,

$$g(q_i) = g(q_i-) \quad \text{and} \quad g(q_i+) = g(q_i) + 2^{-i},$$

i.e. g has a jump discontinuity at every $y = q_i$. Every irrational $y \in [0,1]$ can be approximated by a sequence of rational numbers and since g is monotone, we get $g(y-) = g(y+) = g(y)$, thus g is continuous at every irrational point.

The range R of g is the complement (relative to $[0,1]$) of the set $R^c := \dot{\bigcup}_{i\in\mathbb{N}} I_i$ where $I_i = (g(q_i), g(q_i) + 2^{-i}]$ – these are the 'gaps' caused by the jumps. Since R^c has total length 1, $\lambda(R) = 0$. As g is strictly increasing, it has a continuous generalized inverse, which is given by

$$f(x) = \begin{cases} q_i, & \text{if } x \in I_i,\ i \in \mathbb{N}, \\ y, & \text{if } x = g(y). \end{cases}$$

The function f satisfies $f(0) = 0$, $f(1) = 1$ and it is constant in the interior of each I_i, thus $f'(x)$ exists for almost all $x \in [0,1]$ and $f'(x) = 0$.

Comment A **strictly** increasing function with $f' = 0$ a.e. is constructed in Example 14.5.

14.5 A continuous strictly increasing function with $f' = 0$ Lebesgue almost everywhere

Let $F : [0,1] \to [0,1]$ be the Cantor function [☞Section 2.6], and extend F continuously to $\mathbb{R}$ by setting $F(x) = 0$, $x < 0$, and $F(x) = 1$, $x > 1$. For an enumeration $(q_n)_{n\in\mathbb{N}}$ of $\mathbb{Q} \cap [0,1]$ set

$$f_n(x) := 2^{-n} F(x - q_n) \quad \text{and} \quad f(x) := \sum_{n\in\mathbb{N}} f_n(x), \quad x \in [0,1].$$

As $\|f_n\|_\infty \leqslant 2^{-n}$, the series on the right-hand side converges uniformly, and so $f : [0,1] \to [0,1]$ is continuous as the uniform limit of continuous functions. Moreover, the non-negativity of f_n entails that the series can be differentiated termwise Lebesgue almost everywhere, i.e. $f' = \sum_{n\in\mathbb{N}} f_n'$ almost everywhere, cf. [19, Proposition 5.2.8]. Since $F' = 0$ Lebesgue almost everywhere, this implies

$$f'(x) = \sum_{n\in\mathbb{N}} 2^{-n} F'(x - q_n) = 0 \quad \text{a.e.}$$

In order to prove that f is strictly increasing on $[0,1]$, fix $0 \leqslant x < y \leqslant 1$ and choose $q_k \in \mathbb{Q} \cap [0,1]$ with $x < q_k < y$. Since F, hence f_n, is non-decreasing, it follows that

$$f(y) - f(x) = \sum_{n\in\mathbb{N}} (f_n(y) - f_n(x)) \geqslant f_k(y) - f_k(x) = f_k(y) > 0.$$

Comment 1 An alternative construction can be found in [144, p. 48].

Comment 2 Our example illustrates that the fundamental theorem of calculus may fail even if 'everything is well-defined'. Indeed, f is continuous, f' exists almost everywhere and is integrable, but the fundamental theorem of calculus fails:

$$\int_x^y f'(t)\,dt = 0 < f(y) - f(x), \quad 0 \leqslant x < y \leqslant 1.$$

This phenomenon can be observed for many other increasing functions. The general situation is described by Theorem 14A.

14.6 A continuous function f such that $f' > 1$ a.e. but f is not increasing on any interval

Denote by $F : [0,1] \to [0,1]$ the Cantor function [☞Section 2.6] and set

$$T(x) := F(x)\mathbb{1}_{(0,1)}(x) + F(2 - x)\mathbb{1}_{[1,2]}(x), \quad x \in \mathbb{R}.$$

By the construction of the Cantor function, T is continuous and its derivative is zero Lebesgue almost everywhere. Let $(I_n)_{n\in\mathbb{N}_0}$ be an enumeration of all open intervals (p,q), $0 < p < q < 1$, with rational end points. Define

$$g_k(x) := \frac{1}{3^k} T\left(\frac{x - x_k}{y_k - x_k}\right) \quad \text{and} \quad f_n(x) := 2x - \sum_{k=0}^{n} g_k(x), \quad x \in [0,1],$$

recursively as follows: for $k = 0$ choose any $x_0, y_0 \in I_0$ with $x_0 \neq y_0$. In step $k \geqslant 1$ we consider two cases separately: if $f_{k-1}|_{I_k}$ is not increasing, then we can find $x_k, y_k \in I_k$, $x_k < y_k$ with $f_{k-1}(x_k) > f_{k-1}(y_k)$. If $f_{k-1}|_{I_k}$ is increasing, then we pick $x_k, y_k \in I_k$, $x_k < y_k$ with $f_{k-1}(y_k) \leqslant f_{k-1}(x_k) + 3^{-(k+1)}$; this is possible since f_{k-1} is continuous. The function

$$f(x) := \lim_{n\to\infty} f_n(x) = 2x - \sum_{k=0}^{\infty} g_k(x), \quad x \in [0,1],$$

has all desired properties. First of all, $\|g_k\|_\infty \leqslant 3^{-k}$ implies that the series on the right-hand side converges uniformly for $x \in [0,1]$. Consequently, f is well-defined and continuous as uniform limit of continuous functions. Since $T \geqslant 0$, hence $g_k \geqslant 0$, the series $\sum_{k=0}^{\infty} g_k$ is Lebesgue almost everywhere differentiable and can be differentiated termwise, cf. [19, Proposition 5.2.8]. From $T' = 0$ almost everywhere it follows that $g_k' = 0$ a.e., and hence

$$f'(x) = 2 - \sum_{k=0}^{\infty} g_k'(x) = 2$$

for Lebesgue almost all $x \in [0,1]$. In order to prove that f is not increasing on any interval, fix an interval $I \neq \emptyset$ in $[0,1]$ and choose $n \in \mathbb{N}$ such that $I_n \subseteq I$. Denote by $x_n < y_n$ the points chosen in I_n for the construction of f. Since $g_n(y_n) = 3^{-n}T(1) = 3^{-n}$ and $g_k \geqslant 0$ for all $k \in \mathbb{N}_0$, we have

$$f(y_n) \leqslant f_n(y_n) = f_{n-1}(y_n) - g_n(y_n) = f_{n-1}(y_n) - 3^{-n}.$$

On the other hand, it follows from $g_n(x_n) = 0$ and $\|g_k\|_\infty \leqslant 3^{-k}$ that

$$f(x_n) = f_{n-1}(x_n) - \sum_{k=n+1}^{\infty} g_k(x_n) \geqslant f_{n-1}(x_n) - \sum_{k=n+1}^{\infty} 3^{-k} = f_{n-1}(x_n) - \frac{1}{2}3^{-n}.$$

By our choice of the points x_n, y_n, it holds either that $f_{n-1}(x_n) > f_{n-1}(y_n)$ or that $f_{n-1}(y_n) \leqslant f_{n-1}(x_n) + 3^{-(n+1)}$. In both cases, it is immediate from the previous two inequalities that $f(y_n) < f(x_n)$. Hence, $f|_I$ is not increasing.

14.7 A function which is Lebesgue almost everywhere differentiable but f' does not exist on a dense subset of $\mathbb{R}$

Take an enumeration $(q_n)_{n\in\mathbb{N}}$ of $\mathbb{Q}$ and define

$$f(x) := \sum_{n\in\mathbb{N}} 3^{-n}\mathbb{1}_{[q_n,\infty)}(x), \quad x \in \mathbb{R}.$$

Since f is increasing, it is of bounded variation on compact sets, and so f is Lebesgue almost everywhere differentiable, cf. [19, Theorem 5.2.6]. Yet f is discontinuous at $x \in \mathbb{Q}$. Indeed, if $x = q_n$ then $f(x) \geqslant f(y) + 3^{-n}$ for all $y < x$ and so

$$f(x-) = \lim_{y\uparrow x} f(y) \leqslant f(x) - 3^{-n} < f(x).$$

14.8 $f_n \to f$ and $f_n' \to g$ pointwise does not imply $f' = g$ a.e.

Let $f_n : \mathbb{R} \to \mathbb{R}$, $n \in \mathbb{N}$, be a sequence of continuously differentiable functions such that $f_n \to f$ and $f_n' \to g$ pointwise for some functions f, g. It is relatively easy to see that these assumptions do not imply that $f'(x) = g(x)$ for all $x \in \mathbb{R}$; in fact, f need not be everywhere differentiable. For instance, the sequence $f_n(x) := \sqrt{x^2 + \frac{1}{n}}$ is continuously differentiable but the pointwise limit $f(x) = |x|$ is not differentiable at $x = 0$. Here we will show a much stronger statement: $f'(x) \neq g(x)$ may happen on a set of positive Lebesgue measure, even if f is everywhere differentiable.

Let $C \subseteq [0,1]$ be the geometric Cantor set [☞7.3] with $a_n = \frac{1}{4}\frac{2^n}{4^n}$, that is $C = \bigcap_{n\in\mathbb{N}_0} C_n$ where C_n is obtained by removing from the middle of each of the 2^{n-1} parts of C_{n-1} an open interval of length $2^{-(n-1)} = 4^{-n}$. We will construct a sequence of continuously differentiable functions $f_n : \mathbb{R} \to \mathbb{R}$ such that $f_n(x) \to 0$ and $f_n'(x) \to \mathbb{1}_C(x)$ for all $x \in [0,1]$. Since $\lambda(C) > 0$, this means that the derivative of the limit $f = 0$ differs from the limit of the derivatives on a set of strictly positive Lebesgue measure.

We begin with some preparations. By construction, C_n consists of 2^n intervals of the same length ℓ_n. Take a piecewise linear function $h_n : \mathbb{R} \to \mathbb{R}$ with the following properties, cf. Fig. 14.1:

- $h_n(x) = 1$ for all $0 \leqslant x \leqslant \frac{1}{2}(\ell_{n-1} - 4^{-n})$ and for all $\frac{1}{2}(\ell_{n-1} + 4^{-n}) \leqslant x \leqslant \ell_{n-1}$;
- $h_n(x) = 0$ for all $x \leqslant -\frac{1}{2}4^{-(n+1)}$ and for all $x \geqslant \ell_{n-1} + \frac{1}{2}4^{-(n+1)}$;
- $h_n(\frac{1}{2}\ell_{n-1}) = c_n$ for some $c_n < 0$ which is chosen such that $\int_{\mathbb{R}} h_n(x)\,dx = 0$.

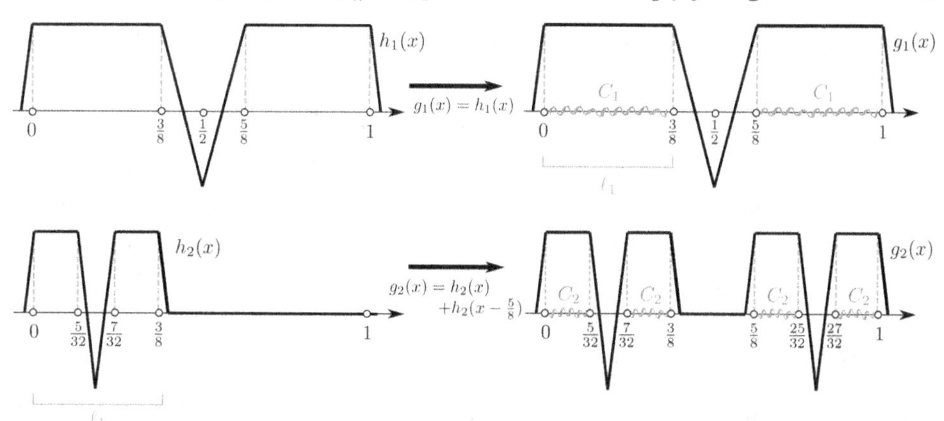

Figure 14.1 The functions h_n and g_n appearing in Example 14.8 for $n = 1, 2$. Mind that the graphs are not true to scale: the peak below zero should be much steeper.

Because of the symmetry of the function h_n with respect to the point $\frac{1}{2}\ell_{n-1}$, we have

$$\int_{-\infty}^{\ell_{n-1}/2} h_n(x)\,dx = \frac{1}{2}\int_{\mathbb{R}} h_n(x)\,dx = 0.$$

If we denote by ξ the (unique) point in the interval $[\frac{1}{2}(\ell_{n-1} - 4^{-n}), \frac{1}{2}\ell_{n-1}]$ such that $h_n(\xi) = 0$, then h_n is positive on $(-\infty, \xi)$ and negative on $[\xi, \frac{1}{2}\ell_{n-1}]$, and so

$$H_n(t) := \int_{-\infty}^{t} h_n(x)\,dx$$

attains its maximum on the interval $(-\infty, \frac{1}{2}\ell_{n-1}]$ at $t = \xi$ and

$$\sup_{t \leqslant \ell_{n-1}/2} |H_n(t)| \leqslant H_n(\xi) \leqslant \xi \leqslant \ell_{n-1}.$$

By the symmetry of h_n, we get $\|H_n\|_\infty \leqslant \ell_{n-1}$. We are now ready to define the sequences of functions with the desired properties. Set

$$g_n(x) := \sum_{k=1}^{2^{n-1}} h_n\left(x - x_k^{(n-1)}\right), \quad x \in \mathbb{R},$$

where $x_k^{(n-1)}$, $k = 1, \ldots, 2^{n-1}$, are the left end points of the 2^{n-1} disjoint intervals which form C_{n-1}, cf. Fig. 14.1. The functions $h_n(x - x_k^{(n-1)})$, $k = 1, \ldots, 2^{n-1}$, have disjoint support, and therefore it follows from our previous considerations that

$$f_n(t) := \int_{-\infty}^{t} g_n(x)\,dx, \quad t \in \mathbb{R},$$

is uniformly bounded in t by ℓ_{n-1}. As $\ell_{n-1} \to 0$, this implies, in particular, $f_n(t) \to 0$ for all $t \in [0,1]$. Since g_n is continuous, f_n is continuously differentiable and $f_n' = g_n$. We finally show that $g_n \to \mathbb{1}_C$ as $n \to \infty$. By construction, we have $g_n = 1$ on C_n and $g_n = 0$ for all x with $d(x, C_n) = \inf_{y \in C_n} |x-y| \geqslant 4^{-n}$. We consider two cases separately for $x \in [0,1]$:

Case 1. $x \in C$. From $x \in \bigcap_{n \in \mathbb{N}_0} C_n$ we find that $g_n(x) = 1$ for all $n \in \mathbb{N}$ and so $\lim_{n\to\infty} g_n(x) = 1 = \mathbb{1}_C(x)$.

Case 2. $x \notin C$. Then x is part of some open interval which has been removed in the construction of the fat Cantor set C. In particular, there are $r > 0$ and $N \in \mathbb{N}$ such that $(x-r, x+r) \notin C_N$. Since the sets C_n are decreasing, $d(x, C_n) \geqslant r$ for all $n \geqslant N$. Hence, $d(x, C_n) \geqslant 4^{-n}$, i.e. $g_n(x) = 0$, for large n.

Comment There is the following classical result from real analysis, cf. [152, Theorem 7.17]: *If $f_n : [0,1] \to \mathbb{R}$ is a sequence of differentiable functions such that $f_n \to f$ pointwise and $f_n' \to g$ uniformly on $[0,1]$, then f is differentiable and $f' = g = \lim_{n\to\infty} f_n'$.* The measure-theoretic version of this result reads as follows: *let $f_n : [0,1] \to \mathbb{R}$ be continuously differentiable functions such that $f_n \to f$ and $f_n' \to g$ pointwise. If $(f_n')_{n\in\mathbb{N}}$ is uniformly integrable* [☞ Definition 1.30], *then f is Lebesgue almost everywhere differentiable and $f' = g$ a.e.* Let us briefly sketch the proof.

By the fundamental theorem of calculus, we have

$$f(t) - f(0) = \lim_{n\to\infty} (f_n(t) - f_n(0)) = \lim_{n\to\infty} \int_0^t f_n'(s)\,ds.$$

Since the sequence $(f_n')_{n\in\mathbb{N}}$ is uniformly integrable, it follows from Vitali's convergence theorem [☞ 1.31] that

$$f(t) - f(0) = \int_0^t g(s)\,ds,$$

and so, by Lebesgue's differentiation theorem [☞ Theorem 14C], $f'(t) = g(t)$ Lebesgue almost everywhere. In our example, uniform integrability of $f_n' = g_n$ fails, and therefore we cannot interchange limit and integration; consequently, $f'(x) = g(x)$ fails on a set of positive Lebesgue measure.

14.9 A function $f(t,x)$ for which $\partial_t \int f(t,x)\,dx$ and $\int \partial_t f(t,x)\,dx$ exist but are not equal

For $t, x \in \mathbb{R}$ set $f(t,x) = t^3 \exp(-t^2 x)$ and $F(t) = \int_{(0,\infty)} f(t,x)\,\lambda(dx)$. Then

$$\frac{F(t) - F(0)}{t} = \int_{(0,\infty)} t^2 \exp(-t^2 x)\,\lambda(dx) = \int_{(0,\infty)} \exp(-y)\,\lambda(dy) = 1$$

and so

$$F'(0) = \lim_{t\to 0} \frac{F(t) - F(0)}{t} = 1,$$

while

$$\int_{(0,\infty)} \partial_t f(t,x)\Big|_{t=0} \lambda(dx) = \int_{(0,\infty)} \exp(-t^2 x)\,(3t^2 - 2xt)\Big|_{t=0} \lambda(dx) = 0.$$

14.10 A function such that $\partial_t \int f(t,x)\,dx$ exists but $\int \partial_t f(t,x)\,dx$ does not

Consider on $((0,\infty), \mathscr{B}(0,\infty))$ Lebesgue measure λ and

$$f(t,x) := \frac{1 - \cos(tx^2)}{x^2}, \quad t, x > 0.$$

As $|1 - \cos y| \leqslant 2\min\{1, |y|^2\}$, the function $x \longmapsto f(t,x)$ is integrable for each $t > 0$, and so

$$F(t) := \int_{(0,\infty)} f(t,x)\,\lambda(dx), \quad t > 0,$$

is well-defined. After a change of variables, $y = tx^2$, it follows that

$$F(t) = \frac{1}{2}\sqrt{t} \int_{(0,\infty)} \frac{1 - \cos y}{y^{3/2}}\,\lambda(dy);$$

in particular, F is differentiable on $(0,\infty)$.

The integral $\int_{(0,\infty)} \partial_t f(t,x)\,\lambda(dx)$ does not exist for any $t > 0$ since

$$\int_{(0,\infty)} |\partial_t f(t,x)|\,\lambda(dx) = \int_{(0,\infty)} |\sin(tx^2)|\,\lambda(dx) = \infty.$$

14.11 A function such that $\int \partial_t f(t,x)\,dx$ exists but $\partial_t \int f(t,x)\,dx$ does not

The function

$$f(t,x) := \begin{cases} \frac{1}{\sqrt{2\pi}} \exp\left(-\frac{x^2}{2t^2}\right), & \text{if } t, x \neq 0, \\ 0, & \text{otherwise}, \end{cases}$$

satisfies

$$F(t) := \int_{\mathbb{R}} f(t,x)\,\lambda(dx) = \frac{1}{\sqrt{2\pi}} \int_{\mathbb{R}} \exp\left(-\frac{x^2}{2t^2}\right) \lambda(dx) = |t|$$

for any $t \neq 0$; for $t = 0$ the identity $F(t) = |t|$ is trivially satisfied since we have $f(0,x) = 0$. This means that $F(t) = \int f(t,x)\,\lambda(dx)$ is not differentiable at $t = 0$. We have

$$\frac{f(t,x) - f(0,x)}{t} = \frac{1}{\sqrt{2\pi}t} \exp\left(-\frac{x^2}{2t^2}\right) \xrightarrow[t\to 0]{} 0$$

for all $x \neq 0$; for $x = 0$ the convergence to zero is trivial. Hence,

$$\left.\frac{\partial}{\partial t} f(t,x)\right|_{t=0} = 0$$

for all $x \in \mathbb{R}$, which implies

$$\int_{\mathbb{R}} \left.\frac{\partial}{\partial t} f(t,x)\right|_{t=0} \lambda(dx) = 0.$$

14.12 A bounded function such that $t \longmapsto f(t,x)$ is continuous but $t \longmapsto \int f(t,x)\,\mu(dx)$ is not continuous

[☞ 12.10]

14.13 An increasing continuous function ϕ and a continuous function f such that $\int_0^1 f(x)\,d\alpha(x) \neq \int_0^1 f(x)\alpha'(x)\,dx$

Let $\alpha : [0,1] \to [0,1]$ be the Cantor function [☞ Section 2.6], and set $f \equiv 1$. By construction, α is continuous, increasing and $\alpha' = 0$ almost everywhere. Hence,

$$\int_0^1 f(x)\alpha'(x)\,dx = 0.$$

On the other hand, it follows from elementary properties of the Riemann–Stieltjes integral that

$$\int_0^1 f(x)\,d\alpha(x) = \int_0^1 d\alpha(x) = \alpha(1) - \alpha(0) = 1.$$

14.14 A nowhere continuous function whose Lebesgue points are dense

Consider Lebesgue measure λ on $(\mathbb{R}, \mathscr{B}(\mathbb{R}))$ and set $f(x) = \mathbb{1}_{\mathbb{Q}}(x)$, $x \in \mathbb{R}$. Since the rationals are dense, it is immediate that f has no continuity points. Yet each $x \in \mathbb{R} \setminus \mathbb{Q}$ is a Lebesgue point:

$$\frac{1}{2r}\int_{(x-r,x+r)} |f(y) - f(x)|\,\lambda(dy) = \frac{1}{2r}\lambda((x-r,x+r)\cap\mathbb{Q}) = 0.$$

14.15 A discontinuous function such that every point is a Lebesgue point

Denote by λ^d Lebesgue measure on $(\mathbb{R}^d, \mathscr{B}(\mathbb{R}^d))$. If $f : \mathbb{R}^d \to \mathbb{R}$ is a measurable function and x a continuity point of f, then x is a Lebesgue point of f, that is,

$$\lim_{r\to 0}\frac{1}{\lambda^d(B_r(x))}\int_{B_r(x)} |f(y) - f(x)|\,\lambda^d(dy) = 0.$$

The converse is not true, i.e. Lebesgue points are, in general, not continuity points [☞ 14.14]. Even if every point $x \in \mathbb{R}^d$ is a Lebesgue point of f, the function f does not need to be continuous.

Consider the measure space $(\mathbb{R}, \mathscr{B}(\mathbb{R}), \lambda)$. Define continuous tent functions

$$t_n(x) := n\left(x + \frac{1}{n}\right)\mathbb{1}_{(-1/n,0]}(x) - n\left(x - \frac{1}{n}\right)\mathbb{1}_{(0,1/n)}(x), \quad x \in \mathbb{R},\ n \in \mathbb{N},$$

and set

$$f(x) := \sum_{n\geqslant 1} t_{4^n}(x - 2^{-n}), \quad x \in \mathbb{R}.$$

Since the support of t_{4^n} is contained in $[-4^{-n}, 4^{-n}]$, it follows that for each $x \neq 0$ only finitely many terms $t_{4^n}(x-2^{-n})$ are non-zero, and therefore f is continuous on $\mathbb{R} \setminus \{0\}$. In particular, each $x \in \mathbb{R} \setminus \{0\}$ is a Lebesgue point.

It remains to consider $x = 0$. Since $f(0) = 0$ and $f(2^{-n}) = 1$ for all $n \in \mathbb{N}$, f is not continuous at $x = 0$. To prove that 0 is a Lebesgue point of f, fix $r \in (0, 1)$

and $k \in \mathbb{N}_0$ such that $2^{-(k+1)} \leqslant r < 2^{-k}$. As $\int_{(0,1)} t_{4^n}(y-2^{-n})\,\lambda(dy) = 4^{-n}$ we have

$$\int_{(-r,r)} |f(y)-f(0)|\,\lambda(dy) \leqslant \sum_{n=k}^{\infty} \int_{(0,1)} t_{4^n}(y-2^{-n})\,\lambda(dy) = \sum_{n=k}^{\infty} 4^{-n} \leqslant 2\cdot 4^{-k},$$

and so

$$\frac{1}{2r}\int_{(-r,r)} |f(y)-f(0)|\,\lambda(dy) \leqslant 2^{k+1}4^{-k}.$$

Letting $r \to 0$ (i.e. $k \to \infty$) we conclude that 0 is a Lebesgue point of f.

14.16 An integrable function f such that $x \longmapsto \int_0^x f(t)\,dt$ is differentiable at $x = x_0$ but x_0 is not a Lebesgue point of f

For $x \in (0,1)$ define

$$f(x) := \begin{cases} 1, & \text{if } \frac{1}{n+1} \leqslant x < \frac{1}{2}\left(\frac{1}{n}+\frac{1}{n+1}\right),\ n \in \mathbb{N}, \\ -1, & \text{if } \frac{1}{2}\left(\frac{1}{n}+\frac{1}{n+1}\right) \leqslant x < \frac{1}{n},\ n \in \mathbb{N}. \end{cases}$$

Extend f to $(-1,1)$ by setting $f(x) := f(-x)$ for $x \in (-1,0)$ and $f(0) := 0$. Since f is with respect to the midpoint of each interval $[\frac{1}{n+1}, \frac{1}{n})$ an odd function, we have

$$\forall n \in \mathbb{N}\ :\ \int_0^{1/n} f(t)\,dt = 0. \tag{14.2}$$

We claim that $x \longmapsto F(x) := \int_0^x f(t)\,dt$ is differentiable at $x = 0$ and $F'(0) = 0$. To prove this, fix $h \in (0,1)$ and choose $n \in \mathbb{N}$ such that $h \in \left[\frac{1}{n+1}, \frac{1}{n}\right)$. From (14.2) we find that

$$\begin{aligned} \left|\frac{1}{h}\int_0^h f(t)\,dt\right| &\leqslant \frac{1}{h}\left|\int_0^{1/(n+1)} f(t)\,dt\right| + \frac{1}{h}\left|\int_{1/(n+1)}^h f(t)\,dt\right| \\ &\leqslant 0 + (n+1)\left(\frac{1}{n} - \frac{1}{n+1}\right) = \frac{1}{n} \xrightarrow[n\to\infty]{} 0. \end{aligned}$$

Because of the symmetry of f, an analogous calculation shows that the left-hand limit exists and is zero. Hence, F is differentiable at $x = 0$ and $F'(0) = 0$. Yet $x = 0$ is not a Lebesgue point of f: as $|f(x)| = 1$ for all $x \in (-1,1)\setminus\{0\}$, we have

$$\frac{1}{2\epsilon}\int_{(-\epsilon,\epsilon)} |f(t)-f(0)|\,dt = \frac{1}{2},$$

which does not converge to 0 as $\epsilon \downarrow 0$.

14.17 Lebesgue points of f need not be Lebesgue points of f^2

Consider on $((-1,1),\mathscr{B}(-1,1))$ Lebesgue measure λ. For $n \geqslant 2$ define disjoint intervals $I_n := \left[\frac{1}{n}, \frac{1}{n} + \frac{1}{n^3}\right)$ and set

$$f(x) := \sum_{n=2}^{\infty} n^{3/4} \mathbb{1}_{I_n}(x), \quad x \in (-1,1).$$

An application of the monotone convergence theorem shows

$$\int_{(-1,1)} |f(x)|^2 \,\lambda(dx) = \sum_{n=2}^{\infty} \int_{I_n} |f(x)|^2 \,\lambda(dx) = \sum_{n=2}^{\infty} n^{3/2} \frac{1}{n^3} < \infty,$$

and so $f \in L^2(-1,1) \subseteq L^1(-1,1)$. We claim that $x = 0$ is a Lebesgue point of f but not of f^2. Fix some $\epsilon \in (0,1)$ and choose $N \in \mathbb{N}$ such that $\epsilon \in \left[\frac{1}{N+1}, \frac{1}{N}\right)$. Then

$$\begin{aligned}\frac{1}{2\epsilon} \int_{(-\epsilon,\epsilon)} |f(x) - f(0)| \,\lambda(dx) &\leqslant \frac{N+1}{2} \sum_{n=N+1}^{\infty} \int_{I_n} |f(x)| \,\lambda(dx) \\ &\leqslant \frac{N+1}{2} \sum_{n=N+1}^{\infty} n^{-9/4} \leqslant \sum_{n=N+1}^{\infty} n^{-5/4} \xrightarrow[N\to\infty]{} 0,\end{aligned}$$

which shows that $x = 0$ is a Lebesgue point of f. From

$$\begin{aligned}\int_{(-1/N,1/N)} |f(x)^2 - f(0)^2| \,\lambda(dx) &= \sum_{n=N+1}^{\infty} \int_{I_n} |f(x)|^2 \,\lambda(dx) \\ &= \sum_{n=N+1}^{\infty} n^{-3/2} \geqslant \sum_{n=N+1}^{2N} n^{-3/2} \geqslant N \cdot (2N)^{-3/2},\end{aligned}$$

we see that

$$\lim_{N\to\infty} \frac{N}{2} \int_{(-1/N,1/N)} |f(x)^2 - f(0)^2| \,\lambda(dx) = \infty,$$

i.e. $x = 0$ is not a Lebesgue point of f^2.

14.18 Functions $f \in L^p$, $0 < p < 1$, without Lebesgue points

Take on $(\mathbb{R}^d, \mathscr{B}(\mathbb{R}^d))$ Lebesgue measure λ^d and set $L^p = L^p(\mathbb{R}^d, \mathscr{B}(\mathbb{R}^d), \lambda^d)$. Write $L(f)$ for the set of Lebesgue points of a measurable function $f : \mathbb{R}^d \to \mathbb{R}$.

Lebesgue's differentiation theorem [☞ 14C] states that $\lambda^d(\mathbb{R}^d \setminus L(f)) = 0$ for any $f \in L^1$. If $f \in L^p$ for some $p \in [1,\infty)$, then $f\mathbb{1}_{B_n(0)} \in L^1$ for all $n \in \mathbb{N}$, and therefore it follows immediately that $\lambda^d(\mathbb{R}^d \setminus L(f)) = 0$ for any $f \in L^p$,

$1 \leqslant p < \infty$. The situation changes drastically for $p \in (0,1)$: a function $f \in L^p$, $0 < p < 1$, may have no Lebesgue points. The reason is that a function $f \in L^p$, $0 < p < 1$, does not need to be locally integrable. There exist $f \in L^p, 0 < p < 1$, such that

$$\forall a < b : \int_{(a,b)} |f(y)| \, \lambda^1(dy) = \infty$$

[☞ 18.12]. Clearly, $L(f) = \emptyset$ for any such function f.

14.19 Lebesgue's differentiation theorem fails for sets which are not shrinking nicely

The classical Lebesgue differentiation theorem states that

$$\lim_{r \to 0} \frac{1}{\lambda^d(B_r(x))} \int_{B_r(x)} |f(y) - f(x)| \, \lambda^d(dy) = 0 \quad \text{Lebesgue a.e.}$$

for every function $f \in L^1(\lambda^d)$. This result remains valid if the open balls $B_r(x)$ are replaced by sets which are shrinking nicely to x. More precisely, choose for each $x \in \mathbb{R}^d$ a sequence $(A_n(x))_{n \in \mathbb{N}} \subseteq \mathscr{B}(\mathbb{R}^d)$ which **shrinks nicely** to x, in the sense that there exists some $r_n = r_n(x) > 0$ such that $r_n \to 0$ as $n \to \infty$, $x \in A_n(x) \subseteq B_{r_n}(x)$ and $\lambda^d(A_n(x)) \geqslant \alpha \lambda^d(B_{r_n}(x))$ for some $\alpha > 0$. If $f \in L^1(\lambda^d)$, then

$$\lim_{n \to \infty} \frac{1}{\lambda^d(A_n(x))} \int_{A_n(x)} |f(y) - f(x)| \, \lambda^d(dy) = 0 \quad \text{a.e.,} \tag{14.3}$$

cf. [150, Theorem 7.10]. The following example shows that (14.3) may break down if the sets $A_n(x)$ do not shrink nicely.

Take a set $P \subseteq [0,1]^2$ with Lebesgue measure $\lambda^2(P) = 1$ such that for every $x \in P$ there exists a line $\ell(x)$ which intersects P only in x, i.e. $\ell(x) \cap P = \{x\}$, cf. [68, Section 5.3] for the existence of such a set. Since Lebesgue measure is inner compact regular, there is a compact set $K \subseteq P$ with $\lambda^2(K) \geqslant \frac{1}{2}$. Fix $x \in K$. As $x \in K \subseteq P$ there exists, by construction, a line $\ell(x)$ such that $P \cap \ell(x) = \{x\}$; hence, $K \cap \ell(x) = \{x\}$. To keep the notation simple, we assume that $\ell(x)$ is the line parallel to the horizontal axis, i.e.

$$\ell(x) = \{x + te_1 \, ; \, t \in \mathbb{R}\}$$

where $e_1 = (1,0)$ is the first unit vector in $\mathbb{R}^2$. Since K^c is open,

$$r(t) := \sup\{r > 0 ; \, B_r(x + te_1) \subseteq K^c\}$$

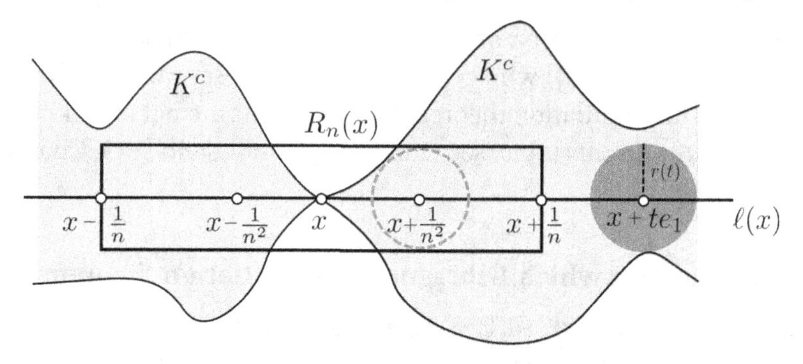

Figure 14.2 The rectangles $R_n(x)$ shrink to x as $n \to \infty$.

is strictly positive for each $t \neq 0$. The mapping $t \longmapsto r(t)$ is continuous and so

$$h_n := \inf_{1/n^2 \leqslant |s| \leqslant 1} r(s) > 0$$

for all $n \in \mathbb{N}$. Moreover, $x \in K$ implies $h_n \to 0$ as $n \to \infty$, and therefore the rectangles

$$R_n(x) := \left(x - \frac{1}{n}, x + \frac{1}{n}\right) \times (-h_n, h_n), \quad n \in \mathbb{N},$$

are shrinking to x; see Fig. 14.2. From

$$\left[\left(x - \frac{1}{n}, x - \frac{1}{n^2}\right) \cup \left(x + \frac{1}{n^2}, x + \frac{1}{n}\right)\right] \times (-h_n, h_n) \subseteq K^c,$$

it follows that

$$\int_{R_n(x)} \mathbb{1}_K(y)\, \lambda^2(dy) \leqslant \lambda^2\left(\left[x - \frac{1}{n^2}, x + \frac{1}{n^2}\right] \times (-h_n, h_n)\right) = \frac{4}{n^2} h_n.$$

Hence,

$$\frac{1}{\lambda^2(R_n(x))} \int_{R_n(x)} \mathbb{1}_K(y)\, \lambda^2(dy) \leqslant \frac{1}{n},$$

which implies that

$$\lim_{n\to\infty} \frac{1}{\lambda^2(R_n(x))} \int_{R_n(x)} \mathbb{1}_K(y)\, \lambda^2(dy) = 0 \neq \mathbb{1}_K(x).$$

In particular,

$$\lim_{n\to\infty} \frac{1}{\lambda^2(R_n(x))} \int_{R_n(x)} |\mathbb{1}_K(y) - \mathbb{1}_K(x)|\, \lambda^2(dy) = 0$$

fails to hold on the set K, which has positive Lebesgue measure.

Comment The construction of the set P goes back to Nikodým [129]. It was Zygmund (cf. [129, p. 168]) who pointed out that the set can be used to show that Lebesgue's differentiation theorem fails if the sets are not shrinking nicely. For further results on this topic, see Guzmán [68] and Stein [173, Chapter 10].

14.20 A measure for which Lebesgue's differentiation theorem fails

Let μ be a measure on $(\mathbb{R}^d, \mathscr{B}(\mathbb{R}^d))$ which is finite on all bounded open sets. If $f : \mathbb{R}^d \to \mathbb{R}$ is measurable and locally integrable, i.e. $\int_K |f|\, d\mu < \infty$ for any compact set $K \subseteq \mathbb{R}^d$, then

$$f(x) = \lim_{r \downarrow 0} \frac{1}{\mu(B_r(x))} \int_{B_r(x)} f(y)\, \mu(dy) \tag{14.4}$$

μ-almost everywhere, cf. [19, Theorem 5.8.8]. This generalizes the classical Lebesgue differentiation theorem [☞14C]. The following examples show that (14.4), in general, fails to hold if μ is not finite on bounded open sets or f is not locally integrable.

(a) *μ is not finite on open balls.* Consider on $(\mathbb{R}, \mathscr{B}(\mathbb{R}))$ the counting measure $\mu = \zeta_{\mathbb{Q}} = \sum_{q \in \mathbb{Q}} \delta_q$ and $f(x) := 1_{\{0\}}(x)$. From $\mu(-r, r) = \infty$ we find that

$$f(0) = 1 \neq 0 = \lim_{r \downarrow 0} \frac{1}{\mu(-r, r)} \int_{(-r,r)} f(y)\, \mu(dy).$$

Since $\mu(\{0\}) = 1 > 0$, this means that (14.4) does not hold μ-almost everywhere. Note that μ is σ-finite and $f \in L^1(\mu)$. Consequently, the example also shows that the assumption of finiteness on open balls cannot be relaxed to σ-finiteness, even if f is integrable.

(b) *f is not locally integrable.* Let λ be Lebesgue measure on $((-1,1), \mathscr{B}(-1,1))$, and pick a positive measurable function $f \geqslant 0$ such that $f(0) = 0$ and $\int_{(-r,r)} f(x)\, \lambda(dx) = \infty$ for all $r > 0$, e.g. $f(x) = |x|^{-1}$ for $x \neq 0$. If we define $\mu := \lambda + \delta_0$, then

$$f(0) = 0 \neq \infty = \lim_{r \downarrow 0} \frac{1}{\mu(-r, r)} \int_{(-r,r)} f(y)\, \mu(dy),$$

i.e. (14.4) fails to hold on the set $\{0\}$ which has strictly positive μ-measure.

Comment Let μ, ν be absolutely continuous measures on the measurable space $(\mathbb{R}^d, \mathscr{B}(\mathbb{R}^d))$, $\nu \ll \mu$, which are finite on all open balls. It follows from (14.4) that the Radon–Nikodým density $f = d\nu/d\mu$ satisfies

$$f(x) = \lim_{r \downarrow 0} \frac{\nu(B_r(x))}{\mu(B_r(x))}$$

for μ-almost every $x \in \mathbb{R}^d$. The above examples show that convergence breaks down if μ or ν is not finite on open balls.

15

Measurability on Product Spaces

Let $(X_i, \mathscr{A}_i)$, $i = 1, 2$, and $(Y, \mathscr{B})$ be measurable spaces. The **product σ-algebra** $\mathscr{A}_1 \otimes \mathscr{A}_2$ is the smallest σ-algebra on $X_1 \times X_2$ containing all 'rectangles' $A_1 \times A_2$ with $A_1 \in \mathscr{A}_1$, $A_2 \in \mathscr{A}_2$, i.e.

$$\mathscr{A}_1 \otimes \mathscr{A}_2 = \sigma(\mathscr{A}_1 \times \mathscr{A}_2).$$

Equivalently, $\mathscr{A}_1 \otimes \mathscr{A}_2$ is the smallest σ-algebra $\mathscr{C}$ on $X_1 \times X_2$ such that the projections

$$\pi_i : (X_1 \times X_2, \mathscr{C}) \to (X_i, \mathscr{A}_i),\ (x_1, x_2) \longmapsto x_i$$

are measurable for $i = 1, 2$. A mapping $f : X_1 \times X_2 \to Y$ is called **jointly measurable** if f is $\mathscr{A}_1 \otimes \mathscr{A}_2/\mathscr{B}$ measurable. Joint measurability implies that $f(x_1, \cdot)$ and $f(\cdot, x_2)$ are $\mathscr{A}_2/\mathscr{B}$, resp. $\mathscr{A}_1/\mathscr{B}$, measurable[1] for each fixed $x_1 \in X_1$, $x_2 \in X_2$, but the converse is false [☞ 15.16, 15.17]. For maps $f : Y \to X_1 \times X_2$ there is the following measurability criterion.

Lemma 15A ([MIMS, p. 149]) *Let $(X_i, \mathscr{A}_i)$, $i = 1, 2$, and $(Y, \mathscr{B})$ be measurable spaces. A function $f : Y \to X_1 \times X_2$ is $\mathscr{B}/\mathscr{A}_1 \otimes \mathscr{A}_2$ measurable if, and only if, the coordinate mappings $\pi_i \circ f : Y \to X_i$ are $\mathscr{B}/\mathscr{A}_i$ measurable for $i = 1, 2$.*

The definition of the product σ-algebra is analogous to the construction of the product topology on the cartesian product of two topological spaces: the product topology is the coarsest (smallest) topology on $X_1 \times X_2$ such that the projections $\pi_i : X_1 \times X_2 \to X_i$, $i = 1, 2$, are continuous. The counterpart to Lemma 15A states that a function $f : Y \to X_1 \times X_2$ is continuous with respect to the product topology if, and only if, the coordinate mappings $\pi_i \circ f$ are continuous. The Borel σ-algebra $\mathscr{B}(X_1 \times X_2)$ associated with the product topology may be strictly larger than the product σ-algebra $\mathscr{B}(X_1) \otimes \mathscr{B}(X_2)$ [☞ 15.6, 15.7].

[1] Since this is part of the assertion of Fubini's theorem [☞ Corollary 1.34], authors often invoke Fubini's theorem for such measurability considerations.

15.1 A function which is Borel measurable but not Lebesgue measurable

Consider the embedding $\iota: \mathbb{R} \to \mathbb{R}^2$, $x \longmapsto (x,0)$. For every $A, B \in \mathscr{B}(\mathbb{R})$ we have

$$\iota^{-1}(A \times B) = \{x\,;\ (x,0) \in A \times B\} = \begin{Bmatrix} A, & \text{if } 0 \in B \\ \emptyset, & \text{if } 0 \notin B \end{Bmatrix} \in \mathscr{B}(\mathbb{R}).$$

Since the rectangles $\mathscr{B}(\mathbb{R}) \times \mathscr{B}(\mathbb{R})$ generate $\mathscr{B}(\mathbb{R}^2)$, this proves Borel measurability of ι.

The function ι is not Lebesgue measurable. To see this, take any $N \subseteq \mathbb{R}$, $N \notin \mathscr{L}(\mathbb{R})$. Since $N \times \{0\}$ is contained in the (Borel measurable) Lebesgue null set $\mathbb{R} \times \{0\}$, it follows from the completeness of the Lebesgue σ-algebra that $N \times \{0\} \in \mathscr{L}(\mathbb{R}^2)$. From

$$\iota^{-1}(N \times \{0\}) = \{x\,;\ (x,0) \in N \times \{0\}\} = N \notin \mathscr{L}(\mathbb{R}),$$

we conclude that ι is not $\mathscr{L}(\mathbb{R})/\mathscr{L}(\mathbb{R}^2)$ measurable.

Comment 1 ι is not $\mathscr{L}(\mathbb{R})/\mathscr{L}(\mathbb{R}^2)$ measurable but it is $\mathscr{L}(\mathbb{R})/\mathscr{L}(\mathbb{R}) \otimes \mathscr{L}(\mathbb{R})$ measurable [☞ Lemma 15A]. In particular, $\mathscr{L}(\mathbb{R}) \otimes \mathscr{L}(\mathbb{R}) \subsetneqq \mathscr{L}(\mathbb{R}^2)$.

Comment 2 By definition, every Borel measurable function $f: \mathbb{R}^k \to \mathbb{R}^d$ satisfies $f^{-1}(\mathscr{B}(\mathbb{R}^d)) \subseteq \mathscr{B}(\mathbb{R}^k)$. Since the Lebesgue σ-algebra is larger than the Borel σ-algebra, this gives $f^{-1}(\mathscr{B}(\mathbb{R}^d)) \subseteq \mathscr{L}(\mathbb{R}^k)$. For Lebesgue measurability we would need $f^{-1}(B) \in \mathscr{L}(\mathbb{R}^k)$ not only for all Borel sets but for all Lebesgue sets, and – as our example shows – this is, in general, not the case. Consequently, Borel measurability does not imply Lebesgue measurability, not even in dimension one [☞ 8.12]. Conversely, Lebesgue measurability does not imply Borel measurability: take $f(x) = \mathbb{1}_L(x)$ for some $L \in \mathscr{L}(\mathbb{R}) \setminus \mathscr{B}(\mathbb{R})$ [☞ 7.20]. Our observations can be summarized in the following abstract way: *If $\mathscr{A} \subsetneqq \mathscr{A}'$ and $\mathscr{C} \subsetneqq \mathscr{C}'$ are σ-algebras, then $\mathscr{A}/\mathscr{C}$ measurability of a function f need not imply $\mathscr{A}'/\mathscr{C}'$ measurability of f – and vice versa.*

15.2 The product of complete σ-algebras need not be complete

Consider the Lebesgue σ-algebra $\mathscr{L}(\mathbb{R}^d)$ on $\mathbb{R}^d$. By definition, $\mathscr{L}(\mathbb{R}^d)$ is the completion of $\mathscr{B}(\mathbb{R}^d)$ with respect to Lebesgue measure.

In particular, $\mathscr{L}(\mathbb{R}) \otimes \mathscr{L}(\mathbb{R})$ is a product of complete σ-algebras. This product is not complete: take any $N \notin \mathscr{L}(\mathbb{R})$, then $N \times \{0\} \in \mathscr{L}(\mathbb{R}^2)$ [☞ 15.1] but $N \times \{0\} \notin \mathscr{L}(\mathbb{R}) \otimes \mathscr{L}(\mathbb{R})$. Indeed, since

$$\iota: (\mathbb{R}, \mathscr{L}(\mathbb{R})) \to (\mathbb{R}^2, \mathscr{L}(\mathbb{R}) \otimes \mathscr{L}(\mathbb{R})),\ x \longmapsto \iota(x) = (x,0)$$

is measurable [☞ Lemma 15A], it follows from $N = \iota^{-1}(N \times \{0\}) \notin \mathscr{L}(\mathbb{R})$ that $N \times \{0\} \notin \mathscr{L}(\mathbb{R}) \otimes \mathscr{L}(\mathbb{R})$.

Comment Since $\mathscr{L}(\mathbb{R}^2)$ is complete, this example also shows that $\mathscr{L}(\mathbb{R}^2)$ is strictly larger than $\mathscr{L}(\mathbb{R}) \otimes \mathscr{L}(\mathbb{R})$.

Comment It happens in many situations that the product of complete measure spaces is not complete: if $(X, \mathscr{A}, \mu)$ and $(Y, \mathscr{B}, \nu)$ are σ-finite, complete measure spaces such that $\mathscr{A} \neq \mathscr{P}(X)$ and $\mathscr{B}$ contains non-empty null sets, then the product space $(X \times Y, \mathscr{A} \otimes \mathscr{B}, \mu \times \nu)$ is not complete.

To prove this, pick some set $Z \in \mathscr{P}(X) \setminus \mathscr{A}$ – because of completeness, Z is not a null set – and some ν-null set $N \in \mathscr{B}$, $N \neq \emptyset$. Denote by $(A_k)_{k\in\mathbb{N}} \subseteq \mathscr{A}$, $A_k \uparrow X$ and $\mu(A_k) < \infty$ some exhausting sequence. Since the measure $\mu \times \nu$ is continuous from below, we have

$$\begin{aligned}(\mu \times \nu)(X \times N) &= \sup_{k\in\mathbb{N}} (\mu \times \nu)(A_k \times N) \\ &= \sup_{k\in\mathbb{N}} \big(\mu(A_k) \cdot \nu(N)\big) = 0,\end{aligned}$$

which means that $Z \times N \subseteq X \times N$ is a subset of a measurable $\mu \times \nu$ null set. If the product space were indeed complete, then $Z \times N$ would be in $\mathscr{A} \otimes \mathscr{B}$, and so the section

$$x \longmapsto \mathbb{1}_{Z\times N}(x, y) = \mathbb{1}_Z(x)\mathbb{1}_N(y) \overset{y\in N}{=} \mathbb{1}_Z(x)$$

would be $\mathscr{A}$ measurable. This is possible only if $Z \in \mathscr{A}$, which is not the case. Thus, the product cannot be complete.

Note that σ-finiteness is essential to conclude that $(\mu \times \nu)(X \times N) = 0$, which leads to the contradiction; σ-finiteness is also essential to ensure the uniqueness of product measure [☞ Theorem 1.33, 16.1].

15.3 $\mathscr{L}(\mathbb{R}) \otimes \mathscr{L}(\mathbb{R}) \subsetneq \mathscr{L}(\mathbb{R}^2)$

[☞ 15.2]

15.4 Sigma algebras $\mathscr{A} = \sigma(\mathscr{G})$ and $\mathscr{B} = \sigma(\mathscr{H})$ such that $\sigma(\mathscr{G} \times \mathscr{H})$ is strictly smaller than $\mathscr{A} \otimes \mathscr{B}$

Let X be a non-empty set. Take $A \subsetneq X$ with $A \neq \emptyset$, and set $\mathscr{G} := \{A\}$ and $\mathscr{H} := \{A^c\}$. Then

$$\sigma(\mathscr{G}) = \sigma(\mathscr{H}) = \{\emptyset, A, A^c, X\} =: \mathscr{A}.$$

The product σ-algebra $\mathscr{A} \otimes \mathscr{A}$ is not generated by $\mathscr{G} \times \mathscr{H}$ since $A \times A \in \mathscr{A} \otimes \mathscr{A}$ but $A \times A \notin \sigma(\mathscr{G} \times \mathscr{H})$.

Comment This pathology cannot happen if the generators contain an exhausting sequence: if $(X, \mathscr{A})$ and $(Y, \mathscr{B})$ are measurable spaces with $\mathscr{A} = \sigma(\mathscr{G})$ and $\mathscr{B} = \sigma(\mathscr{H})$ and there exist sequences $(G_n)_{n \in \mathbb{N}} \subseteq \mathscr{G}$ and $(H_n)_{n \in \mathbb{N}} \subseteq \mathscr{H}$ with $G_n \uparrow X$ and $H_n \uparrow Y$, then $\sigma(\mathscr{G} \times \mathscr{H}) = \mathscr{A} \otimes \mathscr{B}$, cf. [MIMS, p. 138]. As an immediate consequence, it follows that for any two generators $\mathscr{G}$, $\mathscr{H}$ the identity $\sigma(\mathscr{G}' \times \mathscr{H}') = \mathscr{A} \otimes \mathscr{B}$ holds for the enlarged generators $\mathscr{G}' := \mathscr{G} \cup \{X\}$ and $\mathscr{H}' := \mathscr{H} \cup \{Y\}$.

15.5 An example where $\mathscr{P}(X) \otimes \mathscr{P}(X) \neq \mathscr{P}(X \times X)$

We start with the following auxiliary result which can be found in [124].

Lemma 1 *Let $(X, \mathscr{A})$, $(Y, \mathscr{B})$ be measurable spaces and $f : (X, \mathscr{A}) \to (Y, \mathscr{B})$ a measurable function. If the graph of f,*

$$\Gamma_f := \{(x, y) \in X \times Y \,;\; y = f(x)\},$$

belongs to $\mathscr{A} \otimes \mathscr{B}$, then there exist countable families $\mathscr{G} \subseteq \mathscr{A}$, $\mathscr{H} \subseteq \mathscr{B}$ such that $\Gamma_f \in \sigma(\mathscr{G}) \otimes \sigma(\mathscr{H})$ and $\sigma(\mathscr{H})$ contains all singletons $\{f(x)\}$, $x \in X$.

Proof Define

$$\Sigma := \{S \subseteq X \times Y \,;\; \exists (A_i)_{i \in \mathbb{N}} \subseteq \mathscr{A},\ (B_i)_{i \in \mathbb{N}} \subseteq \mathscr{B} \;:\; S \in \sigma(A_i \times B_i, i \in \mathbb{N})\}.$$

It is not difficult to check that Σ is a σ-algebra. Since Σ contains $\mathscr{A} \times \mathscr{B}$, it follows that $\mathscr{A} \otimes \mathscr{B} = \sigma(\mathscr{A} \times \mathscr{B}) \subseteq \sigma(\Sigma) = \Sigma$. By assumption, $\Gamma_f \in \mathscr{A} \otimes \mathscr{B}$, and so there exist $(A_i)_{i \in \mathbb{N}} \subseteq \mathscr{A}$ and $(B_i)_{i \in \mathbb{N}} \subseteq \mathscr{B}$ such that $\Gamma_f \in \sigma(A_i \times B_i, i \in \mathbb{N})$. For fixed $x \in X$ define a family Σ' by

$$\Sigma' := \{S \subseteq X \times Y \,;\; \{y \in Y \,;\; (x, y) \in S\} \in \sigma(B_i, i \in \mathbb{N})\}.$$

Again it is not difficult to see that Σ' is a σ-algebra. Moreover, Σ' contains all sets $A_i \times B_i$, $i \in \mathbb{N}$, since

$$\{y \in Y \,;\; (x, y) \in A_i \times B_i\} = \begin{cases} B_i, & \text{if } x \in A_i, \\ \emptyset, & \text{if } x \notin A_i. \end{cases}$$

Hence, $\sigma(A_i \times B_i, i \in \mathbb{N}) \subseteq \sigma(\Sigma') = \Sigma'$. For $S = \Gamma_f \in \sigma(A_i \times B_i, i \in \mathbb{N})$ we get

$$\{f(x)\} = \{y \in Y \,;\; (x, y) \in \Gamma_f\} \in \sigma(B_i, i \in \mathbb{N}).$$

As $x \in X$ was arbitrary, this shows that $\{f(x)\}$ is contained in the countably generated σ-algebra $\sigma(B_i, i \in \mathbb{N})$ for all $x \in X$. □

To construct an example where $\mathscr{P}(X) \otimes \mathscr{P}(X) \neq \mathscr{P}(X \times X)$ we will apply Lemma 1 for $f(x) = x$ to show that the diagonal does not need to be in $\mathscr{P}(X) \otimes \mathscr{P}(X)$. We need one more auxiliary result which characterizes countably generated σ-algebras.

Lemma 2 *A σ-algebra $\mathscr{A}$ on a set X is countably generated if, and only if, there exists a function $f : X \to \mathbb{R}$ such that*

$$\mathscr{A} = f^{-1}(\mathscr{B}(\mathbb{R})) := \{f^{-1}(B);\ B \in \mathscr{B}(\mathbb{R})\}.$$

Proof Since the Borel σ-algebra $\mathscr{B}(\mathbb{R})$ is countably generated, it is immediate that any σ-algebra of the form $\mathscr{A} = f^{-1}(\mathscr{B}(\mathbb{R}))$ is countably generated. Conversely, assume that $\mathscr{A} = \sigma(A_i, i \in \mathbb{N})$. The function $f(x) := \sum_{i\in\mathbb{N}} 3^{-i}\mathbb{1}_{A_i}(x)$, $x \in X$, is $\mathscr{A}/\mathscr{B}(\mathbb{R})$ measurable, i.e. $f^{-1}(\mathscr{B}(\mathbb{R})) \subseteq \mathscr{A}$. Each A_i can be written in the form $f^{-1}(B)$ for a suitable Borel set B, e.g.

$$A_1 = f^{-1}\left(\left[\tfrac{1}{3}, \tfrac{2}{3}\right]\right) \quad \text{and} \quad A_2 = f^{-1}\left(\left[\tfrac{1}{9}, \tfrac{2}{9}\right] \cup \left[\tfrac{1}{3} + \tfrac{1}{9}, \tfrac{1}{3} + \tfrac{2}{9}\right]\right),$$

and so $\mathscr{A} = \sigma(A_i, i \in \mathbb{N}) \subseteq f^{-1}(\mathscr{B}(\mathbb{R}))$. □

Let X be an arbitrary set, and assume that its diagonal $\Delta = \{(x,x);\ x \in X\}$ belongs to $\mathscr{P}(X) \otimes \mathscr{P}(X)$. Applying Lemma 1 for $f(x) = x$, we find that there exists a countable family $\mathscr{G} \subseteq \mathscr{P}(X)$ such that $\sigma(\mathscr{G})$ contains all singletons $\{x\}$, $x \in X$. Since there are $\mathfrak{c}$ many Borel sets $B \in \mathscr{B}(\mathbb{R})$ [☞ 4.21, Corollary in 4.19], it follows from Lemma 2 that any countably generated σ-algebra contains at most continuum many singletons. Consequently, $\Delta \in \mathscr{P}(X) \otimes \mathscr{P}(X)$ implies that X has at most the cardinality of the continuum. In other words: if the cardinality of X exceeds that of the continuum, then $\Delta \in \mathscr{P}(X \times X)$ but $\Delta \notin \mathscr{P}(X) \otimes \mathscr{P}(X)$, i.e. $\mathscr{P}(X) \otimes \mathscr{P}(X)$ is strictly smaller than $\mathscr{P}(X \times X)$.

Comment Under the continuum hypothesis [☞ p. 43] it can be shown that $\mathscr{P}(\mathbb{R}) \otimes \mathscr{P}(\mathbb{R}) = \mathscr{P}(\mathbb{R}^2)$, cf. [171, Theorem 3.1.24].

15.6 The product of Borel σ-algebras is not always the Borel σ-algebra of the product

Let (X, d) be a metric space whose cardinality is greater than that of the continuum. Since the diagonal $\Delta \subseteq X \times X$ is closed, we have $\Delta \in \mathscr{B}(X \times X)$. On the other hand, $\Delta \notin \mathscr{P}(X) \otimes \mathscr{P}(X)$ [☞ 15.5], and, in particular, $\Delta \notin \mathscr{B}(X) \otimes \mathscr{B}(X)$. Consequently, $\mathscr{B}(X) \otimes \mathscr{B}(X) \neq \mathscr{B}(X \times X)$.

Comment [☞ 15.7] for a further example. There it is shown that the Borel σ-algebra on the Sorgenfrey plane $S \times S$ is strictly larger than the product σ-algebra

$\mathscr{B}(S) \otimes \mathscr{B}(S)$ generated by the Borel σ-algebras on the Sorgenfrey line S.

15.7 Topological spaces X, Y such that $\mathscr{B}(X) = \mathscr{B}(Y)$ but $\mathscr{B}(X \times X) \neq \mathscr{B}(Y \times Y)$

We consider on the real line $\mathbb{R}$ both the Euclidean and the Sorgenfrey topology [☞ Example 2.1.(h)]. Since we need to distinguish between these spaces, we write E for the 'Euclidean real line' and S for the 'Sorgenfrey line'; if we are only interested in the point sets, we write $\mathbb{R}$, $\mathbb{R}^2$, etc. Both topologies lead to the same Borel sets, i.e. $\mathscr{B}(E) = \mathscr{B}(S)$ [☞ 4.18] but the Borel σ-algebras on the products $S \times S$ and $E \times E$ – each with the induced product topology – do not coincide. More precisely, we are going to show that

$$\mathscr{B}(E^2) \subsetneq \mathscr{B}(S^2).$$

Since the half-open rectangles $[a, b) \times [x, y)$ are open in the Sorgenfrey plane S^2, we have

$$(a, b) \times (x, y) = \bigcup_{n \in \mathbb{N}} \left[a + \tfrac{1}{n}, b\right) \times \left[x + \tfrac{1}{n}, y\right) \in \mathscr{B}(S^2).$$

Sets of the form $(a, b) \times (x, y)$ with rational end points are a countable basis for the Euclidean topology on E^2, and so $\mathscr{B}(E^2) \subseteq \mathscr{B}(S^2)$. In order to prove that the inclusion is strict, pick $N \subseteq \mathbb{R}$ such that $N \notin \mathscr{B}(E)$ [☞ 7.20]. We claim that

$$B := \left\{ \frac{1}{\sqrt{2}} \begin{pmatrix} x \\ -x \end{pmatrix} ;\ x \in N \right\}$$

is contained in $\mathscr{B}(S^2)$ but not in $\mathscr{B}(E^2)$. By definition, $N \times \{0\}$ is the pre-image of B under a 45° clockwise rotation about the origin, i.e. $N \times \{0\} = f^{-1}(B)$ where

$$f : \mathbb{R}^2 \to \mathbb{R}^2,\ \begin{pmatrix} x \\ y \end{pmatrix} \longmapsto \frac{1}{\sqrt{2}} \begin{pmatrix} y + x \\ y - x \end{pmatrix},$$

cf. Fig. 15.1. If we set $g(x) := (x, 0)$, $x \in \mathbb{R}$, then

$$N = g^{-1}(N \times \{0\}) = (f \circ g)^{-1}(B).$$

Since $f \circ g$ is continuous in the Euclidean topology, it is $\mathscr{B}(E)/\mathscr{B}(E^2)$ measurable, and $N \notin \mathscr{B}(E)$ implies that $B \notin \mathscr{B}(E^2)$ – otherwise $(f \circ g)^{-1}(B) = N$ would give $N \in \mathscr{B}(E)$. It remains to show that $B \in \mathscr{B}(S^2)$. The half-open rectangles $[x, x+1) \times [-x, -x+1)$, $x \in B$, are open in S^2, and therefore

$$U := \bigcup_{x \in B} [x, x+1) \times [-x, -x+1)$$

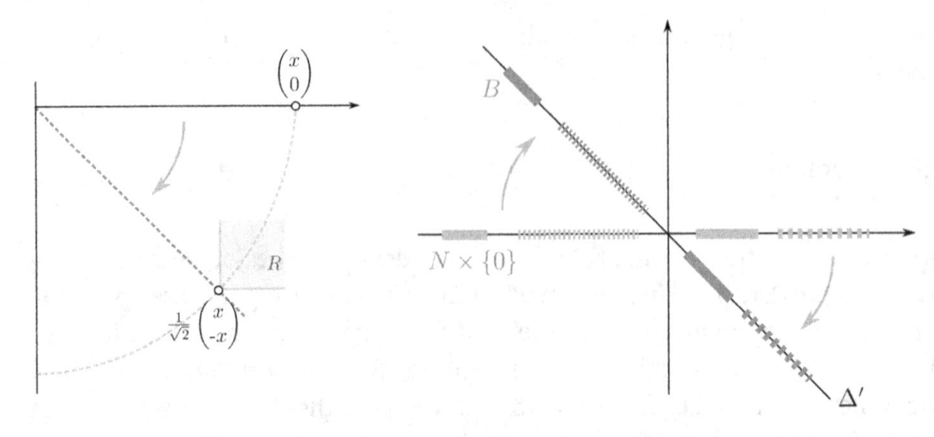

Figure 15.1 Left: A 45° clockwise rotation of the point $(x, 0)$ gives the point $\frac{1}{\sqrt{2}}(x, -x)$ on the antidiagonal. The rectangle $R = \left[\frac{x}{\sqrt{2}}, \frac{x}{\sqrt{2}}\right) \times \left[-\frac{x}{\sqrt{2}}, -\frac{x}{\sqrt{2}}\right)$ is open in the Sorgenfrey topology and intersects the antidiagonal exclusively at the rotated point. Right: The set $B \in \mathscr{B}(S^2) \setminus \mathscr{B}(E^2)$ is obtained by rotation from $N \times \{0\}$ with $N \notin \mathscr{B}(E)$.

is an open subset of the Sorgenfrey plane; in particular, $U \in \mathscr{B}(S^2)$. Since the antidiagonal

$$\Delta' = \left\{\begin{pmatrix} x \\ -x \end{pmatrix};\ x \in \mathbb{R}\right\}$$

is contained in $\mathscr{B}(E^2) \subseteq \mathscr{B}(S^2)$, we conclude that $B = U \cap \Delta' \in \mathscr{B}(S^2)$. This reasoning actually shows that **any** subset of the antidiagonal Δ' is a Borel set in the Sorgenfrey plane.

Comment From $\mathscr{B}(E) = \mathscr{B}(S)$ and $\mathscr{B}(E^2) \subsetneqq \mathscr{B}(S^2)$, we see that

$$\mathscr{B}(S) \otimes \mathscr{B}(S) = \mathscr{B}(E) \otimes \mathscr{B}(E) = \mathscr{B}(E^2) \subsetneqq \mathscr{B}(S^2),$$

i.e. the Borel σ-algebra on the product $S \times S$ is strictly larger than the product σ-algebra $\mathscr{B}(S) \otimes \mathscr{B}(S)$; for a further example of this phenomenon [☞ 15.6].

Our counterexample is also a counterexample to [201, Example 4.12] where it is claimed that $\mathscr{B}(S^2) = \mathscr{B}(E^2)$.

15.8 $\mathscr{B}(X)^{\otimes I}$ is strictly smaller than $\mathscr{B}(X^I)$ for uncountable I

Let X be a topological space containing at least two points and such that all singletons $\{x\}$, $x \in X$, are closed. For an index set I consider X^I with the product topology [☞ Example 2.1.(i)] and the corresponding Borel σ-algebra $\mathscr{B}(X^I)$. If I is uncountable, then $\mathscr{B}(X^I)$ is strictly larger than the product σ-algebra $\mathscr{B}(X)^{\otimes I}$, i.e. the smallest σ-algebra such that all projections $\pi_t : X^I \to X$, $f \longmapsto f(t)$, $t \in I$, are measurable,

$$\mathscr{B}(X)^{\otimes I} = \sigma(\pi_t \,;\; t \in I)$$

[☞ Example 4C]. Pick any $x \in X$ and define an element $f^x \in X^I$ by setting $f^x(t) := x$ for all $t \in I$. We claim that $\{f^x\} \in \mathscr{B}(X^I)$ and $\{f^x\} \notin \mathscr{B}(X)^{\otimes I}$. Since $\{f^x\} = \bigcap_{t\in I} \pi_t^{-1}(\{x\})$ is a closed subset of X^I, we have $\{f^x\} \in \mathscr{B}(X^I)$. To prove that $\{f^x\} \notin \mathscr{B}(X)^{\otimes I}$ we use that for every set $B \in \mathscr{B}(X)^{\otimes I}$ there exists a countable set $S \subseteq I$ such that

$$g \in X^I,\ h \in B,\ g|_S = h|_S \;\Rightarrow\; g \in B \tag{15.1}$$

[☞ 4.5]. Consequently, if $\{f^x\} \in \mathscr{B}(X)^{\otimes I}$, then there exists a countable set $S \subseteq I$ such that

$$g|_S = f^x|_S \;\Rightarrow\; g = f^x. \tag{15.2}$$

This is, however, possible only if I is countable. Indeed, if I is uncountable, pick $y \in X \setminus \{x\}$ and set $g := y\mathbb{1}_{I\setminus S} + x\mathbb{1}_S$; then $g|_S = f^x|_S$ and $g \neq f^x$ because $I \setminus S \neq \emptyset$; hence, (15.2) fails to hold.

Comment 1 The inclusion $\mathscr{B}(X)^{\otimes I} \subseteq \mathscr{B}(X^I)$ holds for any index set I. This follows from the fact that the projections $\pi_t : X^I \to X$ are continuous. Our example shows that the inclusion $\mathscr{B}(X)^{\otimes I} \subseteq \mathscr{B}(X^I)$ is, in general, strict. If I is at most countable and the topology of X has a countable base, then we have $\mathscr{B}(X)^{\otimes I} = \mathscr{B}(X^I)$. Indeed, it suffices to show that $\mathscr{B}(X)^{\otimes I} \supseteq \mathscr{B}(X^I)$. Take a countable basis $\mathscr{U}$ for the topology of X and consider the family of sets $\times_{t\in I} U_t$ where $I_0 := \{t \in I\,;\; U_t \neq X\}$ is finite and $U_t \in \mathscr{U}$ for all $t \in I_0$; in particular,

$$\mathop{\times}_{t\in I} U_t = \bigcap_{t\in I_0} \pi_t^{-1}(U_t) \in \mathscr{B}(X)^{\otimes I}.$$

This family of sets is countable and it is a basis for the product topology on X^I, i.e. any open set O in X^I can be written as a countable union of sets from this family; hence $\mathscr{B}(X^I) \subseteq \mathscr{B}(X)^{\otimes I}$.

Comment 2 The characterization (15.1) shows that $\mathscr{B}(X)^{\otimes I}$ is quite 'small': every set $B \in \mathscr{B}(X)^{\otimes I}$ depends on at most countably many variables. As a consequence, $\mathscr{B}(X)^{\otimes I}$ is too small for many applications, e.g. in the theory of stochastic processes.

For instance, the set of continuous functions $u : [0,1] \to \mathbb{R}$ is **not** contained in $\mathscr{B}(\mathbb{R})^{\otimes[0,1]}$ since we cannot decide whether a given function $u : [0,1] \to \mathbb{R}$ is continuous if we know only its values at countably many points $t \in [0,1]$, cf. [BM, Corollary 4.6]. Moreover, $\mathscr{B}(X)^{\otimes I}$ does not contain all compact subsets of X^I: the set $\{f^x\}$, which we considered above, is compact in X^I – this follows from Tychonoff's theorem – and $\{f^x\} \notin \mathscr{B}(X)^{\otimes I}$.

15.9 The diagonal $\Delta = \{(x, x);\ x \in X\}$ need not be measurable

Consider the σ-algebras $\mathscr{A}_1 := \mathscr{P}(X)$ and $\mathscr{A}_2 := \{\emptyset, X\}$ on $X := \{0, 1\}$. The diagonal

$$\Delta = \{(x, x);\ x \in X\}$$

is not measurable with respect to σ-algebra $\mathscr{A}_1 \otimes \mathscr{A}_2$. Note, however, that the projection $\pi_i(\Delta), i = 1, 2,$ of Δ onto the first and second coordinate, respectively, is X, and therefore $\pi_i(\Delta) \in \mathscr{A}_i$ for $i = 1, 2$.

This example might look a bit artificial since we equip X with two different σ-algebras. One might suspect that this is the actual problem, i.e. that one has $\Delta \in \mathscr{A} \otimes \mathscr{A}$, in general. This, however, is not true. If X has cardinality exceeding that of the continuum $\#X > \mathfrak{c}$, then the diagonal Δ can never be in the product σ-algebra – independently of which σ-algebra is used in X [☞15.5] and [138]. This is sometimes called **Nedoma's pathology**. Here, we give an example of a measurable space $(X, \mathscr{A})$ with $\#X = \mathfrak{c}$ such that $\Delta \notin \mathscr{A} \otimes \mathscr{A}$. We begin with an auxiliary result which is due to Dravecky [47] and which is an immediate consequence of Lemma 1 in Example 15.5.

Lemma (Dravecky) *Let $(X, \mathscr{A})$ be a measurable space. If $\Delta \in \mathscr{A} \otimes \mathscr{A}$, then there exists a countable family $\mathscr{C} \subseteq \mathscr{A}$ such that $\Delta \in \sigma(\mathscr{C}) \otimes \sigma(\mathscr{C})$ and $\{x\} \in \sigma(\mathscr{C})$ for all $x \in X$. In particular, $\sigma(\mathscr{C})$ separates the points of X, i.e. for distinct points $x \neq y$ there exists a set $S \in \sigma(\mathscr{C})$ such that $x \in S$ and $y \notin S$.*

Let X be uncountable and $\mathscr{A}$ be the co-countable σ-algebra [☞Example 4A]. Since X is not countable, there cannot exist a countably generated $\sigma(\mathscr{C}) \subseteq \mathscr{A}$ which contains all singletons $\{x\}$ – indeed, the co-countable σ-algebra $\mathscr{A}$ is the smallest σ-algebra with this property, but $\mathscr{A}$ is not countably generated if X is not countable [☞4.8]. In view of the lemma, Δ is not measurable. Note, however, that all slices $\{x;\ \mathbb{1}_\Delta(x, y_0) = 1\} = \{y_0\}$ and $\{y;\ \mathbb{1}_\Delta(x_0, y) = 1\} = \{x_0\}$ are singletons, hence $\mathscr{A}$ measurable.

Consequence 1. There are measurable functions $f, g : X \times X \to X$ such that the set $\{f = g\} = \{(x, y);\ f(x, y) = g(x, y)\}$ is not measurable. For example, the canonical projections $f(x, y) = x$ and $g(x, y) = y$ are $\mathscr{A} \otimes \mathscr{A} / \mathscr{A}$ measurable, but the set $\{f = g\} = \Delta$ is not.

Consequence 2. If $F : X \times X \to X$ is measurable, then both $x \longmapsto F(x, y)$ and $y \longmapsto F(x, y)$ are measurable. The converse is, in general, false. Take $X = [0, 1]$

with the above σ-algebra and set $F(x, y) = |x - y|$. Then $F^{-1}(\{0\}) = \Delta$ is not measurable.

Comment If the generator $\mathscr{G}$ of $\mathscr{A}$ is countable and separates points, i.e. for any pair (x, y) there is some $G \in \mathscr{G}$ such that $x \in G$ and $y \notin G$, then Δ is measurable. This follows from the fact that

$$\Delta^c = (X \times X) \setminus \Delta = \bigcup \{(G \times G^c) \cup (G^c \times G) \; ; \; G \in \mathscr{G}\}.$$

Dravecky's lemma is a partial converse: if Δ is measurable, then there is a countably generated σ-algebra $\sigma(\mathscr{C}) \subseteq \mathscr{A}$ which separates points in X. Dravecky's motivation was a remark in Rao [138], stating that $\Delta \in \mathscr{A} \otimes \mathscr{A}$ if, and only if, there exists a countably generated σ-algebra $\mathscr{B} \subseteq \mathscr{A}$ such that all singletons are contained in $\mathscr{B}$.

15.10 A metric which is not jointly measurable

Let X be a set with cardinality greater than that of the continuum. Define a metric d on X by $d(x, y) = 0$ if $x = y$ and $d(x, y) = 1$ if $x \neq y$. Then

$$d : (X \times X, \mathscr{B}(X) \otimes \mathscr{B}(X)) \to ([0, 1], \mathscr{B}[0, 1]), \quad (x, y) \longmapsto d(x, y)$$

is not measurable. Indeed: in the metric space (X, d) all singletons $\{x\}$ are open, and therefore every set $A \subseteq X$ is open; thus, $\mathscr{B}(X) = \mathscr{P}(X)$ [☞ 4.14]. Since the cardinality of X exceeds that of the continuum, the diagonal

$$\Delta = \{(x, x) ; \; x \in X\}$$

does not belong to $\mathscr{P}(X) \otimes \mathscr{P}(X) = \mathscr{B}(X) \otimes \mathscr{B}(X)$ [☞ 15.5]. As $\Delta = d^{-1}(\{0\})$ this means that d is not (jointly) measurable.

Comment This argument works, more generally, for **any** metric space (X, d) whose cardinality exceeds that of the continuum, i.e. the metric d always fails to be jointly measurable if $\#X > \mathfrak{c}$.

15.11 A non-measurable set whose projections are measurable

The non-measurable diagonal from Example 15.9 has this property.

15.12 A measurable set whose projection is not measurable

Take a Souslin set $A \subseteq \mathbb{R}$ which is not a Borel set; [☞ 7.19] for the existence of A. Since any Souslin set in $\mathbb{R}$ can be written as the projection of a Borel set in

$\mathbb{R}^2$, there exists some $B \in \mathscr{B}(\mathbb{R}^2)$ such that

$$A = \{x \in \mathbb{R};\ (x, y) \in B \text{ for some } y \in \mathbb{R}\}$$

[☞ 7.19]. Equivalently, $\pi_1(B) = A$ for the projection $\pi_1 : \mathbb{R}^2 \to \mathbb{R}, (x, y) \longmapsto x$. By our choice of A, the projection of the Borel set B onto the first coordinate is not Borel measurable.

Comment Consider $X = \mathbb{R}$ with the Lebesgue σ-algebra $\mathscr{L}(\mathbb{R})$. If we choose $A \in \mathscr{P}(\mathbb{R}) \setminus \mathscr{L}(\mathbb{R})$, then $B := A \times \{0\}$ is a subset of the Lebesgue null set $\mathbb{R} \times \{0\}$, and therefore, by completeness, $B \in \mathscr{L}(\mathbb{R}^2)$. Moreover, $\pi_1(B) = A \notin \mathscr{L}(\mathbb{R})$. However, this counterexample is not completely satisfying since $\mathscr{L}(\mathbb{R}^2)$ is not the 'natural' σ-algebra on the product $\mathbb{R} \times \mathbb{R}$, i.e. $\mathscr{L}(\mathbb{R}^2) \neq \mathscr{L}(\mathbb{R}) \otimes \mathscr{L}(\mathbb{R})$ [☞ 15.2].

Lebesgue [103, p. 191][2] claimed that every coordinate projection of a Borel set $E \subseteq \mathbb{R}^n$ is again Borel measurable; our example shows that this is incorrect. Souslin [170] had discovered this error and, in an attempt to fix the proof, he invented Souslin schemes and discovered analytic sets, which are nowadays also called Souslin sets. Projections of Souslin sets are again Souslin [☞ pp. 138–139, 7.19].

15.13 A non-measurable set whose slices are measurable

Let $(X, \mathscr{A})$ and $(Y, \mathscr{B})$ be measurable spaces. For $C \subseteq X \times Y$ we call

$$C_x := \{y \in Y;\ (x, y) \in C\} \quad \text{and} \quad C^y := \{x \in X;\ (x, y) \in C\}, \quad x \in X,\ y \in Y$$

slices of C. Since the mappings

$$(Y, \mathscr{B}) \ni y \longmapsto (x, y) \in (X \times Y, \mathscr{A} \otimes \mathscr{B}),$$
$$(X, \mathscr{A}) \ni x \longmapsto (x, y) \in (X \times Y, \mathscr{A} \otimes \mathscr{B}),$$

are measurable for each fixed $x \in X$ and $y \in Y$, respectively, it follows that $C \in \mathscr{A} \otimes \mathscr{B}$ implies $C_x \in \mathscr{B}$ and $C^y \in \mathscr{A}$ for all $x \in X$ and $y \in Y$. The converse is not true, i.e. the measurability of the slices does, in general, not imply $C \in \mathscr{A} \otimes \mathscr{B}$.

Let X be an uncountable set and denote by $\mathscr{A}$ the co-countable σ-algebra generated by the singletons [☞ Example 4A],

$$\mathscr{A} = \sigma(\{x\};\ x \in X).$$

Because X is uncountable, $\mathscr{A} \subsetneqq \mathscr{P}(X)$. For any $N \in \mathscr{P}(X) \setminus \mathscr{A}$ the diagonal

[2] Je vais démontrer *que, si E est mesurable B[orel], sa projection l'est aussi* [original italics].

$\Delta N = \{(x,x);\ x \in N\}$ is not contained in $\mathscr{A} \otimes \mathscr{A}$ but all its slices are measurable. Indeed: since the slices $(\Delta N)_x$ and $(\Delta N)^x$, $x \in X$, are either empty or singletons, it is clear that $(\Delta N)_x \in \mathscr{A}$ and $(\Delta N)^x \in \mathscr{A}$ for all $x \in X$. On the other hand, the map $T : (X, \mathscr{A}) \to (X \times X, \mathscr{A} \otimes \mathscr{A})$, $x \longmapsto T(x) = (x,x)$ is measurable:

$$\forall A, B \in \mathscr{A} \ : \ T^{-1}(A \times B) = A \cap B \in \mathscr{A}.$$

Since $N = T^{-1}(\Delta N) \notin \mathscr{A}$, we conclude that $\Delta N \notin \mathscr{A} \otimes \mathscr{A}$.

Comment With more effort, one can show that the diagonal $\Delta = \{(x,x);\ x \in X\}$ is not in $\mathscr{A} \otimes \mathscr{A}$ [☞15.9].

15.14 A measurable function with a non-measurable graph

Let $f : (X, \mathscr{A}) \to (Y, \mathscr{B})$ be a measurable function whose graph is denoted by $\Gamma_f = \{(x, f(x));\ x \in X\} \subseteq X \times Y$. Take $X = Y = [0,1]$, $f = \mathrm{id}$ and $\mathscr{A} = \mathscr{B} = \sigma(\{x\};\ x \in X)$. Since Γ_{id} is the diagonal, Example 15.9 shows that $\Gamma_{\mathrm{id}} \notin \mathscr{A} \otimes \mathscr{B}$.

Comment Let $(X, \mathscr{A})$ and $(Y, \mathscr{B})$ be measurable spaces. If the diagonal satisfies $\Delta = \{(y,y);\ y \in Y\} \in \mathscr{B} \otimes \mathscr{B}$ and if f is an $\mathscr{A}/\mathscr{B}$ measurable function, then $\Gamma_f \in \mathscr{A} \otimes \mathscr{B}$. This follows immediately from the fact that the map $(x,y) \longmapsto (f(x), y)$ is $\mathscr{A} \otimes \mathscr{B}/\mathscr{B} \otimes \mathscr{B}$ measurable.

15.15 A non-measurable function with a measurable graph

Denote by $\mathrm{id} : ([0,1], \mathscr{B}[0,1]) \to ([0,1], \mathscr{L}[0,1])$, $x \longmapsto x$, the identity map. Since $\mathscr{B}[0,1] \subsetneq \mathscr{L}[0,1]$ [☞7.20], it is obvious that id is not measurable. The graph is the diagonal $\{(x,x);\ x \in [0,1]\}$ and this is a Borel set of $[0,1] \times [0,1]$.

15.16 A function $f(x,y)$ which is measurable in each variable but fails to be jointly measurable

Consider $X = \mathbb{R}$ with the co-countable σ-algebra $\mathscr{A}$ [☞Example 4A] and the mapping $f : \mathbb{R}^2 \to \mathbb{R}, (x,y) \longmapsto f(x,y) := x - y$. Since

$$(\mathbb{R}, \mathscr{A}) \to (\mathbb{R}, \mathscr{A}),\ x \longmapsto x \quad \text{and} \quad (\mathbb{R}, \mathscr{A}) \to (\mathbb{R}, \mathscr{A}),\ y \longmapsto -y$$

are measurable, the coordinate mappings $f(x, \cdot)$ and $f(\cdot, y)$ are $\mathscr{A}/\mathscr{A}$ measurable for each fixed $x, y \in \mathbb{R}$. Yet f is not $\mathscr{A} \otimes \mathscr{A}/\mathscr{A}$ measurable: the pre-image of $\{0\} \in \mathscr{A}$ under f is the diagonal $\Delta = \{(x,x);\ x \in \mathbb{R}\}$, which is not in $\mathscr{A} \otimes \mathscr{A}$ [☞15.9].

Comment This example also shows that the sum of measurable mappings need not be measurable: $(x,y) \longmapsto x$ and $(x,y) \longmapsto -y$ are $\mathscr{A} \otimes \mathscr{A}/\mathscr{A}$ measurable but $f(x,y) = x - y$ is not $\mathscr{A} \otimes \mathscr{A}/\mathscr{A}$ measurable. For a more elementary variant of this phenomenon [☞8.15].

15.17 A function $f(x,y)$ which is separately continuous in each variable but fails to be Borel measurable

Let $f : \mathbb{R} \times \mathbb{R} \to \mathbb{R}$ be a function which is separately continuous in each variable, i.e. $f(x,\cdot)$ and $f(\cdot,y)$ are continuous for each fixed $x \in \mathbb{R}$ resp. $y \in \mathbb{R}$. A theorem by Lebesgue [99] states that any such function f is Borel measurable.

More generally, the result holds for metric spaces: if X, Y and Z are metric spaces and $f : X \times Y \to Z$ is separately continuous, then f is Borel measurable, cf. [122]. The following example from Rudin [149] shows that Borel measurability may fail to hold if Z is not a metric space.

Take $X = Y = \mathbb{R}$ with the Euclidean metric and Z the space of bounded Borel measurable functions $u : \mathbb{R} \to \mathbb{R}$ with the topology of pointwise convergence. Define a continuous function $g : \mathbb{R}^2 \to \mathbb{R}$ by

$$g(x,y) := \begin{cases} \frac{2xy}{x^2+y^2}, & \text{if } (x,y) \neq (0,0), \\ 0, & \text{if } (x,y) = (0,0), \end{cases}$$

and set

$$f : \mathbb{R} \times \mathbb{R} \to Z,\ (x,y) \longmapsto \{t \longmapsto f(x,y)(t) := g(x-t, y-t)\}.$$

Since g is separately continuous and Z is equipped with the topology of pointwise convergence, it is immediate that f is separately continuous in each variable. Let us check that f is not Borel measurable. For fixed $t \in \mathbb{R}$, the set

$$U_t := \left\{u \in Z\,;\ |u(t)| < \frac{1}{2}\right\}$$

is open in Z. On the diagonal $\Delta = \{(t,t)\,;\ t \in \mathbb{R}\}$ the function g takes the value 1 except at the origin $(0,0)$, and this gives

$$f^{-1}(U_t) \cap \Delta = \{(t,t)\}, \quad t \in \mathbb{R}.$$

Let $N \subseteq \mathbb{R}$ be non-Borel measurable [☞7.22, 7.20]. The set $U := \bigcup_{t \in N} U_t \subseteq Z$ is a union of open sets, hence open and so $U \in \mathscr{B}(Z)$. Moreover,

$$f^{-1}(U) \cap \Delta = \bigcup_{t \in N} f^{-1}(U_t) \cap \Delta = \{(t,t)\,;\ t \in N\} = N \times N,$$

but the product $N \times N$ is not a Borel set. If it were a Borel set, then it would follow from the Borel measurability of the mapping

$$F : \mathbb{R} \to \mathbb{R}^2, \ x \longmapsto F(x) := (x, x)$$

that $N = F^{-1}(N \times N)$ is Borel measurable, in contradiction to our choice of N. Hence, $f^{-1}(U) \cap \Delta \notin \mathscr{B}(\mathbb{R}^2)$, and as $\Delta \in \mathscr{B}(\mathbb{R}^2)$ this implies $f^{-1}(U) \notin \mathscr{B}(\mathbb{R}^2)$. Consequently, f is not Borel measurable.

15.18 An $\mathcal{A} \otimes \mathcal{B}$ measurable function $f \geqslant 0$ which cannot be approximated from below by simple functions of product form

Let $(X, \mathscr{A})$ and $(Y, \mathscr{B})$ be measurable spaces, and denote by $\mathcal{S}^+(\mathscr{A} \times \mathscr{B})$ the family of all positive simple functions whose steps are rectangles $A \times B \in \mathscr{A} \times \mathscr{B}$:

$$\left\{ (x, y) \longmapsto g(x, y) = \sum_{i=1}^{n} c_i \mathbb{1}_{A_i}(x) \mathbb{1}_{B_i}(y) ; \ n \in \mathbb{N}, A_i \in \mathscr{A}, B_i \in \mathscr{B}, c_i \geqslant 0 \right\}.$$

If $f : X \times Y \to [0, \infty)$ is an $\mathscr{A} \otimes \mathscr{B}$ measurable function, then there need not exist a sequence $(g_k)_{k \in \mathbb{N}} \subseteq \mathcal{S}^+(\mathscr{A} \times \mathscr{B})$ with $g_k \uparrow f$.

Consider $X = Y = \mathbb{R}$ with the Borel σ-algebra $\mathscr{B}(\mathbb{R})$ and Lebesgue measure λ. Since the diagonal

$$\Delta := \{(x, x) ; \ x \in \mathbb{R}\} \subseteq \mathbb{R}^2$$

is a closed subset of $\mathbb{R}^2$, we have $\Delta \in \mathscr{B}(\mathbb{R}^2)$, and so $f := \mathbb{1}_\Delta$ is measurable with respect to $\mathscr{B}(\mathbb{R}^2) = \mathscr{B}(\mathbb{R}) \otimes \mathscr{B}(\mathbb{R})$. Let g be a function of the form

$$g(x, y) = \sum_{i=1}^{n} c_i \mathbb{1}_{A_i}(x) \mathbb{1}_{B_i}(y)$$

for $n \in \mathbb{N}$, $A_i, B_i \in \mathscr{B}(\mathbb{R})$ and $c_i > 0$, and assume that $g \leqslant f$. Since g is strictly positive on $A_i \times B_i$, it follows that $f|_{A_i \times B_i} = 1$, i.e. $A_i \times B_i \subseteq \Delta$. From the definition of the diagonal Δ it is immediate that $A_i = B_i = \emptyset$ or $A_i = B_i = \{x_i\}$ for some $x_i \in \mathbb{R}$. In particular, $\{g \neq 0\}$ is a finite set. This means that for every sequence $(g_k)_{k \in \mathbb{N}} \subseteq \mathcal{S}^+(\mathscr{B}(\mathbb{R}) \times \mathscr{B}(\mathbb{R}))$ with $g_k \leqslant f$, the union

$$\bigcup_{k \in \mathbb{N}} \{g_k \neq 0\}$$

is at most countable. As $\{f \neq 0\}$ is uncountable, this shows that f cannot be approximated from below by simple functions with steps in $\mathscr{B}(\mathbb{R}) \times \mathscr{B}(\mathbb{R})$. Note, however, that there exists a sequence $(g_k)_{k \in \mathbb{N}} \subseteq \mathcal{S}^+(\mathscr{B}(\mathbb{R}) \times \mathscr{B}(\mathbb{R}))$ such that $g_k \uparrow f$ almost everywhere: since Δ is a Lebesgue null set, $g_k := 0$ does the job.

Our next example shows that, in general, even a.e. approximation **from below** may fail.

Consider $X = Y = [0,1]$ with the Borel σ-algebra and Lebesgue measure λ. Take $A \in \mathscr{B}([0,1]^2)$ with strictly positive Lebesgue measure $\lambda^2(A) > 0$ such that

$$I, J \in \mathscr{B}[0,1],\ I \times J \subseteq A \Rightarrow \lambda^2(I \times J) = \lambda^1(I)\lambda^1(J) = 0$$

[☞ 16.5,16.6], and set $f := \mathbb{1}_A$. Let $g \in \mathcal{E}^+(\mathscr{B}[0,1] \times \mathscr{B}[0,1])$ be such that $g \leqslant f$. If we denote by

$$g(x,y) = \sum_{i=1}^{n} c_i \mathbb{1}_{A_i}(x)\mathbb{1}_{B_i}(y), \quad c_i > 0,\ A_i, B_i \in \mathscr{B}[0,1],$$

the canonical representation of g, then $f|_{A_i \times B_i} = 1$, i.e. $A_i \times B_i \subseteq A$ for all $i \in \{1, \dots, n\}$. By our choice of A, this implies $\lambda^2(A_i \times B_i) = 0$. Hence, $g = 0$ Lebesgue almost everywhere. Since $\lambda^2(\{f > 0\}) = \lambda^2(A) > 0$, it follows that there cannot exist a sequence $(g_k)_{k \in \mathbb{N}} \subseteq \mathcal{S}^+(\mathscr{B}[0,1] \times \mathscr{B}[0,1])$ with $g_k \uparrow f$ Lebesgue almost everywhere.

Comment The functions $\mathcal{S}(\mathscr{A} \times \mathscr{B})$ are often denoted by $\mathcal{E}(\mathscr{A}) \otimes \mathcal{E}(\mathscr{B})$ where '$\otimes$' denotes the **tensor product** of functions: $(f \otimes g)(x,y) = f(x)g(y)$ and $\mathcal{E}(\mathscr{A}) \otimes \mathcal{E}(\mathscr{B})$ is the linear span of functions of the form $\mathbb{1}_A \otimes \mathbb{1}_B$ with $A \in \mathscr{A}$ and $B \in \mathscr{B}$. While it is true that the set $\mathcal{E}(\mathscr{A}) \otimes \mathcal{E}(\mathscr{B})$ is dense, w.r.t. convergence almost everywhere, in the $\mathscr{A} \otimes \mathscr{B}$ measurable functions, our example shows that it is in general not possible to approximate, say, positive parts from below with positive functions from $\mathcal{E}(\mathscr{A}) \otimes \mathcal{E}(\mathscr{B})$. A related approximation result in dimension one is the theorem in Example 8.22.

16

Product Measures

Let $(X, \mathscr{A}, \mu)$ and $(Y, \mathscr{B}, \nu)$ be two measure spaces. A **product measure** is any measure ρ on the measurable space $(X \times Y, \mathscr{A} \otimes \mathscr{B})$ that satisfies

$$\forall A \in \mathscr{A},\ B \in \mathscr{B}\ :\ \rho(A \times B) = \mu(A)\nu(B).$$

If both 'marginal' measure spaces are σ-finite, then the product measure exists and is unique [☞ Theorem 1.33], and we write $\rho = \mu \times \nu$.

In Section 1.7 we have constructed ρ using iterated integrals. A close inspection of the argument shows that the existence part, i.e. the formulae (1.26), require only σ-finiteness of the measure appearing in the inner integral.[1]

Based on Carathéodory's extension theorem [☞ Theorem 1.41] we will show a further possibility to extend the set function $(A, B) \longmapsto \mu(A)\nu(B)$ without assuming σ-finiteness.

Theorem 16A *Let $(X, \mathscr{A}, \mu)$ and $(Y, \mathscr{B}, \nu)$ be arbitrary measure spaces. There exist a σ-algebra $\mathscr{F}$ on $X \times Y$ and a measure ρ on $(X \times Y, \mathscr{F})$ such that $\mathscr{A} \otimes \mathscr{B} \subseteq \mathscr{F}$ and*

$$\forall A \in \mathscr{A},\ B \in \mathscr{B}\ :\ \rho(A \times B) = \mu(A)\nu(B).$$

Proof For $F \subseteq X \times Y$ define

$$\rho^*(F) := \inf\left\{ \sum_{n\in\mathbb{N}} \mu(A_n)\nu(B_n)\,;\ (A_n)_n \subseteq \mathscr{A},\ (B_n)_n \subseteq \mathscr{B},\ F \subseteq \bigcup_{n\in\mathbb{N}} A_n \times B_n \right\}.$$

As usual, we agree that $0 \cdot \infty = 0$. We claim that ρ^* is an outer measure, cf. (OM_1)–(OM_3) on p. 28.

[1] More precisely, if we set $\rho(E) = \int_X \left[\int_Y \mathbb{1}_E(x, y)\,\nu(dy)\right] \mu(dx)$ where $(X, \mathscr{A}, \mu)$ is an arbitrary measure space and $(Y, \mathscr{B}, \nu)$ is a σ-finite measure space, then ρ is a product measure with marginals μ and ν. The σ-finiteness of ν is needed only to show that the set $\Sigma := \{E \in \mathscr{A} \otimes \mathscr{B}\,;\ x \longmapsto \int_Y \mathbb{1}_E(x, y)\,\nu(dy) \text{ is measurable}\}$ is stable under the formation of complements: $E \in \Sigma \Rightarrow E^c \in \Sigma$. All other steps in the proof from [MIMS, Theorem 14.5] would go through unchanged.

(OM1) $\rho^*(\emptyset) = 0$.

(OM2) If $F \subseteq G \subseteq X \times Y$, then every covering $G \subseteq \bigcup_{n\in\mathbb{N}} A_n \times B_n$ is also a covering of F, and so $\rho^*(F) \leqslant \rho^*(G)$.

(OM3) Let $(F_k)_{k\in\mathbb{N}} \subseteq X \times Y$ and $F := \bigcup_{k\in\mathbb{N}} F_k$. Without loss of generality, $\rho^*(F_k) < \infty$ for all $k \in \mathbb{N}$; otherwise subadditivity is trivially satisfied. For every $\epsilon > 0$ we choose $(A_{n,k})_{n\in\mathbb{N}} \subseteq \mathscr{A}$ and $(B_{n,k})_{n\in\mathbb{N}} \subseteq \mathscr{B}$ such that $F_k \subseteq \bigcup_{n\in\mathbb{N}} A_{n,k} \times B_{n,k}$ and

$$\sum_{n\in\mathbb{N}} \mu(A_{n,k})\nu(B_{n,k}) \leqslant \rho^*(F_k) + \epsilon 2^{-k}.$$

Since $F \subseteq \bigcup_{k\in\mathbb{N}} \bigcup_{n\in\mathbb{N}} A_{n,k} \times B_{n,k}$ and $\mathbb{N} \times \mathbb{N}$ is countable, it follows that

$$\rho^*(F) \leqslant \sum_{k\in\mathbb{N}} \sum_{n\in\mathbb{N}} \mu(A_{n,k})\nu(B_{n,k}) \leqslant \sum_{k\in\mathbb{N}} (\rho^*(F_k) + \epsilon 2^{-k}) \xrightarrow[\epsilon\downarrow 0]{} \sum_{k\in\mathbb{N}} \rho^*(F_k).$$

By Carathéodory's extension theorem [☞Theorem 1.41],

$$\mathscr{F} := \{F \subseteq X \times Y \,;\, \forall E \subseteq X \times Y \,:\, \rho^*(E) = \rho^*(E \cap F) + \rho^*(E \setminus F)\}$$

is a σ-algebra on $X \times Y$, and the restriction $\rho := \rho^*|_{\mathscr{F}}$ is a measure on $(X \times Y, \mathscr{F})$.

In order to prove $\mathscr{A} \otimes \mathscr{B} \subseteq \mathscr{F}$, we first show that $A \times Y \in \mathscr{F}$ for any $A \in \mathscr{A}$. Fix $A \in \mathscr{A}$ and $E \subseteq X \times Y$ with $\rho^*(E) < \infty$. For $\epsilon > 0$ choose $(A_n)_{n\in\mathbb{N}} \subseteq \mathscr{A}$ and $(B_n)_{n\in\mathbb{N}} \subseteq \mathscr{B}$ such that $E \subseteq \bigcup_{n\in\mathbb{N}} A_n \times B_n$ and $\sum_{n\in\mathbb{N}} \mu(A_n)\nu(B_n) \leqslant \rho^*(E) + \epsilon$. Since

$$E \cap (A \times Y) \subseteq \bigcup_{n\in\mathbb{N}} (A_n \cap A) \times B_n \quad \text{and} \quad E \setminus (A \times Y) \subseteq \bigcup_{n\in\mathbb{N}} (A_n \setminus A) \times B_n,$$

we get

$$\begin{aligned} &\rho^*(E \cap (A \times Y)) + \rho^*(E \setminus (A \times Y)) \\ &\quad \leqslant \sum_{n\in\mathbb{N}} \mu(A_n \cap A)\nu(B_n) + \sum_{n\in\mathbb{N}} \mu(A_n \setminus A)\nu(B_n) \\ &\quad = \sum_{n\in\mathbb{N}} \mu(A_n)\nu(B_n) \leqslant \rho^*(E) + \epsilon \xrightarrow[\epsilon\downarrow 0]{} \rho^*(E). \end{aligned}$$

For $E \subseteq X \times Y$ with $\rho^*(E) = \infty$ the inequality is trivially satisfied; hence,

$$\rho^*(E \cap (A \times Y)) + \rho^*(E \setminus (A \times Y)) \leqslant \rho^*(E)$$

for all $E \subseteq X \times Y$. The reverse inequality holds because of the subadditivity of ρ^*, and so $A \times Y \in \mathscr{F}$. An analogous argument shows $X \times B \in \mathscr{F}$ for all $B \in \mathscr{B}$. Thus,

$$\forall A \in \mathscr{A},\ B \in \mathscr{B} \,:\, A \times B = (A \times Y) \cap (X \times B) \in \mathscr{F}.$$

Since $\mathscr{F}$ is a σ-algebra, this entails $\mathscr{A} \otimes \mathscr{B} = \sigma(\mathscr{A} \times \mathscr{B}) \subseteq \mathscr{F}$.

We still have to show that $\rho(A\times B) = \mu(A)\nu(B)$ for all $A \in \mathscr{A}$ and $B \in \mathscr{B}$. As $A \times B$ is an admissible covering of $A \times B$, the inequality $\rho^*(A\times B) \leqslant \mu(A)\nu(B)$ is immediate. Let $(A_n)_{n\in\mathbb{N}} \subseteq \mathscr{A}$ and $(B_n)_{n\in\mathbb{N}} \subseteq \mathscr{B}$ with $A\times B \subseteq \bigcup_{n\in\mathbb{N}} A_n \times B_n$. Since

$$\sum_{n=1}^{\infty} \mathbb{1}_{A_n}(x)\mathbb{1}_{B_n}(y) = \sup_{k\in\mathbb{N}} \sum_{n=1}^{k} \mathbb{1}_{A_n}(x)\mathbb{1}_{B_n}(y)$$

for each fixed $y \in Y$, it follows from Beppo Levi's theorem [☞ Theorem 1.13] that

$$\begin{aligned}\mu(A)\mathbb{1}_B(y) = \int_X \mathbb{1}_A(x)\mathbb{1}_B(y)\,\mu(dx) &\leqslant \int_X \sum_{n=1}^{\infty} \mathbb{1}_{A_n}(x)\mathbb{1}_{B_n}(y)\,\mu(dx)\\ &= \sup_{k\in\mathbb{N}} \sum_{n=1}^{k} \mathbb{1}_{B_n}(y) \int_X \mathbb{1}_{A_n}(x)\,\mu(dx)\end{aligned}$$

for all $y \in Y$. Another application of Beppo Levi's theorem yields

$$\begin{aligned}\mu(A)\nu(B) = \int_Y \mu(A)\mathbb{1}_B(y)\,\nu(dy) &\leqslant \int_Y \sup_{k\in\mathbb{N}} \sum_{n=1}^{k} \mathbb{1}_{B_n}(y)\mu(A_n)\,\nu(dy)\\ &= \sup_{k\in\mathbb{N}} \sum_{n=1}^{k} \int_Y \mu(A_n)\mathbb{1}_{B_n}(y)\,\nu(dy)\\ &= \sum_{n=1}^{\infty} \mu(A_n)\nu(B_n).\end{aligned}$$

Taking the infimum over all coverings, this gives $\mu(A)\nu(B) \leqslant \rho^*(A\times B)$. Since we already know that $A\times B \in \mathscr{F}$ and $\mu(A)\nu(B) \geqslant \rho^*(A\times B)$, we conclude that

$$\rho(A\times B) = \rho^*(A\times B) = \mu(A)\nu(B). \qquad \square$$

Both approaches, the construction using iterated integrals in Theorem 1.33, and Carathéodory's extension used in Theorem 16A, coincide if both marginal measure spaces $(X, \mathscr{A}, \mu)$ and $(Y, \mathscr{B}, \nu)$ are σ-finite; otherwise [☞ 16.1]. If we interpret $\mathbb{1}_E(x, y)$, $E \in \mathscr{A} \otimes \mathscr{B}$, as a measurable function $f : X \times Y \to \mathbb{R}$, then the non-uniqueness of the product measure in the absence of σ-finiteness is just an outgrowth of the fact that we don't have Tonelli's and Fubini's theorem [☞ 16.12]. For other possibilities when Tonelli and Fubini don't work [☞ 16.10] (lack of positivity), [☞ 16.11, 16.14] (lack of joint measurability) or [☞ 16.16, 16.17] (lack of integrability).

16.1 Non-uniqueness of product measures

Consider $(\mathbb{R}, \mathscr{B}(\mathbb{R}))$ with Lebesgue measure λ and counting measure $\zeta_{\mathbb{R}}$. Let ρ be the product measure on $(\mathbb{R}^2, \mathscr{B}(\mathbb{R}^2))$ constructed in Theorem 16A satisfying

$$\rho(A \times B) = \lambda(A)\zeta_{\mathbb{R}}(B), \quad A, B \in \mathscr{B}(\mathbb{R}).$$

We claim that

$$m(C) := \int_{\mathbb{R}} \int_{\mathbb{R}} \mathbb{1}_C(x,y)\,\lambda(dx)\,\zeta_{\mathbb{R}}(dy), \quad C \in \mathscr{B}(\mathbb{R}^2),$$

is also a product measure, i.e. $m(A \times B) = \lambda(A)\zeta_{\mathbb{R}}(B)$, and that $m \neq \rho$. First we verify that m is a measure on $(\mathbb{R}^2, \mathscr{B}(\mathbb{R}^2))$. Note that $m(C)$ is well-defined since

$$x \longmapsto \mathbb{1}_C(x,y) \quad \text{and} \quad y \longmapsto \int_{\mathbb{R}} \mathbb{1}_C(x,y)\,\lambda(dx)$$

are Borel measurable for each $C \in \mathscr{B}(\mathbb{R}^2)$. If $(C_n)_{n\in\mathbb{N}} \subseteq \mathscr{B}(\mathbb{R}^2)$ is a sequence of pairwise disjoint sets, then $C := \bigcup_{n\in\mathbb{N}} C_n$ satisfies

$$\mathbb{1}_C(x,y) = \sum_{n\in\mathbb{N}} \mathbb{1}_{C_n}(x,y) = \sup_{k\in\mathbb{N}} \sum_{n=1}^{k} \mathbb{1}_{C_n}(x,y), \quad x \in \mathbb{R},\ y \in \mathbb{R},$$

and so Beppo Levi's theorem (applied for Lebesgue measure) gives

$$\int_{\mathbb{R}} \mathbb{1}_C(x,y)\,\lambda(dx) = \sup_{k\in\mathbb{N}} \sum_{n=1}^{k} \int_{\mathbb{R}} \mathbb{1}_{C_n}(x,y)\,\lambda(dx).$$

Integrating with respect to $\zeta_{\mathbb{R}}(dy)$ and applying once more Beppo Levi's theorem (this time for the counting measure), we find that

$$m(C) = \sup_{k\in\mathbb{N}} \sum_{n=1}^{k} \int_{\mathbb{R}} \int_{\mathbb{R}} \mathbb{1}_{C_n}(x,y)\,\lambda(dx)\,\zeta_{\mathbb{R}}(dy) = \sum_{n=1}^{\infty} m(C_n).$$

This finishes the proof showing that m is a measure. If $A, B \in \mathscr{B}(\mathbb{R})$, then it is immediate from $\mathbb{1}_{A\times B}(x,y) = \mathbb{1}_A(x)\mathbb{1}_B(y)$ and the linearity of the integral that

$$m(A \times B) = \lambda(A)\zeta_{\mathbb{R}}(B),$$

and so m is a product measure. In order to prove $m \neq \rho$, we show $m(\Delta) \neq \rho(\Delta)$ for the diagonal $\Delta = \{(x,x);\ x \in \mathbb{R}\}$. Note that $\Delta \in \mathscr{B}(\mathbb{R}^2)$ since Δ is closed. From

$$\int_{\mathbb{R}} \mathbb{1}_\Delta(x,y)\,\lambda(dx) = \lambda(\{y\}) = 0, \quad y \in \mathbb{R},$$

we get $m(\Delta) = 0$. If $(A_n)_{n\in\mathbb{N}}, (B_n)_{n\in\mathbb{N}} \subseteq \mathscr{B}(\mathbb{R})$ are non-empty sets such that $\Delta \subseteq \bigcup_{n\in\mathbb{N}} A_n \times B_n$, then $\bigcup_{n\in\mathbb{N}} A_n \supseteq \mathbb{R}$, and therefore

$$\sum_{n\in\mathbb{N}} \lambda(A_n)\zeta_{\mathbb{R}}(B_n) \geqslant \sum_{n\in\mathbb{N}} \lambda(A_n) = \infty.$$

By the construction of ρ, this implies $\rho(\Delta) = \infty$. Hence, uniqueness of the product measure fails.

Comment As an immediate consequence, neither Tonelli's theorem nor Fubini's theorem extends to non-σ-finite measures; see also Example 16.12.

16.2 A measure on a product space which is not a product measure

Take $X = \{0, 1\}$ and define a measure on the product $(X \times X, \mathscr{P}(X) \otimes \mathscr{P}(X))$ by

$$\mu(\{(x, y)\}) := \begin{cases} 1, & \text{if } x = y, \\ 0, & \text{if } x \neq y. \end{cases}$$

The measure μ cannot be written as product $\rho \times \nu$ of two measures ρ, ν on $(X, \mathscr{P}(X))$. Indeed, if we had $\mu = \rho \times \nu$, then

$$\rho(\{0\})\nu(\{0\}) = \mu(\{0, 0\}) = 1 \quad \text{and} \quad \rho(\{0\})\nu(\{1\}) = \mu(\{0, 1\}) = 0.$$

From the first identity, we see that $\rho(\{0\}) > 0$, and hence, by the second identity, $\nu(\{1\}) = 0$. This, however, contradicts

$$\rho(\{1\})\nu(\{1\}) = \mu(\{1, 1\}) = 1.$$

16.3 The product of complete measure spaces need not be complete

[☞15.2]

16.4 A Lebesgue null set in $[0, 1]^2$ which intersects any set $A \times B$ whose Lebesgue measure is positive

Let λ^d be Lebesgue measure on $[0, 1]^d$ and set $N = \{(x, y) \in [0, 1]^2\,;\ x - y \in \mathbb{Q}\}$. Since the map $(x, y) \longmapsto f(x, y) := x - y$ is Borel measurable, it follows that $N = f^{-1}(\mathbb{Q})$ is a Borel set. By Tonelli's theorem,

$$\begin{aligned} \lambda^2(N) = \iint \mathbb{1}_N(x, y)\,\lambda^1(dx)\,\lambda^1(dy) &= \iint \mathbb{1}_{\mathbb{Q}}(x - y)\,\lambda^1(dx)\,\lambda^1(dy) \\ &= \int \lambda^1(y + \mathbb{Q})\,\lambda^1(dy) = 0. \end{aligned}$$

If $A \times B$ is a set with strictly positive Lebesgue measure, i.e. $\lambda(A) > 0$ and $\lambda(B) > 0$, then $A - B$ contains an open interval $I \neq \emptyset$ [☞comments in 7.28], and therefore

$$(A \times B) \cap N = \{(x, y) \in A \times B;\ x - y \in \mathbb{Q}\} = (A - B) \cap \mathbb{Q} \neq \emptyset.$$

16.5 A set $A \subseteq \mathbb{R}^2$ of positive Lebesgue measure which does not contain any rectangle

Consider Lebesgue measure λ^2 on $([0,1]^2, \mathscr{B}[0,1]^2)$ and define

$$A := \{(x, y) \in [0,1]^2;\ x - y \notin \mathbb{Q}\}.$$

The complement $N = [0,1]^2 \setminus A$ is a Borel set which has Lebesgue measure zero and which satisfies $N \cap (B \times C) \neq \emptyset$ for any two Borel sets B, C with $\lambda(B) > 0$, $\lambda(C) > 0$ [☞16.4]. Consequently, $A \in \mathscr{B}[0,1]^2$ has Lebesgue measure 1, and A does not contain any rectangle $B \times C$ of positive Lebesgue measure.

Comment For a further example [☞16.6]. A simpler example which furnishes a set that does not contain any rectangle (with non-trivial intervals as 'sides') is the Cartesian product of two fat Cantor sets [☞7.3].

16.6 A set $A \subseteq \mathbb{R}^2$ of positive Lebesgue measure such that the intersection of every non-degenerate rectangle with A^c has positive measure

We construct a Borel set $A \in \mathscr{B}(\mathbb{R}^2)$ with strictly positive Lebesgue measure $\lambda^2(A) > 0$ such that $\lambda^2(A^c \cap R) > 0$ for every non-degenerate measurable rectangle R, i.e. $R = S \times T$ for $S, T \in \mathscr{B}(\mathbb{R})$ with $\lambda^1(S) > 0, \lambda^1(T) > 0$. In particular, $A^c \cap R \neq \emptyset$, and so A does not contain any measurable rectangle of strictly positive measure. The idea for the construction of A is from [39].

Take $B \in \mathscr{B}(\mathbb{R})$ such that $\lambda^1(B \cap I) > 0$ and $\lambda^1(B^c \cap I) > 0$ for any open interval $I \neq \emptyset$ [☞7.15]. We claim that

$$A := \{(x, y) \in \mathbb{R}^2;\ x - y \in B\}$$

has the desired properties. As $\lambda^1(B) > 0$, it follows from Tonelli's theorem [☞Theorem 1.34] and the translation invariance of Lebesgue measure that

$$\lambda^2(A) = \int_{\mathbb{R}} \int_{\mathbb{R}} \mathbb{1}_B(x - y)\, \lambda^1(dx)\, \lambda^1(dy) = \int_{\mathbb{R}} \lambda^1(B)\, \lambda^1(dy) = \infty.$$

Let $R = S \times T$ be a non-degenerate measurable rectangle. Since $\lambda^1(S) > 0$ and

$\lambda^1(T) > 0$, Lebesgue's differentiation theorem [☞Theorem 14C] shows that we can find $x_0 \in S$ and $y_0 \in T$ such that

$$\lim_{r\to 0} \frac{1}{2r} \int_{(x_0-r,x_0+r)} \mathbb{1}_S(x)\,\lambda^1(dx) = 1 \text{ and } \lim_{r\to 0} \frac{1}{2r} \int_{(y_0-r,y_0+r)} \mathbb{1}_T(y)\,\lambda^1(dy) = 1.$$

Choose $r > 0$ sufficiently small such that

$$\lambda^1((x_0 - r, x_0 + r) \cap S) > \frac{3}{4}(2r) \quad \text{and} \quad \lambda^1((y_0 - r, y_0 + r) \cap T) > \frac{3}{4}(2r),$$

i.e.

$$\lambda^1(U \cap S) > \frac{3}{4}\lambda^1(U) \quad \text{and} \quad \lambda^1(V \cap T) > \frac{3}{4}\lambda^1(V)$$

for $U := (x_0 - r, x_0 + r)$ and $V := (y_0 - r, y_0 + r)$. Note that this implies

$$\lambda^1(U \cap S^c) < \frac{1}{4}\lambda^1(U) = \frac{r}{2} \quad \text{and} \quad \lambda^1(V \cap T^c) < \frac{1}{4}\lambda^1(V) = \frac{r}{2}. \tag{16.1}$$

Define for $t \in \mathbb{R}$

$$f(t) := \lambda^1((U \cap S) \cap (t + V \cap T)) = \lambda^1(\{x \in U \cap S\,;\ x - t \in V \cap T\}),$$

where we use the shorthand $t + V \cap T := t + (V \cap T) := \{t + y\,;\ y \in V \cap T\}$. From

$$f(t) = \int_{\mathbb{R}} \mathbb{1}_{U\cap S}(x)\mathbb{1}_{V\cap T}(-(t - x))\,\lambda^1(dx),$$

we see that f is the convolution of a bounded function with an integrable function and, hence, f is continuous [MIMS, p. 159]. We will show that f is strictly positive at $t_0 := x_0 - y_0$. As $t_0 + V = U$, we have

$$\begin{aligned} f(t_0) &= \lambda^1((U \cap S) \cap (t_0 + V \cap T)) \\ &= \lambda^1((U \cap S) \cap (U \cap (t_0 + T))) \\ &\geqslant \lambda^1(U) - \lambda^1(U \cap S^c) - \lambda^1(U \cap (t_0 + T)^c). \end{aligned}$$

Using (16.1) and

$$\lambda^1(U \cap (t_0 + T)^c) = \lambda^1((t_0 + V) \cap (t_0 + T)^c) = \lambda^1(t_0 + V \cap T^c) = \lambda^1(V \cap T^c),$$

we conclude that $f(t_0) > r > 0$. Since f is continuous, we can find some $\delta > 0$ such that $f(t) \geqslant r$ for all $t \in I := [t_0 - \delta, t_0 + \delta]$. By Tonelli's theorem,

$$\begin{aligned} \lambda^2(A^c \cap R) &= \lambda^2(\{(x, y) \in S \times T\,;\ x - y \in B^c\}) \\ &\geqslant \lambda^2(\{(x, y) \in \mathbb{R}^2\,;\ x \in S \cap U,\ y \in T \cap V,\ x - y \in B^c \cap I\}) \\ &= \int_{S\cap U} \int_{T\cap V} \mathbb{1}_{B^c\cap I}(x - y)\,\lambda^1(dy)\,\lambda^1(dx). \end{aligned}$$

The change of variables, $t = x - y$, and another application of Tonelli's theorem yields

$$\lambda^2(A^c \cap R) = \int_{S\cap U} \int_{B^c\cap I} \mathbb{1}_{T\cap V}(x-t)\,\lambda^1(dt)\,\lambda^1(dx) = \int_{B^c\cap I} f(t)\,\lambda^1(dt)$$
$$\geqslant r\lambda^1(B^c \cap I).$$

By our choice of B, we have $\lambda^1(B^c \cap I) > 0$, and so $\lambda^2(A^c \cap R) > 0$. This shows that every rectangle R of positive measure intersects A^c in a set of positive Lebesgue measure.

16.7 A set $A \subseteq \mathbb{R}^2$ of positive Lebesgue measure which is not a countable union of rectangles

⌊☞ 16.5, 16.6⌋

16.8 A jointly measurable function such that $x \longmapsto \int f(x,y)\,\mu(dy)$ is not measurable

Consider $X = Y = [0,1]$ with the Borel σ-algebra $\mathscr{B}[0,1]$. Take $A \notin \mathscr{B}[0,1]$ ⌊☞ 7.20⌋ and define

$$\mu(B) := 2\#(A \cap B) + \#(A^c \cap B), \quad B \in \mathscr{B}[0,1].$$

The diagonal $\Delta = \{(x,x);\ x \in [0,1]\}$ is a Borel set in $[0,1]^2$, and so

$$[0,1]^2 \ni (x,y) \longmapsto f(x,y) := \mathbb{1}_\Delta(x,y) \in \mathbb{R}$$

is Borel measurable. Yet

$$F(x) = \int_{[0,1]} f(x,y)\,\mu(dy) = 2\mathbb{1}_A(x) + \mathbb{1}_{A^c}(x) = 1 + \mathbb{1}_A(x)$$

fails to be Borel measurable since $A \notin \mathscr{B}[0,1]$.

Comment This is not a contradiction to Tonelli's theorem since the measure μ is not σ-finite.

16.9 A function $f(x,y)$ such that $f(\cdot,y)$ is $\mathscr{A}$ measurable but $\int f(\cdot,y)\,dy$ is not $\mathscr{A}$ measurable

We construct a σ-algebra $\mathscr{A} \subsetneqq \mathscr{B}(\mathbb{R})$ on $\mathbb{R}$ and a function $f : \mathbb{R}^2 \to \mathbb{R}$ such that $f(x,\cdot) \in L^1(\lambda)$ and $f(\cdot,y)$ is $\mathscr{A}$ measurable for each fixed $x, y \in \mathbb{R}$ but

$$x \longmapsto \int_{\mathbb{R}} f(x,y)\,dy$$

is not $\mathscr{A}$ measurable. This example is taken from [114].

Recall that a set $M \subseteq \mathbb{R}$ is of first category if $M = \bigcup_{n\in\mathbb{N}} A_n$ for a sequence $(A_n)_{n\in\mathbb{N}}$ of nowhere dense sets [☞ Definition 2.3]. Complements of sets of first category are dense in $\mathbb{R}$ [☞ Example 2.6.(d)], subsets of sets of first category are again of first category; conversely if A contains the complement M^c of a set M of first category, then A^c is itself of first category. We claim that

$$\mathscr{A} = \{A \in \mathscr{B}(\mathbb{R})\,;\ A \text{ is of 1st category or } A^c \text{ is of 1st category}\}$$

is a σ-algebra on $\mathbb{R}$. Since the definition of $\mathscr{A}$ treats A and A^c in a symmetric way, $\mathscr{A}$ is closed under complements, and clearly $\emptyset \in \mathscr{A}$. Let $(A_n)_{n\in\mathbb{N}} \subseteq \mathscr{A}$ and set $A = \bigcup_{n\in\mathbb{N}} A_n \in \mathscr{B}(\mathbb{R})$. We consider two cases:

Case 1: All A_n are of first category. It is not difficult to see from the definition that the countable union of sets of first category is of first category, and so $A \in \mathscr{A}$.

Case 2: There exists some $k \in \mathbb{N}$ such that A_k^c is of first category. From $A \supseteq A_k$ and the fact that A_k is the complement of a set of first category, we infer that A^c is of first category, hence $A \in \mathscr{A}$.

By definition, we have $\mathscr{A} \subseteq \mathscr{B}(\mathbb{R})$. Note that this inclusion is strict; for instance, $A = (0,1) \in \mathscr{B}(\mathbb{R})$ is neither of first category [☞ Example 2.6.(c)] nor the complement of a set of first category, since its complement is not dense.

Take $A \in \mathscr{B}(\mathbb{R})$ such that A is of first category and $\lambda(A^c) = 0$ [☞ 7.10]. Define

$$f(x,y) := (x-y)\mathbb{1}_A(x-y)\max\{0, 1-|y|\}, \quad x, y \in \mathbb{R}.$$

For each fixed $y \in \mathbb{R}$, the function $f(\cdot, y)$ vanishes on the $(y+A)^c$ whose complement is of first category. This implies

$$N_c := \{x \in X\,;\ f(x,y) \leqslant c\} \supseteq \{x \in X\,;\ f(x,y) \leqslant 0\} \supseteq (y+A)^c$$

for any $c \geqslant 0$. Thus, $N_c \in \mathscr{A}$ for $c \geqslant 0$. If $c < 0$, then N_c is contained in the set $(y+A)$ of first category, hence of first category, and so $N_c \in \mathscr{A}$. This shows that $f(\cdot, y)$ is $\mathscr{A}/\mathscr{B}(\mathbb{R})$ measurable for each $y \in \mathbb{R}$. Since A^c has Lebesgue measure zero, we have

$$F(x) := \int_{\mathbb{R}} f(x,y)\,\lambda(dy) = \int_{\mathbb{R}} (x-y)\max\{0, 1-|y|\}\,\lambda(dy).$$

The latter integral yields $F(x) = x$ for all $x \in \mathbb{R}$, but the map F is not $\mathscr{A}/\mathscr{B}(\mathbb{R})$ measurable as $F^{-1}(0,1) = (0,1) \notin \mathscr{A}$.

Comment 1 Following the above reasoning, we see that $f(x, \cdot)$ is $\mathscr{A}/\mathscr{B}(\mathbb{R})$ measurable for each $x \in \mathbb{R}$, which means that f is $\mathscr{A}$ (separately) measurable in both components. On the other hand, Fubini's theorem [☞ Corollary 1.35] shows that $(x, y) \longmapsto f(x, y)$ fails to be $\mathscr{A} \otimes \mathscr{A}$ measurable.

Comment 2 By definition, F is the convolution of $h(x) = \max\{0, 1 - |x|\}$ and $g(x) = x\mathbb{1}_A(x)$. As we have seen, g is $\mathscr{A}$ measurable but this measurability property is not inherited by the convolution $F = g * h$.

16.10 Tonelli's theorem fails for non-positive integrands

Equip $(\mathbb{N}, \mathscr{P}(\mathbb{N}))$ with the counting measure $\zeta(A) = \#A$ and define the function

$$f(x, y) := \begin{cases} 2 - 2^{-x}, & \text{if } x = y, \\ -2 + 2^{-x}, & \text{if } x = y + 1, \\ 0, & \text{otherwise,} \end{cases} \quad \text{for} \quad x, y \in \mathbb{N}.$$

On the one hand, we have

$$\begin{aligned} \int_{\mathbb{N}} \int_{\mathbb{N}} f(x, y) \zeta(dx) \zeta(dy) = \sum_{y=1}^{\infty} \sum_{x=1}^{\infty} f(x, y) &= \sum_{y=1}^{\infty} \left((2 - 2^{-y}) + (-2 + 2^{-y-1}) \right) \\ &= -\sum_{y=1}^{\infty} 2^{-y-1} = -\frac{1}{2}, \end{aligned}$$

and, on the other hand,

$$\begin{aligned} \int_{\mathbb{N}} \int_{\mathbb{N}} f(x, y) \zeta(dy) \zeta(dx) &= \sum_{x=1}^{\infty} \sum_{y=1}^{\infty} f(x, y) \\ &= (2 - 2^{-1}) + \sum_{x=2}^{\infty} \left((2 - 2^{-x}) + (-2 + 2^{-x+1}) \right) \\ &= \frac{3}{2} + \sum_{x=2}^{\infty} 2^{-x} = \frac{3}{2} + \frac{1}{2} = 2. \end{aligned}$$

This shows that we cannot interchange the integrals, resp. series. Notice that

$$\sum_{y=1}^{\infty} \sum_{x=1}^{\infty} |f(x, y)| = \sum_{y=1}^{\infty} \left((2 - 2^{-y}) + (2 - 2^{-y-1}) \right) = \infty,$$

i.e. the function $f(x, y)$ is not integrable, resp. jointly summable.

Comment If we use, instead of Tonelli's theorem, Fubini's theorem, then this example shows that we cannot do without **integrability** of the function $f(x, y)$.

16.11 A positive function with $f(x, y) = f(y, x)$ such that the iterated integrals do not coincide

The following example shows that Tonelli's theorem fails if we replace joint measurability by separate measurability.

Consider the product measure space $([0, 1]^2, \mathscr{B}[0, 1]^2, \lambda^2)$. Let us assume that the continuum hypothesis holds[☞ p. 43]. As a consequence, there is a well-ordered set W and a bijection $j : [0, 1] \to W$ such that for each $x \in [0, 1]$ the set $\{w \in W\,;\ w \prec j(x)\} \subseteq W$ is at most countable.

Let $Q = \{(x, y) \in [0, 1]^2\,;\ j(x) \prec j(y)\}$. By construction,

$$Q_x := \{y \in [0, 1]\,;\ \mathbb{1}_Q(x, y) = 1\}, \quad x \in [0, 1],$$

contains, with the possible exception of countably many points, all points from $[0, 1]$. Similarly,

$$Q^y := \{x \in [0, 1]\,;\ \mathbb{1}_Q(x, y) = 1\}, \quad y \in [0, 1],$$

is at most countable. Set $f(x, y) := \mathbb{1}_Q(x, y)$, then

$$y \longmapsto f(x, y) \quad \text{and} \quad x \longmapsto f(x, y) \quad \text{are Borel measurable}$$

and we have

$$\int_{[0,1]} f(x, y)\,\lambda(dy) = 1 \quad \text{and} \quad \int_{[0,1]} f(x, y)\,\lambda(dx) = 0,$$

which implies

$$\int_{[0,1]} \left(\int_{[0,1]} f(x, y)\,\lambda(dy) \right) \lambda(dx) = 1 \neq 0 \int_{[0,1]} \left(\int_{[0,1]} f(x, y)\,\lambda(dx) \right) \lambda(dy).$$

16.12 A positive function $f(x, y)$ whose iterated integrals do not coincide

Consider the measure space $((0, 1), \mathscr{B}(0, 1))$, denote by λ Lebesgue measure and by $\zeta : \mathscr{B}(0, 1) \to [0, \infty]$, $\zeta(B) := \# B$ the counting measure. The indicator function $f(x, y) := \mathbb{1}_\Delta(x, y)$ of the diagonal $\Delta = \{(x, x)\,;\ x \in (0, 1)\}$ is positive and measurable.

For each fixed $y \in (0, 1)$, the set $\{x \in (0, 1)\,;\ f(x, y) \neq 0\} = \{y\}$ is a Lebesgue null set, and so

$$\int_{(0,1)} f(x, y)\,\lambda(dx) = 0, \quad y \in (0, 1),$$

which implies

$$\int_{(0,1)} \int_{(0,1)} f(x,y)\,\lambda(dx)\,\zeta(dy) = 0.$$

If $x \in (0,1)$ is fixed, then

$$\int_{(0,1)} f(x,y)\,\zeta(dy) = \zeta(\{y \in (0,1);\ f(x,y) = 1\}) = \zeta(\{x\}) = 1.$$

Thus,

$$\int_{(0,1)} \int_{(0,1)} f(x,y)\,\zeta(dy)\,\lambda(dx) = 1.$$

Comment This is not a contradiction to Tonelli's theorem since the counting measure $\zeta = \zeta_{\mathbb{R}}$ is not σ-finite. If $\zeta = \zeta_{\mathbb{Q}}$ is the counting measure defined by $\zeta_{\mathbb{Q}}(B) := \#(\mathbb{Q} \cap B)$, then both iterated integrals will be zero – as expected from Tonelli's theorem; note that $\zeta_{\mathbb{Q}}$ is σ-additive.

16.13 A finite measure μ and a Borel set B such that $\iint \mathbb{1}_B(x+y)\,\mu(dx)\,\lambda(dy) \neq \iint \mathbb{1}_B(x+y)\,\lambda(dy)\,\mu(dx)$

Let $X = \mathbb{R}$, denote by λ Lebesgue measure on $(\mathbb{R}, \mathscr{B}(\mathbb{R}))$ and consider a further σ-algebra

$$\mathscr{A} := \{A \in \mathscr{B}(\mathbb{R});\ A \text{ is of 1st category or } A^c \text{ is of 1st category}\}$$

[☞ 16.9]. The set function

$$\mu(A) := \begin{cases} 0, & \text{if } A \text{ is of first category,} \\ 1, & \text{if } A^c \text{ is of first category,} \end{cases}$$

defines a measure on $(X, \mathscr{A})$. Indeed: since there is no set $A \subseteq \mathbb{R}$ which is of first category and has complement of first category, the mapping μ is well-defined. Moreover, $A = \emptyset$ is of first category, and so $\mu(\emptyset) = 0$. If $(A_n)_{n\in\mathbb{N}} \subseteq \mathscr{A}$ are pairwise disjoint sets and $A = \bigcup_{n\in\mathbb{N}} A_n$, then one of the following two cases applies:

Case 1. Each A_n is of first category. In this case, the union $A = \bigcup_{n\in\mathbb{N}} A_n$ is of first category, which implies

$$\mu(A) = 0 = \sum_{n\in\mathbb{N}} \mu(A_n).$$

Case 2. There is some $k \in \mathbb{N}$ such that A_k^c is of first category. From $A \supseteq A_k$ we see that $A^c \subseteq A_k^c$ is of first category, i.e. $\mu(A) = 1$. Since the sets A_n are pairwise disjoint, we have $A_n \subseteq A_k^c$ for all $n \neq k$, which implies that A_n is of first category for $n \neq k$. Hence, $\mu(A_n) = 0$ for all $n \neq k$ and $\mu(A_k) = 1$. This gives

$$\mu(A) = 1 = \mu(A_k) = \sum_{n\in\mathbb{N}} \mu(A_n).$$

Take a set $B \in \mathscr{B}(\mathbb{R})$ of first category such that $\lambda(B^c) = 0$ [☞7.10] and define

$$f(x, y) := \mathbb{1}_B(x + y).$$

Since the translate $y+B$ is of first category for each $y \in \mathbb{R}$, the map $x \longmapsto f(x, y)$ is $\mathscr{A}$ measurable and

$$\int_{\mathbb{R}} f(x, y)\,\mu(dx) = \mu(B - y) = 0, \quad y \in \mathbb{R}.$$

On the other hand, $\lambda(B^c) = 0$ gives

$$\int_{\mathbb{R}} f(x, y)\,\lambda(dy) = \lambda(B - x) = \lambda(B) = \infty, \quad x \in \mathbb{R}.$$

Consequently,

$$\int_{\mathbb{R}}\int_{\mathbb{R}} f(x, y)\,\mu(dx)\,\lambda(dy) = 0 \neq \infty = \int_{\mathbb{R}}\int_{\mathbb{R}} f(x, y)\,\lambda(dy)\,\mu(dx).$$

Comment 1 This example is from [114]. There is no contradiction to Tonelli's theorem since the function f is not $\mathscr{A}\otimes\mathscr{B}(\mathbb{R})/\mathscr{B}(\mathbb{R})$ measurable. Of course, f is $\mathscr{B}(\mathbb{R})\otimes\mathscr{B}(\mathbb{R})/\mathscr{B}(\mathbb{R})$ measurable but this is not good enough since the measure μ 'lives' only on $\mathscr{A}$, which is strictly smaller than $\mathscr{B}(\mathbb{R})$, e.g. $(0, 1) \in \mathscr{B}(\mathbb{R})$ and $(0, 1) \notin \mathscr{A}$.

Comment 2 This example does not use the axiom of choice [☞Section 2.2]. For our reasoning, we need Baire's category theorem only in Euclidean spaces, and in complete separable metric spaces – such as Euclidean space – the Baire category theorem does not require the axiom of choice, cf. [133, p. 95].

16.14 A non-measurable function $f(x, y)$ such that the iterated integral $\iint f(x, y)\,dx\,dy$ exists and is finite

Let λ be Lebesgue measure on $([0, 1], \mathscr{B}[0, 1])$. Pick some non-Lebesgue measurable set L [☞7.22], and take two disjoint sets $A, B \in \mathscr{B}[0, 1]$ with measure

$\lambda(A) = \lambda(B) = \frac{1}{4}$. Define

$$f(x,y) := \mathbb{1}_{A\times L}(x,y) + \mathbb{1}_{B\times([0,1]\setminus L)}(x,y), \quad x,y \in [0,1].$$

We have

$$\int_0^1 f(x,y)\,\lambda(dx) = \int_0^1 \mathbb{1}_A(x)\,\lambda(dx)\cdot \mathbb{1}_L(y) + \int_0^1 \mathbb{1}_B(x)\,\lambda(dx)\cdot \mathbb{1}_{[0,1]\setminus L}(y)$$
$$= \frac{1}{4}\mathbb{1}_{[0,1]}(y),$$

and so $\int_0^1 \int_0^1 f(x,y)\,\lambda(dx)\,\lambda(dy) = \frac{1}{4}$. The function $f(x,y)$ is not measurable in (x,y) or in y, so neither the double integral nor the other iterated integral make sense.

16.15 A function $f(x,y)$ whose iterated integrals exist but do not coincide

Consider Lebesgue measure on $(0,1)$. The function

$$f : (0,1)^2 \to \mathbb{R}, \quad f(x,y) := \frac{x^2 - y^2}{(x^2+y^2)^2},$$

is measurable and the iterated integrals exist but do not coincide:

$$\int_{(0,1)}\int_{(0,1)} \frac{x^2-y^2}{(x^2+y^2)^2}\,\lambda(dx)\lambda(dy) \neq \int_{(0,1)}\int_{(0,1)} \frac{x^2-y^2}{(x^2+y^2)^2}\,\lambda(dy)\lambda(dx).$$

Indeed, we have

$$\frac{d}{dy}\frac{y}{x^2+y^2} = \frac{x^2-y^2}{(x^2+y^2)^2},$$

and so we get

$$\int_{(0,1)}\int_{(0,1)} \frac{x^2-y^2}{(x^2+y^2)^2}\,\lambda(dy)\,\lambda(dx) = \int_{(0,1)} \frac{1}{x^2+1}\,\lambda(dx) = \arctan x\Big|_0^1 = \frac{\pi}{4}.$$

Interchanging the roles of x and y, it follows that

$$\int_{(0,1)}\int_{(0,1)} \frac{y^2-x^2}{(x^2+y^2)^2}\,\lambda(dx)\,\lambda(dy) = \frac{\pi}{4},$$

i.e.

$$\int_{(0,1)}\int_{(0,1)} \frac{x^2-y^2}{(x^2+y^2)^2}\,\lambda(dx)\,\lambda(dy) = -\frac{\pi}{4}.$$

In view of Fubini's theorem, this means that the double integral cannot exist. This can also be seen directly from

$$\begin{aligned}\int_{(0,1)}\int_{(0,1)}\left|\frac{x^2-y^2}{(x^2+y^2)^2}\right|\lambda(dy)\,\lambda(dx) &\geqslant \int_0^1\int_0^x \frac{x^2-y^2}{(x^2+y^2)^2}\lambda(dy)\,\lambda(dx)\\ &= \int_0^1 \frac{x}{x^2+x^2}\lambda(dx)\\ &= \frac{1}{2}\int_0^1 \frac{1}{x}\lambda(dx) = \infty.\end{aligned}$$

16.16 A function $f(x,y)$ which is not integrable but whose iterated integrals exist and coincide

Consider Lebesgue measure on $(-1,1)$. The function

$$f:(-1,1)^2\to\mathbb{R},\ (x,y)\longmapsto f(x,y):=\begin{cases}\dfrac{xy}{(x^2+y^2)^2}, & \text{if } (x,y)\neq(0,0),\\ 0, & \text{if } (x,y)=(0,0),\end{cases}$$

is measurable, the iterated integrals exist and coincide:

$$\int_{(-1,1)}\int_{(-1,1)}\frac{xy}{(x^2+y^2)^2}\lambda(dx)\lambda(dy) = \int_{(-1,1)}\int_{(-1,1)}\frac{xy}{(x^2+y^2)^2}\lambda(dy)\lambda(dx).$$

This follows from the fact that the integrand is an odd function, i.e. we have for $y\neq 0$:

$$\int_{(-1,1)}\frac{xy}{(x^2+y^2)^2}\lambda(dx) = 0.$$

Since $\{0\}$ is a Lebesgue null set, both iterated integrals have common value 0.

The double integral does not exist; the iterated **absolute** integrals yield

$$\begin{aligned}\int_{(-1,1)}\left|\frac{xy}{(x^2+y^2)^2}\right|\lambda(dx) &= \frac{2}{|y|}\int_0^{1/|y|}\frac{\xi}{(\xi^2+1)^2}\lambda(d\xi)\\ &\geqslant \frac{2}{|y|}\underbrace{\int_0^1 \frac{\xi}{(\xi^2+1)^2}\lambda(d\xi)}_{=c\in(0,\infty)}.\end{aligned}$$

Here we use the substitution $x=\xi|y|$ and the fact that $|y|\leqslant 1$, thus $1/|y|\geqslant 1$.

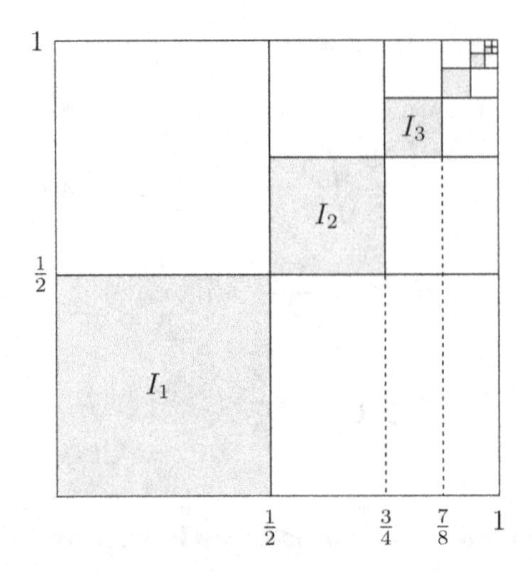

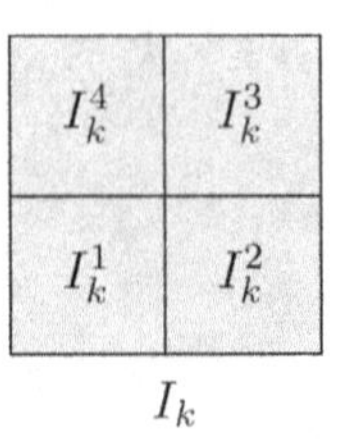

Figure 16.1 A subdivision of $(0,1)^2$ and of the resulting open squares I_k.

Integrating with respect to $y \in (-1,1)$, we conclude that

$$\int_{(-1,1)}\int_{(-1,1)}\left|\frac{xy}{(x^2+y^2)^2}\right|\lambda(dx)\,\lambda(dy) \geqslant c\int_{(-1,1)}\frac{1}{|y|}\,\lambda(dy) = \infty.$$

Comment Fichtenholz [59] has a variant of this example on the measure space $([0,1]^2, \mathscr{B}([0,1]^2), \lambda^2)$ where the iterated integrals of the form

$$\int_A\int_B f(x,y)\,\lambda(dy)\,\lambda(dx) = \int_B\int_A f(x,y)\,\lambda(dx)\,\lambda(dy)$$

exist and coincide for all measurable sets A, B, but the double integral does not exist.

16.17 Yet another example where the iterated integrals exist, but the double integral doesn't

Equip the measurable space $((0,1)^2, \mathscr{B}(0,1)^2)$ with Lebesgue measure λ^2. We subdivide $I_0 = (0,1)^2$ and $I_k \supseteq I_k^1 \cup I_k^2 \cup I_k^3 \cup I_k^4$ into disjoint open intervals as shown in Fig. 16.1, and define a function $f : (0,1) \to \mathbb{R}$ as follows:

$$f(x,y) := \begin{cases} 1/\lambda^2(I_k), & \text{if } (x,y) \in I_k^1 \cup I_k^3,\ k \geqslant 1, \\ -1/\lambda^2(I_k), & \text{if } (x,y) \in I_k^2 \cup I_k^4,\ k \geqslant 1, \\ 0, & \text{otherwise.} \end{cases}$$

By construction, we have

$$\int_{(0,1)} f(x,y)\,\lambda(dy) = 0 \quad \text{and} \quad \int_{(0,1)} f(x,y)\,\lambda(dx) = 0$$

for all $x \in (0,1)$, resp. $y \in (0,1)$. Thus, the iterated integrals exist and coincide:

$$\int_{(0,1)} \left[\int_{(0,1)} f(x,y)\,\lambda(dy) \right] \lambda(dx) = \int_{(0,1)} \left[\int_{(0,1)} f(x,y)\,\lambda(dx) \right] \lambda(dy) = 0$$

whereas the double integral does not exist:

$$\iint_{(0,1)^2} |f(x,y)|\,\lambda^2(dx,dy) = \sum_{k=1}^{\infty} \iint_{I_k} |f(x,y)|\,\lambda^2(dx,dy) = \sum_{k=1}^{\infty} 1 = \infty.$$

16.18 An a.e. continuous function $f(x,y)$ where only one iterated integral exists

Take $X = [-1,1] \times [-1,1]$ together with the Borel sets and Lebesgue measure. The function

$$f(x,y) := \begin{cases} \dfrac{x}{y}, & \text{if } y \neq 0, \\ 0, & \text{if } y = 0, \end{cases}$$

is continuous at $y \neq 0$. Moreover,

$$\int_{-1}^{1} \left(\int_{-1}^{1} \frac{x}{y}\,\lambda(dx) \right) \lambda(dy) = \int_{-1}^{1} 0\,\lambda(dy) = 0$$

exists while the other iterated integral does not exist, since $\int_{-1}^{1} \frac{dy}{|y|} = \infty$.

16.19 Classical integration by parts fails for Lebesgue–Stieltjes integrals

Consider the Dirac measure δ_0 on the measurable space $((-1,1], \mathscr{B}(-1,1])$ and set

$$F(x) := G(x) := \delta_0(-1,x] = \mathbb{1}_{[0,1]}(x).$$

By definition, the Lebesgue–Stieltjes measures $dF(x)$ and $dG(x)$ can be identified with $\delta_0(dx)$, and we get

$$\int_{(-1,1]} F(x)\,dG(x) = \int_{(-1,1]} G(x)\,dF(x) = 1 = F(1)G(1).$$

Since $F(-1)G(-1) = 0$, the familiar integration by parts formula cannot hold:

$$1 = \int_{(-1,1]} F(x)\,dG(x) \neq F(1)G(1) - F(-1)G(-1) - \int_{(-1,1]} G(x)\,dF(x) = 0.$$

Comment The origin of this pathology is the common discontinuity of F and G at $x = 0$. Because of this discontinuity, the integrals $\int G\,dF$ and $\int F\,dG$ do not exist as Riemann–Stieltjes integrals [☞ 3.21, 3.22].

If one allows for joint discontinuities of arbitrary right-continuous and increasing (distribution) functions F and G, then the integration by parts formula must be stated in the following form:

$$\begin{aligned}\int_{(a,b]} F(x-)\,dG(x) &= F(b)G(b) - F(a)G(a) \\ &\quad - \int_{(a,b]} G(x-)\,dF(x) - \sum_{a<x\leqslant b} \Delta F(x)\cdot\Delta G(x);\end{aligned} \tag{16.2}$$

the expression $H(x-) := \lim_{y<x, y\to x} H(y)$ denotes the left-continuous version of $H = F, G$ and $\Delta H(x) := H(x) - H(x-)$ is the jump at x; since H is increasing, there are at most countably many x with $\Delta H(x) \neq 0$, and the sum appearing in (16.2) is well-defined. Alternatively, the integration by parts formula can be written in the form

$$\int_{(a,b]} F(x-)\,dG(x) = F(b)G(b) - F(a)G(a) - \int_{(a,b]} G(x)\,dF(x) \tag{16.3}$$

or in the symmetric form

$$\begin{aligned}\int_{(a,b]} \frac{F(x) + F(x-)}{2}\,dG(x) &= F(b)G(b) - F(a)G(a) \\ &\quad - \int_{(a,b]} \frac{G(x) + G(x-)}{2}\,dF(x).\end{aligned} \tag{16.4}$$

In order to prove (16.2), we observe first that

$$(F(b) - F(a))\,(G(b) - G(a)) = \int_{(a,b]} dF \int_{(a,b]} dG.$$

Writing, in abuse of notation, $d(F \times G) = (dF) \times (dG)$ for the product measure, we can use Tonelli's theorem and conclude

$$\begin{aligned}&(F(b) - F(a))\,(G(b) - G(a)) \\ &\quad = \int_{(a,b]\times(a,b]} d(F \times G)\end{aligned}$$

$$= \int_{(a,b]\times(a,b]} \mathbb{1}_{(a,x]}(y)\, d(F\times G)(x,y) + \int_{(a,b]\times(a,b]} \mathbb{1}_{(x,b]}(y)\, d(F\times G)(x,y)$$
$$= \iint_{(a,b]\times(a,b]} \mathbb{1}_{(a,x]}(y)\, dG(y)\, dF(x) + \iint_{(a,b]\times(a,b]} \mathbb{1}_{(a,y)}(x)\, dF(x)\, dG(y)$$
$$= \int_{(a,b]} (G(x) - G(a))\, dF(x) + \int_{(a,b]} (F(y-) - F(a))\, dG(y)$$
$$= \int_{(a,b]} G(x)\, dF(x) + \int_{(a,b]} F(y-)\, dG(y)$$
$$- G(a)\,(F(b) - F(a)) - F(a)\,(G(b) - G(a)).$$

This identity becomes, after a simple re-arrangement, (16.3). Using

$$\int_{(a,b]} (G(x) - G(x-))\, dF(x) = \int_{(a,b]} \Delta G(x)\, dF(x) = \sum_{a<x\leqslant b} \Delta G(x)\Delta F(x),$$

we obtain (16.2) from (16.3). If we change the roles of F and G in (16.3), add the resulting equality to (16.3) and divide by 2, then we get (16.4).

16.20 A function which is $K(x, dy)$-integrable but fails to be $\mu K(dy)$-integrable

Equip $X = \mathbb{R}$ and $Y = \mathbb{R}^2$ with their Borel σ-algebras, write λ^d for the d-dimensional Lebesgue measure, and define a kernel $K : \mathbb{R}\times\mathscr{B}(\mathbb{R}^2) \to [0,\infty]$ – see the comments further on – by

$$K(x, B) := \lambda^1(\{y\,;\ (x,y)\in B\}) = \int \mathbb{1}_B(x,y)\,\lambda^1(dy), \quad B\in\mathscr{B}(\mathbb{R}^2).$$

The left action of the kernel to the measure λ^1 is

$$\lambda^1 K(B) = \int K(x,B)\lambda^1(dx) = \iint \mathbb{1}_B(x,y)\,\lambda^1(dy)\,\lambda^1(dx) = \lambda^2(B).$$

Consider the function $f : \mathbb{R}^2 \to \mathbb{R}$ defined by

$$f(x,y) = \begin{cases} 1, & \text{if } |y| < |x|,\ y > 0,\\ -1, & \text{if } |y| < |x|,\ y < 0,\\ 0, & \text{otherwise,} \end{cases}$$

see Fig. 16.2. It is obvious that f is measurable and that $f \notin L^1(\lambda^2)$. On the other hand

$$\int_{\mathbb{R}^2} |f(z)| K(x, dz) = \int_{-|x|}^{|x|} |f(x,y)|\,\lambda^1(dy) = 2\int_0^{|x|} (+1)\lambda^1(dy) = 2|x|.$$

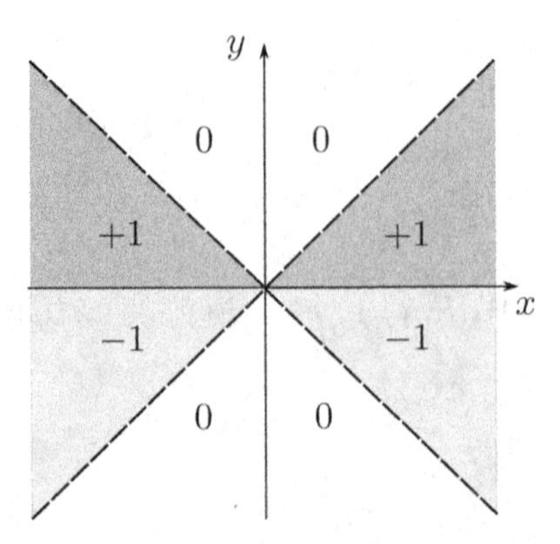

Figure 16.2 The figure shows the different values attained by the function $f(x, y)$ in its domain.

This shows that $f \in L^1(K(x, dy))$ for each x. A similar calculation reveals that

$$\int f(z)K(x, dz) = 0.$$

Comment A **measure kernel** between two measurable spaces $(X, \mathscr{X})$ and $(Y, \mathscr{Y})$ is a mapping $K : X \times \mathscr{Y} \to [0, \infty]$ such that

(a) $x \longmapsto K(x, B)$ is $\mathscr{X}/\mathscr{B}[0, \infty]$ measurable for each $B \in \mathscr{Y}$,
(b) $B \longmapsto K(x, B)$ is a measure for each $x \in X$.

If μ is a measure on $(X, \mathscr{X})$, then the set function

$$\mu K(B) := \int_X K(x, B)\, \mu(dx), \quad B \in \mathscr{Y}$$

defines a measure on $(Y, \mathscr{Y})$. For measurable functions $f : Y \to [0, \infty]$, we have

$$\int_Y f\, d\mu K = \int_X \int_Y f(y)\, K(x, dy)\, \mu(dx).$$

This shows that measure kernels can be used to transport measures from $(X, \mathscr{X})$ to $(Y, \mathscr{Y})$.

A special case of kernels are image measures. Let $\phi : (X, \mathscr{X}) \to (Y, \mathscr{Y})$ be a measurable map. The **image measure** of a measure μ on $(X, \mathscr{X})$ is a measure on $(Y, \mathscr{Y})$ defined by

$$\phi_* \mu(B) = \mu(\phi^{-1}(B)) = \int_X \mathbb{1}_B \circ \phi(x)\, \mu(dx) = \int_X \delta_{\phi(x)}(B)\, \mu(dx), \quad B \in \mathscr{Y}.$$

This means that we can understand image measures as measures obtained from the kernel $K(x,B) := \delta_{\phi(x)}(B)$ and $\phi_*\mu(dy) = \mu\delta_{\phi(\cdot)}(dy)$. For image measures it is well known that $f \in L^1(\phi_*\mu) \iff f\circ\phi \in L^1(\mu)$ [☞ Theorem 1.37], and our example shows that this equivalence breaks down for general measure kernels.

16.21 A consistent family of marginals which does not admit a projective limit

Consider Lebesgue measure λ on $[0,1)$ and let $(A_n)_{n\in\mathbb{N}}$ be a sequence of non-measurable sets with $[0,1) \supseteq A_n \downarrow \emptyset$ and $\lambda^*(A_n) = 1$ [☞ 9.3]. On A_n we consider the σ-algebra $\mathscr{B}(A_n) := A_n \cap \mathscr{B}[0,1)$ and the (trace of) Lebesgue measure

$$\mu_n(A_n \cap B) := \lambda(B), \quad B \in \mathscr{B}[0,1),$$

[☞ Example 1.7.(e), 5.9]. Define

$$j_n : A_n \to \bigtimes_{i=1}^{n} A_i, \; x \longmapsto j_n(x) = (x, \dots, x)$$

and consider the image measure $\mu^{(n)} := j_n(\mu_n) = \mu_n \circ j_n^{-1}$ on the product σ-algebra $\bigotimes_{i=1}^n \mathscr{B}(A_i)$.

By construction, the family $(\mu^{(n)})_{n\in\mathbb{N}}$ is projective (i.e. it satisfies the consistency requirement of Kolmogorov's theorem, see the comments further on), but there is no measure μ on $\bigtimes_{n=1}^\infty A_n$ and $\bigotimes_{n=1}^\infty \mathscr{B}(A_n)$ such that the $\mu^{(n)}$ appear as its marginals. This can be seen as follows.

Assume that there were such a measure μ. The diagonal Δ_n of $A_1 \times \dots \times A_n$ is in the product σ-algebra (since the spaces A_n are separable metric spaces) and we have

$$\mu^{(n)}(\Delta_n) = \mu_n(A_n) = \lambda[0,1) = 1.$$

Thus, the sets $\Gamma_n := \Delta_n \times \bigtimes_{i=n+1}^\infty A_i$ satisfy $\mu(\Gamma_n) = \mu^{(n)}(\Delta_n) = 1$ while – due to the σ-additivity of μ –

$$\bigcap_{n\in\mathbb{N}} A_n = \emptyset \Rightarrow \bigcap_{n\in\mathbb{N}} \Gamma_n = \emptyset \Rightarrow 0 = \lim_{n\to\infty} \mu(\Gamma_n) = 1.$$

Comment Let $(X_t, \mathscr{A}_t)$, $t \in I$, be arbitrarily many measurable spaces and assume that on every finite product $\left(\bigtimes_{t\in H} X_t, \bigotimes_{t\in H} \mathscr{A}_t\right)$, $H \subseteq I$, $\#H < \infty$, there is a probability measure μ_H. **Kolmogorov's existence theorem** for projective limits states that there is – under suitable conditions – a unique probability measure μ on $\left(\bigtimes_{t\in I} X_t, \bigotimes_{t\in I} \mathscr{A}_t\right)$ such that the measures μ_H are the image measures of μ under $\pi_H : \bigtimes_{t\in I} X_t \to \bigtimes_{t\in H} X_t$, i.e. $\mu\circ\pi_H^{-1} = \mu_H$. A necessary condition is that the family $(\mu_H)_H$ is **consistent** or **projective**, meaning that for $H \subseteq K$, $\#K < \infty$, the measure μ_H is the image measure of μ_K under the projection $\pi_H^K : \times_{t\in K} X_t \to \times_{t\in H} X_t$; for sufficiency one needs, in general, further topological conditions. Frequently, one assumes that the spaces X_t are Polish (see e.g. [BM, p. 360]) or Souslin (see e.g. [20, Corollary 7.7.2]), but weaker formulations like the existence of a compact family are possible; see [20, 7.7.1] or [119, Theorem III.T31], [43, III.51, p. 110]. A theorem due to Prohorov gives necessary and sufficient conditions in terms of the tightness of the outer measure [23, IX.4.2, Theorem 1].

If I is countable, e.g. $I = \mathbb{N}$, it is possible to avoid additional topological assumptions, but then the measures μ_H have to be 'transition probabilities', i.e. μ_H, $H = \{1, \dots, n\}$, is of the form

$$\mu_H(A_1 \times \cdots \times A_n) = \int_{A_1} \mu_1(dx_1) \int_{A_2} \mu_2(x_1; dx_2) \cdots \int_{A_n} \mu_n(x_1, \dots, x_n; dx_n)$$

where $A_k \in \mathscr{A}_k$, $\mu_k(x_1, \dots, x_{k-1}, dx_k)$ is a kernel, i.e. $B \longmapsto \mu_k(x_1, \dots, x_{k-1}, B)$ is a probability measure on $(X_k, \mathscr{A}_k)$ for all $x_i \in X_i$, $i = 1, \dots, k$, and the map $(x_1, \dots, x_k) \longmapsto \mu_k(x_1, \dots, x_{k-1}, B)$ is measurable for all $B \in \mathscr{A}_k$; this result is due to Ionescu-Tulcea [128, Proposition V.1.1].

A comprehensive discussion can be found in [139, Chapter VI.4].

17

Radon–Nikodým and Related Results

Let $(X, \mathscr{A})$ be a measurable space. Two measures μ, ν on $(X, \mathscr{A})$ are **(mutually) singular**, $\mu \perp \nu$, if there exists a set $N \in \mathscr{A}$ such that both $\mu(N) = 0$ and $\nu(X \setminus N) = 0$. The measure ν is **absolutely continuous** with respect to μ if

$$N \in \mathscr{A},\ \mu(N) = 0 \Rightarrow \nu(N) = 0,$$

and in this case we write $\nu \ll \nu$. The **Radon–Nikodým theorem** [☞ 1.43] gives a characterization of absolute continuity for σ-finite measures: if μ is σ-finite, then ν is absolutely continuous with respect to μ if, and only if,

$$\nu(A) = \int_A f(x)\,\mu(dx), \quad A \in \mathscr{A}, \tag{17.1}$$

for a μ-a.e. unique function $f : X \to [0, \infty]$. This equivalence breaks down if μ is not σ-finite [☞ 17.1, 17.2]. The function f in (17.1) is the **Radon–Nikodým derivative** of ν with respect to μ, and it is traditionally denoted by $f = \frac{d\mu}{d\nu}$.

By the **Lebesgue decomposition theorem** [☞ 1.44], there exists for any two σ-finite measures μ, ν a unique decomposition $\nu = \nu^\circ + \nu^\perp$ such that $\nu^\circ \ll \mu$ and $\nu^\perp \perp \mu$. Again, σ-finiteness is crucial [☞ 17.11].

A Borel measure μ on $(\mathbb{R}, \mathscr{B}(\mathbb{R}))$ is called **continuous** if $\mu(\{x\}) = 0$ for all $x \in \mathbb{R}$. Continuous measures assign mass zero to each singleton, but they may have atoms [☞ 17.9] and need not be absolutely continuous with respect to Lebesgue measure [☞ 17.7].

17.1 An absolutely continuous measure without a density

We consider on the measurable space $(\mathbb{R}, \mathscr{B}(\mathbb{R}))$ Lebesgue measure λ and the counting measure $\zeta(B) := \#B$. Since $\zeta(N) = 0$ if, and only if, $N = \emptyset$, it is clear that Lebesgue measure is absolutely continuous w.r.t. the counting measure: $\lambda \ll \zeta$.

Assume there were a density $f := \frac{d\lambda}{d\zeta}$, i.e. $\lambda = f \cdot \zeta$. Since $f \not\equiv 0$, there is some $x \in \mathbb{R}$ with $f(x) > 0$, and

$$\lambda(\{x\}) = \int_{\{x\}} f(y)\zeta(dy) = f(x) > 0,$$

which is impossible.

17.2 Another absolutely continuous measure without density

Consider Lebesgue measure λ on $((0,\infty), \mathscr{B}(0,\infty))$ and denote the family of Lebesgue null sets by $\mathscr{N} := \{N \in \mathscr{B}(0,\infty)\,;\ \lambda(N) = 0\}$. The set function

$$\mu(B) := \begin{cases} 0, & \text{if } B \in \mathscr{N}, \\ \infty, & \text{if } B \in \mathscr{B}(0,\infty) \setminus \mathscr{N}, \end{cases}$$

is a measure which has the same null sets as λ. Therefore, λ is absolutely continuous with respect to μ, but we cannot represent λ in the form

$$\lambda(B) = \int_B f(x)\,\mu(dx).$$

This is no contradiction to the Radon–Nikodým theorem since μ is not σ-finite. Note, however, that μ does have a density $f = \frac{d\mu}{d\lambda}$ w.r.t. Lebesgue measure: $f(x) \equiv \infty$.

Comment Examples 17.1 and 17.2 highlight the role of σ-finiteness in the Radon–Nikodým theorem.

17.3 Yet another absolutely continuous measure without density

The following example is from [80, Ex. (19.71)]. Consider on $([0,1], \mathscr{B}[0,1])$ both Lebesgue measure λ and the counting measure ζ and define on the product $([0,1]\times[0,1], \mathscr{B}([0,1]\times[0,1]))$ the following measures:

$$\mu(B) = \int_{[0,1]} \lambda\{y \in [0,1]\,;\ (x,y) \in B\}\zeta(dx) + \int_{[0,1]} \lambda\{x \in [0,1]\,;\ (x,y) \in B\}\zeta(dy),$$

$$\nu(B) = \int_{[0,1]} \lambda\{y \in [0,1]\,;\ (x,y) \in B\}\zeta(dx).$$

It is clear that the iterated integrals – with the non-σ-finite counting measure as outside integral – exist and define measures on the product space [☞ footnote

p. 295]. Moreover, from $\nu(B) \leqslant \mu(B)$ we see that $\nu \ll \mu$, but there does not exist any $\mathscr{B}([0,1]\times[0,1])/\mathscr{B}[0,\infty)$ measurable function $f:\ [0,1]\times[0,1]\to[0,\infty)$ such that $\nu = f\mu$.

Otherwise we would have for all $B \in \mathscr{B}([0,1]\times[0,1])$,

$$\nu(B) = \int \mathbb{1}_B(x,y)\,\nu(d(x,y)) = \iint \mathbb{1}_B(x,y)\,\lambda(dy)\,\zeta(dx),$$

and, using $\nu = f\mu$,

$$\begin{aligned}\nu(B) &= \int \mathbb{1}_B(x,y) f(x,y)\,\mu(d(x,y))\\ &= \iint \mathbb{1}_B(x,y) f(x,y)\,\lambda(dy)\,\zeta(dx) + \iint \mathbb{1}_B(x,y) f(x,y)\,\lambda(dx)\,\zeta(dy).\end{aligned}$$

Now we consider the sets $B = [0,1]\times\{y_0\}$ and $B' = \{x_0\}\times[0,1]$ for arbitrary but fixed $x_0, y_0 \in [0,1]$. From the above formulae we get

$$\nu([0,1]\times\{y_0\}) = 0 = \int_0^1 f(x,y_0)\,\lambda(dx),$$

$$\nu(\{x_0\}\times[0,1]) = 1 = \int_0^1 f(x_0,y)\,\lambda(dy).$$

Since $\lambda(dx)$ and $\lambda(dy)$ are σ-finite measures and since $f \geqslant 0$ is measurable, we get a contradiction to Tonelli's theorem:

$$\int_0^1\int_0^1 f(x,y_0)\,\lambda(dx)\,\lambda(dy_0) = 0 \neq 1 = \int_0^1\int_0^1 f(x_0,y)\,\lambda(dy)\,\lambda(dx_0).$$

Thus, the density f cannot be measurable, after all.

17.4 A not-absolutely continuous measure given by a density

Let $(X,\mathscr{A})$ be a measurable space and μ, ν two measures. Sometimes ν is called **absolutely continuous** with respect to μ if the following condition holds:

$$\forall\epsilon > 0 \quad \exists\delta > 0 \quad \forall A \in \mathscr{A},\ \mu(A) < \delta\ :\ \nu(A) < \epsilon. \tag{17.2}$$

If the measure μ is finite, then this condition is equivalent to the usual definition of absolute continuity

$$\forall N \in \mathscr{A},\ \mu(N) = 0\ :\ \nu(N) = 0. \tag{17.3}$$

If μ is σ-finite, the Radon–Nikodým theorem [☞Theorem 1.43] applies and tells us that under (17.3) the measure ν is of the form $\nu(A) = \int_A f\,d\mu$ for some positive, measurable density.

The following example shows that (17.2) is not equivalent to (17.3) if μ is not finite. Consider Lebesgue measure λ on $((0,\infty),\mathscr{B}(0,\infty))$ and define a further measure

$$\nu(A) := \int_A x\,\lambda(dx).$$

If we take the sequence of sets $A_n := [n, n+1/n)$, we see that

$$\nu(A_n) \geqslant 1 \quad \text{and} \quad \lambda(A_n) = \frac{1}{n} \xrightarrow[n\to\infty]{} 0,$$

which means that (17.2) does not hold. The condition (17.3) is, however, satisfied.

17.5 A measure $\mu \ll \lambda$ such that $\lambda(A_n) \to 0$ does not imply $\mu(A_n) \to 0$

Consider Lebesgue measure λ on $(\mathbb{R},\mathscr{B}(\mathbb{R}))$ and define a measure μ by

$$\mu(B) := \int_B |x|\,\lambda(dx), \quad B \in \mathscr{B}(\mathbb{R}).$$

Clearly, μ is absolutely continuous with respect to one-dimensional Lebesgue measure. For $A_n := [n, n+1/n]$, $n \in \mathbb{N}$, we have $\lambda(A_n) \to 0$ as $n \to \infty$ but

$$\mu(A_n) \geqslant \int_{[n,n+1/n]} n\,\lambda(dx) = 1, \quad n \in \mathbb{N},$$

does not converge to 0 as $n \to \infty$.

17.6 A measure μ which is absolutely continuous w.r.t. Lebesgue measure and $\mu(a,b) = \infty$ for any $(a,b) \neq \emptyset$

Let $\mathbb{Q} = (r_n)_{n\in\mathbb{N}}$ and construct Borel functions $f_n : \mathbb{R} \to [0,\infty)$ such that $\operatorname{supp} f_n \subseteq B_{2^{-n}}(r_n)$ and $\int f_n\,d\lambda = 1$. The function

$$f(x) := \sum_{n=1}^{\infty} f_n(x), \quad x \in \mathbb{R},$$

is positive and a.s. finite. The latter follows from the observation that Lebesgue almost all x are in at most finitely many of the sets $B_{2^{-n}}(r_n)$. We have

$$\int_{\mathbb{R}} \sum_{n=1}^{\infty} \mathbb{1}_{B_{2^{-n}}(r_n)}(x)\,\lambda(dx) = \sum_{n=1}^{\infty} \int_{\mathbb{R}} \mathbb{1}_{B_{2^{-n}}(r_n)}(x)\,\lambda(dx) = \sum_{n=1}^{\infty} 2\cdot 2^{-n} = 2,$$

which means that $\sum_{n=1}^{\infty} \mathbb{1}_{B_{2^{-n}}(r_n)}(x) < \infty$ for almost all x; therefore, $f(x)$ is defined for almost all x by a finite sum.

The measure $\mu := f\lambda$ has a density w.r.t. Lebesgue measure, i.e. it is absolutely continuous. Any open interval (a, b) contains n disjoint open intervals of length $(b-a)/n$, and each of these intervals contains some $B_{2^{-m}}(r_m)$ – just take m large enough. This shows that

$$\mu(a, b) = \int_{\mathbb{R}} \mathbb{1}_{(a,b)}(x) f(x) \lambda(dx) \geqslant n,$$

and since n is arbitrary, μ assigns infinite mass to each non-empty open interval.

17.7 A continuous measure which is not absolutely continuous

Let μ be a measure on the measurable space $([0,1], \mathscr{B}[0,1])$ whose distribution function $F(x) = \mu(0, x]$ is the Cantor function [☞ Section 2.6]. Recall that $F: [0,1] \to [0,1]$ is increasing, continuous and Lebesgue a.e. differentiable with derivative $F' = 0$.

Since $x \longmapsto \mu(0, x] = F(x)$ is continuous, μ assigns measure zero to all singletons $\{x\}$, i.e. μ is continuous.

Suppose there is a measurable function $f \geqslant 0$ such that $F(x) = \int_0^x f(t)\,dt$. By Lebesgue's differentiation theorem [☞ 14C], $F'(x) = f(x)$ a.e. The construction of F yields $F' = 0$ Lebesgue a.e. and so $\int_0^x f(t)\,dt = \int_0^x F'(t)\,dt = 0$. From this we conclude that $F = 0$, which is absurd.

17.8 An absolutely continuous function whose inverse is not absolutely continuous

Let $F: [0,1] \to [0,1]$ be the Cantor function [☞ Section 2.6] and define a function $G(x) := F(x) + x$. Since G is strictly increasing, it has an inverse $g(y) = G^{-1}(y)$. The Cantor function F is not absolutely continuous – we have $F' = 0$ Lebesgue a.e. but F is not constant –, and so G is not absolutely continuous.

Its inverse, however, is an increasing function $g: [0,2] \to [0,1]$ which satisfies for $0 \leqslant y < y' \leqslant 2$ such that $y = G(x)$ and $y' = G(x')$,

$$\frac{y - y'}{g(y) - g(y')} = \frac{G(x) - G(x')}{x - x'} = \frac{F(x) - F(x')}{x - x'} + \frac{x - x'}{x - x'} \geqslant 1$$

and so $|g(y) - g(y')| \leqslant |y - y'|$. Consequently, g is increasing and absolutely continuous but its inverse G is not.

17.9 A continuous measure with atoms

[☞ 6.9]

17.10 The Radon–Nikodým density $f = d\nu/d\mu$ does not necessarily satisfy $f(x) = \lim_{r\downarrow 0} \nu(B_r(x))/\mu(B_r(x))$

[☞14.20]

17.11 Lebesgue's decomposition theorem fails without σ-finiteness

By Lebesgue's decomposition theorem [☞1.44], there exists for any two σ-finite measures μ, ν on $(X, \mathscr{A})$ a decomposition $\nu = \nu^\circ + \nu^\perp$ where ν°, $\nu^\perp$ are measures such that $\nu^\circ \ll \mu$ and $\nu^\perp \perp \mu$. The following example shows that the σ-finiteness plays a crucial role.

Consider $((0,1), \mathscr{B}(0,1))$ with Lebesgue measure λ and the counting measure ζ. Suppose there is a decomposition $\zeta = \zeta^\circ + \zeta^\perp$ such that $\zeta^\circ \ll \lambda$ and $\zeta^\perp \perp \lambda$. Since $\{x\}$ is a Lebesgue null set, we have $\zeta^\circ(\{x\}) = 0$ for all $x \in (0,1)$. Moreover, $\zeta(\{x\}) = 1$ yields $\zeta^\perp(\{x\}) = 1$ for all $x \in (0,1)$. This means that $N = \emptyset$ is the only set N with $\zeta^\perp(N) = 0$. As $\zeta^\perp \perp \lambda$, this gives $\zeta^\perp(N) = 0 = \lambda((0,1) \setminus N)$, i.e. $\lambda((0,1)) = 0$, which is absurd.

17.12 Two mutually singular measures which have the same support

Let $A = \{a_n,\ n \in \mathbb{N}\}$ and $B = \{b_n,\ n \in \mathbb{N}\}$ be countable subsets of $\mathbb{R}$ which are pairwise disjoint and dense in $\mathbb{R}$. Define finite measures on $(\mathbb{R}, \mathscr{P}(\mathbb{R}))$ by

$$\mu := \sum_{n\in\mathbb{N}} 2^{-n}\delta_{a_n} \quad \text{and} \quad \nu := \sum_{n\in\mathbb{N}} 2^{-n}\delta_{b_n}.$$

Since A and B are dense in $\mathbb{R}$, all open balls $B_r(x) \neq \emptyset$ have positive measure with respect to μ and ν, and so $\operatorname{supp}\mu = \mathbb{R}$ and $\operatorname{supp}\nu = \mathbb{R}$. Moreover, $A \cap B = \emptyset$ implies $\nu(A) = 0$. If we set $N := \mathbb{R} \setminus A$, then we see that both $\mu(N) = 0$ and $\nu(N) = \nu(\mathbb{R}) - \nu(A) = \nu(\mathbb{R})$, i.e. μ and ν are mutually singular.

Comment Note that the measure μ is purely atomic, thus singular w.r.t. Lebesgue measure, and still has full support. For further examples of mutually singular measures with full support [☞17.13, 17.15].

17.13 A probability measure μ with full support such that μ and $\mu(c\,\cdot)$ are mutually singular for $c \neq 1$

Consider the infinite product space $X = \mathbb{R}^{\mathbb{N}}$ with the product topology, i.e. the topology with basis

$$\left\{ \bigtimes_{i\in\mathbb{N}} U_i\, ;\ U_i \subseteq \mathbb{R} \text{ open},\ U_i = \mathbb{R} \text{ except for finitely many } i \in \mathbb{N} \right\}.$$

The associated Borel σ-algebra $\mathscr{B}(X)$ is the product σ-algebra $\mathscr{B}(\mathbb{R})^{\otimes\mathbb{N}}$, i.e. the smallest σ-algebra such that all projections, $\pi_i : X \to \mathbb{R}$, $x \longmapsto \pi_i(x) = x(i)$ are measurable [☞ Example 4C]. Denote by

$$p(x) := \frac{1}{\sqrt{2\pi}} \exp\left(-\frac{x^2}{2}\right), \qquad x \in \mathbb{R},$$

the density of the standard Gaussian distribution. By Kolmogorov's existence theorem, cf. [BM, Appendix A.1] or [19, Theorem 3.5.1], there exists a (unique) measure μ on $(X, \mathscr{B}(X))$ such that

$$\mu\left(B_1 \times \cdots \times B_n \times \mathbb{R} \times \mathbb{R} \ldots\right) = \prod_{k=1}^{n} \int_{B_k} p(y)\,\lambda(dy)$$

for all $B_k \in \mathscr{B}(\mathbb{R})$ and $n \in \mathbb{N}$. This implies that

$$\mu(\{\pi_i \in B\}) = \int_B p(y)\,\lambda(dy), \quad i \in \mathbb{N},\ B \in \mathscr{B}(\mathbb{R}),$$

and

$$\mu(\{\pi_{i_1} \in B_1, \ldots, \pi_{i_n} \in B_n\}) = \prod_{k=1}^{n} \mu(\{\pi_{i_k} \in B_k\})$$

for all $1 \leqslant i_1 < \cdots < i_n$, $B_k \in \mathscr{B}(\mathbb{R})$ and $n \in \mathbb{N}$. In other words: the projections π_i, $i \in \mathbb{N}$, are independent random variables, and each π_i has standard Gaussian distribution. By the strong law of large numbers, cf. [MIMS, p. 294],

$$L_c := \left\{x \in X\,;\ \lim_{n\to\infty} \frac{1}{n} \sum_{i=1}^{n} \pi_i(x)^2 = c\right\},$$

satisfies $\mu(L_1) = 1$ and $\mu(L_c) = 0$ for all $c \neq 1$. If we set $X_0 := L_1$, then $\mu(X_0) = 1$ but $\mu(cX_0) = 0$ for all $c \neq 1$ (as usual, $cX_0 = \{cx\,;\ x \in X_0\}$). This shows that the measures $A \longmapsto \mu(A)$ and $A \longmapsto \mu(cA)$ are mutually singular for $c \neq 1$. In order to prove that μ has full support, fix a closed set $F \subsetneqq X$. Since the complement $X \setminus F \neq \emptyset$ is open, there exists a basis set U

$$U = \mathop{\times}_{i\in\mathbb{N}} U_i$$

with $\emptyset \neq U_i \subseteq \mathbb{R}$ open, $I := \{i \in \mathbb{N}\,;\ U_i \neq \mathbb{R}\} < \infty$, such that $U \subseteq X \setminus F$. As p is strictly positive, it follows that

$$\mu(X \setminus F) \geqslant \mu(U) = \prod_{i\in I} \int_{U_i} p(y)\,\lambda(dy) > 0.$$

Hence, $F = X$ is the only closed set with $\mu(X \setminus F) = 0$, and so $\operatorname{supp}\mu = X$.

Comment The measure μ is a Gaussian measure on an infinite-dimensional space. We learnt the above example from V.I. Bogachev; see his monograph [18] for a thorough study of (infinite) Gaussian measures.

This example is a typically infinite-dimensional phenomenon. The following result is known as **Kakutani's theorem** or **Kakutani's alternative**, see Hewitt & Stromberg [80, pp. 452–453] and Jacod & Shiryaev [84, pp. 252–255]. Let $\mu_i, \nu_i, i \in \mathbb{N}$, be probability measures on measurable spaces $(X_i, \mathscr{A}_i)$ and assume that $\mu_i \ll \nu_i$ with Radon–Nikodým density $f_i = \frac{d\mu_i}{d\nu_i}$. We write $\mu = \bigtimes_{i\in\mathbb{N}} \mu_i$ and $\nu = \bigtimes_{i\in\mathbb{N}} \nu_i$ for the product measures in the infinite-dimensional product space $(\bigtimes_{i\in\mathbb{N}} X_i, \bigotimes_{i\in\mathbb{N}} \mathscr{A}_i)$. The following dichotomy holds:

either μ is absolutely continuous w.r.t. ν, i.e. $\mu \ll \nu$;

this is the case if $\prod_{i=1}^{\infty} \int_{X_i} f_i^{\frac{1}{2}}\, d\nu_i > 0$

or μ and ν are mutually singular, i.e. $\mu \perp \nu$;

this is the case if $\prod_{i=1}^{\infty} \int_{X_i} f_i^{\frac{1}{2}}\, d\nu_i = 0$.

17.14 The convolution of two singular measures may be absolutely continuous

Denote by λ Lebesgue measure on $(\mathbb{R}, \mathscr{B}(\mathbb{R}))$, and consider the product measures

$$\mu := \delta_0 \times \lambda \quad \text{and} \quad \nu := \lambda \times \delta_0$$

on $(\mathbb{R}^2, \mathscr{B}(\mathbb{R}^2))$. Since $\lambda^2(\{0\}\times\mathbb{R}) = 0$ and $\mu(\mathbb{R}^2 \setminus (\{0\}\times\mathbb{R})) = 0$, the measure μ is singular with respect to two-dimensional Lebesgue measure λ^2. Analogously, ν is singular with respect to λ^2. We claim that the convolution $\mu * \nu$,

$$(\mu * \nu)(B) = \int_{\mathbb{R}^2}\int_{\mathbb{R}^2} \mathbb{1}_B(x+y)\,\mu(dx)\,\nu(dy), \quad B \in \mathscr{B}(\mathbb{R}^2),$$

is two-dimensional Lebesgue measure λ^2; in particular, $\mu * \nu$ is absolutely continuous with respect to λ^2. If $C, D \in \mathscr{B}(\mathbb{R})$, then

$$\int_{\mathbb{R}^2} \mathbb{1}_{C\times D}(x+y)\,\mu(dx) = \mathbb{1}_C(y_1)\int_{\mathbb{R}} \mathbb{1}_D(x_2+y_2)\,\lambda(dx_2)$$

for any $y = (y_1, y_2) \in \mathbb{R}^2$. Thus,

$$(\mu * \nu)(C \times D) = \left(\int_0^1 \mathbb{1}_C(y_1)\,\lambda(dy_1)\right)\left(\int_0^1 \mathbb{1}_D(x_2)\,\lambda(dx_2)\right) = \lambda^2(C \times D).$$

Since the rectangles $C \times D$, $C, D \in \mathscr{B}(\mathbb{R})$, are a generator of $\mathscr{B}(\mathbb{R}^2)$ which is stable under intersections, the uniqueness of measures theorem [☞ 1.40] yields $\mu * \nu = \lambda^2$.

Comment It is possible to construct a finite measure μ on $([0,1], \mathscr{B}[0,1])$ such that μ is singular with respect to Lebesgue measure but $\mu * \mu$ is absolutely continuous. Take a finite, singular measure μ with Fourier transform $\widehat{\mu}$ satisfying $|\widehat{\mu}(\xi)| \leqslant C|\xi|^{-1/3}$ for $|\xi| \geqslant 1$, cf. [199], then the convolution theorem [MIMS, p. 221] shows that $|\widehat{\mu * \mu}| \leqslant C^2|\xi|^{-2/3}$, $|\xi| \geqslant 1$, which implies that $\widehat{\mu * \mu} \in L^2(dx)$; by Plancherel's theorem, this gives that $\mu * \mu$ has a (square integrable) density with respect to Lebesgue measure and is not singular. See also the paper by Hewitt and Zuckerman [81] for a survey and an extension to Abelian groups.

Wiener and Wintner's paper [199] is one of the first quantitative studies of the problem of how fast the Fourier coefficients of a singular measure μ can decay to zero. In view of the Riemann–Lebesgue lemma, it is clear that an absolutely continuous (with respect to one-dimensional Lebesgue measure, say) measure has a Fourier transform such that $\widehat{\mu}(n) \to 0$. On the other hand, Menchoff [118] showed that the converse is not true: decay does not guarantee absolute continuity. Wiener and Wintner constructed for every $\epsilon > 0$ singular measures on the unit circle such that their Fourier coefficients (or transforms) decay like $n^{-\frac{1}{2}+\epsilon}$ as $n \to \infty$, thus establishing the result that there is a threshold $\varkappa = \frac{1}{2}$ such that there are singular measures whose Fourier coefficients have decay of type $n^{-\varkappa+\epsilon}$ while a measure whose Fourier coefficients show a decay of order $n^{-\varkappa-\epsilon}$ cannot be singular. Subsequently, this line of research attracted a lot of attention. We will continue this remark in the next example [☞ 17.15].

17.15 Singular measures with full support – the case of Bernoulli convolutions

Let C be the classical Cantor middle-thirds set [☞ 2.5] and $F : [0,1] \to [0,1]$ the Cantor function [☞ Section 2.6]. We denote by μ the (Lebesgue–Stieltjes) measure on $(\mathbb{R}, \mathscr{B}(\mathbb{R}))$ whose distribution function is F, i.e.

$$\mu(a,b] = \mu_F(a,b] = F(b) - F(a), \quad 0 \leqslant a < b \leqslant 1.$$

Since $F' = 0$ on $[0,1] \setminus C$, we have $\operatorname{supp}\mu = C$; this means that we may interpret μ as a measure on $([0,1], \mathscr{B}[0,1])$ which is singular to Lebesgue measure λ on $[0,1]$.

The convolution

$$\mu * \mu(B) := \iint \mathbb{1}_B(y+z)\,\mu(dy)\,\mu(dz) = \int \mu(B-z)\,\mu(dz), \quad B \in \mathscr{B}(\mathbb{R})$$

is a measure on $(\mathbb{R}, \mathscr{B}(\mathbb{R}))$ and its support is the closure of

$$\operatorname{supp}\mu + \operatorname{supp}\mu = \{y+z\,;\ y,z \in \operatorname{supp}\mu\}$$

[☞6.6]. Since $\operatorname{supp}\mu = C$ and $C + C = [0,2]$ [☞7.27], we conclude that $\operatorname{supp}\mu * \mu = [0,2]$. In particular, we may interpret $\mu * \mu$ as a measure on $([0,2], \mathscr{B}[0,2])$.

Although $\mu * \mu$ has full support, it is singular with respect to Lebesgue measure on $[0,2]$. For this, we first calculate the inverse Fourier transform of μ,

$$\widecheck{\mu}(\xi) = \int_0^1 e^{i\xi x}\,dF(x) = \int_0^1 e^{i\xi G(y)}\,dy = e^{i\frac{1}{2}\xi} \prod_{k=1}^{\infty} \int_0^1 e^{-i3^{-k}\xi R_k(y)}\,dy,$$

where we use the representation of the inverse G of F in terms of Rademacher functions [☞Section 2.6, Lemma 2.10]. Since R_k takes the values ± 1 on sets of Lebesgue measure $\frac{1}{2}$, we get

$$\widecheck{\mu}(\xi) = e^{i\frac{1}{2}\xi} \prod_{k=1}^{\infty} \frac{e^{i3^{-k}\xi} + e^{-i3^{-k}\xi}}{2} = e^{i\frac{1}{2}\xi} \prod_{k=1}^{\infty} \cos\left(3^{-k}\xi\right).$$

For the sequence $\xi_n = 3^n\pi$, $n \in \mathbb{N}$, we get

$$|\widecheck{\mu}(\xi_n)| = \prod_{k=n+1}^{\infty} \cos\left(3^{n-k}\pi\right) = \prod_{l=1}^{\infty} 1 - \left(1 - \cos\left(3^{-l}\pi\right)\right),$$

and this infinite product converges to a value in $(0,1]$ if, and only if,

$$\sum_{l=1}^{\infty} \left(1 - \cos\left(3^{-l}\pi\right)\right) < \infty;$$

see e.g. [26, Chapter VI.39]. This, however, is easily seen from the elementary estimate $0 \leqslant 1 - \cos x \leqslant \frac{1}{2}x^2$. Thus, $\limsup_{|\xi|\to\infty} |\widecheck{\mu}(\xi)| > 0$. By the convolution theorem [MIMS, p. 221],

$$|\widecheck{\mu * \mu}(\xi)| = |\widecheck{\mu}(\xi)|^2 = \prod_{k=1}^{\infty} \cos^2\left(3^{-k}\xi\right).$$

From this we conclude that $\limsup_{|\xi|\to\infty} |\widecheck{\mu * \mu}(\xi)| > 0$. By the Riemann–Lebesgue lemma, $\mu * \mu$ cannot have an L^1-density with respect to Lebesgue measure.

The Hahn–Jordan decomposition (Theorem 1.47) tells us that $\mu * \mu$ is of the form $\nu^\circ + \nu^\perp$ where $\nu^\circ \ll \lambda$ is absolutely continuous and $\nu^\perp \perp \lambda$ singular with respect to Lebesgue measure. The Jessen–Wintner **law of pure type** [86] entails that $\nu^\circ = 0$, i.e. $\mu * \mu \perp \lambda$.

A modern proof of the Jessen–Wintner law can be found in Breiman [24, Chapter 3.5]. It is based on a probabilistic interpretation of F and G: on the probability space $([0,1], \mathscr{B}[0,1], \lambda)$ the Cantor function F is the distribution function of the random variable (i.e. measurable function) G. From the representation (2.2) we see that G is a series of weighted (weight $2 \cdot 3^{-k}$) independent Bernoulli random variables $\frac{1}{2}(1 - R_k(y))$ taking the values 0 and 1 with equal probability $\frac{1}{2}$; see e.g. [MIMS, pp. 280–283]. Kolmogorov's three series theorem gives necessary and sufficient conditions for the convergence of this series and, in this case, the limit law is an infinite convolution $\pi = \ast_{k=1}^{\infty} \pi_k$ with factors $\pi_k = \frac{1}{2}(\delta_0 + \delta_{2\cdot 3^{-k}})$. The Kolmogorov zero–one law now guarantees that π is of 'pure type', i.e. it is either absolutely continuous or singular with respect to Lebesgue measure.

Comment Starting with Menchoff's 1916 paper [118] and the contributions of Jessen & Wintner [86] and Wiener & Wintner [199] in the 1930s, the study of singular functions 'of Cantor type' and **(infinite) Bernoulli convolutions** has attracted a lot of attention. The historical development and the state of the art is nicely surveyed in the papers [134, 168] by Solomyak and Peres & Solomyak. At this point we want to mention only one key contribution by Salem [155] who fixed some gap in [199] (and its follow-up paper) showing that various singular functions can indeed be chosen to be of Cantor type – this was claimed, but not proved, by Wiener and Wintner. Let us sketch that part of the story which helps to shed more light onto the going ons of the above example.

Consider a symmetric Bernoulli distribution $\beta(1) = \frac{1}{2}(\delta_{-1} + \delta_1)$ which corresponds to a random variable taking values in ± 1 with equal probability; the prototype of such a random variable is any Rademacher function R_k, $k \in \mathbb{N}$, on the probability space $([0,1], \mathscr{B}[0,1], \lambda)$. We will need independent copies and, because of the flip-flop game the Rademacher functions are playing, $(R_k)_{k\in\mathbb{N}}$ is indeed a sequence of independent random variables all having the same law. We are interested in the distributional properties of the series $\sum_{k=1}^{\infty} a^k R_k$ where $a > 0$ is some constant. Note that the scaled random variables $a^k R_k$ are concentrated at the points $\pm a^k$ and have the law $\beta(a^k)$. It is known that

- ▸ $\sum\limits_{k=1}^{\infty} a^k R_k(y)$ converges a.e. and in $L^2(dy)$ if, and only if, $\sum\limits_{k=1}^{\infty} (a^k)^2 < \infty$. In particular, the series either converges or diverges with probability one (for almost all y). This narrows the choice of interesting values of a to $a \in (0, 1)$. There are various ways to prove this, e.g. via **Kolmogorov's three series theorem** [24] or by martingale methods [MIMS, Theorem 28.22]. With yet another proof, this result is due to Wintner (1935); see also [86].

- If the series $\sum_{k=1}^{\infty} a^k R_k$ is convergent, then its law is the infinite convolution $\mu = \mathop{*}_{k=1}^{\infty} \beta(a^k)$ and, as above for $a = \frac{1}{3}$, one can see that the inverse Fourier transform is given by

$$\widehat{\mu}(\xi) = \prod_{k=1}^{\infty} \cos(a^k \xi).$$

- For $a = \frac{1}{3}$, our calculations above show that F_μ is the (shifted) Cantor function on $\left[-\frac{1}{2}, \frac{1}{2}\right]$. If $a = \frac{1}{2}$, then we can use the elementary trigonometric identity $\sin 2x = 2 \sin x \cos x$ to deduce that

$$\frac{\sin \xi}{\xi} = \cos \frac{\xi}{2} \cdot \cos \frac{\xi}{4} \cdot \dots \cdot \cos \frac{\xi}{2^k} \cdot \frac{\sin \frac{\xi}{2^k}}{\frac{\xi}{2^k}} \implies \frac{\sin \xi}{\xi} = \prod_{k=1}^{\infty} \cos \frac{\xi}{2^k},$$

which is the inverse Fourier transform of the measure $\frac{1}{2}\mathbb{1}_{[-1,1]}(x)\,\lambda(dx)$. This shows that the relation of μ with respect to Lebesgue measure depends on $a \in (0, 1)$. In fact, one has

  - ▸▸ μ is either absolutely continuous or singular with respect to Lebesgue measure (Jessen–Wintner law of pure types [86]); if it is absolutely continuous, the support is $\left[-\frac{a}{1-a}, \frac{a}{1-a}\right]$. Note that $\frac{a}{1-a} = \sum_{k=1}^{\infty} a^k$.
  - ▸▸ μ is singular if $a \in (0, \frac{1}{2})$ (Kershner & Wintner, 1935) and absolutely continuous if $a = 2^{-1/n}, n = 1, 2, \dots$ with a differentiable density $d\mu/d\lambda$ having $n - 1$ derivatives (Wintner, 1935), see [86].
  - ▸▸ μ is singular if the reciprocal of $a \in \left(\frac{1}{2}, 1\right)$ is a Pisot–Vijayaraghavan number – an algebraic integer whose conjugates have modulus less than one, e.g. the golden ratio $\frac{1}{2}\left(1 + \sqrt{5}\right)$; this is due to Erdös [54].
  - ▸▸ μ is for Lebesgue almost all $a \in \left[\frac{1}{2}, 1\right)$ absolutely continuous with a density $d\mu/d\lambda \in L^2(\lambda)$ (Solomyak, 1995; see [134], proving a conjecture of Garsia from 1962).

17.16 The maximum of two measures need not be the maximum of its values

Let μ and ν be two (positive) finite measures on a measurable space $(X, \mathscr{A})$. By $\mu \vee \nu$ we denote the smallest (positive) measure which dominates both μ and ν, i.e.

$$\mu \vee \nu = \nu + (\mu - \nu)^+ = \mu + (\nu - \mu)^+,$$

where $\rho^+ - \rho^-$ is the Hahn–Jordan decomposition of the signed measure ρ [☞Theorem 1.47]. In the same way we can define the largest (positive) measure $\mu \wedge \nu$ which is smaller than μ and ν:

$$\mu \wedge \nu = \mu - (\mu - \nu)^+ = \nu - (\nu - \mu)^+.$$

These identities cannot be understood in a setwise sense. Consider, for example, $(X, \mathscr{A}) = (\mathbb{R}, \mathscr{B}(\mathbb{R}))$ and define $\mu = \mathbb{1}_{(-2,0)}\,\lambda$ and $\nu = \mathbb{1}_{(0,2)}\,\lambda$ where λ is Lebesgue measure. We have

$$\mu[-1,1] = 1 \quad \text{and} \quad \nu[-1,1] = 1$$

whereas

$$\begin{aligned} \mu \vee \nu[-1,1] &= \nu[-1,1] + (\mu - \nu)^+[-1,1] \\ &= 1 + \mu[-1,0) + (-\nu)^+(0,1] = 1 + 1 + 0. \end{aligned}$$

This shows that, in general, $\mu \vee \nu(A) \neq \mu(A) \vee \nu(A)$.

Comment By definition, we always have $\mu \vee \nu(A) \geqslant \mu(A) \vee \nu(A)$ as well as $\mu \wedge \nu(A) \leqslant \mu(A) \wedge \nu(A)$. A close look at the proof of the Lebesgue decomposition theorem [☞Theorem 1.44] reveals that

$$\begin{aligned} \mu \wedge \nu(A) &= \inf\{\mu(B) + \nu(A \setminus B);\ B \subseteq A,\ B \in \mathscr{A}\}, \\ \mu \vee \nu(A) &= \sup\{\mu(B) + \nu(A \setminus B);\ B \subseteq A,\ B \in \mathscr{A}\}. \end{aligned}$$

Alternatively, one could use the fact that $\mu, \nu \ll \rho := \mu + \nu$ and that $\mu \wedge \nu$ and $\mu \vee \nu$ have densities with respect to ρ which are given by $\frac{d\mu}{d\rho} \wedge \frac{d\nu}{d\rho}$ and $\frac{d\mu}{d\rho} \vee \frac{d\nu}{d\rho}$, respectively.

18

Function Spaces

Let $(X, \mathscr{A}, \mu)$ be a measure space. For $p \in (0, \infty]$ we denote by $L^p(X, \mathscr{A}, \mu)$ the **Lebesgue space** of pth order, i.e. $f \in L^p(\mu) \iff f : X \to \mathbb{R}$ is measurable and $\|f\|_{L^p} < \infty$ where

$$\|f\|_{L^p} := \left(\int |f|^p \, d\mu\right)^{1/p},$$

$$\|f\|_\infty = \operatorname{esssup} |f| := \inf \{c\, ;\ \mu\{|f| > c\} = 0\};$$

as usual, we identify functions f, g which differ only on a μ-null set [☞ 1.16]. If there is no ambiguity, we use the shorthand L^p or $L^p(\mu)$. **Minkowski's inequality** [☞ Corollary 1.19]

$$\|f + g\|_{L^p} \leqslant \|f\|_{L^p} + \|g\|_{L^p}, \quad f, g \in L^p(\mu),\ p \geqslant 1,$$

shows that $L^p(\mu)$ is a normed linear space if $p \geqslant 1$.

For $p \in (0, 1)$, Minkowski's inequality is reversed [☞ 18.8], and so $L^p(\mu)$ is only a quasi-normed space for $p \in (0, 1)$ [☞ 18.9]. The reversal of the triangle inequality also implies that there are only few convex sets in $L^p(\mu)$, $p \in (0, 1)$, [☞ 18.10]; in particular, the unit ball is not convex. If we consider $X = \mathbb{N}^2$ with the counting measure, we can identify $(L^p(\mu), \|\cdot\|_{L^p})$ with $\mathbb{R}^2$ equipped with the norms $\|x\|_{L^p} = (|x_1|^p + |x_2|^p)^{1/p}$ and $\|x\|_\infty = \max\{|x_1|, |x_2|\}$, and this allows us to plot the unit balls in these spaces for different values of p; see Fig. 18.1. The Lebesgue spaces $L^p(\mu)$, $p \in [1, \infty]$, are Banach spaces [☞ Theorem 1.20]. By definition, the (topological) **dual** $(L^p(\mu))^*$ is the family of all bounded linear functionals $\Lambda : L^p(\mu) \to \mathbb{R}$, and equipped with the norm

$$\|\Lambda\|_{p,*} := \sup_{0 \neq f \in L^p(\mu)} \frac{|\Lambda(f)|}{\|f\|_{L^p}},$$

the dual space is again a Banach space. If $(L^p(\mu))^{**} := ((L^p(\mu))^*)^*$ is isometrically isomorphic to $L^p(\mu)$, then $L^p(\mu)$ is called **reflexive**. For $p \in [1, \infty]$ with

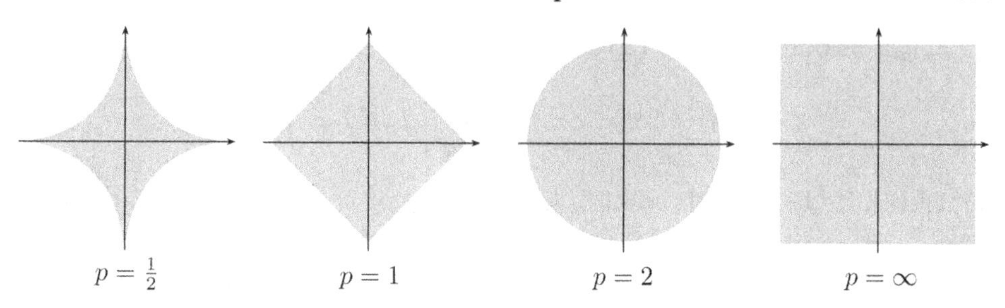

Figure 18.1 The unit balls in the space $\mathbb{R}^2$ equipped with the (quasi-)metric $d_p(x, y) = |x - y|_p = \left(\sum_{i=1}^{2} |x_i - y_i|^p\right)^{1/p}$. Note that the balls are convex if, and only if, $p \in [1, \infty]$. If $p = 1$ and $p = \infty$, then the balls have 'flat' sides, i.e. tangents to the unit ball are not unique – they are not 'rotund' or 'strictly convex'. In fact, the unit balls with $1 < p < \infty$ are 'uniformly rotund' or 'uniformly convex' [☞18.32]. The Milman–Pettis theorem ensures that all uniformly rotund spaces are reflexive, for details see Yosida [202, Section V.2] and Megginson [117, Sections 5.1, 5.2].

conjugate index $q \in [1, \infty]$, $p^{-1} + q^{-1} = 1$, it follows from Hölder's inequality [☞Theorem 1.18] that

$$f \longmapsto \Lambda_g(f) := \int_X fg \, d\mu$$

defines a bounded linear functional on $L^p(\mu)$ for every $g \in L^q(\mu)$. One usually identifies g with Λ_g and writes $L^q(\mu) \subseteq (L^p(\mu))^*$. If $p \in (1, \infty)$, then the dual of $L^p(\mu)$ is isometrically isomorphic to $L^q(\mu)$, cf. [51, Theorem IV.8.1], and, as a consequence, $L^p(\mu)$, $p \in (1, \infty)$, is reflexive. For $p = 1$ and $p = \infty$ identifying the dual of $L^p(\mu)$ is a tricky business [☞18.20–18.25].

A sequence $(f_n)_{n\in\mathbb{N}} \subseteq L^p(\mu)$ **converges weakly in** $L^p(\mu)$ to $f \in L^p(\mu)$ if

$$\forall g \in (L^p(\mu))^* \; : \; \lim_{n\to\infty} \int_X f_n g \, d\mu = \int_X fg \, d\mu, \tag{18.1}$$

and in this case we write $f_n \rightharpoonup f$. For $p \in \{1, \infty\}$ this condition can be cumbersome as it is, in general, hard to identify the duals of $L^1(\mu)$ and $L^\infty(\mu)$. For $L^1(\mu)$ there is an equivalent characterization, which is easier to verify: $f_n \rightharpoonup f$ in $L^1(\mu)$ if, and only if,

$$\forall A \in \mathscr{A} \; : \; \lim_{n\to\infty} \int_A f_n \, d\mu = \int_A f \, d\mu,$$

i.e. it suffices to check (18.1) for $g = \mathbb{1}_A \in L^\infty(\mu)$, $A \in \mathscr{A}$ [☞18.21]. Let us emphasize that this characterization holds for any measure space, it is not needed that $(L^1(\mu))^* = L^\infty(\mu)$. If $p \in (1, \infty)$, then $f_n \rightharpoonup f$ in $L^p(\mu)$ if, and only if, $(f_n)_{n\in\mathbb{N}}$ is bounded in $L^p(\mu)$ and

$$\forall A \in \mathscr{A},\ \mu(A) < \infty : \lim_{n\to\infty} \int_A f_n \, d\mu = \int_A f \, d\mu \tag{18.2}$$

[☞19.18]. If (18.2) holds and $\|f_n\|_{L^p} \to \|f\|_{L^p} < \infty$ for $p \in (1,\infty)$, then we have $f_n \to f$ (strongly) in L^p; the converse is of course also true. This characterization is closely related to the fact that L^p, $p \in (1,\infty)$, is uniformly convex [☞Fig. 18.1, Example 18.32].

Another useful tool is the **Riesz convergence theorem** [MIMS, p. 123], which says that for $p \in [1,\infty)$ and every sequence $(f_n)_{n\in\mathbb{N}} \subseteq L^p(\mu)$ such that the limit $\lim_{n\to\infty} f_n = f$ exists a.e. for some $f \in L^p(\mu)$, we have

$$\lim_{n\to\infty} \|f_n\|_{L^p} = \|f\|_{L^p} \iff \lim_{n\to\infty} \|f_n - f\|_{L^p} = 0.$$

For $p = \infty$ this becomes false [☞12.6].

18.1 Relations between L^r, L^s, L^t if $r < s < t$

On any finite measure space $(X, \mathscr{A}, \mu)$ Hölder's inequality [☞Theorem 1.18] shows

$$\|f\|_{L^p} \leqslant \mu(X)^{p^{-1}-q^{-1}} \|f\|_{L^q}, \quad p \leqslant q,$$

and therefore $L^q \subseteq L^p$ for any $p \leqslant q$. For infinite measure spaces $(X, \mathscr{A}, \mu)$ there is, in general, no direct comparison between any two of the spaces L^p and L^q possible. There are the following characterizations.

Proposition *Let $(X, \mathscr{A}, \mu)$ be a measure space.*

(a) *Let $0 < p < q \leqslant \infty$. If there are sets of arbitrarily small positive measure, i.e.*

$$m_\mu := \inf\{\mu(A);\ A \in \mathscr{A},\ \mu(A) > 0\} = 0,$$

then $L^p \setminus L^q$ is non-empty.

(b) *Let $0 < p < q < \infty$. If there are sets of arbitrarily large finite measure, i.e.*

$$M_\mu := \sup\{\mu(A);\ A \in \mathscr{A},\ \mu(A) < \infty\} = \infty,$$

then $L^q \setminus L^p$ is non-empty.

Proof (a) Fix $\epsilon > 0$. As $m_\mu = 0$ there exists a sequence $(A_n)_{n\in\mathbb{N}} \subseteq \mathscr{A}$ such that $0 < \mu(A_n) \leqslant n^{-2/\epsilon}$. If we set $c_n := \mu(A_n)^{-1/p+\epsilon}$, $n \in \mathbb{N}$, and

$$f(x) := \sum_{n\in\mathbb{N}} c_n \mathbb{1}_{A_n}(x), \quad x \in X,$$

then, by Minkowski's inequality,

$$\|f\|_{L^p} \leqslant \sum_{n\in\mathbb{N}} c_n \mu(A_n)^{1/p} = \sum_{n\in\mathbb{N}} \mu(A_n)^{\epsilon} \leqslant \sum_{n\in\mathbb{N}} \frac{1}{n^2} < \infty,$$

i.e. $f \in L^p$. If $p < q < \infty$, then we obtain

$$\int_X |f|^q \, d\mu \geqslant |c_n|^q \mu(A_n) = \mu(A_n)^{(1-q/p)+q\epsilon} \xrightarrow[n\to\infty]{1-q/p<0} \infty$$

for $\epsilon = \epsilon(p,q)$ sufficiently small. If $p < q = \infty$, we see that

$$f(x) \geqslant c_n \mathbb{1}_{A_n}(x) \geqslant n^{\frac{2}{p\epsilon}-2} \mathbb{1}_{A_n}(x),$$

which implies that $\|f\|_{L^\infty} \geqslant n^{\frac{2}{p\epsilon}-2} \to \infty$ if $\epsilon = \epsilon(p)$ is sufficiently small. Thus, $L^p \setminus L^q \neq \emptyset$.

(b) For fixed $\epsilon > 0$ choose $(A_n)_{n\in\mathbb{N}} \subseteq \mathscr{A}$ such that $n^{2/\epsilon} \leqslant \mu(A_n) < \infty$. Set $c_n := \mu(A_n)^{-1/q-\epsilon}$ and

$$f(x) := \sum_{n\in\mathbb{N}} c_n \mathbb{1}_{A_n}(x), \quad x \in X.$$

By Minkowski's inequality,

$$\|f\|_{L^q} \leqslant \sum_{n\in\mathbb{N}} c_n \mu(A_n)^{1/q} = \sum_{n\in\mathbb{N}} \mu(A_n)^{-\epsilon} \leqslant \sum_{n\in\mathbb{N}} \frac{1}{n^2} < \infty$$

and

$$\int_X |f|^p \, d\mu \geqslant |c_n|^p \mu(A_n) = \mu(A_n)^{(1-p/q)-\epsilon p} \xrightarrow[n\to\infty]{1-p/q>0} \infty$$

for small $\epsilon > 0$, i.e. $f \in L^q \setminus L^p$. □

Corollary *Let $(X, \mathscr{A}, \mu)$ be a measure space and $p, q \in (0,\infty)$. Denote by m_μ and M_μ the quantities defined in the proposition.*

(a) *$L^p \subseteq L^q$ for some (all) $p < q$ if, and only if, $m_\mu > 0$.*
(b) *$L^1 \subseteq L^\infty$ if, and only if, $m_\mu > 0$.*
(c) *$L^p \subseteq L^q$ for some (all) $p > q$ if, and only if, $M_\mu < \infty$.*

Proof (a), (b) *Sufficiency.* Assume that $m_\mu > 0$. If $f \in L^p$ then [☞ 10.5]

$$\sum_{n\in\mathbb{N}} \mu(\{|f|^p > n\}) \leqslant \int_X |f|^p \, d\mu < \infty$$

implies that $\mu(\{|f| > n\}) = 0$ for all n sufficiently large; hence, $f \in L^\infty$. This proves $L^p \subseteq L^\infty$, i.e. $q = \infty$. If $q < \infty$, we get

$$\int_X |f|^q \, d\mu \leqslant \|f\|_{L^\infty}^{q-p} \int_X |f|^p \, d\mu < \infty$$

for all $q \geqslant p$ which shows $L^p \subseteq L^q$, $q \geqslant p$.

Necessity. The converse implication is an immediate consequence of Part (a) of the above proposition.

(c) Assume that $M_\mu < \infty$, and let $f \in L^p$, $f \neq 0$. By Markov's inequality,

$$\mu(\{|f| \geqslant r\}) \leqslant \frac{1}{r^p} \int_X |f|^p \, d\mu < \infty$$

for all $r > 0$. On the other hand, $f \neq 0$ gives $\mu(\{|f| \geqslant r\}) > 0$ for $r > 0$ sufficiently small. By Hölder's inequality, we thus get

$$\left(\int_{\{|f| \geqslant r\}} |f|^q \, d\mu \right)^{1/q} \leqslant \mu(\{|f| \geqslant r\})^{q^{-1}-p^{-1}} \|f\|_{L^p} \leqslant M_\mu^{q^{-1}-p^{-1}} \|f\|_{L^p}$$

for small $r > 0$. Applying the monotone convergence theorem, we conclude that $f \in L^q$. Hence, $L^p \subseteq L^q$. The converse implication follows from the above proposition. □

Lebesgue measure λ on $\mathbb{R}$ satisfies $m_\lambda = 0$ and $M_\lambda = \infty$, and therefore it follows that $L^p(\lambda) \setminus L^q(\lambda) \neq \emptyset$ for any $p, q \in (0, \infty)$. In particular, $L^p(\lambda) \subseteq L^q(\lambda)$ holds if, and only if, $p = q$. For instance, $x^{-\frac{1}{2}} \mathbb{1}_{(0,1)}(x) \in L^1(\lambda) \setminus L^2(\lambda)$ and $(x^2+1)^{-\frac{1}{2}} \in L^2(\lambda) \setminus L^1(\lambda)$.

In arbitrary measure spaces, the L^p-norm is log-convex:

$$\forall s \in (r,t) \;:\; \|u\|_{L^s} \leqslant \|u\|_{L^r}^{\alpha} \|u\|_{L^t}^{1-\alpha} \quad \text{where} \quad \alpha = \frac{\frac{1}{s} - \frac{1}{t}}{\frac{1}{r} - \frac{1}{t}}, \quad 1 - \alpha = \frac{\frac{1}{r} - \frac{1}{s}}{\frac{1}{r} - \frac{1}{t}}.$$

This estimate follows immediately from the Hölder inequality with $p = r/(s\alpha)$ and $q = t/(1-\alpha)s$ applied to $\int |u|^s \, d\mu = \int |u|^{s\alpha} \cdot |u|^{s(1-\alpha)} \, d\mu$. In particular,

$$\forall s \in (r,t) \;:\; L^r(X, \mathscr{A}, \mu) \cap L^t(X, \mathscr{A}, \mu) \subseteq L^s(X, \mathscr{A}, \mu).$$

18.2 One may have $\ell^p(\mu) \subseteq \ell^q(\mu)$, or $\ell^p(\mu) \supseteq \ell^q(\mu)$, or no inclusion at all

Consider a weighted counting measure $\mu = \sum_{n=1}^\infty \mu_n \delta_n$ on $(\mathbb{N}, \mathscr{P}(\mathbb{N}))$ with weights $0 < \mu_n < \infty$. The corresponding L^p-spaces are the 'small-ell-p' spaces

$\ell^p(\mu)$ (we write $\ell^p(\mathbb{N})$ if $\mu_n = 1$), and we define

$$x = (x_n)_{n\in\mathbb{N}} \subseteq \mathbb{R} \ : \ \|x\|_{p,\mu} = \begin{cases} \left(\sum_{n=1}^{\infty} \mu_n |x_n|^p\right)^{\frac{1}{p}}, & \text{if } 1 \leqslant p < \infty, \\ \sup_{n\in\mathbb{N}} |x_n|, & \text{if } p = \infty. \end{cases}$$

The following three cases illustrate the previous Example 18.1:

- *Case 1.* Assume that all weights $\mu_n \geqslant a > 0$, hence $\sum_{n=1}^{\infty} \mu_n = \infty$. If $q > p$, then the following inequality holds:

$$\|x\|_{q,\mu} \leqslant a^{\frac{1}{q}-\frac{1}{p}} \|x\|_{p,\mu} \ \Rightarrow \ \ell^p(\mu) \subseteq \ell^q(\mu).$$

This is, in particular, the case for the classical sequence spaces $\ell^p(\mathbb{N})$ and $\ell^q(\mathbb{N})$. The inequality immediately follows from the observation that for all $m \in \mathbb{N}$,

$$a\,|x_m|^p \leqslant \mu_m\,|x_m|^p \leqslant \sum_{n=1}^{\infty} \mu_n\,|x_n|^p = \|x\|_{p,\mu}^p \ \Rightarrow \ |x_m| \leqslant \frac{\|x\|_{p,\mu}}{a^{1/p}},$$

and so

$$\begin{aligned} \|x\|_{q,\mu}^q &= \sum_{n=1}^{\infty} \mu_n\,|x_n|^q = \sum_{n=1}^{\infty} |x_n|^{q-p} \cdot \mu_n\,|x_n|^p \\ &\leqslant \sum_{n=1}^{\infty} a^{1-\frac{q}{p}} \|x\|_{p,\mu}^{q-p} \cdot \mu_n\,|x_n|^p = a^{1-\frac{q}{p}} \|x\|_{p,\mu}^q. \end{aligned}$$

- *Case 2.* Assume that the weights satisfy $\sum_{n=1}^{\infty} \mu_n < \infty$. In this case, we also have $\lim_{n\to\infty} \mu_n = 0$. If $q > p$, the following inequality holds:

$$\|x\|_{p,\mu} \leqslant \left(\sum_{n=1}^{\infty} \mu_n\right)^{\frac{1}{p}-\frac{1}{q}} \|x\|_{q,\mu} = \mu(\mathbb{N})^{\frac{1}{p}-\frac{1}{q}} \|x\|_{q,\mu} \ \Rightarrow \ \ell^p(\mu) \supseteq \ell^q(\mu),$$

where $\mu(\mathbb{N})$ is the total mass of the measure μ. This follows easily from Hölder's inequality with $s = \frac{q}{p} \geqslant 1$ and its conjugate index $t = \frac{q}{q-p}$:

$$\begin{aligned} \|x\|_{p,\mu}^p &= \sum_{n=1}^{\infty} \mu_n\,|x_n|^p = \sum_{n=1}^{\infty} \mu_n^{1-\frac{p}{q}}\, \mu_n^{\frac{p}{q}}\,|x_n|^p \\ &\leqslant \left(\sum_{n=1}^{\infty} \mu_n\right)^{1-\frac{p}{q}} \left(\sum_{n=1}^{\infty} \mu_n\,|x_n|^q\right)^{\frac{p}{q}}. \end{aligned}$$

▶ *Case 3.* Assume that the weights satisfy $\sum_{n=1}^{\infty} \mu_n = \infty$ and $\lim_{n\to\infty} \mu_n = 0$. In this case 'anything' can happen. For example,

$$\mu_n = \begin{cases} (2k-1)^{-1/2}, & \text{if } n = 2k-1 \text{ odd}, \\ (2k)^{-3/2}, & \text{if } n = 2k \text{ even}. \end{cases}$$

Consider the sequences $x = (x_n)_{n\in\mathbb{N}}$ and $y = (y_n)_{n\in\mathbb{N}}$ with general elements

$$x_n = \begin{cases} 0, & \text{if } n = 2k-1, \\ (2k)^{1/4}, & \text{if } n = 2k, \end{cases} \quad \text{and} \quad y_n = \begin{cases} (2k-1)^{-1/2}, & \text{if } n = 2k-1, \\ 0, & \text{if } n = 2k. \end{cases}$$

Clearly, $x \in \ell^1(\mu) \setminus \ell^2(\mu)$ while $y \in \ell^2(\mu) \setminus \ell^1(\mu)$.

The best one can say, without further conditions on the weight μ, is that one has convexity [☞ last section in 18.1]: $\ell^p(\mu) \cap \ell^q(\mu) \subseteq \ell^\kappa(\mu)$ for all $p \leqslant \kappa \leqslant q$.

18.3 A measure space where $L^p = \{0\}$ for all $0 \leqslant p < \infty$

Let $X \neq \emptyset$, take any σ-algebra $\mathscr{A}$ in X and define

$$\mu(A) := \begin{cases} 0, & \text{if } A = \emptyset, \\ \infty, & \text{if } A \in \mathscr{A},\ A \neq \emptyset. \end{cases}$$

Assume that $f : X \to \mathbb{R}$ is measurable and not zero. Thus, there is some element $c \in f(X)$ such that $c \neq 0$. Set $A := A_c := \{x;\ f(x) = c\}$, then by assumption $\emptyset \neq A \in \mathscr{A}$ and so

$$\int_X |f(x)|^p \,\mu(dx) \geqslant \int_A |c|^p \,d\mu = |c|^p \mu(A) = \infty.$$

This means that $f = 0$ is the only function in L^p, $0 \leqslant p < \infty$.

Comment Note that for $p = \infty$ the space L^∞ is not trivial and consists of all bounded measurable maps $f : X \to \mathbb{R}$.

18.4 A measure space where all spaces L^p, $1 \leqslant p \leqslant \infty$ coincide

Let $X = \{1, 2, \dots, n\}$ for some $n \in \mathbb{N}$, $\mathscr{A} = \mathscr{P}(X)$ and denote by ζ the counting measure on $(X, \mathscr{A})$. Since all functions $f : X \to \mathbb{R}$ are measurable, the spaces L^p are given by

$$L^p(X, \mathscr{A}, \zeta) = \left\{ f : \{1, \dots, n\} \to \mathbb{R};\ \|f\|_{L^p}^p = \sum_{i=1}^n |f(i)|^p < \infty \right\}.$$

The respective norms are given by

$$\|(f_1, \dots, f_n)\|_{L^p} = (|f_1|^p + \dots + |f_n|^p)^{\frac{1}{p}}, \text{ resp. } \|(f_1, \dots, f_n)\|_{L^\infty} = \max_{1 \leqslant i \leqslant n} |f_i|,$$

and so it is clear that we can identify $L^p(X, \mathscr{A}, \zeta)$ with $\mathbb{R}^n$ for any $1 \leqslant p \leqslant \infty$.

18.5 A measure space where $L^1 \subsetneqq L^\infty$

Consider the counting measure ζ on $(\mathbb{N}, \mathscr{P}(\mathbb{N}))$ and observe that the summable sequences of real numbers $\ell^1(\mathbb{N})$ coincide with the space $L^1(\zeta)$; the bounded sequences of real numbers $\ell^\infty(\mathbb{N})$ are $L^\infty(\zeta)$. Since

$$\sum_{n=1}^{\infty} |a_n| < \infty \Rightarrow \lim_{n \to \infty} |a_n| = 0 \Rightarrow \sup_{n \in \mathbb{N}} |a_n| < \infty,$$

it follows that $\ell^1(\mathbb{N}) \subseteq \ell^\infty(\mathbb{N})$. The example $a_n = \frac{1}{n}$ shows that this inclusion is strict.

Comment The converse inclusion, $L^\infty(\mu) \subseteq L^1(\mu)$ holds in any finite measure space $(X, \mathscr{A}, \mu)$. In general, one cannot expect any inclusion [☞ 18.1]. It is possible to characterize those measure spaces where $L^1(\mu) \subseteq L^\infty(\mu)$ holds. The proof of the proposition in [☞ 18.1] shows that our example is, in fact, typical.

18.6 $L^1(\mu) = L^\infty(\mu)$ if, and only if, $1 \leqslant \dim(L^1(\mu)) < \infty$

Let $(X, \mathscr{A}, \mu)$ be a measure space, and write L^p for $L^p(X, \mathscr{A}, \mu)$. If μ is a finite measure, then the inclusion $L^\infty \subseteq L^1$ holds. We are interested under which conditions L^1 coincides with L^∞. If $L^1 = L^\infty$, then there are two immediate consequences:

- $\mathbb{1}_X \in L^\infty = L^1$ shows that μ has to be a finite measure.
- The log-convexity of the norm [☞ 18.1] yields $L^p = L^q$ for all $p, q \in [1, \infty]$.

More is true.

Theorem 1 *Let $(X, \mathscr{A}, \mu)$ be a measure space. If $L^1 = L^\infty$ holds, then one has* $\dim(L^\infty) = \dim(L^1) \in \mathbb{N}$.

Proof We have already seen that $\mathbb{1}_X \in L^\infty = L^1$ entails that $\mu(X) < \infty$. Let us draw three further conclusions from the equality $L^1 = L^\infty$.

1° $m_\mu := \inf\{\mu(A);\ A \in \mathscr{A},\ \mu(A) > 0\} > 0$; this follows from the inclusion $L^1 \subseteq L^\infty$ [☞ 18.1].

2° $\mathscr{A}$ contains at least one atom, i.e. a set A such that $\mu(A) > 0$ and there is no set $B \subseteq A$ with $\mu(B) > 0$ and $\mu(A \setminus B) > 0$.

If there were no atom, then we could find some set $A_1 \subsetneq A_0 := X$ such that $\mu(A_1) < \frac{1}{2}\mu(X) < \infty$. Iterating the procedure we get a sequence $(A_n)_{n\in\mathbb{N}} \subseteq \mathscr{A}$ such that $0 < \mu(A_n) < 2^{-n}\mu(X)$, but this is impossible since we know from the first step that $m_\mu > 0$.

3° $\mathscr{A}$ contains at most finitely many disjoint atoms.

Indeed: suppose there is a sequence $(A_n)_{n\in\mathbb{N}} \subseteq \mathscr{A}$ of disjoint atoms such that $\mu(A_n) = a_n > 0$. There are two possibilities:

Case 1. $\sum_{n=1}^{\infty} a_n = \infty$. Then $f \equiv 1$ is in $L^\infty \setminus L^1$, which is impossible.
Case 2. $\sum_{n=1}^{\infty} a_n < \infty$. In this case there is a subsequence $(a_{n(i)})_{i\in\mathbb{N}}$ such that $\sum_{i=1}^{\infty} a_{n(i)}^{1/2} < \infty$. Then $g(x) := \sum_{i=1}^{\infty} a_{n(i)}^{-1/2}\mathbb{1}_{A_{n(i)}}(x)$ is in $L^1 \setminus L^\infty$, which is impossible.

Thus, there are at most finitely many disjoint atoms.

Write $A_1, \dots, A_N$ for the disjoint atoms in $\mathscr{A}$ and set $A := A_1 \uplus \dots \uplus A_N$. Since $X \setminus A$ is not an atom, Step 1 tells us that $\mu(X \setminus A) = 0$. Any $f \in L^1$ is μ-a.e. constant on each atom [☞ proposition in 8.5], i.e. $f = \sum_{n=1}^{N} c_n \mathbb{1}_{A_n}$ μ-a.e. for suitable constants $c_n \in \mathbb{R}$, and this means that L^1 is N-dimensional. □

Theorem 2 *Let $(X, \mathscr{A}, \mu)$ be a measure space.*

(a) *If $1 \leqslant \dim(L^1) < \infty$, then $L^1 = L^\infty$.*
(b) *If $1 \leqslant \dim(L^\infty) < \infty$, then $L^1 \subseteq L^\infty$. If, in addition, μ has the finite subset property,*

$$A \in \mathscr{A},\ \mu(A) > 0 \Rightarrow \exists A_0 \in \mathscr{A},\ A_0 \subseteq A \ :\ 0 < \mu(A_0) < \infty,$$

then $L^1 = L^\infty$. The finite subset property holds, e.g. for σ-finite measures.

The finite subset property appears naturally in connection with the dual of L^1; see Example 18.21 and the characterization of the finite subset property given there.

Proof **1°** Assume that $\dim(L^1) = N \in \mathbb{N}$. First we show that there is a finite maximal set of linearly independent functions $(\mathbb{1}_{A_n})_{n=1,\dots,N}$ with pairwise disjoint sets $A_n \in \mathscr{A}$ such that $0 < \mu(A_n) < \infty$. Suppose that

$$M := \max\left\{ n \in \mathbb{N} \;\middle|\; \begin{array}{l} \exists B_1, \dots, B_n \in \mathscr{A} \ :\ 0 < \mu(B_i) < \infty \text{ and} \\ (\mathbb{1}_{B_i})_{i\leqslant n} \subseteq L^1 \text{ are linearly independent} \end{array} \right\}$$

is strictly smaller than $\dim(L^1) = N$. Choose measurable sets $B_1, \dots, B_M \in \mathscr{A}$

with $0 < \mu(B_i) < \infty$ such that $(\mathbb{1}_{B_i})_{i\leqslant M} \subseteq L^1$ are linearly independent. Because of the maximality of M, any set $A \in \mathscr{A}$ with $\mathbb{1}_A \in L^1$ satisfies $\mathbb{1}_A = \sum_{i=1}^M c_i \mathbb{1}_{B_i}$ μ-a.e. for suitable constants $c_i \in \mathbb{R}$. If $f \in L^1$, then there exists a sequence of simple functions $(f_k)_{k\in\mathbb{N}} \subseteq L^1 \cap \mathscr{E}(\mathscr{A})$ with $f_k \to f$ in L^1 and μ-a.e. Hence, f coincides μ-a.e. with a function of the form $\sum_{i=1}^M C_i \mathbb{1}_{B_i}$; this shows that we have $\dim(L^1) \leqslant M < \dim(L^1)$, which is impossible.

Consequently, there exists a sequence $(B_i)_{i=1,\dots,N} \subseteq \mathscr{A}$, $0 < \mu(B_i) < \infty$, such that $(\mathbb{1}_{B_i})_{i=1,\dots,N} \subseteq L^1$ is linearly independent. Define

$$\mathscr{G} := \left\{ A \in \mathscr{A}\ ;\ 0 < \|\mathbb{1}_A\|_{L^1} < \infty,\ \exists I \subseteq \{1, \dots, n\} :\ A = \bigcap\nolimits_{i\in I} B_i \cap \bigcap\nolimits_{i\notin I} B_i^c \right\}.$$

The sets in $\mathscr{G}$ are pairwise disjoint and have finite, strictly positive measure; hence, they are linearly independent, and so $\#\mathscr{G} \leqslant N$. On the other hand, it cannot happen that $\#\mathscr{G} < N$. Suppose that, on the contrary, $\mathscr{G} = \{A_1, \dots, A_k\}$ for some $k < N$. As every B_j is of the form $\mathbb{1}_{B_j} = \sum_{i=1}^k c_i \mathbb{1}_{A_i}$ μ-a.e., this would imply that $(\mathbb{1}_{A_j})_{j\leqslant k}$ is a basis for $V := \operatorname{span}\{\mathbb{1}_{B_1}, \dots, \mathbb{1}_{B_N}\}$ in contradiction to our assumption that V has dimension $\dim(L^1) = N$. Therefore, we conclude that $\#\mathscr{G} = N$. The elements $(A_i)_{i\leqslant n}$ in $\mathscr{G}$ have the desired properties: they are pairwise disjoint, linearly independent and satisfy $0 < \|\mathbb{1}_{A_i}\|_{L^1} = \mu(A_i) < \infty$.

Define $A := A_1 \uplus \dots \uplus A_N$. Because of maximality, we get $\mu(X \setminus A) = 0$, otherwise we could find a set $A_0 \subseteq X \setminus A$ such that $(\mathbb{1}_{A_n})_{n=0,\dots,N}$ is still linearly independent, contradicting maximality. Each A_n is an atom: otherwise we could split it into two disjoint sets of positive measure and we would get $N+1$ linearly independent functions – again in contradiction to maximality. Since $\mathscr{A}/\mathscr{B}(\mathbb{R})$ measurable functions are μ-a.e. constant on each atom [☞ proposition in 8.5], it follows that any $\mathscr{A}/\mathscr{B}(\mathbb{R})$ measurable function f satisfies

$$f = \sum_{n=1}^N c_n \mathbb{1}_{A_n} \quad \mu\text{-a.e.}$$

for suitably chosen constants $c_n \in \mathbb{R}$. Since $0 < \mu(A_n) < \infty$, this shows in particular that $L^1 = L^\infty$.

2⁰ Assume that $1 \leqslant \dim(L^\infty) =: N < \infty$. A similar reasoning to that in 1⁰ gives a finite maximal set of linearly independent functions $(\mathbb{1}_{A_n})_{n=1,\dots,N}$ for pairwise disjoint sets with strictly positive measure such that any $\mathscr{A}/\mathscr{B}(\mathbb{R})$ measurable function f satisfies

$$f = \sum_{n=1}^N c_n \mathbb{1}_{A_n} \quad \mu\text{-a.e.} \tag{18.3}$$

for some $c_n \in \mathbb{R}$. This implies that $L^1 \subseteq L^\infty$. Because of the maximality, each A_n

is an atom; otherwise we could split A_n into two disjoint sets of positive measure and obtain $N+1$ linearly independent functions. If μ satisfies the finite subset property, then μ cannot have any atoms of infinite measure, and so $\mu(A_i) < \infty$ for all $i = 1, \dots, N$. From (18.3) we now see that $L^\infty \subseteq L^1$. $\square$

The condition $1 \leqslant \dim L^1 < \infty$ in Theorem 2.(a) cannot be relaxed to become $0 \leqslant \dim L^1 < \infty$ [☞ 18.3]. To see what can go wrong in Theorem 2.(b) if μ does not have the finite subset property, consider $X = (0,1]$ with the σ-algebra generated by the half-open intervals $\left(\frac{k-1}{n}, \frac{k}{n}\right]$, $k = 1, \dots, n$, for $n \in \mathbb{N}$ fixed and the measure

$$\mu(A) := \begin{cases} 0, & \text{if } A = \emptyset, \\ \infty, & \text{otherwise,} \end{cases}$$

then $\dim L^\infty = n$ but $\dim L^1 = 0$; in particular, $L^\infty \neq L^1$.

Comment Our proof of Theorem 2(a) shows that μ can attain only finitely many values, i.e. $\#\{\mu(A);\ A \in \mathscr{A}\} < \infty$, if L^1 has finite dimension. The converse is also true; more generally, L^p has finite dimension for some $p \in [1, \infty)$ if, and only if, $\#\{\mu(A);\ A \in \mathscr{A}\} < \infty$; see also [☞ theorem in 18.15].

18.7 A function where $\sup_{x\in U} |f(x)| \neq \|f\|_{L^\infty(U)}$ for any open set U

Consider Lebesgue measure on $((0,1), \mathscr{B}(0,1))$ and the 'reciprocal' of Thomae's function [☞ Fig. 10.1]

$$f(x) = \begin{cases} 0, & \text{if } x \in (0,1) \setminus \mathbb{Q}, \\ q, & \text{if } x = \frac{p}{q} \in (0,1) \cap \mathbb{Q},\ \gcd(p,q) = 1. \end{cases}$$

Let $U \subseteq (0,1)$ be any open set. Since the rational numbers $\mathbb{Q}$ are a Lebesgue null set, $\|f\|_{L^\infty(U)} = \|f\mathbb{1}_U\|_{L^\infty} = 0$ whereas $\sup_{x\in U} f(x) = \infty$.

Comment If μ is a measure with full topological support, i.e. $\mu(U) > 0$ for every open set $U \neq \emptyset$, then $\|u\|_{L^\infty(\mu)} = \sup_x |u(x)|$ for every **continuous** function. To see what can go wrong, consider $\mu(dx) = \mathbb{1}_{(1/2,1)}(x)\lambda(dx)$ on $((0,1), \mathscr{B}(0,1))$ and the continuous function $u(x) = x^{-1}$. Then $\|u\|_{L^\infty(\mu)} = 2$ while $\sup_x |u(x)| = \|u\|_{L^\infty(\lambda)} = \infty$.

18.8 One cannot compare L^p-norms on $C[0,1]$

Consider Lebesgue measure λ on $[0,1]$ and write $L^p = L^p([0,1], \mathscr{B}[0,1], \lambda)$. For $u \in C[0,1]$ it is not difficult to see that

$$\int_0^1 |u(x)|^p\,dx = 0 \Rightarrow \lambda\{u \neq 0\} = 0 \Rightarrow \forall x \,:\, u(x) = 0,$$

i.e. $u \longmapsto \|u\|_{L^p}$ is indeed a norm on $C[0,1]$. A similar reasoning shows that $\|u\|_{L^\infty} = \sup_x |u(x)|$.

Consider the functions $u_n \in C[0,1]$ given by

$$u_n(x) := \begin{cases} n(1-nx), & \text{if } 0 \leqslant x \leqslant \frac{1}{n}, \\ 0, & \text{if } x > \frac{1}{n}, \end{cases}$$

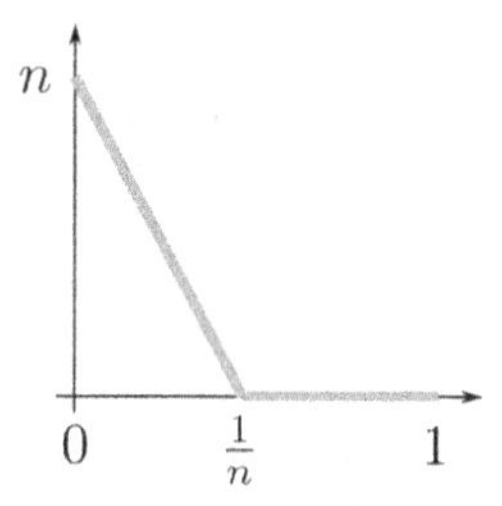

Figure 18.2 Plot of u_n.

cf. Fig. 18.2, and observe that

$$\|u_n\|_{L^p} = n^{1-\frac{1}{p}} \left(\frac{1}{1+p}\right)^{\frac{1}{p}} \quad \text{and} \quad \|u_n\|_{L^\infty} = n.$$

Since $\lim_{n\to\infty} \|u_n\|_{L^p} = \infty$ with a different speed of convergence for each $p \in [0,\infty]$, it is clear that

$$\|u_n\|_{L^q} \leqslant C\|u_n\|_{L^p} \quad p < q$$

cannot hold with a uniform constant C. This means that no pair of norms $(\|\cdot\|_{L^p}, \|\cdot\|_{L^q})$ can be comparable.

Comment The above result shows, in particular, that the topologies in $C[0,1]$ induced by the norms $\|\cdot\|_{L^p}$ are different; see e.g. [160, p. 175, Théorème 2.4.2].

18.9 The spaces L^p with $0 < p < 1$ are only quasi-normed spaces

If one tries to mimic the usual proof that L^p is a normed space [MIMS, Chapter 13], one arrives at the point that the triangle inequality in L^p (Minkowski's inequality) [☞ Corollary 1.19] is deduced from the Hölder inequality [☞ 1.18]. But for $0 < p < 1$ both inequalities are 'the wrong way round'.

Let $(X, \mathscr{A}, \mu)$ be any measure space, $f, g : X \to (0,\infty)$ positive measurable functions and $p^{-1} + q^{-1} = 1$ conjugate exponents such that $p \in (0,1)$; obviously, $q \in (-\infty, 0)$. In this case we have the **reverse Hölder inequality** and the **reverse Minkowski inequality**

$$\forall f \in L^p(\mu),\ g \in L^q(\mu) \,:\, \int fg\,d\mu \geqslant \left(\int f^p\,d\mu\right)^{\frac{1}{p}} \left(\int g^q\,d\mu\right)^{\frac{1}{q}}, \tag{18.4}$$

$$\forall f, h \in L^p(\mu) \;:\; \left(\int (f+h)^p \, d\mu\right)^{\frac{1}{p}} \geqslant \left(\int f^p \, d\mu\right)^{\frac{1}{p}} + \left(\int h^p \, d\mu\right)^{\frac{1}{p}}. \quad (18.5)$$

The reverse Hölder inequality. We want to deduce the reverse Hölder inequality from the 'usual' version. To do so, set

$$r := \frac{1}{p} \in (1, \infty) \quad \text{and} \quad s := -\frac{q}{p} = 1 - q \in (1, \infty) \Rightarrow \frac{1}{r} + \frac{1}{s} = 1.$$

Applying Hölder's inequality with the indices r and s gives

$$\int f^p \, d\mu = \int (fg)^p g^{-p} \, d\mu \leqslant \left(\int (fg)^{pr} \, d\mu\right)^{\frac{1}{r}} \left(\int g^{-ps} \, d\mu\right)^{\frac{1}{s}}$$
$$= \left(\int fg \, d\mu\right)^p \left(\int g^q \, d\mu\right)^{-\frac{p}{q}}.$$

A simple rearrangement gives (18.4). Observe that the above calculation needs that $\int f^p \, d\mu < \infty$ and $\int g^q \, d\mu < \infty$. □

The reverse Minkowski inequality. As in the classical proof we begin with

$$(f+h)^p = f(f+h)^{p-1} + h(f+h)^{p-1}.$$

Using the reverse Hölder inequality for the conjugate exponents $p^{-1} + q^{-1} = 1$ gives

$$\int (f+h)^p \, d\mu = \int f(f+h)^{p-1} \, d\mu + \int h(f+h)^{p-1} \, d\mu$$
$$\geqslant \left[\left(\int f^p \, d\mu\right)^{\frac{1}{p}} + \left(\int h^p \, d\mu\right)^{\frac{1}{p}}\right] \left(\int (f+h)^p \, d\mu\right)^{\frac{1}{q}}.$$

This inequality can easily be rearranged to become (18.5), since the only 'exceptional' case where $\int (f+h)^p \, d\mu = \infty$ is trivial. □

We can still see that L^p, $0 < p < 1$, is a metric space which is even quasi-normed. This follows from the elementary inequalities

$$(a+b)^p \leqslant a^p + b^p, \quad a, b \geqslant 0, \; 0 < p \leqslant 1$$

and

$$(a+b)^r \leqslant 2^{r-1}(a^r + b^r), \quad a, b \geqslant 0, \; r \geqslant 1$$

when applied to the L^p-norm.

The first inequality shows that for any $f, g, h \in L^p(X, \mathscr{A}, \mu)$

$$d_p(f,g) := \|f-g\|_{L^p}^p = \int |f-g|^p \, d\mu \leqslant \int |f-h|^p \, d\mu + \int |h-g|^p \, d\mu$$
$$= d_p(f,h) + d_p(h,g),$$

i.e. $(f,g) \longmapsto d_p(f,g)$ is a translation-invariant metric on $L^p(X, \mathscr{A}, \mu)$.

The second inequality with $r = \frac{1}{p}$ shows that

$$\|f+g\|_{L^p} = \left(\int |f+g|^p \, d\mu\right)^{\frac{1}{p}} \leqslant \left(\int |f|^p \, d\mu + \int |g|^p \, d\mu\right)^{\frac{1}{p}}$$
$$\leqslant 2^{\frac{1}{p}-1} \left(\|f\|_{L^p} + \|g\|_{L^p}\right),$$

which means that $(L^p, \|\cdot\|_{L^p})$ is a **quasi-normed** space.

18.10 The spaces L^p with $0 < p < 1$ are not locally convex

Let λ be Lebesgue measure on $([0,1], \mathscr{B}[0,1])$. We show that the only open and convex sets of $L^p = L^p([0,1], \mathscr{B}[0,1], \lambda)$, $p \in (0,1)$, equipped with the metric $d_p(\cdot,\cdot)$ from the previous example, are $\emptyset$ and L^p.

Let $U \neq \emptyset$ be some convex neighbourhood of $0 \in L^p$ and pick any $f \in L^p$. Since U is an open set, there is some $r > 0$ such that the (metric) open ball $B_r(0) = \{u \in L^p \,;\, d_p(u,0) < r\}$ is contained in U.

Fix some $n \in \mathbb{N}$ such that

$$\frac{1}{n^{1-p}} \int |f|^p \, d\lambda < r, \quad \text{thus,} \quad nf \in B_{nr}(0).$$

Since λ has no non-empty atoms, we find a partitioning $B_1 \uplus \dots \uplus B_n = [0,1]$ consisting of disjoint measurable sets $B_1, \dots, B_n \in \mathscr{B}[0,1]$ such that

$$\int_{B_i} |f|^p \, d\lambda = \frac{1}{n} \int |f|^p \, d\lambda, \quad i = 1, \dots, n.$$

The functions $g_i := nf \mathbb{1}_{B_i}$ satisfy

$$\int |g_i|^p \, d\lambda = n^p \int_{B_i} |f|^p \, d\lambda < r \implies g_i \in B_r(0)$$

for every $i = 1, \dots, n$. Since U is convex, $f = \frac{1}{n}(g_1 + \dots + g_n) \in U$, and this shows that $U = L^p$.

Comment This result is due to Day [41]. A close inspection of this argument shows that it goes through in an arbitrary measure space such that the map

$$B \longmapsto \int |f|^p \mathbb{1}_B \, d\mu$$

attains all values in the interval $[0, \|f\|_{L^p}^p]$. Let us show that this property holds for any measure μ which has no atoms [☞ p. 123] with finite measure. Define for fixed $f \in L^p$ a finite measure ν on $(X, \mathscr{A})$ by

$$\nu(A) := \int_A |f(x)|^p \, \mu(dx), \quad A \in \mathscr{A}.$$

We claim that ν is atomless. In order to prove this, we note that $\nu(A) = \nu(\tilde{A})$ for $\tilde{A} := A \cap \{|f| > 0\}$, $A \in \mathscr{A}$, and consider several cases separately:

Case 1. $\tilde{A}$ is an atom of μ. Then $\mu(\tilde{A}) = \infty$, and therefore $\int_{\tilde{A}} |f(x)|^p \, \mu(dx) < \infty$ implies

$$0 = \mu(\tilde{A} \cap \{|f| > 0\}) = \mu(A \cap \{|f| > 0\}).$$

Hence, $\nu(A) = \nu(\tilde{A}) = 0$, which means that A is not an atom of ν.

Case 2. $\tilde{A}$ is not an atom of μ and $\mu(\tilde{A}) = 0$. Then $\nu(A) = \nu(\tilde{A}) = 0$, i.e. A is not an atom of ν.

Case 3. $\tilde{A}$ is not an atom of μ and $\mu(\tilde{A}) > 0$. This implies that there exists a set $B \in \mathscr{A}$ such that $B \subseteq \tilde{A}$, $\mu(B) > 0$ and $\mu(\tilde{A} \setminus B) > 0$. Because of $B \subseteq \tilde{A} \subseteq \{|f| > 0\}$ we have $\mu(B) = \mu(B \cap \{|f| > 0\}) > 0$, which implies $\nu(B) = \int_B |f(x)|^p \, \mu(dx) > 0$. Analogously, it follows that $\nu(B \setminus A) > 0$. This shows that A is not an atom of ν.

It is known that the range of any atomless finite measure is convex, cf. [19, Corollary 1.12.10],[1] we can find for any $\alpha \in [0, \nu(X)] = [0, \|f\|_{L^p}^p]$ a set $A \in \mathscr{A}$ such that $\nu(A) = \alpha$.

18.11 The dual of $L^p(\lambda)$ with $0 < p < 1$ is trivial

For any $p \in (0,1)$ the dual space of $L^p = L^p([0,1], \mathscr{B}[0,1], \lambda)$ is trivial, i.e. the only continuous linear functional $T : L^p([0,1]) \to \mathbb{R}$ is 0.

Indeed: let $T : L^p \to \mathbb{R}$ be a continuous linear functional. As $0 \in T(L^p)$ we have $A_\epsilon := T^{-1}(-\epsilon, \epsilon) \neq \emptyset$ for all $\epsilon > 0$. Moreover, $A_\epsilon \subseteq L^p$ is for each $\epsilon > 0$ a convex and open set. The only open convex and non-empty set in L^p, $p \in (0,1)$, is the whole space [☞ 18.10], so we find that $A_\epsilon = L^p$, i.e. $T(L^p) \subseteq (-\epsilon, \epsilon)$ for all $\epsilon > 0$. Hence, $T = 0$.

[1] We define atoms differently than [19] but for finite measures the definitions are equivalent [☞ Lemma 6A].

Comment Consider, more generally, a measure space $(X, \mathscr{A}, \mu)$. If μ has no atoms of finite measure, then the dual of $L^p(X, \mathscr{A}, \mu)$ is trivial for any $p \in (0, 1)$; this is an immediate consequence of the above reasoning and the comment in Example 18.10. If $(X, \mathscr{A}, \mu)$ is σ-finite, then L^p is non-trivial if, and only if, μ has at least one atom with finite measure, cf. Day [41].

18.12 Functions $f \in L^p$, $0 < p < 1$, need not be locally integrable

Let λ^d be Lebesgue measure on $(\mathbb{R}^d, \mathscr{B}(\mathbb{R}^d))$ and $L^p = L^p(\mathbb{R}^d, \mathscr{B}(\mathbb{R}^d), \lambda^d)$ for $p > 0$. If $f \in L^p$ for some $p \geqslant 1$, then

$$\forall x \in \mathbb{R}^d \quad \forall r > 0 : \int_{B_r(x)} |f(x)| \, \lambda^d(dx) < \infty;$$

this is a consequence of the fact that the inclusion $L^p(X, \mathscr{A}, \mu) \subseteq L^1(X, \mathscr{A}, \mu)$, $p \geqslant 1$, holds for finite measure spaces $(X, \mathscr{A}, \mu)$. In particular, any function $f \in L^p$, $p \geqslant 1$, is locally integrable [☞ p. 203]. The following example shows that local integrability breaks down for $f \in L^p$ with $0 < p < 1$.

Fix $0 < p < 1$. For an enumeration $(q_n)_{n \in \mathbb{N}}$ of $\mathbb{Q}$ define a sequence of open intervals by

$$I_n := \left(q_n - \frac{1}{n^{2\left(\frac{1}{p}-1\right)}}, \; q_n + \frac{1}{n^{2\left(\frac{1}{p}-1\right)}} \right)$$

and set

$$f(x) := \sum_{n \in \mathbb{N}} \frac{1}{\lambda(I_n)} \mathbb{1}_{I_n}(x), \quad x \in \mathbb{R}.$$

By Beppo Levi, we have

$$\|f\|_{L^p} \leqslant \sum_{n \in \mathbb{N}} \left(\int_{I_n} |f(x)|^p \, \lambda(dx) \right)^{1/p} = \sum_{n \in \mathbb{N}} \lambda(I_n)^{\frac{1}{p}-1} = 2^{\frac{1}{p}-1} \sum_{n \in \mathbb{N}} n^{-2} < \infty.$$

Hence, $|f| < \infty$ almost everywhere and $f \in L^p$. On the other hand, any open interval $(a, b) \neq \emptyset$ contains infinitely many I_n, and so

$$\int_{(a,b)} |f(x)| \, \lambda(dx) \geqslant \sum_{n \in \mathbb{N} : I_n \subseteq (a,b)} \frac{1}{\lambda(I_n)} \lambda(I_n) = \sum_{n \in \mathbb{N} : I_n \subseteq (a,b)} 1 = \infty.$$

This shows that f is not locally integrable.

18.13 The spaces L^q with $q < 0$ are not linear spaces

Let $(X, \mathscr{A}, \mu)$ be an arbitrary measure space. If $0 < p < 1$, the conjugate index q is negative; this explains the interest in the family L^q with $q < 0$. Formally we can define L^q as follows: $f \in L^q$ if f is measurable and

$$\int |f|^q \, d\mu = \int \frac{1}{|f|^{\frac{p}{1-p}}} \, d\mu < \infty, \quad \frac{1}{p} + \frac{1}{q} = 1.$$

Note, however, that L^q is not a linear space any longer.

Assume that μ has infinite mass, and let $f \in L^q$ be such that $f > 0$, e.g. $f(x) := |x|^{-3/q} + 1$ on $(\mathbb{R}, \mathscr{B}(\mathbb{R}), \lambda)$. Then

$$\int_X |f(x) + 1|^q \, \mu(dx) \leqslant \int_X |f(x)|^q \, \mu(dx) < \infty,$$

i.e. $g := f + 1 \in L^q$. However, $1 = g - f \notin L^q$ as $\mu(X) = \infty$.

18.14 A measure space where L^p is not separable

It is well known that in a σ-finite measure space $(X, \mathscr{A}, \mu)$ such that $\mathscr{A}$ is generated by a countable family, the spaces $L^p(X, \mathscr{A}, \mu)$, $1 \leqslant p < \infty$, are separable [MIMS, pp. 362–363] and these conditions are close to being optimal [MIMS, p. 364]. The following example shows that we cannot avoid σ-finiteness.

Take $X = [0, 1]$ with the natural Euclidean metric $d(x, y) = |x - y|$ and let ζ be the counting measure on $([0, 1], \mathscr{B}[0, 1])$, i.e. $\zeta(B) := \#B$. Obviously, ζ is not σ-finite. The pth power ζ-integrable simple functions are of the form

$$\mathcal{E}(\mathscr{B}[0,1]) \cap L^p(\zeta) = \left\{ \sum_{n=1}^{N} y_n \mathbb{1}_{A_n} \,;\; N \in \mathbb{N},\ y_n \in \mathbb{R},\ A_n \in \mathscr{B}[0,1],\ \#A_n < \infty \right\},$$

which entails that the space $L^p(\zeta)$ is given by

$$\left\{ u : [0,1] \to \mathbb{R} \,;\; \exists (x_n)_n \subseteq [0,1]\ \forall x \neq x_n \,:\, u(x) = 0 \text{ and } \sum_{n=1}^{\infty} |u(x_n)|^p < \infty \right\}.$$

Obviously, $(\mathbb{1}_{\{x\}})_{x \in [0,1]} \subseteq L^p(\zeta)$, but no countable system can approximate this family since

$$\|\mathbb{1}_{\{x\}} - \mathbb{1}_{\{y\}}\|^p_{L^p(\zeta)} = 0 \text{ or } 2$$

according to whether $x = y$ or $x \neq y$.

Comment Let $(X, \mathscr{A}, \mu)$ be a measure space. If $L^p(\mu)$ is separable for some $p \in [1, \infty)$, then $L^p(\mu)$ is separable for all $p \in [1, \infty)$, cf. [MIMS, p. 133, Problem 13.15].

18.15 Separability of the space L^∞

Let $(X, \mathscr{A}, \mu)$ be a measure space. If μ is a purely atomic measure [☞ p. 123] which has only finitely many disjoint atoms, say $A_1, \dots, A_n$, then

$$L^\infty(X, \mathscr{A}, \mu) = \left\{ f : X \to \mathbb{R};\ f \text{ is measurable and } \max_{x \in A_1 \cup \dots \cup A_n} |f(x)| < \infty \right\}.$$

This means that we can identify $L^\infty(X, \mathscr{A}, \mu)$ with (a subspace of) $\mathbb{R}^n$ where $n = \#\operatorname{supp}\mu$, which is obviously separable. We will show that this is the only situation where L^∞ is separable; see the theorem further on.

Our analysis uses the following lemma, which is interesting in its own right.

Lemma *Let B be a Banach space which contains an uncountable family of non-empty open and pairwise disjoint subsets $\mathscr{U} = (U_i)_{i \in I}$. Then B cannot be separable.*

Proof Assume there is a countable dense subset $(u_n)_{n \in \mathbb{N}} \subseteq B$. Since the sets $U_i \in \mathscr{U}$ are open and non-empty, we have $(u_n)_{n \in \mathbb{N}} \cap U_i \neq \emptyset$ for all $i \in I$, and we find for each $i \in I$ some index $n(i) \in \mathbb{N}$ such that $u_{n(i)} \in U_i$. Since the sets U_i, $i \in I$, are pairwise disjoint, this means that we can count the sets in $\mathscr{U}$, which is a contradiction to I being uncountable. □

Returning to the Banach space $B = L^\infty(X, \mathscr{A}, \mu)$, let us assume that there is an uncountable family $(A_i)_{i \in I} \subseteq \mathscr{A}$ such that their symmetric differences have strictly positive measure, $\mu(A_i \vartriangle A_k) = \mu((A_i \setminus A_k) \cup (A_k \setminus A_i)) > 0$ for all $i \neq k$. Typically, this is the case if μ is a measure with full topological support (i.e. charging every non-empty open set) on a metric space (X, d); a precise characterization is given below. Define a family of open sets $\mathscr{U} = (U_i)_{i \in I}$ via

$$U_i := \left\{ f \in L^\infty;\ \|f - \mathbb{1}_{A_i}\|_{L^\infty} < \tfrac{1}{2} \right\}.$$

The sets in $\mathscr{U}$ are open and pairwise disjoint: suppose that $f \in U_i \cap U_k$ for some $i \neq k$. The condition $\mu((A_i \setminus A_k) \cup (A_k \setminus A_i)) > 0$ ensures that

$$1 = \|\mathbb{1}_{A_i} - \mathbb{1}_{A_k}\|_{L^\infty} \leqslant \|\mathbb{1}_{A_i} - f\|_{L^\infty} + \|f - \mathbb{1}_{A_k}\|_{L^\infty} < \frac{1}{2} + \frac{1}{2} = 1,$$

which is a contradiction. Thus, $\mathscr{U}$ satisfies the conditions of the lemma, and we conclude that $L^\infty(X, \mathscr{A}, \mu)$ is not separable.

For our reasoning we assumed that there exists an uncountable family of sets $(A_i)_{i \in I} \subseteq \mathscr{A}$ such that $\mu((A_i \setminus A_k) \cup (A_k \setminus A_i)) > 0$ for all $i \neq k$. It turns out that such a family exists if, and only if, $\dim L^\infty = \infty$. It is clear that $\dim L^\infty = \infty$ is necessary for the existence; sufficiency will be shown in (the proof of) the following theorem.

Theorem *Let $(X, \mathscr{A}, \mu)$ be a measure space. The following statements are equivalent.*

(a) *L^∞ is separable.*
(b) $\dim(L^\infty) < \infty$.
(c) *μ is purely atomic and has at most finitely many disjoint atoms.*

Proof Without loss of generality, we may assume that $\mu(X) > 0$. We are going to show the implications (a)⇒(c)⇒(b)⇒(a).

(a)⇒(c). We prove this implication by contradiction. For every measure μ, there exist a purely atomic measure μ_a and a non-atomic measure μ_d on $(X, \mathscr{A})$ such that $\mu = \mu_a + \mu_d$, cf. [87, Theorem 2.1]. If (c) does not hold, then μ is either not purely atomic or there are infinitely many disjoint atoms. We claim that there exists in both cases a sequence $(A_k)_{k\in\mathbb{N}} \subseteq \mathscr{A}$ of pairwise disjoint sets with $\mu(A_k) > 0$ for all $k \in \mathbb{N}$. We consider the two cases separately:

Case 1. μ is not purely atomic, i.e. $\mu_d(X) > 0$. Since μ_d is non-atomic, the set X is not an atom of μ_d, and so there exists a set $A_1 \in \mathscr{A}$ with $A_1 \subseteq X$ such that $\mu_d(A_1) > 0$ and $\mu_d(X \setminus A_1) > 0$. By the same argument, $X \setminus A_1$ is not an atom, i.e. there is some $A_2 \subseteq X \setminus A_1$ with $\mu_d(A_2) > 0$ and $\mu_d((X \setminus A_1) \setminus A_2) > 0$. By iteration, we get a sequence $(A_k)_{k\in\mathbb{N}} \subseteq \mathscr{A}$ of pairwise disjoint sets with $\mu(A_k) \geqslant \mu_d(A_k) > 0$.

Case 2. μ is purely atomic and has infinitely many disjoint atoms. Since atoms have, by definition, positive measure, this gives immediately the desired sequence of pairwise disjoint sets with positive measure.

For the thus constructed sequence $(A_k)_{k\in\mathbb{N}}$, define a map $T : \mathscr{P}(\mathbb{N}) \to \mathscr{A}$ by

$$T(B) := \bigcup_{k\in B}^{\bullet} A_k, \quad B \subseteq \mathbb{N}.$$

Since the sets A_k, $k \in \mathbb{N}$, are pairwise disjoint and have strictly positive measure, it follows easily that the uncountable family $(T(B))_{B\subseteq\mathbb{N}}$ satisfies

$$\mu((T(B) \setminus T(B')) \cup (T(B') \setminus T(B))) > 0 \quad \text{for all } B \neq B'.$$

The considerations preceding the theorem show that L^∞ is not separable.

(c)⇒(b). By assumption, μ has finitely many disjoint atoms, say, $A_1, \dots, A_n$; set $X_0 := \bigcup_{i=1}^n A_i$. Since μ is purely atomic and $X \setminus X_0$ does not contain any atom, it follows that $\mu(X \setminus X_0) = 0$. In particular, $f|_{X\setminus X_0} = 0$ μ-a.e. for any measurable function f. Moreover, f is μ-a.e. constant on each atom, which means that we can find constants $c_1, \dots, c_n \in \mathbb{R}$ such that

$$f = \sum_{i=1}^{n} c_i \mathbb{1}_{A_i} \quad \mu\text{-a.e.}$$

Hence, $\dim(L^\infty) \leqslant n < \infty$.

(b)⇒(a). If $\dim(L^\infty) < \infty$, then there exists some $n \in \mathbb{N}$ and $f_1, \dots, f_n \in L^\infty$ such that every function $f \in L^\infty$ can be written as

$$f = \sum_{i=1}^{n} c_i f_i \quad \mu\text{-a.e.}$$

for suitable constants $c_i \in \mathbb{R}$. Consequently, the countable family

$$\left\{ \sum_{k=1}^{n} q_k f_k \,;\; q_k \in \mathbb{Q} \right\}$$

is dense in L^∞. □

18.16 $C_b(X)$ need not be dense in $L^p(\mu)$

Let (X, d) be a metric space and consider $L^p = L^p(X, \mathscr{B}(X), \mu)$, $1 \leqslant p < \infty$, for a Borel measure μ. If μ is outer regular [☞ p. 185] or $\mu(B_r(0)) < \infty$ for all $r > 0$, then $C_b(X) \cap L^p(\mu)$ is dense in L^p, cf. [MIMS, p. 189]. If neither of the two conditions is satisfied, then $C_b(X) \cap L^p(\mu)$ need not be dense in $L^p(\mu)$.

Let $\zeta_{\mathbb{Q}} = \sum_{q \in \mathbb{Q}} \delta_q$ be the counting measure on $(\mathbb{R}, \mathscr{B}(\mathbb{R}))$. For any $1 \leqslant p < \infty$ the space L^p contains the family $\{\mathbb{1}_{\{q\}}; q \in \mathbb{Q}\}$ of linearly independent elements, and therefore L^p is an infinite-dimensional space. On the other hand,

$$L^p \cap C_b(X) = \{0\}.$$

Indeed: let $f \in L^p \cap C_b(X)$, and suppose that $f(x) > 0$ for some $x \in \mathbb{R}$. Since f is continuous, there is some $\delta > 0$ such that $f(y) \geqslant \frac{1}{2} f(x) > 0$ for all $y \in B_\delta(x)$. Hence,

$$\int_{\mathbb{R}} |f|^p \, d\zeta_{\mathbb{Q}} \geqslant \int_{B_\delta(x)} |f|^p \, d\zeta_{\mathbb{Q}} \geqslant \frac{f(x)^p}{2^p} \zeta_{\mathbb{Q}}(B_\delta(x)) = \infty,$$

in contradiction to $f \in L^p$. Analogously, it follows that $f \in C_b(X) \cap L^p$ cannot attain any values strictly smaller than zero. Hence, $f = 0$ is the only function in $C_b(X) \cap L^p$, and so $C_b(X) \cap L^p$ is not dense in L^p.

Note that any open set U satisfies $\zeta_{\mathbb{Q}}(U) = \infty$; this implies that $\zeta_{\mathbb{Q}}$ is not outer regular and does not satisfy $\zeta_{\mathbb{Q}}(B_r(0)) < \infty$, i.e. the above-mentioned sufficient conditions for denseness of $C_b(X) \cap L^p$ are not satisfied.

18.17 A subset of L^p which is dense in L^r, $r < p$, but not dense in L^p

We will use the following lemma, cf. [151, Theorem 1.18].

Lemma *Let $(X, \|\cdot\|)$ be a normed space and $\Lambda : X \to \mathbb{R}$ a non-zero linear functional. Λ is not continuous if, and only if, its kernel*

$$\ker(\Lambda) := \Lambda^{-1}(0) = \{x \in X\,;\ \Lambda x = 0\}$$

is dense in X.

Fix $p \in (1, \infty)$ and denote the conjugate exponent by $q \in (1, \infty)$. Take some $g \in L^q := L^q(\mathbb{R}, \mathscr{B}(\mathbb{R}), \lambda)$ such that $g \notin L^s$ for all $s > q$ [☞ 10.16], and set

$$D := \left\{ f \in L^p \cap L^\infty\,;\ \int_{\mathbb{R}} fg\,d\lambda = 0 \right\}.$$

As $g \in L^q$, an application of Hölder's inequality shows that the linear functional

$$(L^p, \|\cdot\|_{L^p}) \to (\mathbb{R}, |\cdot|),\ f \longmapsto \int_{\mathbb{R}} fg\,d\lambda$$

is continuous, and by the above lemma its kernel is not dense in $(L^p, \|\cdot\|_{L^p})$. Since D is a subset of the kernel, this implies that D is not dense in $(L^p, \|\cdot\|_{L^p})$. Now let $r \in (1, p)$ with conjugate exponent $s > q$. By construction, $g \notin L^s$, and therefore the functional

$$\Lambda : (L^r \cap L^\infty, \|\cdot\|_{L^r}) \to (\mathbb{R}, |\cdot|),\ f \longmapsto \Lambda f := \int_{\mathbb{R}} fg\,d\lambda$$

is not continuous. Indeed: from

$$\|g\|_{L^s} = \sup\left\{ \int_{\mathbb{R}} fg\,d\lambda\,;\ f \in L^r,\ \|f\|_{L^r} \leqslant 1 \right\}$$

[☞ (1.16), 18.19], and the fact that $(L^r \cap L^\infty, \|\cdot\|_{L^r})$ is dense in $(L^r, \|\cdot\|_{L^r})$, it follows that

$$\|g\|_{L^s} = \sup\left\{ \int_{\mathbb{R}} fg\,d\lambda\,;\ f \in L^r \cap L^\infty,\ \|f\|_{L^r} \leqslant 1 \right\}.$$

As $g \notin L^s$, the left-hand (hence, the right-hand) side equals $+\infty$, which means that the functional Λ is not continuous. Applying once more the lemma, we obtain that

$$\ker(\Lambda) = \left\{ f \in L^r \cap L^\infty ;\ \int_{\mathbb{R}} fg \, d\lambda = 0 \right\}$$

is dense in $(L^r \cap L^\infty, \|\cdot\|_{L^r})$. Since $(L^r \cap L^\infty, \|\cdot\|_{L^r})$ is dense in $(L^r, \|\cdot\|_{L^r})$, we conclude that $\ker(\Lambda)$ is dense in $(L^r, \|\cdot\|_{L^r})$. Finally, $L^r \cap L^\infty \subseteq L^p \cap L^\infty$ implies that $\ker(\Lambda) \subseteq D$, and so D is dense in $(L^r, \|\cdot\|_{L^r})$.

18.18 L^p is not an inner product space unless $p = 2$ or $\dim(L^p) \leqslant 1$

Let $(X, \mathscr{A}, \mu)$ be a measure space and consider $L^p = L^p(X, \mathscr{A}, \mu)$ for $p \in [1, \infty]$. The mapping

$$\langle f, g \rangle := \int_X fg \, d\mu, \quad f, g \in L^2$$

defines a scalar product on L^2 which induces the L^2 norm $\|f\|_{L^2} = \sqrt{\langle f, f \rangle}$. If $p \neq 2$ then L^p is not an inner product space, i.e. there does not exist a scalar product $\langle \cdot, \cdot \rangle : L^p \times L^p \to \mathbb{R}$ with $\|f\|_{L^p} = \sqrt{\langle f, f \rangle}$, $f \in L^p$. The only exception is the trivial case $\dim(L^p) \leqslant 1$ for which it can easily be seen that L^p is indeed an inner product (hence, a Hilbert) space.

Recall that a norm $\|\cdot\|$ on a linear space V is induced by a scalar product if, and only if, the **parallelogram law** holds:

$$\|f + g\|^2 + \|f - g\|^2 = 2\|f\|^2 + 2\|g\|^2, \quad f, g \in V, \tag{18.6}$$

see [MIMS, p. 338, Exercise 26.2] or [202, I.5, Theorem 1] (Fréchet–von Neumann–Jordan theorem).

Assume that $\dim(L^p) > 1$ and fix $p \in [1, \infty)$. We claim that there are disjoint sets $A, B \in \mathscr{A}$ with $0 < \mu(A), \mu(B) < \infty$. Indeed: since $\dim(L^p) > 1$ there exist $f \in L^p$ and $c > 0$ such that $A := \{|f| > c\}$ and $A^c = \{|f| \leqslant c\}$ have positive measure. By Markov's inequality, $\mu(A) < \infty$. Consider several cases separately:

Case 1. $\mu(A^c) < \infty$. In this case, A and $B := A^c$ do the job.

Case 2. $\mu(A^c) = \infty$. If there exists some $B \subseteq A^c$ with $0 < \mu(B) < \infty$, then we are done. If there does not exist any such set B, then $g|_{A^c} = 0$ μ-a.e. for every $g \in L^p$. As $\dim(L^p) > 1$ this means that A cannot be an atom, and therefore there exists some $F \in \mathscr{A}$, $F \subseteq A$ such that $\mu(F) \in (0, \mu(A))$ and $\mu(A \setminus F) \in (0, \mu(A))$. Consequently, F and $A \setminus F$ are disjoint sets with non-zero finite measure.

Let $A, B \in \mathscr{A}$ be two disjoint sets of non-zero finite measure. For the functions $f := \mu(A)^{-1/p}\mathbb{1}_A$ and $g := \mu(B)^{-1/p}\mathbb{1}_B$ we have

$$\|f+g\|_{L^p}^2 + \|f-g\|_{L^p}^2 = 2 \cdot 2^{2/p} \quad \text{and} \quad 2\|f\|_{L^p}^2 + 2\|g\|_{L^p}^2 = 4,$$

which means that the parallelogram law (18.6) fails if $p \neq 2$. Consequently, we have shown that L^p is not an inner product space for $p \in [1, \infty) \setminus \{2\}$ if $\dim(L^p) > 1$.

It remains to consider the case $p = \infty$. We claim that there exists a function $h \in L^\infty$ such that $\mu(\{h > c\}) > 0$ and $\mu(\{h < -c\}) > 0$ for some $c > 0$. Since $\dim(L^p) \geqslant 2$ there exists a function $f \in L^\infty(\mu)$ such that $f \geqslant 0$ and f is not constant μ-almost everywhere. In particular, we can choose $r \in (0, \|f\|_{L^\infty})$ such that $\{f \leqslant r\}$ and $\{f > r\}$ both have positive measure. The function

$$h(x) := \|f\|_{L^\infty}\mathbb{1}_{\{f>r\}}(x) - \|f\|_{L^\infty}\mathbb{1}_{\{f\leqslant r\}}(x), \quad x \in X,$$

is clearly in L^∞ and satisfies $\mu(\{h > c\}) > 0$ and $\mu(\{h < -c\}) > 0$ for the constant $c := \frac{1}{2}\|f\|_{L^\infty}$.

For a function h with the above properties, consider the positive part $f := h^+$ and the negative part $g := h^-$. As $f - g = h$ and $f + g = |h|$, we have

$$\|f+g\|_{L^\infty}^2 + \|f-g\|_{L^\infty}^2 = 2\|h\|_{L^\infty}^2.$$

Since either $\|h^+\|_{L^\infty} = \|h\|_{L^\infty}$ or $\|h^-\|_{L^\infty} = \|h\|_{L^\infty}$, it follows from $\|h^+\|_{L^\infty} > 0$ and $\|h^-\|_{L^\infty} > 0$ that

$$\|f+g\|_{L^\infty}^2 + \|f-g\|_{L^\infty}^2 = 2\|h\|_{L^\infty}^2 < 2\|h^-\|_{L^\infty}^2 + 2\|h^+\|_{L^\infty}^2 = 2\|f\|_{L^\infty}^2 + 2\|g\|_{L^\infty}^2,$$

i.e. the parallelogram law (18.6) does not hold.

18.19 The condition $\sup_{\|g\|_{L^q}\leqslant 1} \int |fg| \, d\mu < \infty$ need not imply that $f \in L^p(\mu)$

Let $(X, \mathscr{A}, \mu)$ be a measure space, $1 \leqslant p \leqslant \infty$ and $\frac{1}{p} + \frac{1}{q} = 1$. Throughout this example all functions are tacitly assumed to be measurable. Hölder's inequality shows that

$$\int_X |fg| \, d\mu \leqslant \|f\|_{L^p}\|g\|_{L^q};$$

in particular, $f \in L^p$ and $g \in L^q$ implies $fg \in L^1$. The converse direction

$$\sup_{\|g\|_{L^q}\leqslant 1} \int_X |fg| \, d\mu < \infty \Rightarrow f \in L^p$$

trivially holds if $p = 1$ – just take $g \equiv 1 \in L^\infty$.

If $1<p<\infty$ we need, in addition, the finite subset property,

$$\forall A\in\mathscr{A},\ \mu(A)>0\quad \exists A_0\subseteq A,\ A_0\in\mathscr{A}\ :\ 0<\mu(A_0)<\infty. \tag{18.7}$$

This condition is certainly fulfilled if the measure space is σ-finite.

To see what can go wrong if (18.7) does not hold, consider $(\mathbb{R},\mathscr{B}(\mathbb{R}))$ with Lebesgue measure λ and set

$$\mu(A):=\begin{cases}\lambda(A), & \text{if } A\in\mathscr{B}(\mathbb{R}), A\subseteq[0,\infty),\\ \infty, & \text{for all other } A\in\mathscr{B}(\mathbb{R}).\end{cases}$$

Every function $h\in L^r(\mu)$, $r<\infty$, must satisfy $h\mathbb{1}_{(-\infty,0)}=0$ μ-a.e. Therefore, $f:=\mathbb{1}_{(-\infty,0)}$ is not in $L^p(\mu)$ while $fg=0$ a.e. for all $g\in L^q(\mu)$, i.e.

$$\sup_{\|g\|_{L^q(\mu)}\leqslant 1}\int_{\mathbb{R}}|fg|\,d\mu=0\quad\text{but}\quad f\notin L^p(\mu).$$

In order to prove the sufficiency of condition (18.7), fix a measurable function f with $M:=\sup_{\|g\|_{L^q}\leqslant 1}\int|fg|\,d\mu<\infty$. Let $A_0\subseteq X$ be such that $0<\mu(A_0)<\infty$ and set $f_n:=\min\{|f|,n\}\mathbb{1}_{A_0}$. Clearly, $f_n\in L^p$ and so

$$G_n(g):=\int f_n g\,d\mu,\quad g\in L^q,$$

is a bounded linear functional with $\|G_n\|_{L^q\to\mathbb{R}}=\|f_n\|_{L^p}$. Since

$$|G_n(g)|\leqslant\int|fg|\,d\mu\leqslant M\|g\|_{L^q},\quad n\in\mathbb{N},\ g\in L^q,$$

it follows that $\|f_n\|_{L^p}=\|G_n\|_{L^q\to\mathbb{R}}\leqslant M$ uniformly for all $n\in\mathbb{N}$. Letting $n\to\infty$ yields that $f\mathbb{1}_{A_0}\in L^p$.

Now we use the above argument for all $A\in\mathscr{A}_0=\{A\in\mathscr{A}\,;\ \mu(A)<\infty\}$, and we conclude, as before, that $\|f\mathbb{1}_A\|_{L^p}\leqslant M<\infty$ uniformly for $A\in\mathscr{A}_0$. Set $c:=\sup_{A\in\mathscr{A}_0}\|f\mathbb{1}_A\|_{L^p}<\infty$. By the definition of the supremum, there is a sequence $A_k\in\mathscr{A}_0$ such that $A_k\uparrow X_0\subseteq X$ and $c=\lim_{k\to\infty}\|f\mathbb{1}_{A_k}\|_{L^p}$.

We claim that $f=f\mathbb{1}_{X_0}$ a.e. If $\mu(X\setminus X_0)=0$, there is nothing to show. Suppose that $\mu(X\setminus X_0)>0$ and that f is not a.e. zero on $X\setminus X_0$. Because of (18.7), there is some $A\in\mathscr{A}_0$ with $A\subseteq X\setminus X_0$ and $f\mathbb{1}_A\neq 0$. In particular,

$$\|f\mathbb{1}_{X_0\cup A}\|_{L^p}^p=\|f\mathbb{1}_{X_0}\|_{L^p}^p+\|f\mathbb{1}_A\|_{L^p}^p>c^p,$$

which is impossible. Thus, $f=f\mathbb{1}_{X_0}$ a.s. and $\|f\|_{L^p}=\|f\mathbb{1}_{X_0}\|_{L^p}=c<\infty$.

Our argument can easily be adapted to the situation where $p=\infty$ and $q=1$.

Comment If (18.7) holds, then we can obtain the norm of the Banach space $L^p, 1 \leqslant p \leqslant \infty$, by the following variational principle:

$$\|f\|_{L^p} = \sup_{\|g\|_{L^q} \leqslant 1} \int_X fg \, d\mu = \sup_{\|g\|_{L^q} \leqslant 1} \int_X |fg| \, d\mu.$$

Notice that this formula works for all measurable functions f and characterizes whether f belongs to L^p or not.

If (18.7) is not satisfied, then the above formula is still valid **provided we already know that** $f \in L^p$, i.e. it is a representation of the L^p-norm but it does not guarantee that a measurable function f is indeed in L^p.

18.20 Identifying the dual of L^p with L^q is a tricky business

Let $(X, \mathscr{A}, \mu)$ be an arbitrary measure space and $p, q \in [1, \infty]$ conjugate exponents, $p^{-1} + q^{-1} = 1$. Hölder's inequality [☞Theorem 1.18] shows that

$$\forall g \in L^q \ : \ f \longmapsto \Lambda_g(f) := \int fg \, d\mu$$

defines a bounded linear functional Λ_g on $L^p(\mu)$; that is $\Lambda_g \in (L^p)^*$. Usually, one identifies Λ_g and g and writes $L^q \subseteq (L^p)^*$. This identification may, sometimes, be misleading as the following example shows. If $1 < p < \infty$, there is an isometric isomorphism between $(L^p)^*$ and L^q, and this is still true for $(L^1)^*$ and L^∞ if μ is σ-finite [☞18.21] but $(L^\infty)^*$ cannot, in general, be represented by L^1 [☞18.24].

Consider the measure space $X = \{0, 1\}$ with $\mathscr{A} = \mathscr{P}(X)$ and the measure $\mu(\{0\}) = 1$ and $\mu(\{1\}) = \infty$. The measure μ is infinite and not σ-finite. Moreover, for $1 \leqslant p < \infty$

$$u \in L^p(\mu) \iff u(0) \in \mathbb{R}, \ u(1) = 0 \Rightarrow \|u\|_{L^p} = |u(0)|,$$

and

$$u \in L^\infty(\mu) \iff u(0), u(1) \in \mathbb{R} \Rightarrow \|u\|_{L^\infty} = \max\{|u(0)|, |u(1)|\}.$$

For $p \in (1, \infty)$ every bounded linear functional Λ on $L^p(\mu)$ is of the form

$$\Lambda(u) = \Lambda_{a,b}(u) = au(0) + bu(1) = au(0) \quad \text{for} \quad a, b \in \mathbb{R}.$$

Obviously, $\Lambda_{a,b} = \Lambda_{a,0}$, i.e. $(L^p(\mu))^*$ can be identified with $\mathbb{R} \times \{0\}$ and this shows that for $1 < p < \infty$ and the conjugate $1 < q = p(p-1)^{-1} < \infty$

$$(L^p(\mu))^* \longleftrightarrow L^q(\mu) \quad \text{via the map} \quad g \simeq \Lambda_g, \ \Lambda_g(u) = \int ug \, d\mu = u(0)g(0).$$

Let us now consider $p = 1$ and $q = \infty$. The above consideration is still valid for $p = 1$ and shows that $(L^1(\mu))^*$ is isomorphic to $\mathbb{R} \times \{0\}$. On the other hand, $L^\infty(\mu)$ can be identified with $\mathbb{R} \times \mathbb{R}$. Notice that the above construction gives an imbedding $\iota : L^\infty(\mu) \to (L^1(\mu))^*$ and that we actually do not identify $L^\infty(\mu)$ with $(L^1(\mu))^*$ but $\iota(L^\infty(\mu))$ and $(L^1(\mu))^*$. The map ι is surjective, and this means that $\iota(L^\infty(\mu))$ identifies all L^∞-functions which lead to the same functional. This explains that $(L^1(\mu))^*$ is to be identified with the quotient space $L^\infty(\mu)/\sim$, where $f \sim g$ if $\Lambda_f = \Lambda_g$. In other words, $\iota(L^\infty(\mu))$ is isomorphic to $\mathbb{R} \times \{0\}$.

18.21 The dual of L^1 can be larger than L^∞

Equip $\mathbb{R}$ with the co-countable σ-algebra $\mathscr{A}$ [☞Example 4A], and let ζ be counting measure on $(\mathbb{R}, \mathscr{A})$. The space of integrable functions $u : X \to \mathbb{R}$ is given by

$$u \in L^1(\mathbb{R}, \mathscr{A}, \zeta) \iff \#\{u \neq 0\} \leqslant \#\mathbb{N} \text{ and } \sum_{x\in\mathbb{R}} |u(x)| = \sum_{x\in\{u\neq 0\}} |u(x)| < \infty.$$

The functional $\Lambda(u) := \sum_{x>0} u(x)$ is a bounded linear functional on L^1 but it is not induced by any L^∞-function. Indeed, if there were some $g \in L^\infty$ such that

$$\Lambda(u) = \int u g \, d\zeta = \sum_{x\in\mathbb{R}} u(x)g(x) \text{ for all } u \in L^1(\mathbb{R}, \mathscr{A}, \zeta),$$

then we would have $g(x) = \mathbb{1}_{(0,\infty)}(x)$, but this function is not measurable for the co-countable σ-algebra. Thus, $(L^1)^* \supsetneqq L^\infty$.

Comment 1 If the measure space $(X, \mathscr{A}, \mu)$ is σ-finite, then there is always an isometric isomorphism between the dual $(L^1(X, \mathscr{A}, \mu))^*$ and $L^\infty(X, \mathscr{A}, \mu)$, see [MIMS, p. 241] or Dunford & Schwartz [51, Theorem IV.8.5], each linear functional $\Lambda \in (L^1(X, \mathscr{A}, \mu))^*$ is of the form

$$\exists g \in L^\infty \quad \forall u \in L^1 \ : \ \Lambda(u) = \Lambda_g(u) = \int_X u \, g \, d\mu \quad \text{and} \quad \|\Lambda_g\| = \|g\|_{L^\infty}.$$

If the measure space $(X, \mathscr{A}, \mu)$ has the finite subset property,

$$\forall A \in \mathscr{A},\ \mu(A) > 0 \quad \exists A_0 \subseteq A,\ A_0 \in \mathscr{A} \ : \ 0 < \mu(A_0) < \infty, \tag{18.8}$$

then there is still an isometric embedding $L^\infty \subseteq (L^1)^*$, i.e. $\Lambda_g(u) = \int_X u \, g \, d\mu$ defines an element in $(L^1)^*$ such that $\|\Lambda_g\| = \|g\|_{L^\infty}$ In fact, the finite subset property is equivalent to $\|\Lambda_g\| = \|g\|_{L^\infty}$.

Lemma *Let $(X, \mathscr{A}, \mu)$ be a measure space, $g \in L^\infty$ and $\Lambda = \Lambda_g : L^1 \to \mathbb{R}$ the linear functional defined by*

$$\Lambda_g(f) = \int gf \, d\mu.$$

The functional Λ_g is continuous and its operator norm satisfies $\|\Lambda_g\| \leqslant \|g\|_{L^\infty}$.

Moreover, $\|\Lambda_g\| = \|g\|_{L^\infty}$ holds for all $g \in L^\infty$ if, and only if, the finite subset property (18.8) is satisfied.

Proof Recall that

$$\|\Lambda_g\| = \sup_{u \in L^1, u \neq 0} \frac{|\Lambda_g(u)|}{\|u\|_{L^1}} = \sup_{f \in L^1, \|f\|_{L^1}=1} |\Lambda_g(f)|.$$

Therefore, $\|\Lambda_g\| \leqslant \|g\|_{L^\infty}$ follows from Hölder's inequality. Considering $-f$ instead of f allows us to replace, under the supremum, $|\Lambda_g(f)|$ by $\Lambda_g(f)$. If, in addition, $g \geqslant 0$, then Λ_g preserves positivity and it is enough to consider $f \geqslant 0$.

Assume now that the property (18.8) holds and consider, for any $\epsilon > 0$, the set $A := \{|g| > \|g\|_{L^\infty} - \epsilon\}$. Since $\mu(A) > 0$, we can use the finite subset property to get some $A_0 \subseteq A$ such that $0 < \mu(A_0) < \infty$. Consequently, the function $f := \mu(A_0)^{-1} \operatorname{sgn}(g) \mathbb{1}_{A_0}$ is integrable with $\|f\|_{L^1} = 1$, and we get

$$\|\Lambda_g\| \geqslant \Lambda_g(f) = \int fg \, d\mu = \frac{1}{\mu(A_0)} \int_{A_0} |g| \, d\mu \geqslant \|g\|_{L^\infty} - \epsilon.$$

Letting $\epsilon \to 0$ proves $\|\Lambda_g\| \geqslant \|g\|_{L^\infty}$, hence $\|\Lambda_g\| = \|g\|_{L^\infty}$.

Conversely, assume that $\|\Lambda_g\| = \|g\|_{L^\infty}$ holds for all $g \in L^\infty$. Fix any $A \in \mathscr{A}$ with strictly positive measure $\mu(A) > 0$ and set $g := \mathbb{1}_A \in L^\infty$; by construction, we have $\|g\|_{L^\infty} = 1$. Since $\|\Lambda_g\| = \|g\|_{L^\infty} = 1$ and $g \geqslant 0$, there is some $f \in L^1$, $f \geqslant 0$, $\|f\|_{L^1} = 1$, such that $\int_A f \, d\mu = \int fg \, d\mu \geqslant \frac{1}{2}$; in particular, the set $\{f > 0\} \cap A$ has strictly positive μ-measure. Using the fact that we can approximate the integrable function $f\mathbb{1}_A$ and its integral from below by elementary functions, we infer that there is some measurable set $A_0 \subseteq A$ such that $0 < \mu(A_0) < \infty$. □

If, in addition to (18.8), μ is **localizable** (also known as **decomposable**), then $(L^1(X, \mathscr{A}, \mu))^*$ and $L^\infty(X, \mathscr{A}, \mu)$ are isometrically isomorphic, cf. [80, Theorem 20.20], [139, Theorem 5.5.8] or [205, Theorem 35.3, p. 256]. The situation is different if $1 < p < \infty$ and, consequently, $1 < q < \infty$: in this case $(L^p(X, \mathscr{A}, \mu))^*$ is isometrically isomorphic to $L^q(X, \mathscr{A}, \mu)$ for any type of measure space, cf. [51, Theorem IV.8.1].

Comment 2 Let $(X, \mathscr{A}, \mu)$ be a measure space and $f_n, f \in L^1 = L^1(X, \mathscr{A}, \mu)$. By definition, $f_n \rightharpoonup f$ weakly in L^1 if

$$\forall \Lambda \in (L^1)^* \ : \ \lim_{n\to\infty} \Lambda(f_n) = \Lambda(f). \tag{18.9}$$

This convergence is often denoted as $\sigma(L^1, (L^1)^*)$-convergence. As we know, the dual of L^1 can be strictly larger than L^∞, and this makes it often hard to verify (18.9). However, it turns out that it actually suffices to check (18.9) for a much smaller class of functionals, namely, $f_n \rightharpoonup f$ weakly in L^1 if, and only if,

$$\forall A \in \mathscr{A} \ : \ \lim_{n\to\infty} \int_A f_n \, d\mu = \int_A f \, d\mu,$$

cf. [51, Theorem IV.8.7] or [202, Theorem V.1.4]. In the language of functional analysis, this means that $\sigma(L^1, (L^1)^*)$-convergence is equivalent to $\sigma(L^1, L^\infty)$-convergence. This result holds for **any** measure space, it is not needed that $(L^1)^* = L^\infty$.

18.22 The dual of L^1 can be isometrically isomorphic to a space which is strictly smaller than L^∞

Consider $X = \mathbb{R}$ with $\mathscr{A} = \mathscr{P}(\mathbb{R})$ and the measure

$$\mu(A) := \begin{cases} 0, & \text{if } A = \emptyset \text{ or } A \text{ is a countable set with } 0 \notin A, \\ 1, & \text{if } A \text{ is countable and } 0 \in A, \\ \infty, & \text{if } A \text{ is uncountable.} \end{cases}$$

The space of integrable functions $L^1(\mu)$ consists of all functions f which are zero except for countably many points and satisfy $|f(0)| < \infty$. If $f \in L^1(\mu)$, then $\int f \, d\mu = f(0)$ and

$$\mu(\{x \in \mathbb{R};\ f(x) \neq f(0)\}) = 0,$$

i.e. each $f \in L^1(\mu)$ can be identified with the function $f(0)\mathbb{1}_{\{0\}}$. This means that

$$\mathbb{R} \ni c \longmapsto c\mathbb{1}_{\{0\}} \in L^1(\mu)$$

defines an isometric isomorphism between $\mathbb{R}$ and $L^1(\mu)$. In particular,

$$(L^1(\mu))^* \cong \mathbb{R}^* \cong \mathbb{R} \cong L^1(\mu).$$

On the other hand, $L^\infty(\mu)$ contains all bounded functions $f : \mathbb{R} \to \mathbb{R}$ and so there is no isomorphism between $\mathbb{R} \cong (L^1(\mu))^*$ and $L^\infty(\mu)$; in this sense $(L^1(\mu))^*$ is strictly smaller than $L^\infty(\mu)$.

Comment The example is from McShane [115], another one is [☞18.3]. The pathology happens because the measure space does not have the finite subset property; see also Comment 1 to the previous Example 18.21.

18.23 A measure space such that the dual of L^1 is L^1

[☞18.3, 18.22].

18.24 The dual of L^∞ can be larger than L^1

Let $(X, \mathscr{A}, \mu)$ be a measure space and assume that there is a sequence of pairwise disjoint sets $(A_n)_{n\in\mathbb{N}} \subseteq \mathscr{A}$ such that $0 < \mu(A_n) < \infty$; moreover, denote by $A_0 = \biguplus_{n\in\mathbb{N}} A_n$ the disjoint union.

Define a linear functional on the subspace $M := \operatorname{span}\{\mathbb{1}_{A_n}, n \in \mathbb{N}_0\} \subseteq L^\infty$. Note that $u \in M$ if, and only if, $u = \sum_{n=0}^\infty a_n \mathbb{1}_{A_n}$ such that only finitely many of the coefficients a_n are non-zero. Consider the sequence of integrable functions $f_n := \mu(A_n)^{-1}\mathbb{1}_{A_n}$, $n \geqslant 1$. Because of the structure of M it is clear that for all $u \in M$,

$$\Lambda(u) := \lim_{n\to\infty} \int f_n u \, d\mu = a_0 \quad \text{and} \quad |\Lambda(u)| \leqslant \|u\|_\infty = \max_{n\geqslant 0} |a_n|.$$

In particular, Λ is a bounded linear functional on a linear subspace of L^∞. By the Hahn–Banach theorem, we can extend Λ to L^∞ without changing its norm.

Suppose now that we can identify $(L^\infty)^*$ with L^1. This means that the functional Λ can be written in the form

$$\Lambda(u) = \int u \cdot f \, d\mu$$

for a suitable function $f \in L^1$. If we take $u := \mathbb{1}_{A_0} = \sum_{n=1}^\infty \mathbb{1}_{A_n}$ we get, on the one hand, with the dominated convergence theorem

$$\Lambda(u) = \int \sum_{n=1}^{\infty} \mathbb{1}_{A_n} \cdot f \, d\mu = \sum_{n=1}^{\infty} \int \mathbb{1}_{A_n} \cdot f \, d\mu = \sum_{n=1}^{\infty} \Lambda(\mathbb{1}_{A_n}) = 0,$$

while, on the other hand, $\mathbb{1}_{A_0} \in M$ and so

$$\Lambda(u) = \Lambda(\mathbb{1}_{A_0}) = 1.$$

This means that we have reached a contradiction and we conclude that $(L^\infty)^*$ is strictly larger than L^1.

Comment Our example is a variant of the 'standard' counterexample: let λ be Lebesgue measure on $([0,1], \mathscr{B}[0,1])$ and set $M := C[0,1] + \mathbb{R} \cdot g$ for some $g \in L^\infty[0,1] \setminus C[0,1]$ such that $g \geqslant 0$ and $\|g\|_{L^\infty} = 1$, e.g. $g = \mathbb{1}_{[0,\frac{1}{2}]}$. Define

$$\Lambda(u) := \alpha \quad \text{for all } u = \phi + \alpha g \in M;$$

this is clearly a continuous linear functional on the linear space $M \subsetneq L^\infty[0,1]$ such that $\Lambda|_{C[0,1]} = 0$. Using the Hahn–Banach theorem we can extend Λ to $L^\infty[0,1]$. If Λ were of the form

$$\Lambda(u) = \int_0^1 u(x) \cdot f(x)\, dx \quad \text{for all } u \in L^\infty[0,1]$$

for some $f \in L^1[0,1]$, then $\int_0^1 \phi(x) f(x)\, dx = \Lambda(\phi) = 0$ for all $\phi \in C[0,1]$ would imply that $f \equiv 0$. But this cannot be the case as Λ is not identically zero.

A result due to Yosida & Hewitt [204, Theorem 2.3] (see also [202, Example IV.5] or [51, table on p. 378 and Theorem IV.8.16]) shows that we can identify $(L^\infty(X, \mathscr{A}, \mu))^*$ with the family of finitely additive set functions on $(X, \mathscr{A})$.

It is known that $\Lambda \in (L^\infty(X, \mathscr{A}, \mu))^*$ has a representation with a function from $L^1(X, \mathscr{A}, \mu)$ if (and only if) the set function $A \longmapsto \Lambda(\mathbb{1}_A)$ is σ-additive; see Bogachev [19, Proposition 4.4.2].

Still, $L^1(X, \mathscr{A}, \mu)$ is always isometrically embedded in $(L^\infty(X, \mathscr{A}, \mu))^*$. Using Hölder's inequality for a given $f \in L^1(X, \mathscr{A}, \mu)$ we see that

$$\Lambda_f(u) := \int u \cdot f\, d\mu, \quad |\Lambda_f(u)| \leqslant \|f\|_{L^1} \cdot \|u\|_{L^\infty}$$

defines a bounded linear functional $\Lambda \in (L^\infty(X, \mathscr{A}, \mu))^*$.

Moreover, $\Lambda_f(\operatorname{sgn} f) = \|f\|_{L^1}$, i.e. we have the isometric embedding

$$L^1(X, \mathscr{A}, \mu) \hookrightarrow (L^\infty(X, \mathscr{A}, \mu))^*.$$

This imbedding is strict, if $X = \bigcup_{n\in\mathbb{N}} A_n$ and $0 < \mu(A_n) < \infty$. Moreover, if μ is purely discontinuous with at most finitely many (disjoint) atoms of finite measure, then $L^1 = (L^\infty)^*$, cf. [205, Theorem 50.2, p. 362]. Note that the space L^∞ of a purely discontinuous measure with finitely many atoms is finite-dimensional [☞ 18.15].

18.25 A measure space where the dual of L^∞ is L^1

See Example 18.4 and the comment to Example 18.24, which shows that this is essentially the only situation where $(L^\infty)^*$ can be identified with L^1.

18.26 Non-uniqueness in the Riesz representation theorem

Let $(X, \mathcal{O})$ be a locally compact Hausdorff space. The Riesz representation theorem [☞Theorem 1.22] asserts that every positive linear form Λ on the space of continuous functions with compact support $C_c(X, \mathbb{R})$ is given by an integral of the form $\Lambda(u) = \int u \, d\mu$ with a unique regular Borel measure μ on $(X, \mathscr{B}(X))$.

The following example shows that the regularity of the representing measure is essential for the uniqueness. Let $X = \mathbb{R}$ be equipped with the discrete metric $d(x, y) = 0$ if $x = y$ and $d(x, y) = \infty$ if $x \neq y$ [☞Example 2.1.(b)]. In the corresponding discrete topology, singletons $\{x\}$ are both open and compact, thus all subsets $U \subseteq \mathbb{R}$ are open and all compact sets have finitely many elements. In particular, the only dense subset of X is $\mathbb{R}$, i.e. (X, d) is not separable.

The compact sets of (X, d) are exactly the sets with finitely many elements, and so $C_c(X, \mathbb{R}) = \{u : X \to \mathbb{R};\ \#\{u \neq 0\} < \infty\}$. Thus, $\Lambda(u) := \sum_{n \in \mathbb{N}} u(n)$ is a positive linear functional on $u \in C_c(X, \mathbb{R})$.

Equip X with the Borel sets $\mathscr{B}(X)$. Any subset of X is open, and therefore $\mathscr{B}(X) = \mathscr{P}(X)$. Denote by $\zeta_{\mathbb{N}}(A) := \#(A \cap \mathbb{N})$ the counting measure on $\mathbb{N}$ and set

$$\mu(A) = \zeta_{\mathbb{N}}(A) + \nu(A \setminus \mathbb{N}) \quad \text{and} \quad \nu(A) = \begin{cases} 0, & \text{if } A \text{ is countable}, \\ \infty, & \text{if } A \text{ is not countable}. \end{cases}$$

Note that $\zeta_{\mathbb{N}}(A) = \mu(A)$ if $\#A < \infty$, i.e. $\zeta_{\mathbb{N}} = \mu$ on all compact sets of X. Thus,

$$\forall u \in C_c(X, \mathbb{R}) \ : \ \Lambda(u) = \int_X u \, d\zeta = \int_X u \, d\mu,$$

which means that the representing measure of Λ is not unique.

Comment Note that $\zeta(\mathbb{R} \setminus \mathbb{N}) = 0$ while $\mu(\mathbb{R} \setminus \mathbb{N}) = \infty$. Therefore, we have for the open set $\mathbb{R} \setminus \mathbb{N}$

$$\mu(\mathbb{R} \setminus \mathbb{N}) \neq \sup\{\Lambda(u);\ u \in C_c(X, \mathbb{R}),\ u \leqslant \mathbb{1}_{\mathbb{R} \setminus \mathbb{N}}\} = 0,$$

which means that μ does not satisfy the regularity condition of Riesz's theorem.

18.27 Non-uniqueness in the Riesz representation theorem II

This example continues 18.26 and contains a further situation where the representing measure in Riesz's theorem is not unique. Let Ω be the ordinal space [☞Section 2.4] and consider the following σ-algebra [☞Example 4B]

$$\mathscr{A} := \{A \subseteq \Omega;\ A \cup \{\omega_1\} \text{ or } A^c \cup \{\omega_1\} \text{ contains an uncountable compact set}\}$$

the Dieudonné measure [☞ Example 5B]

$$\mu(A) := \begin{cases} 1, & \text{if } A \cup \{\omega_1\} \text{ contains an uncountable compact set } C, \\ 0, & \text{if } A^c \cup \{\omega_1\} \text{ contains an uncountable compact set } C. \end{cases}$$

The measure μ is not regular since every open neighbourhood $U(\omega_1)$ of $\{\omega_1\}$ has mass 1 while $\mu(\{\omega_1\}) = 0$. On the other hand,

$$\forall f \in C(\Omega) \ : \ \Lambda f := \int_\Omega f(x)\,\mu(dx) = f(\omega_1).$$

This follows from the fact that f is finally constant, i.e. there exists some countable ordinal number $\alpha_f \prec \omega_1$ such that $f|_{[\alpha_f,\omega_1]} \equiv f(\omega_1)$ (cf. [☞ Section 2.4]) and $\mu([\alpha_f,\omega_1]) = 1$.

Since μ is not regular, it cannot be the representing measure of Λ in the sense of Riesz's representation theorem. In fact, we can write $\Lambda f = \int_\Omega f(x)\,\delta_{\omega_1}(dx)$, i.e. the Dirac measure δ_{ω_1} is indeed the regular measure guaranteed by Riesz's theorem.

18.28 A measure space where L^∞ is not weakly sequentially complete

A metric space (X,d) with (topological) dual X^* is said to be **weakly sequentially complete** if for every sequence $(x_n)_{n\in\mathbb{N}} \subseteq X$ for which

$$\forall x^* \in X^* \ : \ (\langle x^*, x_n\rangle)_{n\in\mathbb{N}} \text{ is a Cauchy sequence,} \tag{18.10}$$

there is an element $x \in X$ such that $x_n \rightharpoonup x$ weakly, i.e.

$$\forall x^* \in X^* \ : \ \lim_{n\to\infty} \langle x^*, x_n\rangle = \langle x^*, x\rangle.$$

Because of the completeness of $\mathbb{R}$, (18.10) is equivalent to

$$\forall x^* \in X^* \ : \ \lim_{n\to\infty} \langle x^*, x_n\rangle \in \mathbb{R} \text{ exists.}$$

Let $(X,\mathscr{A},\mu)$ be a measure space. If $1 \leqslant p < \infty$, then $L^p = L^p(X,\mathscr{A},\mu)$ is weakly sequentially complete, cf. [51, Corollary IV.8.8, Theorem IV.8.6]. For $1 < p < \infty$ the dual of L^p is L^q, where $p^{-1} + q^{-1} = 1$, and so weak sequential completeness means that for every sequence $(f_n)_{n\in\mathbb{N}} \subseteq L^p$ for which

$$\forall g \in L^q \ : \ \lim_{n\to\infty} \int_X f_n g\,d\mu \in \mathbb{R} \text{ exists,}$$

we can find some $f \in L^p$ such that

$$\forall g \in L^q \ : \ \lim_{n\to\infty} \int_X f_n g\,d\mu = \int_X f g\,d\mu.$$

Although the dual of L^1 can be strictly larger than L^∞ [☞ 18.21], this characterization also holds for $p = 1$, cf. [202, Theorem V.1.4] or [205, Theorem 52.1, p. 373].

We will show that, in general, weak sequential completeness fails for $p = \infty$.

Consider $(\mathbb{N}, \mathscr{P}(\mathbb{N}))$ with the counting measure $\zeta = \zeta_{\mathbb{N}}$. We can identify $L^\infty(\zeta)$ with ℓ^∞, the space of bounded real-valued sequences $(u(k))_{k\in\mathbb{N}}$ with norm $\|u\|_{\ell^\infty} = \sup_{k\in\mathbb{N}} |u(k)|$. We will show that the sequence $(f_n)_{n\in\mathbb{N}} \subseteq \ell^\infty$ defined by

$$f_n(k) := I_{[1,n]}(k), \quad k, n \in \mathbb{N},$$

is weakly Cauchy but not weakly convergent in ℓ^∞. Denote by

$$c_0 := \left\{(u(k))_{k\in\mathbb{N}} \subseteq \mathbb{R};\ \lim_{k\to\infty} u(k) = 0\right\}$$

the space of real-valued sequences converging to zero. This space is continuously embedded into ℓ^∞ by $\iota\colon c_0 \to \ell^\infty, u \longmapsto u$. If Λ is a continuous linear functional on ℓ^∞, then the composition $\Lambda\circ\iota$ is a continuous linear functional on c_0, i.e. $\Lambda\circ\iota \in c_0^*$. Since c_0^* is isometrically isomorphic to $\ell^1 = \ell^1(\mathbb{N})$, [51, p. 375, Table IV.A], it follows that

$$(\Lambda\circ\iota)(u) = \sum_{k\in\mathbb{N}} a(k)u(k), \quad u \in c_0,$$

for some sequence $(a(k))_{k\in\mathbb{N}} \in \ell^1$. As $f_n \in c_0$, this gives

$$\Lambda f_n = \sum_{k=1}^{n} a(k), \quad n \in \mathbb{N}.$$

It follows from $(a(k))_{k\in\mathbb{N}} \in \ell^1$ that $\lim_{n\to\infty} \Lambda f_n$ exists and is finite and hence, $(\Lambda f_n)_{n\in\mathbb{N}}$ is a Cauchy sequence. This proves that $(f_n)_{n\in\mathbb{N}}$ is a weak Cauchy sequence in ℓ^∞. Now suppose that there is some $f \in \ell^\infty$ such that $f_n \rightharpoonup f$ weakly in ℓ^∞. Since $\mathbb{1}_{\{k\}} \in \ell^1$, this gives

$$f(k) = \lim_{n\to\infty} f_n(k) = 1$$

for all $k \in \mathbb{N}$, i.e. $f = 1_{\mathbb{N}}$. But f_n cannot converge weakly to $f = 1_{\mathbb{N}}$. To see this, define a linear functional

$$\Lambda u := \lim_{n\to\infty} u(n)$$

on the subspace M comprising all $u \in \ell^\infty$ with $u = a(0)\mathbb{1}_{\mathbb{N}} + \sum_{n\in\mathbb{N}} a(n)\mathbb{1}_{\{n\}}$ and $a(n) \neq 0$ for only finitely many $n \in \mathbb{N}_0$. Obviously, $|\Lambda u| \leqslant \|u\|_{\ell^\infty}$ and $\Lambda u = a_0$ for all $u \in M$. Consequently, Λ is a bounded linear functional on M, and, by the Hahn–Banach theorem, we can extend Λ to a bounded linear functional on ℓ^∞. As $\Lambda f_n = 0$ but $\Lambda \mathbb{1}_{\mathbb{N}} = 1$, we conclude that f_n does not converge weakly to $\mathbb{1}_{\mathbb{N}}$. Since $f = \mathbb{1}_{\mathbb{N}}$ was the only possible candidate, this means that the sequence does not converge weakly.

Comment A similar reasoning shows that, more generally, $L^\infty(X, \mathscr{A}, \zeta)$ fails to be weakly sequentially complete if there is a sequence $(A_k)_{k\in\mathbb{N}} \subseteq \mathscr{A}$ of pairwise disjoint sets with $0 < \zeta(A_k) < \infty$.

18.29 Uniform boundedness does not imply weak compactness in L^1

Let $(X, \mathscr{A}, \mu)$ be a measurable space, and write $L^p = L^p(X, \mathscr{A}, \mu)$ for $1 \leqslant p \leqslant \infty$. If $p \in (1, \infty)$ and if $(f_n)_{n\in\mathbb{N}}$ is a sequence of measurable functions such that $\sup_{n\in\mathbb{N}} \|f_n\|_{L^p} < \infty$, then there exists a function $f \in L^p$ such that $f_n \xrightarrow[n\to\infty]{} f$, i.e. weak (topologically) convergence holds:

$$\forall g \in L^q \ : \ \int f_n g \, d\mu \xrightarrow[n\to\infty]{} \int f g \, d\mu;$$

here $q \in (1, \infty)$ is the conjugate index, $p^{-1} + q^{-1} = 1$. This is a consequence of the Banach–Alaoglu theorem (see e.g. Rudin [151, Theorem 3.15]) and the fact that L^p, $p \in (1, \infty)$, is a reflexive Banach space, e.g. [MIMS, p. 243] and Examples 18.20 *et seq.* The following example shows that the statement fails to hold for $p = 1$.

Consider on $((0, \infty), \mathscr{B}(0, \infty))$ Lebesgue measure λ and $f_n := \mathbb{1}_{[n,n+1)}$ for all $n \in \mathbb{N}$. Clearly, $(f_n)_{n\in\mathbb{N}}$ is bounded in L^1. Suppose that $f_n \xrightarrow[n\to\infty]{} f$ for some $f \in L^1$. For $g := \mathbb{1}_{(a,b)} \in L^\infty$ with $0 < a < b < \infty$ this entails

$$\int_{(a,b)} f \, d\lambda = \lim_{n\to\infty} \int_{(a,b)} f_n \, d\lambda = 0.$$

Since the open intervals (a, b), $0 < a < b < \infty$, are a generator of $\mathscr{B}(0, \infty)$, this implies $\int_B f \, d\lambda = 0$ for any $B \in \mathscr{B}(0, \infty)$, and so $f = 0$. This, however, is impossible since we may take $g = \mathbb{1}_{(0,\infty)}$ and get

$$\int_{(0,\infty)} f \, d\lambda = \lim_{n\to\infty} \int_{(0,\infty)} f_n \, d\lambda = 1.$$

Comment The proper (necessary and sufficient) condition for sequential weak compactness in L^1 is uniform integrability [☞ Definition 1.30]. In L^p, uniform integrability is ensured by uniform boundedness. A discussion on various forms of uniform integrability can be found in [MIMS, Chapter 22].

18.30 The algebra $L^1(\lambda^d)$ does not have a unit element

Let $(\mathbb{R}^d, \mathscr{B}(\mathbb{R}^d), \lambda^d)$ be the d-dimensional Lebesgue measure space and consider on the space of integrable functions $L^1 = L^1(\lambda^d)$ the convolution 'product'

$$(f * g)(x) := \int_{\mathbb{R}^d} f(y)g(x-y)\,\lambda^d(dy), \quad x \in \mathbb{R}^d.$$

By Tonelli's theorem, $f, g \in L^1$ implies $f * g \in L^1$ and $\|f * g\|_{L^1} \leqslant \|f\|_{L^1}\|g\|_{L^1}$. The space $(L^1, *)$ is an algebra, but there is no unit element. Assume there were some $\epsilon \in L^1$ such that $\epsilon * g = g$ for all $g \in L^1$. Since g is integrable, the Fourier transform

$$\widehat{g}(\xi) := \frac{1}{(2\pi)^d} \int_{\mathbb{R}^d} e^{-ix\cdot\xi} g(x)\,dx$$

and the inverse Fourier transform

$$\check{g}(\xi) := \int_{\mathbb{R}^d} e^{ix\cdot\xi} g(x)\,dx$$

are well-defined. It follows from the convolution theorem [MIMS, p. 221] that

$$(2\pi)^d\,\widehat{\epsilon} \cdot \widehat{g} = \widehat{g}.$$

Since there is some $g \in L^1$ such that $\widehat{g} \not\equiv 0$ – e.g. $g(x) = (2\pi)^{-d/2} e^{-|x|^2/2}$ –, we get $\widehat{\epsilon} \equiv (2\pi)^{-d}$, which is impossible since $\lim_{|\xi|\to\infty} \widehat{\epsilon}(\xi) = 0$ by the Riemann–Lebesgue lemma [MIMS, p. 222].

Comment If we consider the convolution on the set of finite Borel measures on $(\mathbb{R}^d, \mathscr{B}(\mathbb{R}^d))$,

$$(\mu * \nu)(B) := \int_{\mathbb{R}^d} \int_{\mathbb{R}^d} \mathbb{1}_B(x+y)\,\mu(dx)\,\nu(dy), \quad B \in \mathscr{B}(\mathbb{R}^d),$$

then there **is** a unit element; namely, the Dirac measure δ_0 centred at 0. Note that if $\mu = f\lambda^d$ and $\nu = g\lambda^d$ are absolutely continuous measures, then the convolution is given by $\mu * \nu = (f * g)\,\lambda^d$ with $f * g$ defined at the beginning of this example.

18.31 The algebra $L^1(\lambda^d)$ contains non-trivial divisors of zero

Consider on $(\mathbb{R}^d, \mathscr{B}(\mathbb{R}^d), \lambda^d)$ the algebra $(L^1, *)$ [☞ 18.30]. Pick two bounded disjoint open sets $U, V \subseteq \mathbb{R}^d$ and two non-trivial functions $\phi, \psi \in C_c^\infty(\mathbb{R}^d)$ with support in U and V, respectively. If we denote by $f := \widehat{\phi}$ and $g := \widehat{\psi}$ their Fourier transforms [☞ 18.30], then $f, g \in L^1$ and $f \not\equiv 0$, $g \not\equiv 0$. However,

$$\widecheck{f * g} = \phi \cdot \psi \equiv 0 \implies f * g \equiv 0.$$

18.32 Uniform convexity/rotundity of L^p

Let $(X, \mathscr{A}, \mu)$ be a measure space. For $1 < p < \infty$ the L^p space is **uniformly convex** or **uniformly rotund**, i.e. for any $\epsilon > 0$ there is some $\delta > 0$ such that for any two functions $f, g \in L^p$ with $\|f\|_{L^p} = 1 = \|g\|_{L^p}$ the implication

$$\|f - g\|_{L^p} \geqslant \epsilon \Rightarrow \left\| \frac{f+g}{2} \right\|_{L^p} \leqslant 1 - \delta \qquad (18.11)$$

holds, cf. [19, Theorem 4.7.15]. For $p = 1$ and $p = \infty$ uniform convexity does, in general, not hold.

Consider Lebesgue measure λ on $((0,1), \mathscr{B}(0,1))$.

L^1 is not uniformly convex. The functions

$$f(x) := 1 + \sin(2\pi x) \quad \text{and} \quad g(x) := 1, \quad x \in (0,1)$$

satisfy $\|f\|_{L^1} = 1 = \|g\|_{L^1}$, $\|f - g\|_{L^1} = \frac{2}{\pi}$ and $\|\frac{1}{2}(f + g)\|_{L^1} = 1$. Consequently, uniform convexity of L^1 fails (choose $\epsilon = \frac{2}{\pi}$ in (18.11)).

L^∞ is not uniformly convex. The functions

$$f(x) := \mathbb{1}_{(0,1)}(x) \quad \text{and} \quad g(x) := \mathbb{1}_{(1/2,1)}(x), \quad x \in (0,1),$$

satisfy $\|f\|_{L^\infty} = 1 = \|g\|_{L^\infty}$, $\|f - g\|_{L^\infty} = 1$ and $\|\frac{1}{2}(f + g)\|_{L^\infty} = 1$, and so L^∞ is not uniformly convex.

Comment An elegant, and at the same time simple, proof of the uniform convexity of the spaces L^p, $1 < p < \infty$, can be found in Lieb & Loss [106, pp. 42–45]; it is based on Hanner's inequalities; see also [MIMS, p. 131].

In uniformly convex Banach spaces it is possible to characterize strong (i.e. norm) convergence in terms of weak convergence. As a consequence of L^p, $1 < p < \infty$, being uniformly convex, one has the following characterization of L^p-convergence for $1 < p < \infty$: $\|f_n - f\|_{L^p} \to 0$ if, and only if, $f_n \rightharpoonup f$ weakly in L^p and $\|f_n\|_{L^p} \to \|f\|_{L^p}$, cf. [19, Corollary 4.7.16].

Geometrically, uniform convexity of the unit ball means that the mid-point $\frac{1}{2}(x + y)$ of any chord with end points $\|x\| = \|y\| = 1$ which are 'far away' $\|x - y\| > \epsilon$ is also 'far away' from the perimeter of the unit ball. In infinite dimensions there can be unit balls which are strictly convex (i.e. they have exactly one tangent functional) but still there are points $\|x_n\| = \|y_n\| = 1$ such that $\inf_{n\in\mathbb{N}} \|x_n - y_n\| > 0$ and $\sup_{n\in\mathbb{N}} \|\frac{1}{2}(x_n + y_n)\| = 1$. This is a typical infinite-dimensional phenomenon, as in finite dimensions the notions of rotundity (strict convexity) and uniform rotundity (uniform convexity) coincide [117, Proposition 5.2.14].

By the **Milman–Pettis theorem**, uniformly convex Banach spaces are reflexive; see [202, V.2, Theorem 2] or [117, Theorem 5.2.15].

18.33 An absolutely continuous measure such that the translation operator is not continuous in L^1

Given a function $f: \mathbb{R} \to \mathbb{R}$ define the translation $\tau_h f(x) := f(x-h)$, $x \in \mathbb{R}$, $h \in \mathbb{R}$. For one-dimensional Lebesgue measure λ on $(\mathbb{R}, \mathscr{B}(\mathbb{R}))$ the translation operator is continuous in $L^1(\lambda)$, i.e.

$$\forall f \in L^1(\lambda) \ : \ \lim_{h\to 0} \|\tau_h f - f\|_{L^1(\lambda)} = 0,$$

cf. [MIMS, p. 159]. The situation is different for general measures μ. For instance, if we consider the Dirac measure $\mu = \delta_0$, then

$$\|\tau_h \mathbb{1}_{\{0\}} - \mathbb{1}_{\{0\}}\|_{L^1(\mu)} = \int |\mathbb{1}_{\{h\}}(x) - \mathbb{1}_{\{0\}}(x)| \, \delta_0(dx) = 1$$

does not converge to 0 as $h \to 0$, and therefore τ_h is not continuous in $L^1(\delta_0)$. The following example shows that continuity in $L^1(\mu)$ may fail to hold even if μ is absolutely continuous with respect to Lebesgue measure.

Consider the measure

$$\mu(dx) := \phi(x)\lambda(dx) := \sum_{n=1}^{\infty} \frac{1}{n^3} g\left(x - \tfrac{1}{n}\right) \lambda(dx)$$

for the function $g(x) := x^{-\frac{1}{2}} \mathbb{1}_{(0,1)}(x) \in L^1(\lambda)$. It follows from the Beppo Levi theorem and the translation invariance of Lebesgue measure that μ is a finite measure:

$$\mu(\mathbb{R}) = \sum_{n=1}^{\infty} \frac{1}{n^3} \int_{\mathbb{R}} g\left(x - \tfrac{1}{n}\right) \lambda(dx) = \left(\int_{\mathbb{R}} g(x)\lambda(dx)\right) \sum_{n=1}^{\infty} \frac{1}{n^3} < \infty.$$

This shows that $\phi \in L^1(\lambda)$; in particular, $\phi < \infty$ Lebesgue almost everywhere. The function $f(x) := x^{-1}\mathbb{1}_{(0,1)}(x)$ is in $L^1(\mu)$ since

$$\int_{\mathbb{R}} |f(x)| \, \mu(dx) \leqslant \sum_{n=1}^{\infty} \frac{1}{n^3} \int_{\left(\frac{1}{n}, 1+\frac{1}{n}\right)} \frac{1}{x} \frac{1}{\left(x - \frac{1}{n}\right)^{1/2}} \lambda(dx)$$

$$\leqslant \sum_{n=1}^{\infty} \frac{n}{n^3} \int_{(0,1)} \frac{1}{x^{1/2}} \lambda(dx)$$
$$= 2 \sum_{n=1}^{\infty} \frac{1}{n^2} < \infty.$$

The estimate

$$\int_{\mathbb{R}} \left| f\left(x - \tfrac{1}{n}\right) \right| \mu(dx) \geqslant \frac{1}{n^3} \int_{\left(\frac{1}{n}, 1+\frac{1}{n}\right)} \frac{1}{x - \frac{1}{n}} \frac{1}{\left(x - \frac{1}{n}\right)^{1/2}} \lambda(dx)$$
$$= \frac{1}{n^3} \int_{(0,1)} \frac{1}{x^{3/2}} \lambda(dx) = \infty$$

shows that $\tau_{1/n} f \notin L^1(\mu)$ for all $n \in \mathbb{N}$. In particular, $\tau_{1/n} f$ does not converge to f in $L^1(\mu)$.

18.34 There is no Bochner integral in spaces which are not locally convex

Let $(X, \mathscr{A}, \mu)$ be a measure space and $(E, \|\cdot\|)$ a Banach space. Throughout, vector-valued quantities will be denoted by Latin letters $f, g, h, \ldots$ and scalars by lowercase Greek letters $\gamma, \kappa, \phi, \ldots$.

The **Bochner integral** is an integral for Banach-space-valued functions $f : X \to E$. Such a function is called **strongly measurable** if there exists a sequence of E-valued simple functions $f_n(x) := \sum_{i=1}^{n} g_i \mathbb{1}_{A_i}(x)$ where $n \in \mathbb{N}$, $g_i \in E$ and $A_i \in \mathscr{A}$ have finite μ-measure, i.e. $\mu(A_i) < \infty$, such that for μ almost all $x \in X$

$$\lim_{n \to \infty} \|f_n(x) - f(x)\|_E = 0.$$

We denote the E-valued simple functions by $\mathcal{E}_E(\mathscr{A})$ and the (equivalence classes of μ-a.e. coinciding) E-valued strongly measurable functions by $L_E^0(\mathscr{A})$.

Pettis' theorem [202, p. 131, V.4 theorem] tells us that a Banach-space-valued function f is strongly measurable if, and only if, the following two conditions are satisfied:

(a) there is a μ-null set $N \in \mathscr{A}$ such that the range $\{f(x);\ x \in X \setminus N\} \subseteq E$ is separable, i.e. it contains a countable dense subset;

(b) f is weakly measurable, i.e. $x \longmapsto \langle g^*, f(x) \rangle$ is a measurable (real or complex) function for each $g^* \in E^*$.

A strongly measurable f is said to be **Bochner integrable** if there exists a sequence of E-valued simple functions $(f_n)_{n\in\mathbb{N}} \subseteq \mathcal{E}_E(\mathscr{A})$ such that

$$\lim_{n\to\infty} \int_X \|f(x) - f_n(x)\|_E \, \mu(dx) = 0.$$

In this case, $\int_X f(x)\,\mu(dx) := \lim_{n\to\infty} \int_X f_n(x)\,\mu(dx)$. One can show that f is Bochner integrable if, and only if, the real-valued function $x \longmapsto \|f(x)\|_E$ is in $L^1(X, \mathscr{A}, \mu)$. A full discussion can be found in [202, pp. 130–136, Chapters V.4, 5] or in [120].

The following theorem due to Albiac and Ansorena [5] shows that the existence of a Bochner integral already entails that E is locally convex. In particular, there is no Bochner integral in quasi-Banach spaces such as $E = L^p$ with $0 < p < 1$ [☞18.9, 18.10]. Recall that a quasi-Banach space is a complete linear space $(E, \|\cdot\|_E)$ where $f \longmapsto \|f\|_E$ is a **quasi-norm**, i.e. absolutely homogeneous, positive definite, and $\|f + g\|_E \leqslant \gamma(\|f\|_E + \|g\|_E)$ for some fixed constant $\gamma \geqslant 1$. The spaces $\mathcal{E}_E(\mathscr{A})$ and $L^0_E(\mathscr{A})$ defined above still make sense in quasi-Banach spaces; we will also need the spaces

$$L^1_E(\mathscr{A}) := \left\{ f \in L^0_E(\mathscr{A});\ \int_X \|f(x)\|_E \, \mu(dx) < \infty \right\},$$

$$L^\infty_E(\mathscr{A}) := \left\{ f \in L^0_E(\mathscr{A});\ \operatorname*{esssup}_{x\in X} \|f(x)\|_E < \infty \right\},$$

and $L^1_{\mathbb{R}}(\mathscr{A})$ refers to the scalar-valued L^1-functions.

Theorem (Albiac, Ansorena) *Let $(X, \mathscr{A}, \mu)$ be a measure space such that μ is a not purely atomic, σ-finite[2] measure and let $(E, \|\cdot\|_E)$ be a quasi-Banach space. If there exists a continuous linear operator $I : L^1_E(\mathscr{A}) \to E$ such that for every E-valued simple function*

$$I\left(\sum_{i=1}^n \mathbb{1}_{A_i} \cdot g_i\right) = \sum_{i=1}^n \mu(A_i) g_i$$

holds, then E is locally convex, hence isomorphic to a Banach space.

Proof Let $g \in L^\infty_E(\mathscr{A})$ and $\phi \in L^1_{\mathbb{R}}(\mathscr{A})$. The function $x \longmapsto \phi(x)g(x)$ satisfies $\|\phi(x)g(x)\|_E \leqslant |\phi(x)| \operatorname{esssup} \|g\|_E$ and is, therefore, an element of $L^1_E(\mathscr{A})$. Consequently, the bilinear operator

$$T : L^\infty_E(\mathscr{A}) \times L^1_{\mathbb{R}}(\mathscr{A}) \to L^1_E(\mathscr{A}), \quad (g, \phi) \longmapsto \phi g$$

[2] σ-finiteness is only needed in order to ensure that all standard definitions of 'atom' coincide [☞(6.1) and Lemma 6A]; if we use the condition (6.2) from Lemma 6A as the definition of an atom, then we can drop σ-finiteness.

is well-defined and continuous.

Since μ is not purely atomic, there is some $A \in \mathscr{A}$ such that $0 < \mu(A) < \infty$ and A does not contain any atom. Denote by $\mathbb{K} = \{g \in L^\infty_E(\mathscr{A});\ \operatorname{esssup}\|g\| \leqslant 1\}$ the closed unit ball in $L^\infty_E(\mathscr{A})$ and set $\phi(x) := \mathbb{1}_A(x)$. Since T is continuous, the set

$$T(\mathbb{K} \times \{\mathbb{1}_A\}) = \{\mathbb{1}_A \cdot g;\ \operatorname{esssup}\|g\| \leqslant 1\}$$

is a bounded set in $L^1_E(\mathscr{A})$. By assumption, the operator I is continuous, and so $I(T(\mathbb{K} \times \{\mathbb{1}_A\}))$ is bounded, i.e.

$$\exists \kappa > 0 \quad \forall g \in L^\infty_E(\mathscr{A}),\ \operatorname{esssup}\|g\| \leqslant 1 \ :\ \|I(\mathbb{1}_A \cdot g)\|_E \leqslant \kappa.$$

Pick any $n \in \mathbb{N}$, elements $g_1, \dots, g_n \in E$ with $\|g_i\| = 1$ and $\theta_1, \dots, \theta_n \in [0,1]$ such that $\sum_{i=1}^n \theta_i = 1$. Since A is not an atom, we know from Sierpiński's theorem on the range of non-atomic measures, cf. Sierpiński [165, p. 241] or Bogachev [19, Corollary 1.12.10], that there is a partition $A_1 \uplus \dots \uplus A_n = A$ of the set A such that $\mu(A_i) = \theta_i \mu(A)$.

The simple function $g(x) = \sum_{i=1}^{n} \mathbb{1}_{A_i}(x) g_i$ satisfies $\|g(x)\|_E \leqslant \max_{1\leqslant i\leqslant n}\|g_i\|_E = 1$, hence $\|I(g)\|_E = \|I(\mathbb{1}_A \cdot g)\|_E \leqslant \kappa$. Thus,

$$\left\| \sum_{i=1}^{n} \mu(A_i) g_i \right\|_E \leqslant \kappa \Rightarrow \left\| \sum_{i=1}^{n} \theta_i g_i \right\|_E \leqslant \frac{\kappa}{\mu(A)},$$

which means that the origin of E has a convex neighbourhood, i.e. the linear space E is locally convex. □

Comment The paper by Gal & Goldstein [63] is a predecessor of the paper by Albiac & Ansorena [5]. These authors show that in quasi-Banach spaces $(E, \|\cdot\|_E)$ with the additional property that $\|\kappa f\|_E = |\kappa|^p \|f\|_E$ (this includes the spaces L^p, $0 < p < 1$) the Riemann integral of a continuous E-valued function $f : [a,b] \to E$ does not satisfy the 'triangle inequality'

$$\left\| \int_a^b f(x)\,dx \right\|_E \nleqslant \int_a^b \|f(x)\|_E\,dx.$$

The reason for this failure is, of course, that a true quasi-norm does not satisfy the triangle inequality with constant 1.

19

Convergence of Measures

Up to now, we have studied fixed measures μ on a measurable space $(X, \mathscr{A})$ and their integrability properties, i.e. the spaces L^p. We will now change our point of view and investigate the **space of measures** over a fixed measurable space $(X, \mathscr{A})$

$$\begin{aligned} \mathcal{M}^+ = \mathcal{M}^+(X) &= \{\mu\,;\ \mu \text{ is a positive measure on } (X, \mathscr{A})\} \\ \text{and}\quad \mathcal{M}_b^+ = \mathcal{M}_b^+(X) &= \left\{\mu \in \mathcal{M}^+\,;\ \mu(X) < \infty\right\}. \end{aligned}$$

Since sums and positive scalar multiples of (finite) measures are again (finite) measures, $\mathcal{M}^+$ and $\mathcal{M}_b^+$ are positive cones, and the space of bounded signed measures $\mathcal{M}_b := \mathcal{M}_b^+ - \mathcal{M}_b^+ = \{\mu - \nu\,;\ \mu, \nu \in \mathcal{M}_b^+\}$ is a vector space. We want to introduce a topology in $\mathcal{M}^+$ and $\mathcal{M}_b^+$. One would expect that **setwise** convergence, i.e.

$$\forall A \in \mathscr{A}\ :\ \mu(A) := \lim_{n\to\infty} \mu_n(A), \tag{19.1}$$

is a natural choice, but this leads immediately to problems. First of all, it is not obvious that the limit (19.1) preserves σ-additivity – it does, this is the famous **Nikodým convergence theorem** [19, Theorem 4.6.3] or [51, Section III.7.4]. On the other hand, (19.1) is so strong that it does not lead to a rich theory. Very natural examples such as $\mu_n = \delta_{x_n}$ with $x_n = 1/n$ do not converge in this sense: Take $A = \{0\}$, 0 being the limit of the points x_n, and observe that

$$\lim_{n\to\infty} \delta_{1/n}(\{0\}) = 0 \neq \delta_0(\{0\}) = 1,$$

[☞ 19.5, 19.6, 19.9] for further examples.

In order to get a sufficiently interesting theory, we restrict ourselves to measurable spaces which are also topological spaces $(X, \mathscr{O})$, and we consider only the Borel sets $\mathscr{A} = \mathscr{B}(X)$ and Borel measures. In order to ensure that compact sets are closed, we assume that X is Hausdorff, i.e. $\mathcal{O}$ is rich enough to separate points.

A sequence of finite measures $(\mu_n)_{n\in\mathbb{N}} \subseteq \mathcal{M}_b^+$ converges **weakly** to a measure μ, if

$$\forall f \in C_b(X) : \lim_{n\to\infty} \int f\,d\mu_n = \int f\,d\mu. \tag{19.2}$$

For weak convergence of a sequence $(\mu_n)_{n\in\mathbb{N}}$, it is not sufficient that the limit $I(f) := \lim_{n\to\infty} \int f\,d\mu_n$ exists and is finite – even in topologically good situations: let X be a normal topological space ([53, Section 1.5], [200, Chapter 5]), then $f \longmapsto I(f)$ defines a positive, bounded linear functional on $(C_b(X), \|\cdot\|_\infty)$, i.e. a positive element $I \in (C_b(X))^*$, but it is known that the dual space $(C_b(X))^*$ consists of regular **finitely additive** set functions, and so $I(f)$ cannot be of the form $\int f\,d\mu$ with a (σ-additive!) measure μ; see [51, IV.6.2, Theorem 2]; for another such example [☞ 19.7].

In order to avoid this kind of pathology, the notion of vague convergence is helpful. Since this is based on compact sets, we have to ensure that the space $(X, \mathcal{O})$ has sufficiently many compact sets. Typically, X is assumed to be locally compact[1] and all measures should be locally finite, hence finite on compact sets.

A sequence of locally finite measures $(\mu_n)_{n\in\mathbb{N}} \subseteq \mathcal{M}^+(X)$ converges **vaguely** to a locally finite measure μ, if

$$\forall f \in C_c(X) : \lim_{n\to\infty} \int f\,d\mu_n = \int f\,d\mu. \tag{19.3}$$

On a locally compact space, **Riesz's representation theorem** [☞ 1.22] shows that every linear functional $I : C_c(X) \to \mathbb{R}$ which is positive ($I(f) \geqslant 0$ for all $f \geqslant 0$) is of the form $I(f) = \int f\,d\mu$ for some unique, (inner compact and outer) regular $\mu \in \mathcal{M}^+$. Thus, it is enough to require in (19.3) that the limit $I(f) := \lim_{n\to\infty} \int f\,d\mu_n$ exists and is finite for all $f \in C_c(X)$; the representing measure μ will automatically be regular [☞ p. 185].

A further consequence of Riesz's theorem is that the regular (signed) measures $\mathcal{M}_{\text{reg}}(X) = \mathcal{M}_{\text{reg}}^+(X) - \mathcal{M}_{\text{reg}}^+(X)$ are the topological dual of $C_c(X)$; a functional analyst would, therefore, call vague convergence the **weak*** **convergence** in $\sigma(\mathcal{M}_{\text{reg}}(X), C_c(X))$.

To keep things simple, we assume from now on that X is a metric space, and μ, μ_n are locally finite Borel measures defined on $\mathcal{B}(X)$. The proof of the follow-

[1] For a normed linear space X this means, effectively, that X is finite-dimensional; see [97, Section II.3, Corollary 3].

ing theorem can be found in [MIMS, p. 250]; it is stated there for regular Borel measures, but it holds, without alterations, for locally finite Borel measures.

Theorem 19A (portmanteau theorem; [MIMS, p. 250]) *Let (X, d) be a locally compact metric space and $\mu, \mu_n \in \mathcal{M}^+$ locally finite Borel measures on $\mathscr{B}(X)$. The following assertions are equivalent:*

(a) $\mu_n \to \mu$ *vaguely;*
(b) $\limsup_{n\to\infty} \mu_n(K) \leqslant \mu(K)$ *for all compact sets* $K \subseteq X$ *and* $\liminf_{n\to\infty} \mu_n(U) \geqslant \mu(U)$ *for all relatively compact open sets* $U \subseteq X$*;*
(c) $\lim_{n\to\infty} \mu_n(B) = \mu(B)$ *for all relatively compact Borel sets* $B \subseteq X$ *such that* $\mu(\partial B) = 0$ *where* $\partial B := \overline{B} \setminus B^\circ$ *is the topological boundary.*

An immediate consequence is the following characterization of weak convergence in terms of vague convergence.

Corollary 19B *Let (X, d) be a locally compact metric space and $\mu, \mu_n \in \mathcal{M}_b^+(X)$ (inner compact) regular Borel measures on $\mathscr{B}(X)$. One has*

$$\mu_n \xrightarrow{\text{weakly}} \mu \iff \mu_n \xrightarrow{\text{vaguely}} \mu \quad \text{and} \quad \lim_{n\to\infty} \mu_n(X) = \mu(X).$$

Without the mass preservation condition $\mu_n(X) \to \mu(X)$, the corollary is false [☞ 19.1]. It is also not enough to require that $\lim_{n\to\infty} \mu_n(X)$ exists [☞ 19.2].

A slightly different version of the portmanteau theorem holds for weak convergence.

Theorem 19C (portmanteau theorem; [MINT, Satz 25.3]) *Let (X, d) be any metric space and $\mu, \mu_n \in \mathcal{M}_b^+(X)$ bounded Borel measures on $\mathscr{B}(X)$. The following assertions are equivalent:*

(a) $\mu_n \to \mu$ *weakly;*
(b) $\lim_{n\to\infty} \int f \, d\mu_n = \int f \, d\mu$ *for all uniformly continuous* $f \in C_b(X)$*;*
(c) $\limsup_{n\to\infty} \mu_n(F) \leqslant \mu(F)$ *for all closed sets* $F \subseteq X$ *and* $\lim_{n\to\infty} \mu_n(X) = \mu(X)$*;*
(d) $\liminf_{n\to\infty} \mu_n(U) \geqslant \mu(U)$ *for all open sets* $U \subseteq X$ *and* $\lim_{n\to\infty} \mu_n(X) = \mu(X)$*;*
(e) $\lim_{n\to\infty} \mu_n(B) = \mu(B)$ *for all Borel sets* $B \subseteq X$ *such that* $\mu(\partial B) = 0$ *where* $\partial B := \overline{B} \setminus B^\circ$ *is the topological boundary.*

Remark 19D The bottleneck in the proofs of the above results is the problem to find a sequence of continuous functions $(f_k)_{k\in\mathbb{N}}$ which are sandwiched between a closed (or compact) set F and an open set $U \supseteq F$, such that $f_k \downarrow \mathbb{1}_F$. This is straightforward in metric spaces, see e.g. [MIMS, p. 417]; in general topological spaces one needs further separation properties (normal or completely regular spaces) in order to prove Urysohn's lemma – see Engelking [53, Theorem 1.5.11 and Corollary 3.3.3] or Willard [200, Theorem 15.3, Exercise 20C]. This still does not guarantee the existence of a sequence $f_k \downarrow \mathbb{1}_F$, but only of a downward filtering family $f_i \downarrow \mathbb{1}_F$, and further assumptions such as (inner compact) regularity of the measures μ_n, μ or τ-additive (⌊☞comment to 12.9⌋ implying the validity of monotone convergence for nets, cf. [20, Section 7.2] or [43, III.49, p. 108]) are needed. Let us briefly mention what can be found in the literature.

- Theorem 19A holds for regular (hence, τ-additive, cf. [20, Proposition 7.2.2]) measures on a locally compact Hausdorff space X; see [13, Theorems 30.2, 30.8, 30.10].
- Theorem 19C (without part (b)) holds for τ-smooth bounded measures on a completely regular topological space X; see [192, Theorem 3.5] or [20, Corollary 8.2.10] if X is perfectly normal. These results are also known as **Alexandroff's theorem**.
- A generalization of the concept of vague convergence to general Hausdorff spaces – essentially taking the property $\liminf_{n\to\infty} \mu_n(U) \geqslant \mu(U)$ for all open sets $U \subseteq X$ as the definition – is discussed in Topsoe [186, Sections 8–11], Schwartz [159, Appendix 1] and Gänssler & Stute [64, Section 8.4].

Since vague convergence is in many situations the weak* convergence in the space of signed regular measures $\mathcal{M}_{\text{reg}}$, one might ask what the **strong** or **norm convergence** might be. This role is played by the **total variation norm**

$$\|\mu\|_{TV} := \mu^+(X) + \mu^-(X) = |\mu|(X), \quad \mu \in \mathcal{M}_b(X), \tag{19.4}$$

where $\mu^\pm$ are the summands appearing in the Hahn–Jordan decomposition ⌊☞Theorem 1.47⌋. One can show that

$$\|\mu\|_{TV} = \sup\left\{\sum_{n=1}^{\infty} |\mu(A_n)|\,;\ X = \bigcupdot_{n=1}^{\infty} A_n,\ A_n \in \mathscr{A}\right\}$$

and that the space $\mathcal{M}_b(X)$ is complete with respect to this norm [36, Proposition 4.1.7]. Moreover, the total variation norm $\|\mu\|_{TV}$ is comparable to the norm $\|\mu\| := \sup\{|\mu(A)|\,;\ A \in \mathscr{A}\}$, cf. [19, p. 177].

19.1 Classical counterexamples related to vague and weak convergence

Consider $\mathbb{R}$ with the Euclidean metric. The space $C_\infty(\mathbb{R})$ of continuous functions vanishing at infinity is defined as the closure of $C_c(\mathbb{R})$ with respect to the uniform norm.

Equivalently, $u \in C_\infty(\mathbb{R})$ if u is continuous and $\lim_{|x|\to\infty} u(x) = 0$. A sequence $(\mu_n)_{n\in\mathbb{N}}$ of Borel measures is C_∞**-convergent** if

$$\forall u \in C_\infty(\mathbb{R}) \; : \; \lim_{n\to\infty} \int u(x)\,\mu_n(dx) = \int u(x)\,\mu(dx) \in \mathbb{R}$$

for a measure μ. Clearly, weak convergence implies C_∞-convergence which, in turn, implies vague convergence. The following examples show that the converse is not true.

(a) The sequence $\mu_n(dx) = \mathbb{1}_{[-n,n]}(x)\,dx$ converges vaguely to Lebesgue measure since for each $u \in C_c(\mathbb{R})$ with $\operatorname{supp} u \subseteq [-N,N]$ we have

$$\lim_{n\to\infty} \int u(x)\,\mu_n(dx) = \lim_{n\to\infty} \int_{-n}^{n} u(x)\,dx = \int_{-N}^{N} u(x)\,dx = \int u(x)\,dx.$$

C_∞-convergence – hence weak convergence – fails because we can find a $u \in C_\infty(\mathbb{R})$, $u \geqslant 0$, such that $\int u(x)\,dx = \infty$ while $\int u(x)\,\mu_n(dx) < \infty$: just take $u(x) = \min\{1, |x|^{-1}\}$.

(b) The sequence $\mu_n = \delta_n$ is C_∞-convergent, but does not converge weakly to $\mu = 0$. In fact, for $u \in C_\infty(\mathbb{R})$

$$\int u(x)\,\delta_n(dx) = u(n) \xrightarrow[n\to\infty]{} 0,$$

while $u(x) := \cos(\pi x)$ is continuous and bounded, but

$$\int \cos(\pi x)\,\delta_n(dx) = \cos(\pi n)$$

does not converge as $n \to \infty$.

(c) The sequence $\mu_n = \delta_{\frac{1}{n}}$ converges weakly to $\mu = \delta_0$ since

$$\int u(x)\,\mu_n(dx) = u(\tfrac{1}{n}) \xrightarrow[n\to\infty]{} u(0) = \int u(x)\,\delta_0(dx),$$

while convergence in total variation norm [☞ (19.4)] fails:

$$\|\mu_n - \delta_0\|_{TV} = 2.$$

Comment Let X be a topological space and μ, μ_n, $n \in \mathbb{N}$, Borel measures on $(X, \mathscr{B}(X))$. If $\sup_{n\in\mathbb{N}} \mu_n(X) < \infty$, then vague and C_∞-convergence coincide since $C_c(X)$ is – by definition – dense in $C_\infty(X)$. Vague and weak convergence coincide, if the sequence μ_n preserves mass, i.e. if $\lim_{n\to\infty} \mu_n(X) = \mu(X)$ [☞ Corollary 19B].

19.2 Vague convergence does not preserve mass

The sequence of probability measures $\mu_n := \delta_n$ on $(\mathbb{R}, \mathscr{B}(\mathbb{R}))$ converges vaguely to $\mu = 0$ and mass is not preserved:

$$\lim_{n\to\infty} \mu_n(\mathbb{R}) = 1 \neq 0 = \mu(\mathbb{R}).$$

Comment If $\mu_n \to \mu$ vaguely, then mass is preserved if, and only if, $\mu_n \to \mu$ weakly [☞ Corollary 19B].

19.3 Vague convergence of positive measures $\mu_n \to \mu$ does not imply $|\mu_n - \mu| \to 0$

Let μ be Lebesgue measure on $[0,1]$ and define $g_n(x) := \sin(n\pi x)$. The measures $\mu_n := (1-g_n)\lambda$ are positive measures and we get for any continuous function $u \in C[0,1] = C_b[0,1] = C_c[0,1]$,

$$\left| \int_0^1 u(x)\,\mu_n(dx) - \int_0^1 u(x)\,\mu(dx) \right| = \left| \int_0^1 u(x)\sin(n\pi x)\,dx \right| \xrightarrow[n\to\infty]{} 0$$

by the Riemann–Lebesgue lemma [MIMS, p. 222]. For $u \equiv 1$ we have

$$\int_0^1 u(x)\,|\mu - \mu_n|(dx) = \int_0^1 |\sin(n\pi x)|\,dx = \frac{1}{n\pi}\int_0^{n\pi} |\sin y|\,dy = \frac{1}{\pi}.$$

Comment [☞ 19.4] for a further example.

19.4 Vague convergence $\mu_n \to 0$ does not entail vague convergence $|\mu_n| \to 0$

Define on $([0,1], \mathscr{B}[0,1])$ a sequence of signed measures

$$\mu_n := \delta_0 - \delta_{\frac{1}{n}} \quad \text{and} \quad |\mu_n| = \delta_0 + \delta_{\frac{1}{n}}.$$

Since $C_c[0,1] = C_b[0,1] = C[0,1]$, we get for any $u \in C[0,1]$,

$$\int u\,d\mu_n = u(0) - u(1/n) \xrightarrow[n\to\infty]{} 0,$$

whereas

$$\int u\,d|\mu_n| = u(0) + u(1/n) \xrightarrow[n\to\infty]{} 2u(0).$$

19.5 Vague convergence does not imply $\mu_n(B) \to \mu(B)$ for all Borel sets

Let (X, d) be a locally compact metric space and μ_n, μ finite measures. The portmanteau theorem [☞Theorem 19A] states that $\mu_n \to \mu$ vaguely if, and only if, $\mu_n(B) \to \mu(B)$ for all relatively compact Borel sets B with $\mu(\partial B) = 0$. The following example shows that we cannot drop the condition $\mu(\partial B) = 0$.

Consider $(\mathbb{R}, \mathscr{B}(\mathbb{R}))$ with Lebesgue measure λ and define a sequence of measures $\mu_n = f_n \lambda$ with density

$$f_n(x) := \sqrt{\frac{n}{2\pi}} \exp\left(-n\frac{x^2}{2}\right), \quad x \in \mathbb{R},\ n \in \mathbb{N}.$$

Then $\mu_n \to \delta_0$ weakly [☞19.13], hence $\mu_n \to \delta_0$ vaguely. Yet $\mu_n(\{0\}) = 0$ does not converge to $\delta_0(\{0\}) = 1$.

Comment [☞12.8] for another example.

19.6 A sequence of absolutely continuous measures which converges weakly to λ on $[0,1]$ but $\mu_n(B) \to \lambda(B)$ fails for some Borel set $B \subseteq [0,1]$

[☞12.8]

19.7 A sequence of measures μ_n such that $\lim_{n\to\infty} \int f\,d\mu_n$ exists, but is not of the form $\int f\,d\mu$

Let $X = \mathbb{N}$, consider the discrete topology, i.e. $\mathscr{O} = \mathscr{P}(\mathbb{N}) = \mathscr{B}(X)$, and let $\mu_n := \delta_n$ be a sequence of Dirac measures. For all functions from the space

$$D := \left\{ f : \mathbb{N} \to \mathbb{R};\ \lim_{n\to\infty} f(n) \text{ exists and is finite} \right\},$$

the limit

$$I(f) := \lim_{n\to\infty} \int f\,d\delta_n, \quad f \in D,$$

exists and defines a continuous linear functional: $|I(f)| \leqslant \|f\|_\infty$ if $f \in D$. Since $I(\mathbb{1}_A) = 0$ for every finite $A \subseteq \mathbb{N}$, it is not possible that $I(f) = \int f\,d\mu$ for $f \in D$: we have $\mu(\{n\}) = I(\mathbb{1}_{\{n\}}) = 0$ for all $n \in \mathbb{N}$ and so, by σ-additivity, $\mu = 0$.

Note that I can be extended to a linear functional on $C_b(\mathbb{N})$: use $D \subseteq C_b(\mathbb{N})$ along with the Hahn–Banach theorem; [☞5.4] for a related construction.

19.8 Weakly convergent sequences need not be tight

Let (X, d) be a metric space with Borel σ-algebra $\mathscr{B}(X)$. A family $\mathcal{M}$ of measures on $(X, \mathscr{B}(X))$ is **(uniformly) tight** if there exists for every $\epsilon > 0$ a compact set $K = K_\epsilon \subseteq X$ such that $\sup_{\mu \in \mathcal{M}} \mu(K^c) \leqslant \epsilon$.

For complete separable metric spaces (X, d), **Prohorov's theorem** states that a family $\mathcal{M}_1$ of probability measures is tight if, and only if, $\mathcal{M}_1$ is weakly sequentially relatively compact, i.e. every sequence $(\mu_n)_{n\in\mathbb{N}} \subseteq \mathcal{M}_1$ has a weakly convergent subsequence, cf. [20, Theorem 8.6.2] or [15, Theorem 5.2]. This equivalence breaks down if (X, d) is not complete: it is possible to construct a metric space (X, d) which is not complete and a probability measure μ on $(X, \mathscr{B}(X))$ such that μ is not tight [☞ 5.26]. If we set $\mathcal{M}_1 := \{\mu\}$, then $\mathcal{M}_1$ is relatively compact but not tight. Equivalently, the sequence $\mu_n := \mu$, $n \in \mathbb{N}$, is weakly convergent but not tight.

Comment The other implication of Prohorov's theorem remains valid for general metric spaces, i.e. if a family $\mathcal{M}_1$ of probability measures is tight, then $\mathcal{M}_1$ is relatively compact, cf. [15, Theorem 5.1]. There are versions of Prohorov's theorem for not necessarily finite measures and for signed measures; see [20, Section 8.6].

19.9 Signed measures μ_n such that $\int f \, d\mu_n \to \int f \, d\mu$ for all $f \in C(\mathbb{R})$ but $\mu_n(B) \to \mu(B)$ fails for sets with $\mu(\partial B) = 0$

If $\mu, \mu_n, n \in \mathbb{N}$, are (positive) measures on $(\mathbb{R}, \mathscr{B}(\mathbb{R}))$ such that $\mu_n \to \mu$ weakly, i.e. $\int f \, d\mu_n \to \int f \, d\mu$ for all $f \in C_b(\mathbb{R})$, then

$$\lim_{n\to\infty} \mu_n(B) = \mu(B)$$

for every $B \in \mathscr{B}(\mathbb{R})$ with $\mu(\partial B) = 0$. This implication of the portmanteau theorem breaks down for signed measures.

Consider the measures $\mu_n := \delta_0 - \delta_{1/n}$ and $\mu = 0$ on $(\mathbb{R}, \mathscr{B}(\mathbb{R}))$. For every $f \in C(\mathbb{R})$ we have

$$\int f \, d\mu_n = f(0) - f\left(\frac{1}{n}\right) \xrightarrow[n\to\infty]{} 0 = \int f \, d\mu.$$

If we take $B = \{0\}$, then trivially $\mu(\partial B) = 0$ but $\mu_n(B) = 1$ does not converge to $\mu(B) = 0$.

19.10 Signed measures μ_n such that $\int f\,d\mu_n \to \int f\,d\mu$ for all bounded uniformly continuous functions f but μ_n does not converge weakly to μ

Let (X, d) be a metric space with Borel σ-algebra $\mathscr{B}(X)$. If μ_n,μ are (positive) bounded measures on $(X, \mathscr{B}(X))$, then $\mu_n \to \mu$ weakly if, and only if,

$$\lim_{n\to\infty} \int_X f\,d\mu_n = \int_X f\,d\mu$$

for all bounded uniformly continuous functions $f : X \to \mathbb{R}$ [☞ Theorem 19C]. This equivalence breaks down for signed measures.

Consider $\mathbb{R}$ with the Euclidean metric and the sequence of signed measures

$$\mu_n := \delta_n - \delta_{n+n^{-1}}, \quad n \in \mathbb{N}.$$

Since

$$\int_{\mathbb{R}} f\,d\mu_n = f(n) - f\left(n + \frac{1}{n}\right), \quad n \in \mathbb{N},$$

it follows that $\int_{\mathbb{R}} f\,d\mu_n \to 0$ for all uniformly continuous functions f. However, μ_n does not converge weakly to $\mu \equiv 0$: pick a bounded continuous function f such that $f(n) = 1$ and $f(n + n^{-1}) = 0$ for all $n \in \mathbb{N}$, then

$$\lim_{n\to\infty} \int_{\mathbb{R}} f\,d\mu_n = 1 \neq 0 = \int_{\mathbb{R}} f\,d\mu.$$

19.11 A sequence of measures which does not converge weakly but whose Fourier transforms converge pointwise

Lévy's continuity theorem states that a sequence $(\mu_n)_{n\in\mathbb{N}}$ of finite measures converges weakly if, and only if, $\widehat{\mu}_n \to \phi$ for a function ϕ which is continuous at $\xi = 0$; see e.g. [MIMS, pp. 255–256, Problem 21.3]. The following example shows that the equivalence breaks down if the limit function ϕ is not continuous at $\xi = 0$.

Consider on the measurable space $(\mathbb{R}, \mathscr{B}(\mathbb{R}))$ the measure $\mu = \frac{1}{2}\delta_{-1} + \frac{1}{2}\delta_1$. Denote by μ_n the $2n$-fold convolution of μ [☞ Definition 1.39], i.e.

$$\mu_n := \underbrace{\mu * \cdots * \mu}_{2n \text{ times}}, \quad n \in \mathbb{N}.$$

Since the Fourier transform $\widehat{\mu}$ of μ is given by

$$\widehat{\mu}(\xi) = \frac{1}{2\pi} \int_{\mathbb{R}} e^{-ix\xi}\,\mu_n(dx) = \frac{1}{2\pi} \cos(\xi), \quad \xi \in \mathbb{R},$$

it follows from the convolution theorem [MIMS, p. 221] that

$$\widehat{\mu}_n(\xi) = (2\pi)^{2n-1}\widehat{\mu}(\xi)^{2n} = \frac{1}{2\pi}\cos(\xi)^{2n}, \quad \xi \in \mathbb{R}.$$

Letting $n \to \infty$ we find that

$$\lim_{n\to\infty} \widehat{\mu}_n(\xi) = \phi(\xi) := \begin{cases} \frac{1}{2\pi}, & \text{if } \xi = k\pi,\ k \in \mathbb{Z}, \\ 0, & \text{otherwise,} \end{cases}$$

for all $\xi \in \mathbb{R}$. The sequence $(\mu_n)_{n\in\mathbb{N}}$ does not converge weakly. If it were weakly convergent, say, $\mu_n \to \mu$, then $\widehat{\mu}_n \to \widehat{\mu}$ pointwise, and so $\widehat{\mu} = \phi$. But this is impossible since $\widehat{\mu}$ is continuous at $\xi = 0$ and ϕ is discontinuous at $\xi = 0$.

19.12 Lévy's continuity theorem fails for nets

Lévy's continuity theorem says that a sequence $(\mu_n)_{n\in\mathbb{N}}$ of probability measures converges vaguely[2] to a probability measure μ if, and only if, the inverse Fourier transforms $\check{\mu}_n(\xi) := \int e^{ix\xi}\,\mu_n(dx)$ converge pointwise to a limit $\phi(\xi)$ which is continuous at the point $\xi = 0$. In this case $\phi = \check{\mu}$ is the inverse Fourier transform of the bounded measure μ.

The following example is from Berg & Forst [14, Example 3.15]. Denote by $\mathbb{S}^{\mathbb{R}}$, $\mathbb{S} = \{e^{i\phi}\,;\ \phi \in [0, 2\pi)\}$, the infinite-dimensional torus. Since $\mathbb{S}$ is a compact group, $\mathbb{S}^{\mathbb{R}}$ is again compact by Tychonoff's theorem, and the map $J : \mathbb{R} \to \mathbb{S}^{\mathbb{R}}$, $J(x) := (e^{-i\xi x})_{\xi\in\mathbb{R}}$, is a continuous and injective homomorphism. Notice that J is not an isomorphism $\mathbb{R} \to J(\mathbb{R})$ since this would mean that $J(\mathbb{R})$ is a locally compact subgroup of the compact group $\mathbb{S}^{\mathbb{R}}$, hence itself compact; see e.g. [79, p. 35].

This means that there is a net $(y_\alpha)_\alpha$ in $J(\mathbb{R})$ such that $\lim_\alpha y_\alpha = J(0)$, but the pre-image $(x_\alpha)_\alpha \subseteq \mathbb{R}$, $x_\alpha := J^{-1}(y_\alpha)$ does not converge to 0.

Now consider the probability measures $(\delta_{x_\alpha})_\alpha$. By definition, $\delta_{x_\alpha} \to \delta_0$ vaguely if, and only if, $x_\alpha \to 0$, and so $(\delta_{x_\alpha})_\alpha$ does not converge to δ_0. On the other hand,

$$\check{\delta}_{x_\alpha}(\xi) = e^{i\xi x_\alpha} = J(x_\alpha) = J(J^{-1})(y_\alpha) = y_\alpha \to J(0) = 1 = \check{\delta}_0(\xi)$$

for all $\xi \in \mathbb{R}$.

[2] Since we are dealing with probability measures, vague convergence is equivalent to weak convergence [☞Corollary 19B].

19.13 A sequence of non-atomic measures converging weakly to a purely atomic measure

Denote by λ Lebesgue measure on $(\mathbb{R}, \mathscr{B}(\mathbb{R}))$, and let p_t be the heat kernel

$$p_t(x) := \frac{1}{\sqrt{2\pi t}} \exp\left(-\frac{x^2}{2t}\right), \quad x \in \mathbb{R},\ t > 0.$$

Define a sequence of measures μ_n, $n \in \mathbb{N}$, on $(\mathbb{R}, \mathscr{B}(\mathbb{R}))$ by

$$\mu_n(B) := \int_B p_{1/n}(x)\,\lambda(dx), \quad B \in \mathscr{B}(\mathbb{R}),\ n \in \mathbb{N}.$$

A change of variables shows that

$$\int_{\mathbb{R}} f(x)\,\mu_n(dx) = \int_{\mathbb{R}} f(x) p_{1/n}(x)\,\lambda(dx) = \int_{\mathbb{R}} f(x/\sqrt{n}) p_1(x)\,\lambda(dx),$$

which implies that

$$\lim_{n\to\infty} \int_{\mathbb{R}} f(x)\,\mu_n(dx) = f(0) = \int_{\mathbb{R}} f(x)\,\delta_0(dx)$$

for any $f \in C_b(\mathbb{R})$, i.e. $\mu_n \to \delta_0$ weakly. This means that absolute continuity of μ_n (with respect to Lebesgue measure) is not preserved under weak limits.

19.14 A sequence of purely atomic measures converging weakly to a non-atomic measure

Define a sequence $(\mu_n)_{n\in\mathbb{N}}$ of (probability) measures on $([0,1], \mathscr{B}([0,1]))$ by

$$\mu_n := \frac{1}{n} \sum_{k=0}^{n-1} \delta_{k/n}.$$

If $f : [0,1] \to \mathbb{R}$ is a continuous and bounded function, then

$$\int_{[0,1]} f(x)\,\mu_n(dx) = \frac{1}{n} \sum_{k=0}^{n-1} f\left(\frac{k}{n}\right).$$

Since f is Riemann integrable and since the expression on the right-hand side is a Riemann sum, we find that

$$\lim_{n\to\infty} \int_{[0,1]} f(x)\,\mu_n(dx) = \int_{[0,1]} f(x)\,dx,$$

i.e. μ_n converges weakly to Lebesgue measure $\lambda|_{[0,1]}$.

Comment In fact, the set of purely atomic probability measures on a locally compact metric space X is dense, w.r.t. weak convergence of measures, in the set of all probability measures; see [13, Corollary 30.5] and note that weak convergence $\mu_n \to \mu$ is equivalent to vague convergence $\mu_n \to \mu$ for probability measures μ_n, μ [☞ Corollary 19B].

19.15 A net of Dirac measures converging weakly to a non-Dirac measure

Consider the ordinal space $\Omega_0 = [0, \omega_1)$ equipped with the canonical order topology [☞ Section 2.4]. Take the co-countable σ-algebra $\mathscr{A}$ [☞ Example 4A] and the measure [☞ Example 5A]

$$\mu(A) := \begin{cases} 0, & \text{if } A \text{ is countable,} \\ 1, & \text{if } A^c \text{ is countable,} \end{cases} \qquad A \in \mathscr{A}.$$

The net δ_α, $\alpha < \omega_1$, converges weakly to μ. Indeed: if $f : \Omega_0 \to \mathbb{R}$ is a continuous function, then f is eventually constant, i.e. there are $\alpha_f < \omega_1$ and $c \in \mathbb{R}$ such that $f(\alpha) = c$ for all $\alpha > \alpha_f$ [☞ Section 2.4]. Since $[0, \alpha_f]$ is countable, this gives, in particular, that $f : (\Omega_0, \mathscr{A}) \to (\mathbb{R}, \mathscr{B}(\mathbb{R}))$ is measurable. Moreover,

$$\forall \alpha > \alpha_f \ : \int_{\Omega_0} f \, d\delta_\alpha = f(\alpha) = c.$$

Invoking, once more, that $[0, \alpha_f]$ is countable, we find from the definition of μ that $\mu([0, \alpha_f]) = 0$ and $\mu(\Omega_0 \setminus [0, \alpha_f]) = 1$. Hence,

$$\int_{\Omega_0} f \, d\mu = \int_{\Omega_0 \setminus [0, \alpha_f]} f \, d\mu = c\mu(\Omega_0 \setminus [0, \alpha_f]) = c.$$

Consequently, the net $\int f \, d\delta_\alpha$ converges to $\int f \, d\mu$ for any continuous function f, and so δ_α converges weakly to μ.

Comment The example is taken from Bogachev [20, Chapter 8].

19.16 $f_n \mu \to f \mu$ weakly does not imply $f_n \to f$ in probability

Consider Lebesgue measure λ on $((0, 1), \mathscr{B}(0, 1))$ and

$$\mu_n(B) := \int_B f_n(x) \, \lambda(dx), \quad B \in \mathscr{B}(0, 1), \ n \in \mathbb{N},$$

with density function

$$f_n(x) := 1 + \sin(2\pi n x), \quad x \in (0, 1), \ n \in \mathbb{N}.$$

Since f_n is positive and

$$\mu_n(0,1) = 1 + \int_{(0,1)} \sin(2\pi n x)\,\lambda(dx) = 1,$$

each μ_n, $n \in \mathbb{N}$, is a probability measure. Denote by $\widehat{\mu}_n$ its Fourier transform,

$$\widehat{\mu}_n(\xi) = \frac{1}{2\pi}\int_{(0,1)} e^{-ix\xi} f_n(x)\,\lambda(dx), \quad \xi \in \mathbb{R}.$$

Since $x \longmapsto e^{-ix\xi}\mathbb{1}_{(0,1)}(x) \in L^1(\lambda)$ for each fixed $\xi \in \mathbb{R}$, it follows from the Riemann–Lebesgue lemma [MIMS, p. 222] that

$$\lim_{n\to\infty} \frac{1}{2\pi}\int_{(0,1)} e^{-ix\xi} \sin(2\pi n x)\,\lambda(dx) = 0, \quad \xi \in \mathbb{R}.$$

Thus,

$$\lim_{n\to\infty} \widehat{\mu}_n(\xi) = \frac{1}{2\pi}\int_0^1 e^{-ix\xi}\,\lambda(dx) = \widehat{\mu}(\xi), \quad \xi \in \mathbb{R},$$

where $\mu := \mathbb{1}_{(0,1)}\,\lambda$ is Lebesgue measure restricted to $(0,1)$. By Lévy's continuity theorem, pointwise convergence $\widehat{\mu}_n \to \widehat{\mu}$ implies weak convergence $\mu_n \to \mu$, i.e. $f_n\lambda \to \mathbb{1}_{(0,1)}\,\lambda$ weakly. Yet f_n does not converge in probability to $\mathbb{1}_{(0,1)}$. Indeed, $f_n \to \mathbb{1}_{(0,1)}$ in probability would imply the existence of a subsequence converging almost everywhere to $\mathbb{1}_{(0,1)}$, but such a subsequence cannot exist [☞ 11.9].

Comment As a consequence of the above example, we obtain that weak convergence $f_n\mu \to f\mu$ does not imply any of the following modes of convergence: $f_n \to f$ in measure, $f_n \to f$ in probability, $f_n \to f$ almost everywhere, $f_n \to f$ in L^p. For an example showing that $f_n\mu \to f\mu$ weakly does not entail $f_n \xrightharpoonup[n\to\infty]{} f$ weakly in L^p, $1 \leqslant p < \infty$ [☞ 19.17].

19.17 $f_n\mu \to f\mu$ weakly does not imply $f_n \rightharpoonup f$ weakly in $L^1(\mu)$

Let X be a metric space and μ a Borel measure on $(X, \mathscr{B}(X))$. If $(f_n)_{n\in\mathbb{N}} \subseteq L^1(\mu)$ converges weakly in L^1 to some f, i.e. $f_n \xrightharpoonup[n\to\infty]{} f$ [☞ p. 331], then it follows from $\mathbb{1}_A \in L^\infty$ that

$$\forall A \in \mathscr{A} \;:\; \int_X \mathbb{1}_A f_n\,d\mu \xrightarrow[n\to\infty]{} \int_X \mathbb{1}_A f\,d\mu;$$

since we can take any measurable set A whose boundary has measure zero, $(f\mu)(\partial A)$, the portmanteau theorem [☞ 19C] shows that the measures $f_n\mu$

converge weakly to $f\mu$. The converse is, in general, false even if μ is a finite measure and $(f_n)_{n\in\mathbb{N}}$ is bounded in $L^1(\mu)$.

Define on $((-1,1),\mathscr{B}(-1,1))$ a finite measure μ by $\mu = \lambda + \delta_0$ and set

$$f_n(x) := \frac{n}{2}\mathbb{1}_{\left(-\frac{1}{n},\frac{1}{n}\right)\setminus\{0\}}(x), \quad x \in (-1,1),\ n \in \mathbb{N}.$$

If $g \in C_b(-1,1)$, then

$$\left|\int_{(-1,1)} g(x)f_n(x)\,\mu(dx) - g(0)\right| = \frac{n}{2}\int_{(-1/n,1/n)} |g(x)-g(0)|\,\lambda(dx)$$
$$\leqslant \sup_{|x|\leqslant 1/n} |g(x)-g(0)| \xrightarrow[n\to\infty]{} 0,$$

and this shows that $f_n\mu \to \delta_0 = 1_{\{0\}}\mu$ weakly. However, f_n does not converge to $1_{\{0\}}$ weakly in $L^1(\mu)$ since

$$\lim_{n\to\infty}\int_{(-1,1)} g(x)f_n(x)\,\mu(dx) = \int_{(-1,1)} g(x)\mathbb{1}_{\{0\}}(x)\,\mu(dx)$$

fails to hold for $g = \mathbb{1}_{\{0\}} \in L^\infty(\mu)$.

Comment In fact, the sequence $(f_n)_{n\in\mathbb{N}}$ does not converge weakly in $L^p(\mu)$ to $f = 1_{\{0\}}$ for any $1 \leqslant p < \infty$. For $p = 1$ we have shown this above, for $p > 1$ note that $(f_n)_{n\in\mathbb{N}} \subseteq L^p(\mu)$ is not bounded in $L^p(\mu)$, i.e.

$$\sup_{n\in\mathbb{N}} \|f_n\|_{L^p(\mu)} = \infty,$$

and therefore it cannot be weakly convergent in $L^p(\mu)$ [☞ theorem in 19.18]. Consequently, weak convergence of the measures $f_n\mu \to f\mu$ does not imply weak convergence in L^p of the densities $f_n \rightharpoonup f$ for any $p \in [1,\infty)$.

19.18 $f_n \rightharpoonup f$ weakly in $L^p(\mu)$ for $p > 1$ does not imply $f_n\mu \to f\mu$ weakly

Consider Lebesgue measure λ on $(\mathbb{R},\mathscr{B}(\mathbb{R}))$ and set

$$f_n(x) := \mathbb{1}_{[n,n+1]}(x), \quad x \in \mathbb{R}.$$

We have $\|f_n\|_{L^p} = 1$ for all $n \in \mathbb{N}$ and $p \geqslant 1$, and therefore $(f_n)_{n\in\mathbb{N}}$ is bounded in L^p. If $q \in (1,\infty)$ is the conjugate index of $p \in (1,\infty)$ and $g \in L^q(\lambda)$, then by Jensen's inequality [☞ Theorem 1.23]

$$\left|\int \mathbb{1}_{[n,n+1]}(x)g(x)\,\lambda(dx)\right|^q \leqslant \int \mathbb{1}_{[n,n+1]}(x)|g(x)|^q\,\lambda(dx).$$

Since $f_n(x) = \mathbb{1}_{[n,n+1]}(x) \to 0$ for all $x \in \mathbb{R}$, it follows from the dominated convergence theorem that $f_n \rightharpoonup 0$ weakly in L^p for $1 < p < \infty$. On the other hand, $\mu_n := f_n \lambda$ does not converge weakly to $\mu := 0$ since

$$\lim_{n\to\infty} \mu_n(\mathbb{R}) = 1 \neq 0 = \mu(\mathbb{R}).$$

Comment The sequence $\mu_n = f_n \lambda$ does not converge weakly to 0, but it converges vaguely to 0. It is, therefore, natural to ask whether for sequences $(f_n)_{n\in\mathbb{N}}$ of positive functions the implication

$$f_n \rightharpoonup f \text{ in } L^p(\mu) \text{ for some } 1 < p < \infty \Rightarrow f_n \mu \to f \mu \text{ vaguely}$$

holds. The answer is affirmative for many measure spaces, for instance for finite measures and for Borel measures with the property that $\mu(K) < \infty$ for any compact set K. This follows from the portmanteau theorem, and the following characterization of weak convergence in L^p, $1 < p < \infty$, cf. [202, Theorem V.1.3] or [205, Theorem 53.4, p. 382].

Theorem *Let $(X, \mathscr{A}, \mu)$ be a measure space and let $1 < p < \infty$. A sequence $(f_n)_{n\in\mathbb{N}} \subseteq L^p$ converges weakly to $f \in L^p$ if, and only if,* $\sup_{n\in\mathbb{N}} \|f_n\|_{L^p} < \infty$ *and*

$$\forall A \in \mathscr{A},\ \mu(A) < \infty :\ \lim_{n\to\infty} \int_A f_n \, d\mu = \int_A f \, d\mu. \tag{19.5}$$

For $p = 1$ and $p = \infty$, L^p-boundedness and (19.5) are necessary but not sufficient for weak convergence in L^p, cf. [205, Theorem 53.4, p. 382].

References

Each reference is followed, in parentheses, by a list of page numbers where it is cited.

[BM] Schilling, R.L. and Partzsch, L.: *Brownian Motion*. De Gruyter, Berlin 2012.

[MIMS] Schilling, R.L.: *Measures, Integrals and Martingales*. Cambridge University Press, Cambridge 2017 (2nd ed.).

[MINT] Schilling, R.L.: *Maß und Integral*. De Gruyter, Berlin 2015.

[4] Akhiezer, N.I.: *The Classical Moment Problem and Some Related Questions in Analysis*. Oliver and Boyd, Edinburgh 1965.

[5] Albiac, F. and Ansorena, J.L.: Integration in quasi-Banach spaces and the fundamental theorem of calculus. *Journal of Functional Analysis* **264** (2013) 2059–2076.

[6] Arzelà, C.: Sulla integrazione per serie. I, II. *Accademia dei Lincei, Rendiconti, IV. Serie* **1** (1885) 532–537 & 566–569.

[7] Balcerzak, M. and Kharazishvili, A.: On uncountable unions and intersections of measurable sets. *Georgian Mathematical Journal* **6** (1999) 201–212.

[8] Ball, J.M. and Murat, F.: Remarks on Chacon's biting lemma. *Proceedings of the American Mathematical Society* **107** (1989) 655–663.

[9] Banach, S.: *Oeuvres avec des commentaires* (2 vols., ed. St. Hartman and E. Marczewski). PWN, Warsaw 1967, 1979.

[10] Banach, S.: Sur le problème de la mesure. *Fundamenta Mathematicae* **4** (1923) 7–33 (reprinted in [9, pp. 66–89]).

[11] Banach, S. and Kuratowski, C.: Sur une généralisation du problème de la mesure. *Fundamenta Mathematicae* **14** (1929) 127–131 (reprinted in [9, pp. 182–186]).

[12] Banach, S. and Tarski, A.: Sur la décomposition des ensembles de points en parties respectivement congruentes. *Fundamenta Mathematicae* **6** (1924) 244–277 (reprinted in [9, pp. 118–148]).

[13] Bauer, H.: *Measure and Integration Theory*. De Gruyter, Berlin 2001 (translated from the German ed. *Maß- und Integrationstheorie*. De Gruyter, Berlin 1990).

[14] Berg, C. and Forst, G.: *Potential Theory on Locally Compact Abelian Groups*. Springer, Berlin 1975.

[15] Billingsley, P.: *Convergence of Probability Measures*. Wiley, New York 1999 (2nd ed.; 1st ed. Wiley 1968).

[16] Bingham, N.H., Goldie, C.M. and Teugels, J.L.: *Regular Variation*. Cambridge University Press, Cambridge 1989 (second corrected printing).

[17] Bingham, N.H. and Ostaszewski, D.: Set theory and the analyst. *European Journal of Mathematics* **5** (2019) 2–48.

[18] Bogachev, V.I.: *Gaussian Measures*. American Mathematical Society, Providence, RI 1998.

[19] Bogachev, V.I.: *Measure Theory, Vol. 1*. Springer, Berlin 2007.

[20] Bogachev, V.I.: *Measure Theory, Vol. 2*. Springer, Berlin 2007.

[21] Bourbaki, N.: *Elements of Mathematics – General Topology Chapters 5–10*. Springer, Berlin 1989 (translated from the French: *Topologie générale Chapitres 5 à 10, Hermann, Paris 1974*).

[22] Bourbaki, N.: *Elements of Mathematics – Integration I, Chapters 1–6*. Springer, Berlin 2004. (translated from the French: *Intégration, Hermann, Paris 1965, 1967, 1959*).

[23] Bourbaki, N.: *Elements of Mathematics – Integration II, Chapters 7–9*. Springer, Berlin 2004. (translated from the French: *Intégration, Hermann, Paris 1963, 1969*).

[24] Breiman, L.: *Probability*. Addison-Wesley, Reading, MA 1968 (several reprintes by SIAM, Philadelphia 1992 ff).

[25] Brézis, H.: How to recognize constant functions. Connections with Sobolev spaces. *Russian Mathematical Surveys* **57** (2002) 693–708.

[26] Bromwich, T.J.I'a.: *An Introduction to the Theory of Infinite Series*. AMS Chelsea Publishing, Providence, RI 1991 (Reprint of the 2nd ed., London 1926).

[27] Broughton, A. and Huff, B.W.: A comment on unions of σ-fields. *American Mathematical Monthly* **84** (1977) 553–554.

[28] Cajori, F.: *A History of Mathematical Notations (2 vols.)*. Open Court, Chicago, IL 1929.

[29] Cantor, G.: *Gesammelte Abhandlungen mathematischen und philosophischen Inhalts*. Springer, Berlin 1932 (reprinted by G. Olms, Hildesheim 1962).

[30] Cantor, G.: Über unendliche, lineare Punktmannigfaltigkeiten (5. Fortsetzung). *Mathematische Annalen* **21** (1883) 545–591 (reprinted in [29, pp. 165–204, commentaries pp. 204–208 and 208–209]).

[31] Cantor, M.: *Vorlesungen über Geschichte der Mathematik. Band 3: von 1668 bis 1758*. B.G. Teubner, Leipzig 1901 (2nd ed).

[32] Capek, P.: The atoms of a countable sum of set functions. *Mathematica Slovaca* **1** (1989) 81–89.

[33] Carathéodory, C.: *Gesammelte Mathematische Schriften* (5 vols., ed. the Bayerische Akademie der Wissenschaften). C.H. Beck, München 1956.

[34] Carathéodory, C.: Über das lineare Maß von Punktmengen – eine Verallgemeinerung des Längenbegriffs. *Nachrichten der königlichen Gesellschaft der Wissenschaften zu Göttingen, math.-phys. Klasse* (1914) 404–425 (reprinted in [33, vol. 4, pp. 249–275, addendum on the history of measurability pp. 276–277]).

[35] Carathéodory, C.: *Vorlesungen über reelle Funktionen*. Teubner, Leipzig 1927 (2nd ed.; reprinted by Chelsea, New York 1968).

[36] Cohn, D.L.: *Measure Theory*. Birkhäuser, Boston 1980.

[37] Darboux, G.: Mémoire sur la théorie des fonctions discontinues. *Annales de l'École Normale, Paris* **4** (1875) 57–112.

[38] Darst, R.B.: C^∞-functions need not be bimeasurable. *Proceedings of the American Mathematical Society* **27** (1971) 128–132.

[39] Darst, R. and Goffman, C.: A Borel set which contains no rectangles. *The American Mathematical Monthly* **77** (1970) 728–729.

[40] Davies, R.O.: Measures not approximable or not specifiable by means of balls. *Mathematika* **18** (1971) 157–160.

[41] Day, M.M.: The spaces L^p with $0 < p < 1$. *Bulletin of the American Mathematical Society* **46** (1940) 816–823.

[42] Dekker, T.J. and deGroot, J.: Decompositions of a sphere. *Fundamenta Mathematicae* **43** (1956) 185–194.

[43] Dellacherie, C. and Meyer, P.-A.: *Probabilités et potentiel. Chapitres I à IV*. Hermann, Paris 1975 (English translation *Probabilities and Potential Pt. A*, North Holland 1979).

[44] Diestel, J. and Uhl, J.J. (Jr): *Vector Measures*. American Mathematical Society, Providence, RI 1977.

[45] Dieudonné, J.: *Choix d'oeuvres mathématiques* (2 vols). Hermann, Paris 1981.

[46] Dieudonné, J.: Un exemple d'espace normal non susceptible d'une structure uniforme d'espace complet. *Comptes Rendus Hebdomadaires de l'Academie des Sciences* **209** (1939) 145–147 (reprinted in [45, pp. 159-161]).

[47] Dravecký, J.: Spaces with measurable diagonal. *Matematický časopis* **25** (1975) 3–9.

[48] du Bois-Reymond, P.: Der Beweis des Fundamentalsatzes der Integralrechnung. *Mathematische Annalen* **16** (1880) 115–128.

[49] Dudley, R.M.: *Real Analysis and Probability*. Wadsworth & Brooks/Cole, Pacific Grove, CA 1989.

[50] Dugundji, J.: *Topology*. Allyn & Bacon, Boston 1968.

[51] Dunford, N. and Schwartz, J.T.: *Linear Operators. Part I: General Theory*. Interscience Publishers, New York 1958 (several reprints).

[52] Dyer, R.H. and Edmunds, D.E.: *From Real to Complex Analysis*. Springer, London 2014.

[53] Engelking, R.: *General Topology* (revised and completed ed.). Heldermann Verlag, Berlin 1989.

[54] Erdös, P.: On a family of symmetric Bernoulli convolutions. *American Journal of Mathematics* **61** (1939) 974–976.

[55] Erdös, P. and Kakutani, S.: On a perfect set. *Colloqium Mathematicum* **4** (1957) 195–196.

[56] Erdös, P. and Stone, A.H.: On the sum of two Borel sets. *Proceedings of the American Mathematical Society* **25** (1970) 304–306.

[57] Evans, L.C. and Gariepy, R.F.: *Measure Theory and Fine Properties of Functions*. CRC Press, Boca Raton, FL 1992.

[58] Fan, K.: Entfernung zweier zufälligen Größen und die Konvergenz nach Wahrscheinlichkeit. *Mathematische Zeitschrift* **40** (1943/44) 681–683.

[59] Fichtenholz, G.: Sur une fonction de deux variables sans intégrale double. *Fundamenta Mathematicae* **6** (1924) 30–36.

[60] Folland, G.B.: *Real Analysis - Modern Techniques and their Applications*. Wiley, New York 1984.

[61] Foran, J.: A note on convergence in measure. *Real Analysis Exchange* **22** (1996/97) 802–805.

[62] Fréchet, M.: Sur divers modes de convergence d'une suite de fonctions d'une variable. *Bulletin of the Calcutta Mathematical Society* **11** (1921) 187–206.

[63] Gal, S.G. and Goldstein, J.A.: Semigroups of linear operators on p-Fréchet spaces, $0 < p < 1$. *Acta Mathematica Hungarica* **114** (2007) 13–36.

[64] Gänssler, P. and Stute, W.: *Wahrscheinlichkeitstheorie*. Springer, Berlin 1977.

[65] Gelbaum, B.R. and Olmsted, J.M.H.: *Counterexamples in Analysis*. Holden–Day, San Francisco, CA 1964 (unabridged and corrected reprint by Dover, Mineola, NY 2003).
[66] Gordon, R.A.: *The Integrals of Lebesgue, Denjoy, Perron, and Henstock*. American Mathematical Society, Providence, RI 1994.
[67] Grafakos, L.: *Classical Fourier Analysis*. Springer, Berlin 2008 (2nd ed.).
[68] Guzmán M. de: *Differentiation of Integrals in* $\mathbb{R}^n$. Lecture Notes in Mathematics **481** Springer, Berlin 1975.
[69] Hahn, H. and Rosenthal, A.: *Set Functions*. The University of New Mexico Press, Albuquerque , NM 1948.
[70] Halmos, P.R.: On the set of values of a finite measure. *Bulletin of the American Mathematical Society* **53** (1947) 138–141.
[71] Halmos, P.R.: *Naive Set Theory*. Springer, New York 1974.
[72] Halperin, I.: Discontinuous functions with the Darboux property. *Canadian Mathematical Bulletin* **2** (1959) 111–118.
[73] Hankel, H.: Untersuchungen über die unendlich oft oscillirenden und unstetigen Functionen. *Mathematische Annalen* **20** (1882) 63–112. [Posthumous reprint of a presentation given in 1870 at Tübingen University]
[74] Hankel, H.: Grenze. In: J.S. Ersch, J.G. Gruber: *Allgemeine Encyclopädie der Wissenschaften und Künste. Erste Section A–G* **Part 90** (ed. H. Brockhaus). F.A. Brockhaus, Leipzig 1871, pp. 189–211.
[75] Hausdorff, F.: *Hausdorff – Gesammelte Werke Bd. II* (ed. E. Brieskorn et al.). Springer, Berlin 2002.
[76] Hausdorff, F.: *Grundzüge der Mengenlehre*. Veit & Comp., Leipzig 1914 (reprinted in [75] and by Chelsea, New York 1949, 1965, 1978. The 2nd and 3rd 'editions' from 1927/1935 are substantially different; the 3rd ed. was translated into English).
[77] Hawkins, T.: *Lebesgue's Theory of Integration*. Chelsea, New York 1979.
[78] Henle, J. and Wagon, S.: A translation invariant measure. *The American Mathematical Monthly* **90** (1983) 62–63.
[79] Hewitt, E. and Ross, K.A.: *Abstract Harmonic Analysis – 1*. Springer, Berlin 1979 (2nd ed.).
[80] Hewitt, E. and Stromberg, K.: *Real and Abstract Analysis*. Springer, New York 1975.
[81] Hewitt, E. and Zuckerman, H.S.: Singular measures with absolutely continuous convolution squares. *Mathematical Proceedings of the Cambridge Philosophical Society* **62** (1966) 399–420.
[82] Hoffmann-Jørgensen, J.: *The Theory of Analytic Spaces*. Various Publications Series **10**. Matematisk Institut, Aarhus Universitet, Aarhus 1970.
[83] Ionescu Tulcea, A. and Ionescu Tulcea, C.: *Topics in the Theory of Lifting*. Springer, 1969.
[84] Jacod, J. and Shiryaev, A.N.: *Limit Theorems for Stochastic Processes*. Springer, Berlin 2003 (2nd ed.).
[85] Jech, T.: *Set Theory*. Academic Press, New York 1978.
[86] Jessen, B. and Wintner, A.: Distribution functions and the Riemann zeta function. *Transactions of the American Mathematical Society* **38** (1935) 48–88.
[87] Johnson, R.A.: Atomic and nonatomic measures. *Proceedings of the American Mathematical Society* **25** (1970) 650–655.
[88] Kánnai, Z.: A one-to-one popcorn function. *The American Mathematical Monthly* **124** (2017) 746–748.

[89] Kechris, A.S.: *Classical Descriptive Set Theory*. Springer, New York 1994.

[90] Kestelman, H.: *Modern Theories of Integration*. Dover, New York 1960 (reprint of the Oxford University Press ed., 1937).

[91] Kharazishvili, A.B.: *Strange Functions in Real Analysis*. Chapman & Hall/CRC, Boca Raton, FL 2006 (2nd ed.).

[92] Kharazishvili, A.B.: *Set Theoretical Aspects of Real Analysis*. CRC Press, Boca Raton, FL 2015.

[93] Kuratowkski, K.: *Topology I*. Academic Press, New York 1968.

[94] Lakatos, I.: *Proofs and Refutations. The Logic of Mathematical Discovery*. Cambridge University Press, Cambridge 1976.

[95] Lang, R.: A note on the measurability of convex sets. *Archiv der Mathematik* **47** (1986) 90–92.

[96] Lang, R.: A simple example of a non-Borel σ-field. *Statistics & Decisions* **4** (1986) 97–98.

[97] Lang, S.: *Real Analysis*. Addison-Wesley, Reading , MA 1969 (originally published as *Analysis II*).

[98] Lebesgue, H.: *Oeuvres Scientifiques* (5 vols., ed. F. Châtelet and G. Choquet). L'Enseignement Mathématique, Institut de Mathématiques Université de Genève, Geneva 1972.

[99] Lebesgue, H.: Sur l'approximation des fonctions. *Bulletin des Sciences Mathématiques* **22** (1898) 278–287 (reprinted in [98, vol. 3, pp. 11–20]).

[100] Lebesgue, H.: Sur une généralisation de l'intégrale définie. *Comptes Rendus Hebdomadaires de l'Academie des Sciences* **132** (1901) 1–3. (reprinted in [98, vol. 1, pp. 197–199]; English translation in the appendix to Chapter 1).

[101] Lebesgue, H.: Intégrale, Longeur, Aire. *Annali di Matematica Pura et Applicata* **7** (1902) 129 pp (also published as: Thèse de Doctorat, Faculté des Sciences de Paris Juin 1902; reprinted in [98, vol. 1, pp. 201–331]).

[102] Lebesgue, H.: *Leçons sur l'intégration*. Collection Borel, Gauthier–Villars, Paris 1904. (reprinted in [98, vol. 2, pp. 11–154]; 2nd revised and enlarged ed., Gauthier-Villars, Paris 1928).

[103] Lebesgue, H.: Sur les fonctions représentables analytiquement. *Journal de Mathématiques Pures et Appliquées* (1905) 139–216 (reprinted in [98, vol. 3, pp. 103–180]).

[104] Lebesgue, H.: Remarques sur la définition de l'intégrale. *Bulletin des Sciences Mathématiques* **29** (1905) 272–275 (reprinted in [98, vol. 2, pp. 155–158]).

[105] Lebesgue, H.: Sur les transformations ponctuelles transformant les plans en plans qu'on peut définir par des procédés analytiques. Extrait d'une lettre adressée par Mr. H. Lebesgue à Mr. C. Segre. *Atti Accademia Reale delle Science di Torino* **42** (1907) 3–10. (reprinted in [98, vol. 3, pp. 219–226]).

[106] Lieb, E.H. and Loss, M.: *Analysis*. American Mathematical Society, Providence, RI 1997.

[107] Lukacs, E.: *Characteristic Functions*. Griffin, London 1970 & Hafner, New York 1970 (revised and enlarged 2nd ed.; 1st ed. Griffin, London 1960).

[108] Luxemburg, W.A.J.: Arzelà's dominated convergence theorem for the Riemann integral. *The American Mathematical Monthly* **78** (1971) 970–979.

[109] Lyapunov, A.A.: Sur les fonctions-vecteurs complètement additives. *Izvestija Akademii Nauk SSSR Seriya Matematicheskaya* **4** (1940) 465–478.

[110] Malliavin, P.: *Integration and Probability*. Springer, New York 1995.

[111] Maraham, D.: On a theorem of Von Neumann. *Proceedings of the American Mathematical Society* **9** (1958) 987–994.

[112] Marczewski, R.: *Collected Mathematical Papers*. Polish Academy of Sciences, Institute of Mathematics, Warsaw 1996.

[113] Marczewski, R.: Remarks on the convergence of measurable sets and measurable functions. *Colloqium Mathematicum* **3** (1955) 118–124 (reprinted in [112, pp. 475–481]).

[114] Mattner, L.: Product measurability, parameter integrals, and a Fubini counter-example. *L'Enseignement Mathématique* **45** (1999) 271–279.

[115] McShane, E.J.: Linear functionals on certain Banach spaces. *Proceedings of the American Mathematical Society* **1** (1950) 402–408.

[116] Medvedev, F.A.: *Scenes From the History of Real Functions*. Birkhäuser, Basel 1991 (translated from the Russian ed., Moscow, Nauka 1975).

[117] Megginson, R.E.: *An Introduction to Banach Space Theory*. Springer, New York 1998.

[118] Menchoff, D.E.: Sur l'unicité du développement trigonométrique. *Comptes Rendus de l'Academie des Sciences, Paris* **163** (1916) 433–436.

[119] Meyer, P.-A.: *Probability and Potentials*. Blaisdell, Waltham, MA 1966 (translation from the French ed. *Probabilités et potentiel*, Hermann, Paris 1966).

[120] Mikusiński, J.: *The Bochner Integral*. Birkhäuser, Basel 1978.

[121] Mikusiński, J. and Ryll-Nardzweski, C.: Sur le produit de composition. *Studia Mathematica* **12** (1951) 51–57.

[122] Montgomery, D.: Non-separable metric spaces. *Fundamenta Mathematicae* **25** (1935) 527–533.

[123] Munroe, M.: *Introduction to Measure and Integration*. Addison-Wesley, Reading, MA 1953.

[124] Musiał, K.: Projective limits of perfect measure spaces. *Fundamenta Mathematicae* **110** (1980) 163–189.

[125] Natanson, I.P.: *Theorie der Funktionen einer reellen Veränderlichen*. Akademie-Verlag, Berlin 1975 (4th ed.).

[126] Neumann, J. von: *Collected Works* (6 vols., ed. A.H. Taub). Pergamon Press, New York 1961.

[127] Neumann, J. von: Über die analytischen Eigenschaften von Gruppen linearer Transformationen und ihrer Darstellungen. *Mathematische Zeitschrift* **30** (1929) 3–42 (reprinted in [126, pp. 509–548]).

[128] Neveu, J.: *Bases Mathématiques du Calcul des Probabilités*. Masson, Paris 1964.

[129] Nikodym, O.: Sur les ensembles accessibles. *Fundamenta Mathematicae* **10** (1927) 116–168.

[130] Nussbaum, A.E.: On a theorem by I. Glicksberg. *Proceedings of the American Mathematical Society* **13** (1962) 645–646.

[131] Osgood, W.F.: Non-uniform convergence and the integration of series term by term. *American Journal of Mathematics* **19** (1897) 155–190.

[132] Overdijk, D.A., Simons, F.H. and Thiemann, J.G.F.: A comment on unions of rings. *Indagationes Mathematicae* **41** (1979) 439–441.

[133] Oxtoby, J.C.: *Measure and Category*. Springer, New York 1980 (2nd ed.).

[134] Peres, Y. and Solomyak, B.: Absolute continuity of Bernoulli convolutions, a simple proof. *Mathematical Research Letters* **3** (1996) 231–239.

[135] Pesin, I.N.: *Classical and Modern Integration Theories*. Academic Press, New York 1970 (translated from the Russian ed., Nauka, Moscow 1966).

[136] Pfeffer, W.F.: *Derivation and Integration*. Cambridge University Press, Cambridge 2001.

[137] Purves, R.: Bimeasurable functions. *Fundamenta Mathematicae* **58** (1966) 149–157.
[138] Rao, B.: Remarks on analytic sets. *Fundamenta Mathematicae* **66** (1970) 237–239.
[139] Rao, M.M.: *Measure Theory and Integration*. Marcel Dekker, New York 2004 (2nd ed.).
[140] Rao, K.P.S.B. and Rao, B.V.: Borel spaces. *Dissertationes Mathematicae* **190** (1981).
[141] Ridder, J.: Maß- und Integrationstheorie in Strukturen. *Acta Mathematica* **73** (1941) 131–173.
[142] Riemann, B.: *Gesammelte Mathematische Werke, Wissenschaftlicher Nachlaß und Nachträge – Collected Papers* (ed. H. Weber and R. Dedekind, new ed. R. Narashiman). Springer and BSB B.G. Teubner, Berlin and Leipzig 1990.
[143] Riemann, B.: Ueber die Darstellbarkeit einer Function durch eine trigonometrische Reihe. *Abhandlungen der königlichen Gesellschaft der Wissenschaften, Göttingen* **13** (1867) 227–271. Posthumous publication of Riemann's habilitation thesis, submitted to Göttingen university 1854 (reprinted in [142, Kap. XII, pp. 259–271]).
[144] Riesz, F. and Sz.-Nagy, B.: *Functional Analysis*. Ungar, New York, 1955 (unabridged reprint by Dover, Mineola, NY 1990).
[145] Robinson, R.M.: On the decompositions of spheres. *Fundamenta Mathematicae* **34** (1947) 246–260.
[146] Ross, K.A. and Stone, A.H.: Products of separable spaces. *The American Mathematical Monthly* **71** (1964) 398–403.
[147] Royden, H.L.: *Real Analysis*. Macmillan, New York 1988 (3rd ed.).
[148] Rubel, L.A.: A pathological Lebesgue-measurable function. *Journal of the London Mathematical Society* **38** (1963) 1–4.
[149] Rudin, W.: Lebesgue's first theorem. In: L. Nachbin (ed.): *Mathematical Analysis and Applications*, Advances in Math. Supplementary Studies, vol. **7B**, Academic Press, New York 1981, pp. 741–747.
[150] Rudin, W.: *Real and Complex Analysis*. McGraw-Hill, New York 1986 (3rd ed.).
[151] Rudin, W.: *Functional Analysis*. McGraw-Hill, New York 1991 (2nd ed.).
[152] Rudin, W.: *Principles of Mathematical Analysis*. McGraw-Hill, New York 1976 (3rd ed.).
[153] Saks, S.: *Theory of the Integral*. Hafner, New York 1937 (unabridged reprint Dover, New York 1965).
[154] Salem, R.: *Oeuvres mathématiques* (ed. A. Zygmund). Hermann, Paris 1967.
[155] Salem, R.: On singular monotonic functions of the Cantor type. *Journal of Mathematics and Physics* **21** (1942) 69–82 (reprinted in [154, pp. 239–251]).
[156] Scheinberg, S.: Topologies which generate a complete measure algebra. *Advances in Mathematics* **7** (1971) 231–239.
[157] Schoenberg, I.J.: *Selected Papers* (2 vols., ed. C. de Boor). Birkhäuser, Boston 1988.
[158] Schoenberg, I.J.: On the Peano curve of Lebesgue. *Bulletin of the American Mathematical Society* **44** (1938) 519 (reprinted in [157, vol. 1, p. 197]).
[159] Schwartz, L.: *Radon Measures on Arbitrary Topological Spaces and Cylindrical Measures*. Oxford University Press, London 1973.
[160] Schwartz, L.: *Analyse I. Théorie des ensembles et topologie*. Hermann, Paris 1991.
[161] Schwartz, L.: *Analyse III. Calcul intégral*. Hermann, Paris 1998 (nouvelle ed.).
[162] Sierpiński, W.: *Oeuvres choisies* (3 vols). PWN Warsaw 1975.

[163] Sierpiński, W.: Sur la question de la mesurabilité de la base de M. Hamel. *Fundamenta Mathematicae* **1** (1920) 105–111 (reprinted in [162, vol. 2, pp. 322–327]).
[164] Sierpiński, W.: Sur un problème concernant les ensembles mesurables superficiellement. *Fundamenta Mathematicae* **1** (1920) 112–115 (reprinted in [162, vol. 2, pp. 328–330]).
[165] Sierpiński, W.: Sur les fonctions d'ensemble additives at continues. *Fundamenta Mathematicae* **3** (1922) 240–246 (reprinted in [162, vol. 2, pp. 457–463]).
[166] Simmons, S.M.: A converse Steinhaus theorem for locally compact groups. *Proceedings of the American Mathematical Society* **49** (1975) 383–386.
[167] Smith, H.J.S.: On the integration of discontinuous functions. *Proceedings of the London Mathematical Society* **6** (1875) 140–153.
[168] Solomyak, B.: Notes on Bernoulli convolutions. In: M.L. Lapidus, M. Van Frankenhuysen (eds.): *Fractal Geometry and Applications: A Jubilee of Benoıt Mandelbrot. Part 1*. American Mathematical Society, Providence, RI 2004, 207–230.
[169] Solovay, R.M.: A model of set theory in which every set of reals is Lebesgue measurable. *Annals of Mathematics* **92** (1970) 1–56.
[170] Souslin, M.: Sur une définition des ensembles mesurables B sans nombres transfinis. *Comptes Rendus Hebdomadaires de l'Academie des Sciences Paris* **164** (1917) 88–91.
[171] Srivastava, S.M.: *A Course on Borel Sets*. Springer, New York 1998.
[172] Steen, L.A. and Seebach, J.A. Jr.: *Counterexamples in Topology*. Springer, New York 1978 (2nd ed.; unaltered reprint by Dover, New York 1995).
[173] Stein, E.M.: *Harmonic Analysis: Real-Variable Methods, Orthogonality, and Oscillatory Integrals*. Princeton University Press, Princeton, NJ 1993.
[174] Stein, E.M. and Shakarchi, R.: *Real Analysis* (*Princeton Lectures in Analysis III*). Princeton University Press, Princeton, NJ 2005.
[175] Steinhaus, H.: *Selected Papers* (ed. St. Hartman et al.) PWN Warsaw 1985.
[176] Steinhaus, H.: A new property of the Cantor set (in Polish). *Wektor* (1917) 1–3 (English translation in [175, pp. 205–207]).
[177] Steinhaus, H.: Sur les distances des points dans les ensembles de mesure positive. *Fundamenta Mathematicae* **1** (1920) 93–104 (reprinted in [175, pp. 296–304]).
[178] Stieltjes, T.-J.: *Oeuvres Complètes – Collected Papers* (2 vols., ed. G. van Dijk). Springer, Berlin 1993.
[179] Stieltjes, T.-J.: Recherches sur les fractions continues. *Annales de la Faculté des Sciences de Toulouse* **8** (1894) 1–122 (reprinted and translated in [178, vol. 2, pp. 406–566 and pp. 609–745]).
[180] Stoyanov, J.M.: *Counterexamples in Probability*. Dover, Mineola, NY 2013 (3rd ed.; 1st ed. Wiley, Chichester 1987).
[181] Stromberg, K.: The Banach–Tarski paradox. *The American Mathematical Monthly* **86** (1979) 151–161.
[182] Thielman, H.P.: Types of functions. *The American Mathematical Monthly* **60** (1953) 156–161.
[183] Thomae, K.J.: *Einleitung in die Theorie der bestimmten Integrale*. Nebert, Halle (Saale) 1875.
[184] Thomas, J.: A regular space, not completely regular. *The American Mathematical Monthly* **76** (1969) 181–182.
[185] Titchmarsh, E.C.: *Introduction to the Theory of Fourier Integrals*. Oxford University Press, Oxford 1948 (2nd ed).

[186] Topsoe, F.: *Topology and Measure.* Lecture Notes in Mathematics **133**, Springer, Berlin 1970.
[187] Triebel, H.: *Theory of Function Spaces I, II, III, IV.* Birkhäuser, Basel 1983, 1992, 2006 and 2020.
[188] Uhl, J.J. (Jr.): The range of a vector-valued measure. *Proceedings of the American Mathematical Society* **23** (1969) 158–163.
[189] Ulam, S.: *Sets, Numbers, and Universes. Selected Works* (ed. W.A. Beyer et al.). MIT Press, Cambridge, MA 1974.
[190] Ulam, S.: Zur Masstheorie der allgemeinen Mengenlehre. *Fundamenta Mathematicae* **16** (1930) 140–150 (reprinted in [189, pp. 9–19]).
[191] Ushakov, N.G.: *Selected Topics in Characteristic Functions.* VSP, Utrecht 1999 (reprinted by De Gruyter, Berlin).
[192] Vakhania, N.N., Tarieladze, V.I. and Chobanyan, S.A.: *Probability Distributions on Banach Spaces.* D. Reidel, Dordrecht 1987.
[193] Volterra, V.: *Opere Matematiche* (5 vols., ed. Accademia Nazionale dei Lincei). Accademia Nazionale dei Lincei, Rome 1954.
[194] Volterra, V.: Alcune osservazioni sulle funzioni punteggiate discontinue. *Giornale di Matematiche (Battaglini)* **19** (1881) 76–86 (reprinted in [193, vol. 1, pp. 7–15]).
[195] Volterra, V.: Sui principii del calcolo integrale. *Giornale di matematiche (Battaglini)* **19** (1881) 340–372 (reprinted in [193, vol. 1, pp. 16–48]).
[196] Wagon, S.: *The Banach–Tarski Paradox.* Cambridge University Press, Cambridge 1985.
[197] Walter, W.: A counterexample in connection with Egorov's theorem. *The American Mathematical Monthly* **84** (1977) 118–119.
[198] Wiener, N.: *Collected Works With Commentaries* (4 vols., ed. P. Masani). MIT Press, Cambridge, MA 1976–1985.
[199] Wiener, N. and Wintner, A.: Fourier–Stieltjes transforms and singular infinite convolutions. *American Journal of Mathematics* **60** (1938) 513–522 (reprinted in [198, vol. 2, pp. 677–686]).
[200] Willard, S.: *General Topology.* Addison-Wesley, Reading, MA 1970 (reprinted by Dover, New York 2004).
[201] Wise, G.L. and Hall, E.B.: *Counterexamples in Probability and Analysis.* Oxford University Press, Oxford 1993.
[202] Yosida, K.: *Functional Analysis.* Springer, Berlin 1980 (6th ed.).
[203] Yosida, K.: *Collected Papers* (ed. K. Itô). Springer, Tokyo 1992.
[204] Yosida, K., Hewitt, E.: Finitely additive measures. *Transactions of the American Mathematical Society* **72** (1952) 46–66 (reprinted in [203, pp. 344–364]).
[205] Zaanen, A.D.: *Integration.* North-Holland, Amsterdam 1967 (completely revised ed. of *An Introduction to the Theory of Integration*, North-Holland, Amsterdam 1958).
[206] Zygmund, A.: *Trigonometrical Series.* Warszawa-Lwow 1935 (reprinted by Dover, New York 1955).
[207] Zygmund, A.: *Trigonometric Series* (2 vols bound in 1). Cambridge University Press, Cambridge 1968.

Index

This should be used in conjunction with the references, the table of contents, the lists of symbols and the list of topics. Generally, index entries are theorems, concepts and methods, but not concrete counterexamples. These can be found using the table of contents or the list of topics. Numbers following entries are either chapter numbers (§n) or page numbers (n), bold numbers refer to definitions and important statements. The symbols ◉ and ⊞ refer to pictures and tables. Unless otherwise stated, 'integral', 'integrability', etc. always mean the (abstract) Lebesgue integral. Within the index we use 'L…' and 'R…' as a shorthand for '(abstract) Lebesgue…' and 'Riemann…'.

Printed in the United States
by Baker & Taylor Publisher Services